Environmental Science and Engineering

The ultimate goal of this series is to contribute to the protection of our environment, which calls for both profound research and the ongoing development of solutions and measurements by experts in the field. Accordingly, the series promotes not only a deeper understanding of environmental processes and the evaluation of management strategies, but also design and technology aimed at improving environmental quality. Books focusing on the former are published in the subseries Environmental Science, those focusing on the latter in the subseries Environmental Engineering.

Nitish Kumar
Editor

Mercury Toxicity

Challenges and Solutions

Springer

Editor
Nitish Kumar
Department of Biotechnology
Central University of South Bihar
Gaya, Bihar, India

ISSN 1863-5520 ISSN 1863-5539 (electronic)
Environmental Science and Engineering
ISBN 978-981-99-7718-5 ISBN 978-981-99-7719-2 (eBook)
https://doi.org/10.1007/978-981-99-7719-2

This Springer imprint is published by the registered company Springer Nature Singapore Pte Ltd.
The registered company address is: 152 Beach Road, #21-01/04 Gateway East, Singapore 189721, Singapore

Paper in this product is recyclable.

Preface

Mercury (Hg) is a toxic heavy metal which is widely dispersed in nature. Mercury is well-known for being a bio-accumulative heavy metal and a persistent atmospheric contaminant. The WHO determined that 0.002 mg/L of mercury was the permitted level. The zero oxidation state of mercury (Hg^0) can be found as a vapor or as a liquid metal. The mercuric state (Hg^{2+}) can either produce inorganic salts or organomercury compounds. All of these result in toxicity or death, with the exception of elemental liquid mercury.

Mercury mostly comes from various natural sources such as soils, forest fires, volcanoes, mercury-rich geological zones, submarine vents, as well as from fresh water rivers, lakes, and seas. Human activity has, however, enhanced the content of mercury in the environment in various ways, including through a range of industrial and combustion operations like coal-fired power generation, metal smelting and mining, and incineration of waste. Multiple mercury emissions sources have the potential to penetrate the food chain and endanger human health. The main ways that humans are exposed to mercury are either by inhaling mercury from dental amalgam or by eating fish and seafood that have been tainted with mercury. High quantities of mercury in the human diet can cause brain abnormalities that can be long-lasting and even irreversible. The risks are particularly serious for young children since they could experience neurological abnormalities. To counter the poisonous effects of mercury, many plants and microorganisms have evolved detoxification processes.

This global environmental problem is clarified in the book, which also suggests multidisciplinary techniques for addressing contamination. There are three sections in this book. The various sources and locations of mercury in soil and plant ecosystems are discussed in the first section. The second portion goes through the health dangers of mercury toxicity. Sustainable mitigation tactics and solutions are covered in the third section. This book gives a general review of mercury-polluted areas' potential for bioremediation treatments supported by fungi, bacteria, or plants. These restoration techniques benefit from being economical and environmentally beneficial because they use plants to absorb and immobilize toxins from soil and water, and fungi and bacteria to break them down. Students, professors, researchers, and environmental specialists working on mercury poisoning around the world will find this book to be a useful resource.

Gaya, Bihar, India Nitish Kumar

Acknowledgments

First and foremost, I would like to praise and thank God, the Almighty, who has granted countless blessings, knowledge, and opportunity to accomplish the project work. Thanks to all the authors of the various chapters for their contributions. It had been a bit of a long process from the initial outlines to developing the full chapters and then revising them in the light of reviewer's comments. We sincerely acknowledge the author's willingness to go through this process. I also acknowledge the work and knowledge of the members of our review panels, many of which had to be done at short notice. Thanks to all the people at Springer Nature, especially Yosuke Nishida and Mr. Sivananth Sivachandran with whom we corresponded for their advice and facilitation in the production of this book. I am grateful to my family members Mrs. Kiran (Wife), Miss Kartika Sharma and Laavanya Sharma (Daughters), and parents for their incredible and selfless support all the time.

Gaya, Bihar, India Nitish Kumar

Contents

About the Editor

Dr. Nitish Kumar is Senior Assistant Professor at the Department of Biotechnology, Central University of South Bihar, Gaya, Bihar, India. Dr. Kumar completed his doctoral research at the Council of Scientific and Industrial Research—Central Salt & Marine Chemicals Research Institute, Bhavnagar, Gujarat, India. He has more than 12 years of research and teaching experience in the field of plant and microbial biotechnology, and bioremediation. He has published more than 70 research articles in leading international and national journals, more than 20 chapters, and 10 books with Springer and Taylor & Francis. Dr. Kumar is Recipient of the Young Scientist Award from the Science and Engineering Research Board (SERB) in 2014. He has received many awards/fellowships/projects from various organizations, for example, the CSIR, DBT, ICAR and SERB-DST, and BRNS-BARC, among others. He is an Active Reviewer for journals, including Biotechnology Reports, Aquatic Botany, Industrial Crops and Products, PLoS One, Plant Biochemistry and Biotechnology, and 3Biotech. He also serves as Associate Editor of the journal Gene (Elsevier).

Part I
Source and Distribution of Mercury in Soil and Plant

Chapter 1
Environmental Mercury Exposure—A Continuing Challenge

Shihab Uddin, Sumona Khanom, and Md. Rafiqul Islam

1.1 Introduction

The environment and human health are significantly at risk from mercury (Hg), a naturally occurring heavy metal that is also a very toxic neurotoxin (Wang et al. 2023a; Singh et al. 2023). Natural processes like volcanic eruptions, geological processes, reemission from ocean, and anthropogenic activities like gold mining activities, cement production, combustion of fuels, other industrial processes, and waste incineration all contribute to the environmental Hg release (Streets et al. 2019; Huang et al. 2017; Liu et al. 2012). Three main types of Hg that predominate in the environment are methylmercury (Me-Hg), elemental Hg (Hg^0), and inorganic Hg (Hg^{2+}) (Gworek et al. 2020). Among the types, the most hazardous form of Hg, Me-Hg, is created by the conversion of Hg^{2+} by certain microbes in aquatic settings (Peterson et al. 2023). On the other hand, Hg^{2+} is the result of the oxidation of Hg^0 in the environment (Zhang et al. 2022). When Hg is released, it can travel a great distance in the atmosphere before eventually settling in water or soil, where it can persist and continue to endanger ecosystems and human populations. Mercury can travel far from its point of origin due to this atmospheric movement, having an impact in places that are remote from the points of emission (Dastoor et al. 2022). This ultimately results in the increased possibility of environmental Hg exposure to a greater number of wildlife and human populations.

Environmental Hg exposure is the term used to describe the presence and buildup of Hg in various environmental compartments, such as the air, water, soil, and food

S. Uddin (✉) · Md. R. Islam
Department of Soil Science, Faculty of Agriculture, Bangladesh Agricultural University, Mymensingh 2202, Bangladesh
e-mail: shihab.ss@bau.edu.bd

S. Khanom
Department of Plant Pathology, Faculty of Agriculture, Bangladesh Agricultural University, Mymensingh 2202, Bangladesh

N. Kumar (ed.), *Mercury Toxicity*, Environmental Science and Engineering,
https://doi.org/10.1007/978-981-99-7719-2_1

chain, which may contaminate them and cause exposure to other living things, such as humans. Methylmercury frequently bioaccumulates and biomagnifies in the food chain, particularly in fish and other predatory seafood, which ultimately creates an exposure route for humans who consume such contaminated food (Chen and Dong 2022). Humans can also be exposed to environmental Hg by inhalation, dermal contact, and consumption of other contaminated foods, while plants can be exposed to atmospheric deposition, irrigation with contaminated water, and background soil Hg. Exposure to environmental Hg can have negative effects on wildlife, especially in aquatic habitats where it can impair neurological and reproductive processes, causing population losses and weakening ecosystem health (Batchelar et al. 2013). Mercury exposure can have negative neurological and developmental impacts on people, particularly in sensitive populations like fetuses and young children (Lin et al. 2023). This is especially true if people consume fish that has been methylated. International cooperation and efficient strategies to reduce Hg emissions and their negative consequences are needed to address the issue of environmental Hg exposure. Controlling Hg pollution and protecting human health and the environment are made possible by programs like the Minamata Convention (Zhao et al. 2023). To design focused strategies for the mitigation and protection of ecosystems and public health, it is essential to comprehend the origins, distribution, and consequences of environmental Hg exposure, given the global nature of Hg emissions. Special research activities need to be conducted to explore the Hg exposure pathways.

1.2 Environmental Mercury Exposure Pathways

Environmental Hg exposure pathways refer to the different routes that an organism might encounter Hg and potentially run the risk of developing health problems. Plants, animals, and humans may experience environmental Hg exposure in several ways, which results in an imbalance in the biological functions of that organism. The details of the possible environmental Hg exposure pathways will be described in this section.

1.2.1 Plants

Mercury is present everywhere in the soil, water, or air around plants, which results in easy exposure of this deadly material to both terrestrial and aquatic plants. Terrestrial plants can be exposed through atmospheric deposition, soil contamination, or the biomagnification process, whereas aquatic plants can also be exposed by these processes, water contamination, and sediment uptake (Fig. 1.1) (Zhou et al. 2021). Numerous natural and human activities result in the emission of Hg into the atmosphere. It can travel a great distance once in the sky before being deposited onto land by rainfall and dry deposition. Mercury can be absorbed by terrestrial plants through their leaves. Like terrestrial plants, aquatic plants can also uptake atmospheric Hg through their submerged leaves and stems in water bodies, where Hg is deposited by dry deposition and rainfall. Due to geological processes, Hg can be found naturally in soil and sediment, but human activities like mining, industrial emissions, and agricultural practices can greatly raise the amount of Hg in the soil and water bodies. Mercury can be absorbed by the roots of both terrestrial and aquatic plants that are grown in contaminated soil and sediments. Like soil, Hg is also naturally present in water, and aquatic plants can absorb Hg present in contaminated water through their roots. Both terrestrial and aquatic plants are not part of the biomagnification process. However, plants may be indirectly impacted by biomagnification. Herbivores that eat Hg-contaminated plants absorb the Hg when they do so. Mercury is also accumulated by predators that eat these herbivores. Biomagnification can have an impact on terrestrial predators, including several species of mammals and birds. The process continues until the top of the food chain, leading to higher concentrations of Hg in aquatic predators in both terrestrial and aquatic ecosystems. As soon as Hg enters a plant, it can travel throughout its tissues and even gather in specific areas of the plant. A few of the factors that impact how much Hg plants absorb and accumulate are the type of plant, the soil and water quality, the concentration of Hg, and the amount of environmental exposure. Plants can occasionally act as bioindicators of pollution, showing when and how much Hg pollution is present in each ecosystem. The growth, development, and overall health of plants can also be impacted by Hg exposure. Careful monitoring and knowledge of the amount of Hg present in plants are necessary to assess the health of ecosystems and potential risks to both plant and animal life.

Fig. 1.1 Environmental mercury exposure pathways for terrestrial and aquatic plants

1.2.2 Animals

Like plants, animals can also be exposed to environmental Hg through various processes. These processes include inhalation, dermal contact, direct ingestion of contaminated plants, direct or indirect ingestion through predation, ingestion of soil and sediments, absorption through gills, biomagnification, and transfer from mother to offspring. Animals that live on land may be directly exposed to Hg if they eat plants that have absorbed it from the soil or environment. For instance, insects, birds, and herbivorous mammals that feed on contaminated vegetation may also consume Hg in the food they eat. Similarly, by eating aquatic plants that have absorbed Hg from the water and sediment, aquatic herbivores, such as specific fish species and aquatic invertebrates, can be exposed to Hg. Mercury exposure by biomagnification is a possibility for predators in terrestrial habitats (Fig. 1.2). The levels of Hg in carnivorous animals can rise when they eat other animals (herbivores) that have accumulated Hg from plants. Aquatic carnivores, such as fish, turtles, and birds that eat fish, can also be exposed to Hg by ingesting contaminated prey. Fish at the base of the food chain may consume Hg directly from water and sediment, and this Hg is passed on to their predators, causing increased Hg concentrations in top-level predators (Fig. 1.3). Additionally, while scavenging for food, some animals, especially small mammals and birds, may absorb soil or sediment. These animals may consume Hg directly if the soil or sediment is contaminated with Hg. Likewise, by consuming Hg-containing sediment particles, benthic species, such as bottom-feeding fish and invertebrates, can be exposed to Hg in aquatic ecosystems. Furthermore, certain aquatic organisms, particularly fish and amphibians, can directly absorb Hg from water through their gills (Barboza et al. 2018). Young animals can also become exposed to Hg in some species because it can be transferred from mother to child during reproduction or through breast milk in mammals (Chen et al. 2014a, b). Animals exposed to Hg may experience negative impacts on their well-being, temperament, and ability to reproduce. High Hg concentrations in predators, including fish, birds of prey, and marine mammals, can have detrimental effects on their populations and even put the health of the animals and people who eat them at risk.

1.2.3 Humans

Mercury exposure in humans can happen both directly and indirectly in several different ways (Fig. 1.4). Humans are most frequently exposed to Hg through the ingestion of contaminated fish and seafood (Rodrigues et al. 2023). There are many ways that Hg can get into water bodies, including industrial discharges and atmospheric deposition. Once in the water, Hg undergoes a transformation into Me-Hg, which is extremely poisonous and can bioaccumulate in fish and other aquatic life. Mercury can gradually build up in the bodies of people who consume fish on a regular basis, particularly predatory species higher up the food chain. Some land animals,

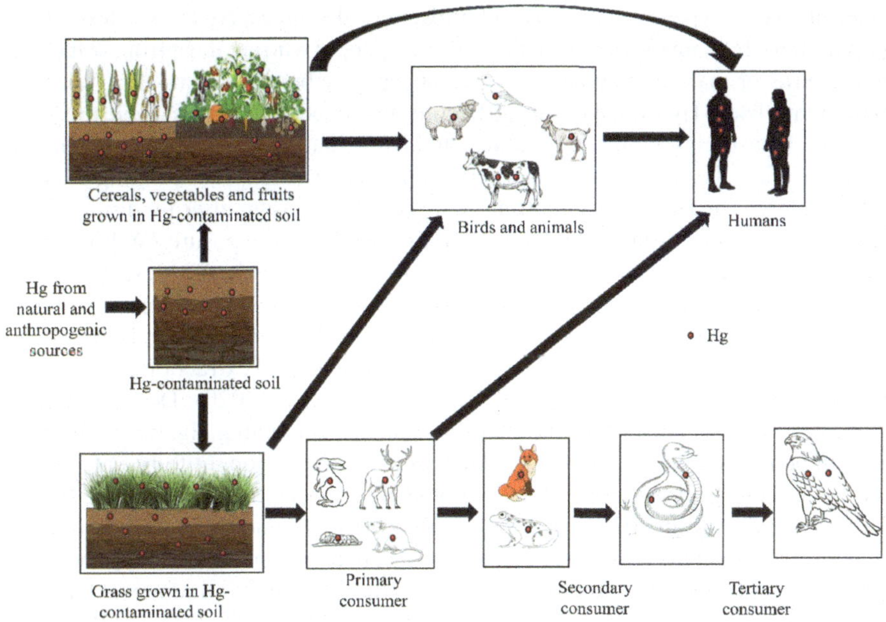

Fig. 1.2 Mercury biomagnification process in terrestrial environment

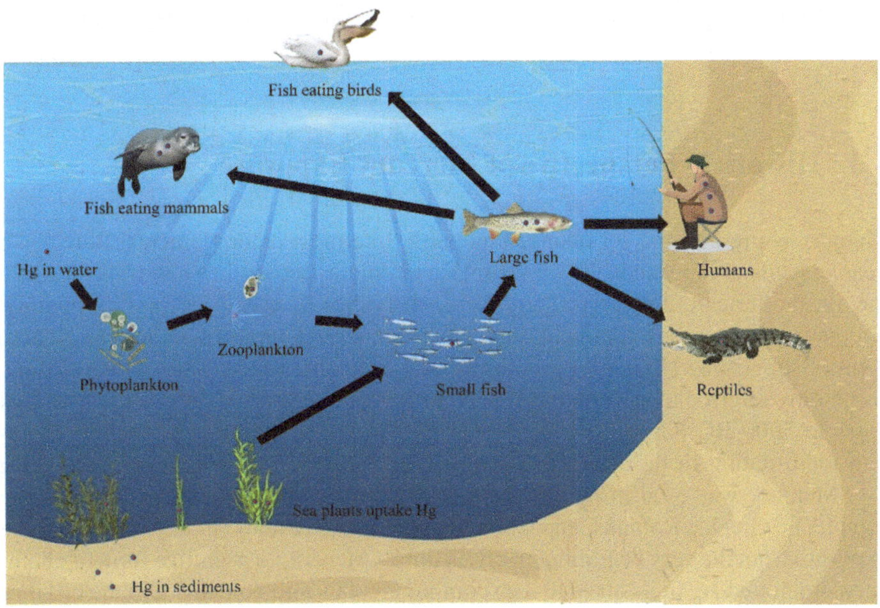

Fig. 1.3 Mercury biomagnification process in aquatic environment

especially some types of mammals and birds, may also ingest Hg in their food. If the animals have built up the poison in their tissues, people who eat meat from some fish-eating birds or marine mammals, beef, or sheep may be exposed to Hg. Terrestrial plants may absorb Hg from the soil and become poisoned in locations where the levels are high. Consuming grains, vegetables, and fruits that have been polluted with Hg can pose a risk of exposure in areas where traditional or subsistence farming methods are prevalent. The two most important routes of Hg exposure after food sources are thought to be inhalation of contaminated air and dermal exposure (Wilman et al. 2023). Exposure to Hg may result from specific professional activities. Workers in sectors where Hg-containing items, such as batteries, electrical equipment, and thermometers, are produced may be at risk (Wilson et al. 2018). The use of Hg to extract gold from ore in small-scale gold mining and artisanal gold extraction methods can expose workers to high quantities of Hg (Mambrey et al. 2020). Dental amalgam fillings, which are made of a combination of metals containing Hg, have long been used in dentistry. Even while dental amalgam is usually thought to be safe to use, some people might be worried about possible Hg exposure from this source (Berlin 2020; Joy and Qureshi 2020; Tibau and Grube 2019). Thiomersal, a preservative that contains Hg, was once employed in various vaccinations and medicinal goods. The exposure risk from this source is currently very low, though, as thiomersal has been extensively phased out of vaccinations since the early 2000s (Counter and Buchanan 2004). Cosmetics are widely used to make oneself more lucrative. In recent years, the presence of Hg in cosmetics has been reported in many countries (Podgórska et al. 2021; Wang and Fang 2021). It is also a potential source of Hg exposure for humans.

1.3 Historical Perspective of Mercury Exposure

Mercury is present in a set quantity on earth, like most other elements. Between a year and tens of thousands of years, this Hg circulates across the subterranean earth, the air, the terrestrial reservoir, and different bodies of water. Despite its toxicity, Hg contains numerous chemical qualities that are advantageous to humans. There is proof suggesting Hg has been continuously used since ancient times. Vermillion, commonly referred to as cinnabar (HgS), was found covering a human skeleton that dates to 5000 BC (Czaika and Edwards 2013). A ceremonial cup from an Egyptian tomb dating to the fifteenth century BC is another historical instance of Hg use. Mercury was used in numerous ways by all ancient civilizations, such as the Egyptians, Greeks, Romans, and Chinese. These uses resulted in environmental Hg exposure from the very beginning of civilization. Alchemy, a primitive form of chemistry, employed Hg in its quest to turn common metals into gold and find the elixir of life. In the belief that it might treat specific illnesses, it was also used in traditional medicine. It was employed in rituals and to cure skin diseases. Around 500 BC, it was employed as a medical treatment and an aphrodisiac in China and India. Mercury was reportedly used as a contraceptive by Chinese women 4,000 years ago (Czaika

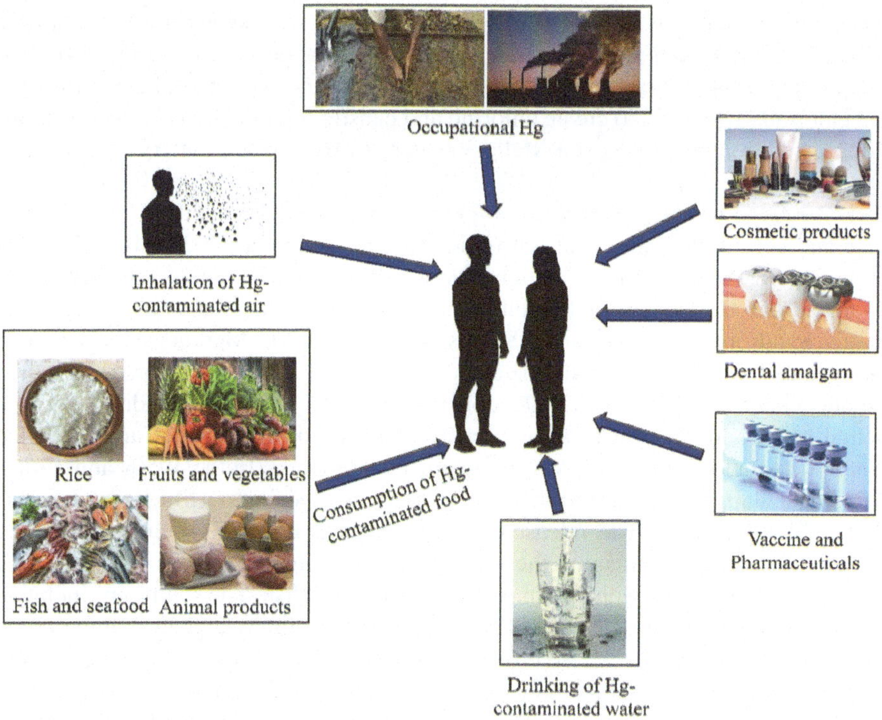

Fig. 1.4 Environmental mercury exposure pathways for humans

and Edwards 2013). In conventional Chinese medicine, HgS is still employed as a sedative. Mercury compounds were employed in cosmetics and religious ceremonies in ancient Egypt. Mercury had been used to amalgamate gold by the year 1000 AD (Czaika and Edwards 2013). Mercury forms gleaming pellets around the gold, which employees then burn. The refined gold remains after the Hg has evaporated. Today, artisanal small-scale gold mining businesses continue to use this method, exposing over 10 million workers to the deadly substance and discharging 650–1000 tons of Hg annually into the environment. During the Roman era, people extracted tungsten ore from deep inside the earth. The Romans used slave and jail labor to run a Hg quarry in Spain. They utilized Hg as a pigment; Hg-containing paint was discovered in Roman homes that were covered by Mount Vesuvius' volcanic ash in 79 BC. Although recently Hg was used in a wide range of products due to its chromatic qualities, the usage of Hg in paint has persisted into the current era. The inventions of the barometer by Torricelli in 1643 and the Hg thermometer by Fahrenheit in 1720 led to the widespread usage of Hg in science. A few measurement tools, including blood-pressure meters, are still produced in China, even though the health care industry no longer uses Hg in thermometers. Many new technologies made during the Industrial Revolution raised consumer appetites for Hg. Mercury fulminate was initially employed as an explosive detonator in 1799. Polyvinyl chloride, whose initial synthesis depended on Hg as a catalyst, was created for the first time in 1835. Mercury was used in Thomas Edison's incandescent lamp in 1891; it is still used in small fluorescent light bulbs today. Mercury might have been utilized in the chlor-alkali process to create chlorine and caustic soda in 1894 (Ito and Manabe 2022). Additionally, the Ruben-Mallory battery (a Hg dry-cell battery) was created and extensively employed during World War II. Mercury was primarily used in the early 1900s to produce scientific equipment, recover gold and silver, make fulminate and vermilion, and create felt (Guerrero and Schneider 2023a, b). When hat manufacturers, referred to as hatters, were exposed to Hg fumes and dust while using Hg-based chemicals in the felting process of producing hats, the phrase "Mad Hatter Syndrome" first appeared (Soares and Preto 2020). Mercury exposure over an extended period led to neuropsychiatric symptoms such as tremors, irritability, mood swings, and difficulties with coordination, hence the word "madness." These actual hatters are thought to have served as inspiration for Lewis Carroll's "Mad Hatter" character in "Alice's Adventures in Wonderland." In the 1950s and 1960s, Minamata, Japan, saw an outbreak of the Minamata sickness (Murata and Karita 2022). Methylmercury was released into Minamata Bay by the chemical company Chisso Corporation, damaging the local fish. Major Hg poisoning in the populace was caused by the ingestion of tainted seafood and shellfish. The symptoms included loss of coordination, muscle weakness, sensory abnormalities, and, in extreme cases, paralysis and death. The severe effects of industrial Hg pollution on human health were symbolized by the Minamata illness on a global scale. All these events historically led to human exposure to Hg. Today, people are still using Hg for different purposes, which has increased the risk of Hg exposure to humans and wildlife. This will ultimately threaten plant, animal, and human life.

1.4 Toxicological Implications

The term "toxicological implication" refers to the results or implications of exposure to a toxic material like Hg on living things, such as plants, animals, and humans. It entails researching the damaging effects of Hg on biological systems and comprehending the methods through which Hg impairs living things. Depending on the form of Hg, its concentration, the length of exposure, and the susceptibility of the exposed organisms, toxicological implications might include a wide range of health and environmental effects. The creation of safety standards, laws, and best practices is dependent on this information to safeguard people, ecosystems, and the environment from dangerous exposures. Once Hg enters the body of an organism, it disrupts normal functioning at molecular and cellular levels. In this section, we will discuss the adverse impacts of Hg on plants, animals, and humans.

1.4.1 Toxicological Effects in Plants

Since organic forms (mostly Me-Hg) of Hg are more toxic compared to the inorganic form (Hg^{2+}), they have a substantial adverse effect on plants. Most crop species are susceptible to Hg when it is present beyond the tolerable limit. Due to its propensity to accumulate in the roots, roots usually exhibit phytotoxic symptoms (Chen et al. 2014a, b). When plants uptake excessive Hg from the contaminated soil, all the vital physiological processes like photosynthesis, respiration, transpiration, and cell division are impaired (Fig. 1.5). When photosynthesis is affected, plants can't produce energy from sunlight, resulting in stunted growth, low biomass production, and decreased total plant health. It was also reported that Hg^{2+} can substitute the integral component (Mg^{2+}) of chlorophyll, which ultimately reduces the photosynthesis capacity of plants (Azevedo and Rodriguez 2012). Like chlorophyll, aquaporins (water channels in plants) are similarly affected by Hg, which inhibits transpiration and the subsequent uptake of water by vascular tissues. In addition, Hg purposefully interferes with the antioxidant defense enzymes in plants, particularly catalase, glutathione reductase, ascorbate peroxidase, and superoxide dismutase. Furthermore, it influences additional antioxidant components, including glutathione and non-protein thiols. All these combined result in oxidative stress in plants by generating reactive oxygen species and leading to damage to cell membranes and proteins, affecting the plants' antioxidant defense systems (Zhou et al. 2008; Israr et al. 2006). The propensity of Hg to interact with sulfhydryl (SH) groups present in enzymes and proteins, as well as its great selectivity for phosphate groups in lipids, nucleotides, and compounds with plenty of energy, is a likely mechanism for Hg toxicity (Sapre et al. 2019). Furthermore, Hg may prevent plants from absorbing vital elements including calcium, potassium, and magnesium. This disruption may result in nutritional deficits and imbalances, which could have an impact on the growth and development of plants. Changes in plants' physiological processes due to Hg exposure can result in wilting, diminished root growth, and leaf chlorosis (yellowing). These may result in reduced stunted growth and plant productivity, which ultimately reduce plants' ability to compete with other species and survive in their natural habitat. Aquatic plant communities might change in terms of distribution and composition due to Hg exposure. Sensitive species may become extinct or be replaced by invading or Hg-tolerant species, changing the composition and operation of the ecosystem. Mercury from sediments and water can be absorbed and accumulated by aquatic plants, and herbivorous animals and other creatures farther up the food chain may be exposed. Mercury slows down photosynthesis in aquatic plants, causing them to grow slower and produce less than terrestrial plants. High amounts of Hg exposure can also interfere with aquatic plants' ability to reproduce and spread by interfering with their reproductive processes. Overall, a comprehensive strategy is needed to address the toxicological effects of Hg on terrestrial and aquatic plants, including limiting Hg emissions, cleaning up contaminated areas, encouraging sustainable land use, and preserving water quality. We can maintain biodiversity, ecosystem services, and the

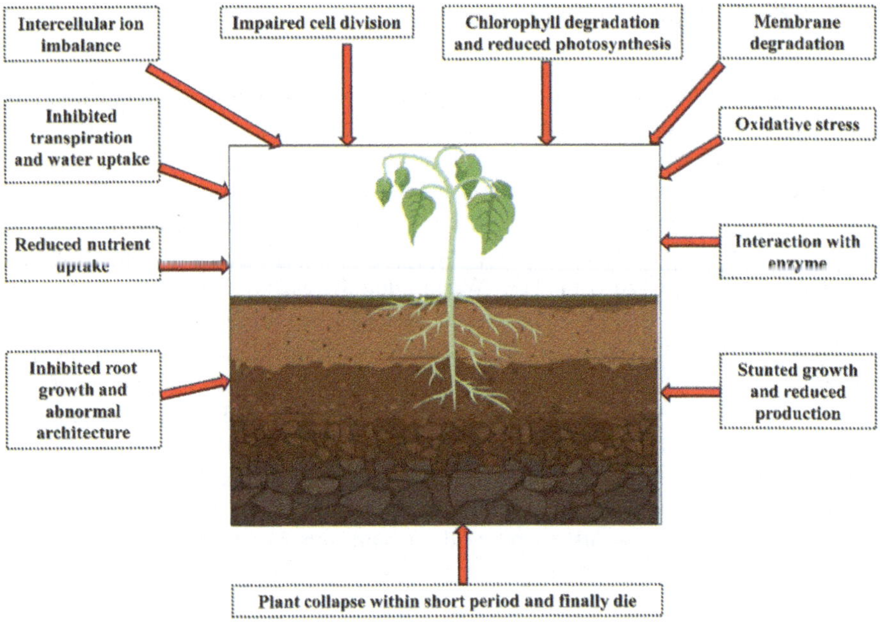

Fig. 1.5 Toxicological effects of environmental mercury in plants

general health of the environment by safeguarding plants and ecosystems from Hg exposure.

1.4.2 Toxicological Effects in Animals

Animals, excluding humans, are not exempt from the toxicological effects of Hg, which can be severe and vary based on species, exposure levels, and sensitivity. Animals can be exposed to Hg in several ways, such as by ingesting foods containing Hg, being exposed to polluted water or soil, or breathing in Hg vapor. Once Hg enters the animal's body, it interacts with the animal, leading to some adverse health effects. Freshwater and marine fish are at high risk of Hg exposure and its adverse effects. Low levels of Hg in the water can nevertheless alter fish genetically, biochemically, physiologically, morphologically, and behaviorally (Huang et al. 2011). Speciation, bioavailability, as well as absorption and transformation of Hg, are some of the variables that affect Hg toxicity. The consequences of Hg toxicity on multiple individuals, species, and stages of life might differ substantially (Morcillo et al. 2016), and it is also recognized that these parameters differ between different species. Methylmercury can build up in their tissues, which can result in decreased metabolism and liver function, behavioral changes, hampered reproduction, deformities, harm to the gills and olfactory organs, and even death (Zheng et al. 2019; Barboza et al. 2018;

O'Bryhim et al. 2017). As a result of Hg exposure, fish may feed less frequently or not at all, which could result in a lack of vital nutrients for larval growth (Liu et al. 2016). Additionally, it is known that exposure to inorganic Hg causes oxidative stress, which results in histological damage to fish gonads, like the thickening of tubule walls. Male testicular testosterone levels are much lower after Hg exposure, and Sertoli cell enlargement and interstitial irritation can also be seen, along with spermatogenic degradation and apoptosis (Zhang et al. 2016). Aquatic ecosystems and animals that use fish as a food source may be negatively impacted by fish exposure to Hg.

Birds, especially those that predate Hg-contaminated marine fish, are at greater risk of Hg exposure. Mercury exposure has several deleterious effects on birds, like depressed reproductive success (Costa et al. 2014), hatching time alteration (Yu et al. 2016), reduced fledging and fledglings (Costa et al. 2014), altered behavioral pattern (Tartu et al. 2015), neurotoxicity (Wolf et al. 2017), endocrine disruption (Franceschini et al. 2017), immune dysfunction (Whitney and Cristol 2017), oxidative stress (Costantini et al. 2014), chromosomal damage (Kenow et al. 2008), altered metabolism (Hoffman et al. 2005), and reduced growth conditions in terms of liver-to-kidney mass (Fort et al. 2015). These effects greatly hamper overall growth and even death, resulting in the loss of biodiversity. Reptiles like turtles and lizards that are exposed to Hg may exhibit altered behavior, decreased foraging prowess, and diminished reproductive success (Hopkins et al., 2013). Exposure to Hg can be harmful to amphibians as well. As a result of the effects of Me-Hg on their growth, metamorphosis, and immune system, they are more vulnerable to other environmental stresses (Todd et al. 2012). Mercury exposure in mammals, notably marine mammals like dolphins and seals, can have negative effects on the nervous system, interfere with reproduction, and alter behavior and feeding habits (UNEP 2023). Finally, the total biodiversity of ecosystems can be impacted by Hg exposure, which can also change community structures and cause decreases in some animal species.

1.4.3 Toxicological Effects in Humans

Humans who are exposed to Hg may experience serious toxicological effects that affect different body systems and result in a variety of health problems (Fig. 1.6). The chemical form of Hg, the exposure route, the duration and intensity of the exposure, the individual's age, and general health all affect how dangerous Hg is. The physiological processes of humans can be disturbed by Hg exposure. Mercury has a negative impact on the brain and nervous system, decreasing cognitive abilities like memory, concentration, and the capacity to learn at relatively modest concentrations (Wang and Matsushita 2021; Cariccio et al. 2019). Various other neurological conditions can also manifest, including speech impairment, hearing loss, limb muscle paralysis, and ultimately death (Wang et al. 2021). Chronic exposure to Hg causes elevated oxidative stress and heart attack by boosting the generation of free radicals and decreasing the function of enzymes that fight oxidative damage (glutathione

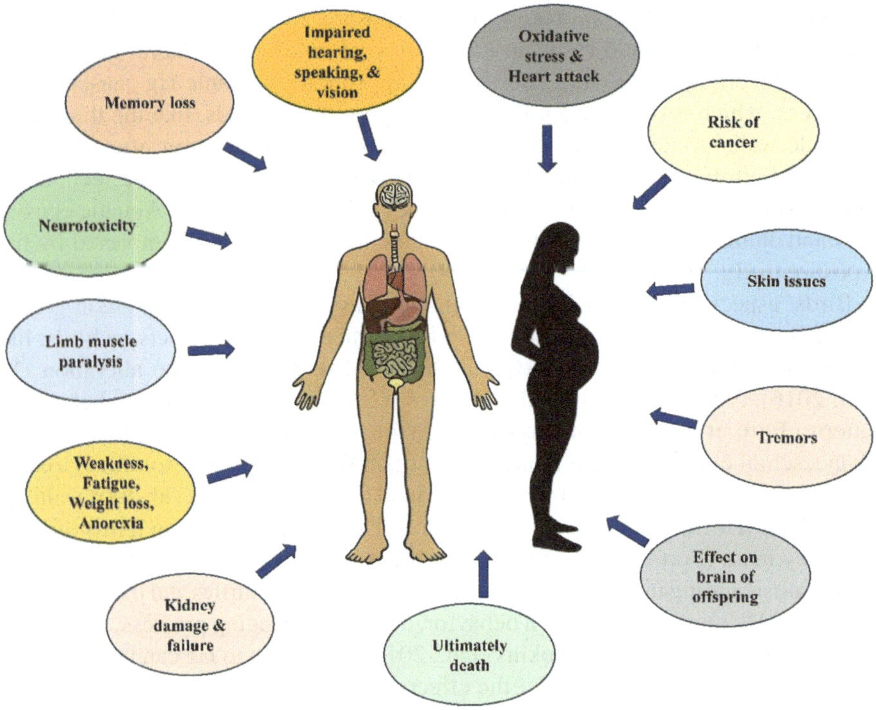

Fig. 1.6 Toxicological effects of environmental mercury in humans

peroxidase, catalase, and superoxide dismutase) (Hu et al. 2021; Pamphlett et al. 2021). Despite conflicting evidence, Hg exposure is linked to an increased risk of cancer. Through the control of cell proliferation, Hg may increase the risk of developing cancer. Cancer may be influenced by genotoxicity, oxidative DNA damage, and Hg's impact on epigenetics (Skalny et al. 2022).

Additionally, the well-known neurotoxic Me-Hg has a negative effect on the growing human brain. The blood—brain barrier and the placental barrier are both easily crossed by it, making any form of contact during pregnancy extremely dangerous. The International Agency for Research on Cancer has classified Me-Hg as category 2B and has determined that it may be carcinogenic. Congenital cases, or children born to exposed parents, manifested more symptoms than did the parents. Extremely delayed developing skills and significant nervous system disturbance were among the symptoms. Ataxia, sensory problems with the hands and feet, hearing problems, balance issues, speech problems, trembling in the hands and feet, and problems with eye movement are only a few of the disease's typical symptoms. According to both the National Academy of Science and the Environmental Protection Agency, 8–10% of American women are thought to have Hg levels that would cause neurological abnormalities in any offspring they may have (Matta and Gjyli 2016). Mercury still has the brain as its preferred target organ, but it can harm any organ and cause problems with the kidneys, muscles, and nerves. Intracellular calcium homeostasis may be disturbed, and the membrane potential may be disturbed. Given that stability constants are exceptionally high, Hg attaches to readily accessible thiols (Matta and Gjyli 2016). Children who are exposed to Hg during pregnancy, especially Me-Hg, may experience developmental delays, learning difficulties, and intellectual impairments. Mercury exposure can influence reproductive health, resulting in irregular menstrual cycles in women and men, as well as decreased fertility and poorer sperm quality in men. Heart attacks, hypertension, and an increased risk of cardiovascular disease can all be brought on by prolonged exposure to high Hg levels. People with weakened immune systems are more prone to infections and diseases because of Hg exposure. Renal damage from Hg exposure can result in compromised renal function and even kidney failure. Mercury vapor inhalation can result in chest pain, shortness of breath, coughing, and other respiratory issues. Constipation, diarrhea, and nausea are gastrointestinal symptoms that can be brought on by ingesting specific types of Hg. Skin rashes and dermatitis can develop after coming into contact with specific Hg compounds. Alterations in mood, irritability, anxiety, and sadness can result from prolonged Hg exposure.

1.5 Why Environmental Mercury Exposure is a Continuous Challenge?

Environmental Hg exposure is still a significant problem with complex effects on both the environment and human health. Several reasons contribute to the persistent worry about Hg exposure, despite greater knowledge and regulatory actions. When it comes to effective regulation and mitigation, Hg presents special challenges because of its widespread global distribution, long-distance transit, and bioaccumulation in the food chain. The factors responsible for making it more challenging are as follows.

1.5.1 Global Nature of Mercury Pollution

The global environmental problem of Hg pollution impacts ecosystems and human populations everywhere and cuts beyond country boundaries. Both natural and man-made activities, such as industrial processes, coal-fired power plants, and artisanal gold mines, release Hg into the atmosphere. Mercury can travel great distances through the sky after being released, eventually settling in remote areas far from its initial source (Yuan et al. 2023). This occurrence is referred to as "global Hg transport." As a result, Hg contamination can have an effect in areas that are far away from the sources of pollution (Soares et al. 2023). The extensive dispersion of Hg in the environment makes it difficult and complex to handle, necessitating global collaboration and sustained efforts to lessen its detrimental impacts on ecosystems and human health.

1.5.2 Persistent and Bioaccumulative Nature of Mercury

Mercury is a persistent and bioaccumulative contaminant, which means it is difficult to break down in the environment and can build up over time in living things. When Hg is introduced into the atmosphere, it can circulate between the air, water, and soil for a considerable amount of time. It persists and changes into more hazardous forms due to some natural processes, such as the transformation of Hg^0 to Me-Hg by specific bacteria in water bodies (Mandor 2023). When an organism absorbs Hg more quickly than its body can remove it, this is known as bioaccumulation (Hu et al. 2023). The organic form of Hg (Me-Hg) is easily absorbed by plankton in aquatic habitats and is later ingested by smaller fish. Mercury builds up and concentrates as larger predatory fish eat these smaller fish, increasing levels of the toxin in top predators. Humans who consume fish and seafood are especially at risk from this bioaccumulation process because they may be exposed to higher levels of Me-Hg. Since Hg is persistent and bioaccumulative, managing it is a constant problem because reducing emissions is insufficient to eliminate completely the threats posed by Hg that are already present in the environment (Joy and Qureshi 2023).

1.5.3 Multiple Sources of Mercury

Different sources, both natural and man-made, release Hg into the environment. The details of the sources and distribution of Hg are discussed in Chapter 1. Volcanic eruptions, geothermal emissions, and reemissions from the ocean are examples of natural sources of Hg emissions, whereas waste incineration, industrial processes like cement production, artisanal gold mining, and coal combustion are major human sources of Hg emissions. Humans can control anthropogenic emissions, but they have no control over natural emissions. Natural emissions can also affect humans, although they are far from human residential areas due to the global cycling process. Hg-containing products still exist despite efforts to decrease their use in many industries, and their incorrect disposal can cause Hg to leak into the environment. Numerous items, including batteries, switches, and some types of electrical equipment, may contain Hg, creating concerns both during their useful lives and after they are discarded.

1.5.4 Legacy Contamination

When we talk about legacy contamination, we're talking about the long-term environmental damage and health concerns brought on by past human activities that have left dangerous compounds in the environment. Even decades or centuries after their initial release, these pollutants frequently linger for long periods of time. Mercury has been left in soil and sediments because of legacy contamination brought on by previous industrial activity and poor waste management (Guerrero and Schneider 2023a, b). The process of addressing old Hg pollution can be drawn out and difficult. Legacy pollution poses major environmental and public health risks since it can harm soil, water bodies, air, and living things.

1.5.5 Global Trade and Consumption

Mercury consumption and global trade have a huge impact on the environment and human health. The regulation of the international movement of Hg continues to face difficulties despite international accords like the Minamata Convention on Hg, which tries to regulate Hg commerce and use. Waste containing Hg is frequently exported from developed countries to poor countries, potentially contaminating the environment and posing health hazards to populations who engage in informal recycling. Global commerce and consumption of Hg must be reduced to protect both the environment and human health. This can be done by enforcing stricter laws, raising public awareness, and implementing sustainable alternatives.

1.5.6 Reliance on Mercury-Containing Products

Some industries and sectors continue to extensively rely on Hg-containing products despite the acknowledged hazards to human health and the environment. For instance, Hg is used in some industrial operations to produce caustic soda and chlorine, as well as in small-scale artisanal gold mining to recover gold. Additionally, a variety of applications still use Hg-containing products like, skin care products, thermometers, fluorescent lighting, and some electrical switches and relays (Qureshi 2021). Dependence on these goods is frequently a result of things like cost-effectiveness, ignorance of better alternatives, and lax laws in some areas. Comprehensive efforts are needed to reduce the use of products containing Hg, including the promotion of safer alternatives, education about the dangers of Hg, the implementation of regulations, and the support of technological advancements that eliminate the need for Hg in various fields and applications.

1.5.7 Limited Infrastructure and Resources

The mitigation of Hg pollution faces substantial obstacles due to inadequate infrastructure and resources. Many developing nations lack the essential tools and knowledge to efficiently monitor and manage Hg emissions, particularly those with small industrial capacities. Additionally, for cash-strapped governments and enterprises, the expense of implementing Hg reduction measures may be prohibitive. Lack of effective waste management practices may result in the inappropriate disposal of products containing Hg, worsening environmental contamination. To help these countries develop the capacity to successfully combat Hg pollution, international cooperation and support are essential.

1.5.8 Complex Regulation and Monitoring Challenges

As a result of its many forms and exposure sources, Hg regulation is difficult to implement efficiently. International collaboration and coordination are also necessary to manage Hg emissions from various businesses and enforce regulations globally. Monitoring Hg levels in the environment, food, and human populations is a difficult and expensive process. It is necessary to use expensive technology and specialist knowledge, which may not be present in all areas, to obtain reliable and representative data on Hg levels. As a result of underreporting and delayed responses, Hg poisoning's symptoms are often vague and easy to confuse with those of other disorders. To combat ongoing Hg exposure, coordinated international efforts are required to set strict restrictions, expand monitoring capacities, and improve public health surveillance systems to quickly identify and treat instances. To protect people's health and the environment from the damaging impacts of ongoing Hg exposure, governments, organizations, and communities must work together.

1.5.9 Lack of Awareness

Addressing Hg exposure still faces a huge issue due to a lack of knowledge. The risks caused by Hg and its many routes of exposure are unknown to many people, communities, and even some industries. Due to a lack of knowledge, Hg-containing products, such as broken fluorescent bulbs or electronic debris, may be improperly handled and disposed of, which can release deadly substances into the environment. Because Hg is frequently used in artisanal gold mining, miners may not fully comprehend the dangers of handling it, which can result in direct exposure by inhalation or skin contact. Furthermore, consumers might not be aware of safer alternatives to products containing Hg, which would increase the demand for such goods. Public awareness of the risks of exposure, safer substitutes, and the importance of efficient waste management is crucial if we are to lessen the detrimental impacts of Hg exposure on human health and the environment. To address the lack of awareness and promote responsible behaviors to limit Hg exposure internationally, education programs, community outreach, and collaboration with relevant stakeholders are crucial.

Political and Economic Factors

Making considerable adjustments to industrial processes is frequently required to address Hg exposure, and these changes can be affected by political and economic reasons. It might be difficult to strike a balance between commercial interests and environmental concerns.

1.6 Vulnerable Populations

Groups of people are considered vulnerable populations if they are more likely to experience negative health impacts from Hg exposure for a variety of reasons. Because Hg is such a strong neurotoxin, some people are more vulnerable to its negative effects. Pregnant women, newborns, and young children are at risk for Hg exposure because Hg can disrupt a person's growing nervous system and brain, which can result in cognitive deficits and neurodevelopmental problems (Li et al. 2023; Lee et al. 2023). Mercury exposure during pregnancy can be extremely dangerous to the developing fetus, as nutrition is passed from the mother. Infants who are breastfed may also be at risk of Hg exposure if their moms are (Lin et al. 2023). Fish can bioaccumulate Me-Hg, making subsistence fishermen and populations who depend largely on fish from polluted waters for their nutrition particularly vulnerable. Indigenous groups are at risk if the fish they eat is contaminated, especially those whose traditional diets include fish from nearby bodies of water (Wang et al. 2023b). During the gold extraction process, artisanal small-scale miners and their families are directly exposed to Hg, and these miners are considered a vulnerable group (Molina Ramirez 2023). Additionally, some occupational groups may be more exposed to Hg than others, such as those who work in coal-fired power plants and garbage incineration. In addition, there may be greater risks of Hg exposure for certain occupational groups, including those who work in dental offices, labs, and businesses that utilize or manufacture products containing Hg. Due to poor nutrition, limited access to healthcare, and exposure to environmental pollutants, those who live in poorer areas may be more at risk of being exposed to Hg. Furthermore, pre-existing medical disorders, particularly those that affect the brain system and kidneys, may make a person more vulnerable to the harmful effects of Hg exposure. However, regulations, public health programs, and awareness campaigns that lower Hg emissions, encourage safe handling techniques, and teach the public about the consequences of Hg exposure are essential if we are to safeguard these vulnerable groups.

1.7 Regulatory Framework and Guidelines

Mercury regulations and standards are essential for addressing the threats to the environment and human health that come with this harmful element. Various national and international organizations have created rules, norms, and recommendations to reduce exposure to and emissions of Hg in response to these worries. Industry, healthcare, waste management, and emissions from power plants and other sources are all covered by these frameworks. Governments and businesses may lessen the negative impacts of Hg and support a safer and more sustainable future by putting certain regulatory measures into place and adhering to them. Some of the regulatory framework and guidelines are described in this section.

1.7.1 Minamata Convention on Mercury

Over 120 nations have ratified the Minamata Convention, an international agreement that addresses Hg pollution and was the first step toward safeguarding the next generation (Kessler 2013). The convention covers all facets of Hg, including mining, trading, use, and emissions. Throughout its existence, it seeks to regulate and lower exposure to and discharges of Hg. The Minamata Convention's main goals are to protect human health and the environment by reducing Hg exposure, particularly for vulnerable groups like pregnant women, children, and indigenous communities. The treaty supports Hg use reduction or phase-out in goods and procedures. It promotes the creation and application of Hg-free substitutes across numerous industries. The convention establishes rules for regulating and lowering Hg emissions from industrial operations like artisanal small-scale gold mining, cement manufacturing, waste incineration, and coal-fired power plants (Zhao et al. 2023). The agreement aims to advance safer and more environmentally friendly methods of gold extraction by addressing the effects of Hg use in artisanal gold mining on the environment and human health (Keane et al. 2023; Schwartz et al. 2023). The pact creates procedures to control Hg trade, export, and import, preventing its unlawful trafficking and ensuring responsible management throughout the supply chain. The convention talks about cleaning up Hg-contaminated areas to lessen dangers to the environment and human health. It also encourages public awareness and education on the dangers of Hg exposure and the significance of Hg reduction measures. In addition, it promotes information exchange between nations to aid in the application of best practices. Furthermore, the Minamata Convention offers help to developing nations in putting its provisions into practice, including technical support and financial support for technology transfer and capacity building. Parties are required by the agreement to report on their Hg emissions, the steps they have taken to put the convention's provisions into effect, and their progress toward meeting its goals.

1.7.2 Occupational Safety and Health Administration (OSHA)

When it comes to protecting workers who can be exposed to Hg at work, the Occupational Safety and Health Administration (OSHA) in the United States is crucial. To prevent employees from inhaling Hg vapor or coming into contact with Hg-containing materials, OSHA establishes and enforces permissible exposure limits (e.g., 50 μg Hg vapor m^{-3} of the breathing zone air) (Dudeja et al. 2023). The use of engineering controls, personal protective equipment, and appropriate training are just a few examples of the controls and safety measures that employers are expected to use to prevent or eliminate Hg exposure. In addition to conducting inspections to determine if these rules and guidelines are being followed, OSHA also provides educational materials and other support to employers and employees in an effort to raise knowledge of safe handling procedures for Hg and other hazardous materials in the workplace.

1.7.3 Environmental Protection Agency (EPA)

To protect both human health and the environment, the Environmental Protection Agency (EPA) of the United States plays a critical role in Hg regulation. The EPA establishes emission standards (Mercury and Air Toxics Standards, MATS) for Hg from industrial sources, regulates Hg in water discharges, handles hazardous waste containing Hg, addresses Hg in products, and implements the Minamata Convention to reduce global Hg emissions under several federal laws, including the Clean Air Act, Clean Water Act, and Toxic Substances Control Act (EPA 2023a, b). The EPA aims to reduce Hg exposure, mitigate its negative effects, and encourage responsible practices to ensure a safer and healthier future for everyone. This is done through research, risk assessments, technical assistance, and enforcement.

1.7.4 World Health Organization (WHO)

The World Health Organization (WHO) highlights the need for efficient methods to limit Hg pollution and exposure and acknowledges Hg emissions as a significant worldwide health hazard. According to the WHO, exposure to Hg can have serious health repercussions, especially for vulnerable groups like children and pregnant women. WHO supports the execution of the Minamata Convention on Hg, which strives to regulate and decrease Hg discharges and emissions, as part of its efforts. WHO also promotes awareness of the dangers of Hg exposure through research, advice, and public health activities. In addition, it urges nations to establish strict policies to control emissions and safeguard both human health and the environment from the damaging impacts of this dangerous metal.

1.7.5 Food and Agriculture Organization (FAO) and World Health Organization (WHO) Joint Expert Committee on Food Additives (JECFA)

By assessing the safety of food additives and contaminants, including Hg, and establishing acceptable levels in food and drinking water, the Food and Agriculture Organization (FAO) and WHO Joint Expert Committee on Food Additives (JECFA) play a crucial role in Hg regulation. To establish safe limits for Hg exposure through food intake, JECFA conducts thorough risk assessments, considering the potential effects on various demographic groups' health. Their scientific knowledge drives global standards and recommendations, ensuring that Hg levels in food are maintained within acceptable ranges to safeguard public health and encourage ethical production and consumption of food practices everywhere. JECFA set the provisional tolerable weekly intake for Me-Hg of 1.6 $\mu g\ kg^{-1}$ body weight per week (Yaginuma-Sakurai et al. 2012).

1.7.6 European Union (EU) Regulations

The European Union (EU) has put in place a comprehensive set of regulations to control pollution and Hg exposure. These rules address a variety of issues, including banning the export of metallic Hg, limiting the use of Hg in particular goods, establishing emission standards for industrial operations, phasing out the use of dental amalgam, and governing how to dispose of Hg-containing lights. The EU wants to protect human health from the harmful consequences of Hg exposure by reducing Hg emissions, preventing Hg leakage into the environment, and taking these actions. These rules demonstrate the EU's dedication to encouraging safer practices, lowering Hg-related risks, and protecting its people and the environment from the dangers that Hg poses. The EU has also fixed a set of permissible Hg limits in different environments and food items (EU 2023).

1.7.7 National Regulations

Regarding Hg exposure, numerous nations each have their own distinct laws and recommendations. Depending on the nation's priorities for public health and the environment, these may differ in breadth and strictness.

1.7.8 Safer Alternatives and Best Practices

International organizations and regulatory bodies actively encourage the use of safer substitutes for Hg in a variety of applications. To reduce the danger of exposure, they also share the best handling and management techniques for garbage and items containing Hg.

Public Health Initiatives

Governmental organizations and non-government organizations engage in public health initiatives to raise awareness of the risks of Hg exposure and its effects on health. These initiatives, outreach programs, and instructional campaigns may target vulnerable populations, such as expectant mothers, children, and subsistence fishermen who rely heavily on fishing for their livelihood.

1.8 Mitigation and Remediation of Mercury Exposure

Significant dangers to the environment and human health come with Hg exposure. To reduce the negative consequences of this hazardous heavy metal, mitigation and remediation activities are extremely important. We can effectively minimize Hg exposure and protect the well-being of ecosystems and communities by putting in place comprehensive policies and focused initiatives. To build a more secure and sustainable future for everyone, we'll showcase the numerous techniques and technologies used in this debate to mitigate and clean up Hg contamination.

1.8.1 Emission Controls

Emissions from coal combustion in industries are the most significant source of Hg in the environment, accounting for about 45% of the anthropogenic emissions (Thepanondh and Tunlathorntham 2020; Dziok et al. 2015). Thus, it is essential to put strong emission controls in place for commercial operations and other operations that release Hg into the atmosphere. In addition, choosing the most relevant and effective technology to reduce Hg emissions can be successful in limiting Hg emissions (Burmistrz et al. 2016). Before flue gases are released into the atmosphere, Hg can be captured and removed from them using technologies including activated carbon injection, scrubbers, and catalytic converters (Hadi et al. 2015; Di Natale et al. 2011). In this method, powdered activated carbon is introduced into the flue gas stream just before it enters the emission control system. The increased surface area of activated carbon particles allows for the adsorption of Hg compounds as well as Hg vapor. These methods are extensively practiced in countries in Europe and America to reduce Hg emissions.

1.8.2 Regulations and Bans

Mercury is widely used in many products, especially electronic and electrical devices and medical equipment. The amount of Hg released into the environment can be considerably reduced by enforcing laws and restrictions on these goods containing Hg. For instance, restrictions on the use of Hg in thermometers, batteries, and certain industrial applications All the developed countries, especially the exporters of such kinds of Hg-containing products, like countries in North America, Europe, Japan, and China, have already signed and executed laws regarding the restrictions and bans on using Hg-containing products (Rhee 2015). This will significantly reduce the risk of Hg exposure.

1.8.3 Waste Management

Mercury should never be released into the environment, so it must be handled and disposed of properly. Mercury contamination can be reduced through recycling and the secure trash disposal of items containing Hg. Counties all over the world have established technologies for efficient management of Hg-containing waste, like solidification or stabilization (Eckley et al. 2020; Zhang et al. 2009) and thermal treatment (e.g., batch retorting, ex-situ thermal desorption, in-situ vitrification) (Rumayor et al. 2016; Kunkel et al. 2006).

1.8.4 Clean-Up of Contaminated Sites

It is essential to clean up Hg-contaminated areas, such as old industrial sites and places where artisanal gold mining occurs. To stop further Hg migration, remediation techniques may include excavation or dredging with removal, containment in-place, ex-situ soil washing, solidification/stabilization, and thermal treatment (Fontaine 2023; Randall and Chattopadhyay 2013; Liu et al. 2018; Xu et al. 2015; Subires-Munoz et al. 2011; Zhang et al. 2009). In the United States, within the framework of the multimillion-dollar cleanup of a defunct chlor-alkali facility, approximately 5 million kg of Hg-contaminated soil were hauled over 1,000 km to a toxic waste disposal facility in 2015. Such initiatives have the potential to significantly lower Hg flows and levels at a site. Alternative methods are frequently used, or off-site removals are concentrated primarily on locations with the greatest levels of contamination or the possibility of runoff due to the high costs.

1.8.5 Mercury-Free Gold Mining Practices

Mercury exposure for miners and nearby communities can be decreased by encouraging and promoting Hg-free or low-Hg gold extraction methods in artisanal and small-scale gold mining (Keane et al. 2023). In the twentieth century, these methods have become popular across the world. Some examples of these methods are panning, sluicing, shaking tables, spiral concentrators, vortex concentrators, and centrifuges (EPA 2023a, b).

1.8.6 Public Awareness and Mass Education

Better methods for handling Hg and decreasing exposure can result from raising awareness of the risks of exposure and educating communities, workers, and vulnerable groups. This is most effective in areas where fish and seafood consumption are more popular. For instance, in Surabaya, Indonesia, education on Hg exposure from fish and fish products was provided to schoolchildren in the Kenjeran Beach Area. The result showed that education significantly improved knowledge and Hg exposure in the intervention group compared to those who were not educated (Mahmudiono et al. 2023).

1.8.7 Mercury-Free Alternatives

Alternatives free of Hg are now more common than ever for a variety of products. These substitutes include solid-state or electronic switches and relays, composite resin or ceramic dental fillings in place of amalgam, Hg-free UV lamps and light sources, Hg-free batteries, digital and infrared thermometers in place of Hg-based models, gallium or Galinstan thermometers, and replacements for traditional fluorescent lamps like LED and CFL. Industries have been working hard to provide safer and more environmentally friendly options as knowledge of environmental and health issues rises. As a result, there is a declining demand for items that contain Hg and a rising need for sustainable replacements. This would be the most significant approach if it could be implemented all over the world, as it would reduce the possibility of Hg exposure to humans.

1.8.8 Fish Consumption Advisories

Fish consumption advisories are recommendations that are made by public health organizations to educate the public about the potential dangers of eating species of fish that are captured in particular bodies of water. The advisories are generally determined by the concentrations of pollutants like Hg and other pollutants that are detected in fish. These warnings are primarily intended to safeguard vulnerable groups from the risks associated with consuming tainted fish. Giving warnings about eating fish in places with Hg-contaminated water bodies can help communities learn how much fish is safe to eat, especially for vulnerable groups like children and pregnant women. The advisories may differ from state to state, country to country, and from time to time. For instance, American states have more freedom to set rules for regional water bodies compared to Japan (Sei and Watanabe 2012).

1.8.9 Health Interventions

Different chemical forms of Hg exist, and each form has a unique impact on the health of the body. For those who have been exposed to Hg, early diagnosis and treatment of Hg poisoning are crucial. Chelating agents, chelating agent combinations, plasma exchange, hemodialysis, and plasmapheresis are all forms of treatment for Hg poisoning symptoms in patients (Ye et al. 2016). However, the mechanism of chelating agents in the detoxification of Hg is still unclear. Extensive research is needed to explore the mechanisms.

1.8.10 Monitoring and Research

Monitoring and research are essential steps in reducing Hg exposure and creating solutions for effective mitigation. It is essential to continuously monitor Hg levels in the environment, food supply, and human population to pinpoint probable sources of pollution, weigh the hazards of exposure, and monitor the success of corrective actions. International partnerships create worldwide monitoring networks that track Hg contamination across international borders, making it easier to determine how it is distributed globally and how it is transported over great distances. Real-time monitoring of Hg concentrations in the air and water is now possible because of technological advancements, enabling more precise data collection and quick responses to pollution surges. Additionally, continuing research examines the short- and long-term consequences of Hg exposure on human health. Additionally, by improving the precision of risk assessments relating to Hg exposure, these research initiatives guarantee that regulatory choices are supported by the most recent research findings.

1.9 Outlook and Recommendations

The Minamata Convention and other current agreements and regulations must be strengthened by the governments of all the countries and organizations working on Hg. These entail enforcing emission controls, keeping track of compliance, and prosecuting offenders of Hg-related legislation. Promoting the use of safer substitutes for Hg in diverse industries and applications should continue to be a priority. Safer and more environmentally friendly practices may result from the research and development of Hg-free technologies. Customized awareness efforts should concentrate on informing communities, employees, and at-risk groups about the dangers of Hg exposure and how to reduce them. This is especially important in areas with a high prevalence of artisanal gold mining, cement production, and subsistence fishing. To combat global Hg contamination, governments, organizations, and scientific communities must collaborate internationally and share knowledge. Sharing research findings and best practices can result in more efficient tactics. It is crucial to make investments in the cleanup of hazardous locations, like old industrial zones and mining districts. Proper cleanup procedures can safeguard nearby populations and stop further Hg exposure. It is essential to support the development of poor nations' capacity to address Hg exposure. This includes developing frameworks for monitoring and regulation, encouraging Hg-safe practices, and training healthcare workers. The sources, movement, and destiny of Hg in the environment require ongoing study. Monitoring Hg levels in the environment on a regular basis can assist in identifying developing exposure issues and evaluating the success of mitigating measures. International cooperation is necessary to solve this issue because artisanal and small-scale gold mining as well as cement production are key sources of Hg exposure. Hg-free mining practices have become more widespread in recent years. Sustainable mining techniques and such Hg-free alternatives can lessen the impact on ecosystems and vulnerable populations. Industries that use or emit Hg should also adopt ethical standards and make investments in Hg emission-reducing technology. Promoting corporate social responsibility can be extremely important for reducing Hg emissions. Lastly, encouraging neighborhood groups to take an active role in Hg exposure reduction initiatives can result in more long-lasting and regionally specific solutions. All these initiatives will enable us to live in an environment with less Hg.

References

Azevedo R, Rodriguez E (2012) Phytotoxicity of mercury in plants: a review. J Bot 2012:848614. https://doi.org/10.1155/2012/848614

Barboza LGA, Vieira LR, Branco V, Carvalho C, Guilhermino L (2018) Microplastics increase mercury bioconcentration in gills and bioaccumulation in the liver, and cause oxidative stress and damage in *Dicentrarchus labrax* juveniles. Sci Rep 8(1):15655. https://www.nature.com/articles/s41598-018-34125-z

Batchelar KL, Kidd KA, Drevnick PE, Munkittrick KR, Burgess NM, Roberts AP, Smith JD (2013) Evidence of impaired health in yellow perch (*Perca flavescens*) from a biological mercury hotspot in northeastern North America. Environ Toxicol Chem 32(3):627–637

Berlin M (2020) Mercury in dental amalgam: a risk analysis. Neurotoxicol 81:382–386

Burmistrz P, Kogut K, Marczak M, Zwoździak J (2016) Lignites and subbituminous coals combustion in Polish power plants as a source of anthropogenic mercury emission. Fuel Process Technol 152:250–258

Cariccio VL, Samà A, Bramanti P, Mazzon E (2019) Mercury involvement in neuronal damage and in neurodegenerative diseases. Biol Trace Elem Res 187:341–356

Chen B, Dong S (2022) Mercury contamination in fish and its effects on the health of pregnant women and their fetuses, and guidance for fish consumption—a narrative review. Int J Environ Res Pub Health 19(23):15929. https://doi.org/10.3390/ijerph192315929

Chen YA, Chi WC, Trinh NN, Huang LY, Chen YC, Cheng KT et al (2014a) Transcriptome profiling and physiological studies reveal a major role for aromatic amino acids in mercury stress tolerance in rice seedlings. PLoS ONE 9:e95163. https://doi.org/10.1371/journal.pone.0095163

Chen Z, Myers R, Wei T, Bind E, Kassim P, Wang G, Ji Y, Hong X, Caruso D, Bartell T, Gong Y (2014b) Placental transfer and concentrations of cadmium, mercury, lead, and selenium in mothers, newborns, and young children. J Expos Sci Environ Epidemiol 24(5):537–544

Costa RA, Eeva T, Eira C, Vaqueiro J, Medina P, Vingada JV (2014) Great tits breeding performance and mercury contamination from the paper and pulp industry in the west coast of Portugal. Chem Ecol 30:206–215

Costantini D, Meillère A, Carravieri A, Lecomte V, Sorci G, Faivre B, Weimerskirch H, Bustamante P, Labadie P, Budzinski H, Chastel O (2014) Oxidative stress in relation to reproduction, contaminants, gender and age in a long-lived seabird. Oecologia 175:1107–1116

Counter SA, Buchanan LH (2004) Mercury exposure in children: a review. Toxicol Appl Pharmacol 198(2):209–230

Czaika E, Edwards B (2013) History of mercury use in products and processes. Retrieved from: https://mercurypolicy.scripts.mit.edu/blog/?p=367. Accessed on 30 July 2023

Dastoor A, Angot H, Bieser J, Christensen JH, Douglas TA, Heimbürger-Boavida LE, Jiskra M, Mason RP, McLagan DS, Obrist D, Outridge PM (2022) Arctic mercury cycling. Nat Rev Earth Environ 3(4):270–286

Di Natale F, Erto A, Lancia A, Musmarra D (2011) Mercury adsorption on granular activated carbon in aqueous solutions containing nitrates and chlorides. J Hazard Mat 192:1842–1850

Dudeja P, Dudeja K, Grover S, Singh H, Jabin Z (2023) Pathway to mercury-free dentistry: an insight into past, present, and future. Eur Oral Res 57(2):60–67

Dziok T, Strugała A, Rozwadowski A, Macherzyński M (2015) Studies of the correlation between mercury content and the content of various forms of sulfur in Polish hard coals. Fuel 159:206–213

Eckley CS, Gilmour CC, Janssen S, Luxton TP, Randall PM, Whalin L, Austin C (2020) The assessment and remediation of mercury contaminated sites: a review of current approaches. Sci Total Environ 707:136031. https://doi.org/10.1016/j.scitotenv.2019.136031

EPA (2023a) Artisanal and Small-Scale Gold Mining Without Mercury. Retrieved from: https://www.epa.gov/international-cooperation/artisanal-and-small-scale-gold-mining-without-mercury. Accessed on 29 July 2023

EPA (2023b) What EPA is doing to reduce mercury pollution, and exposures to mercury. Retrieved from: https://www.epa.gov/mercury/what-epa-doing-reduce-mercury-pollution-and-exposures-mercury. Accessed on 30 July 2023

EU (2023) Mercury. Retrieved from: https://eur-lex.europa.eu/legal-content/EN/TXT/PDF/?uri=CELEX:32022R0617. Accessed on 30 July 2023

Fontaine BL Jr (2023) Comprehensive occupational and environmental risk assessment of elemental mercury at the Edison national historic site in West Orange, NJ. Collections 19(2):243–260

Fort J, Lacoue-Labarthe T, Nguyen HL, Boué A, Spitz J, Bustamante P (2015) Mercury in wintering seabirds, an aggravating factor to winter wrecks? Sci Total Environ 527:448–454. https://doi.org/10.1016/j.scitotenv.2015.05.018

Franceschini MD, Evers DC, Kenow KP, Meyer MW, Pokras M, Romero LM (2017) Mercury correlates with altered corticosterone but not testosterone or estradiol concentrations in common loons. Ecotoxicol Environ Saf 142:348–354

Guerrero S, Schneider L (2023) The global roots of pre-1900 legacy mercury. Proc Natl Acad Sci 120(31):e2304059120. https://doi.org/10.1073/pnas.2304059120

Guerrero S, Schneider L (2023) The global roots of pre-1900 legacy mercury. Proc Natl Acad Sci 120(31):e2304059120. https://doi.org/10.1073/pnas.2304059120

Gworek B, Dmuchowski W, Baczewska-Dąbrowska AH (2020) Mercury in the terrestrial environment: a review. Environ Sci Eur 32(1):1–19

Hadi P, To MH, Hui CW, Lin CSK, McKay G (2015) Aqueous mercury adsorption by activated carbons. Water Res 73:37–55

Hoffman DJ, Spalding MG, Frederick PC (2005) Subchronic effects of methylmercury on plasma and organ biochemistries in great egret nestlings. Environ Toxicol Chem 24:3078–3084

Hopkins BC, Willson JD, Hopkins WA (2013) Mercury exposure is associated with negative effects on turtle reproduction. Environ Sci Technol 47(5):2416–2422

Hu XF, Lowe M, Chan HM (2021) Mercury exposure, cardiovascular disease, and mortality: a systematic review and dose-response meta-analysis. Environ Res 193:110538. https://doi.org/10.1016/j.envres.2020.110538

Hu H, Gao Y, Yu H, Xiao H, Chen S, Tan W, Tang J, Xi B (2023) Mechanisms and biological effects of organic amendments on mercury speciation in soil–rice systems: a review. Ecotoxicol Environ Saf 251:114516. https://doi.org/10.1016/j.ecoenv.2023.114516

Huang W, Cao L, Shan X, Lin L, Dou S (2011) Toxicity testing of waterborne mercury with red sea bream (*Pagrus major*) embryos and larvae. Bull Environ Contam Toxicol 86:398–405

Huang Y, Deng M, Li T, Japenga J, Chen Q, Yang X, He Z (2017) Anthropogenic mercury emissions from 1980 to 2012 in China. Environ Pollut 226:230–239

Israr M, Sahi S, Datta R, Sarkar D (2006) Bioaccumulation and physiological effects of mercury in Sesbania drummonii. Chemosphere 65:591–598

Ito H, Manabe A (2022) Chlor–alkali electrolysis. In: Smolinka T, Garche J (eds) Electrochemical power sources: fundamentals, systems, and applications. Elsevier, pp 281–304

Joy A, Qureshi A (2020) Mercury in dental amalgam, online retail, and the Minamata convention on mercury. Environ Sci Technol 54(22):14139–14142

Joy A, Qureshi A (2023) Reducing mercury emissions from coal-fired power plants in India: possibilities and challenges. Ambio 52(1):242–252

Keane S, Bernaudat L, Davis KJ, Stylo M, Mutemeri N, Singo P, Twala P, Mutemeri I, Nakafeero A, Etui ID (2023) Mercury and artisanal and small-scale gold mining: review of global use estimates and considerations for promoting mercury-free alternatives. Ambio 52(5):833–852

Kenow KP, Hoffman DJ, Hines RK, Meyer MW, Bickham JW, Matson CW, Stebbins KR, Montagna P, Elfessi A (2008) Effects of methylmercury exposure on glutathione metabolism, oxidative stress, and chromosomal damage in captive-reared common loon (*Gavia immer*) chicks. Environ Pollut 156:732–738

Kessler R (2013) The Minamata convention on Mercury: a first step toward protecting future generations. Environ Health Perspect 121(10):304–309

Kunkel AM, Seibert JJ, Elliott LJ, Kelley R, Katz LE, Pope GA (2006) Remediation of elemental mercury using in situ thermal desorption (ISTD). Environ Sci Technol 40:2384–2389

Lee S, Kim JH, Moon HB, Park J, Choi G, Kim S (2023) Effects of mercury exposure on fetal body burden and its association with infant growth. Environ Res 217:114780. https://doi.org/10.1016/j.envres.2022.114780

Li J, Tian X, Zhao J, Cui L, Wei L, Gao Y, Li B, Li YF (2023) Temporal changes of blood mercury concentrations in Chinese newborns and the general public from 1980s to 2020s. J Trace Elem Med Biol 127126. https://doi.org/10.1016/j.jtemb.2023.127126

Lin YC, Chang WH, Li TC, Iwata O, Chen HL (2023) Health risk of infants exposed to lead and mercury through breastfeeding. Expos Health 15(1):255–267

Liu Q, Kingler RH, Wimpee B, Dellinger M, King-Heiden T, Grzybowski J, Gerstenberger SL, Weber DN, Carvan MJ (2016) Maternal methylmercury from a wild-caught walleye diet induces developmental abnormalities in zebrafish. Reprod Toxicol 65:272–282

Liu LW, Li W, Song WP, Guo MX (2018) Remediation techniques for heavy metal-contaminated soils: principles and applicability. Sci Total Environ 633:206–219

Liu G, Cai Y, O'Driscoll N, Feng X, Jiang G (2012) Overview of mercury in the environment. In: Liu G, Cai Y, O'Driscoll N (eds) Environmental chemistry and toxicology of mercury, pp 1–12. https://doi.org/10.1002/9781118146644.ch1

Mahmudiono T, Wijanarko MAV, Elkarima E, Fikrah'Arifah D, Indriani D, Setyaningtyas SW, Chen HL (2023) Education on mercury exposure from fish and its processed products among school children in the Kenjeran Beach Area, Surabaya. J Pub Health Africa 14(s2):2623. https://www.publichealthinafrica.org/jphia/article/view/2623

Mambrey V, Rakete S, Tobollik M, Shoko D, Moyo D, Schutzmeier P, Steckling-Muschack N, Muteti-Fana S, Bose-O'Reilly S (2020) Artisanal and small-scale gold mining: a cross-sectional assessment of occupational mercury exposure and exposure risk factors in Kadoma and Shurugwi, Zimbabwe. Environ Res 184:109379. https://doi.org/10.1016/j.envres.2020.109379

Mandor M (2023) Elemental mercury removal from natural gas using nano-TiO_2. J Chem Environ 2(1):98–108

Matta G, Gjyli L (2016) Mercury, lead and arsenic: impact on environment and human health. J Chem Pharm Sci 9(2):718–725

Molina Ramirez EK (2023) Remediation of soils contaminated with mercury during artisanal and small-scale gold mining operations in Colombia using silver-coated plates. https://ore.exeter.ac.uk/repository/handle/10871/133553

Morcillo P, Meseguer J, Esteban MA, Cuesta A (2016) In vitro effects of metals on isolated head-kidney and blood leucocytes of the teleost fish *Sparus aurata* L. and *Dicentrarchus labrax* L. Fish Shellfish Immunol 54:77–85

Murata K, Karita K (2022) Minamata disease. In: Nakajima T, Nakamura K, Nohara K, Kondoh A (eds) Overcoming environmental risks to achieve sustainable development goals. Curr Top Environ Health Prev Med. Springer, Singapore, pp 9–19. https://doi.org/10.1007/978-981-16-6249-2_2

O'Bryhim JR, Adams DH, Spaet JLY, Mills G, Lance SL (2017) Relationships of mercury concentrations across tissue types, muscle regions and fins for two shark species. Environ Pollut 223:323–333

Pamphlett R, Doble PA, Bishop DP (2021) The prevalence of inorganic mercury in human kidneys suggests a role for toxic metals in essential hypertension. Toxics 9:67. https://doi.org/10.3390/toxics9030067

Peterson BD, Krabbenhoft DP, McMahon KD, Ogorek JM, Tate MT, Orem WH, Poulin BA (2023) Environmental formation of methylmercury is controlled by synergy of inorganic mercury bioavailability and microbial mercury-methylation capacity. Environ Microbiol. https://doi.org/10.1111/1462-2920.16364

Podgórska A, Puścion-Jakubik A, Grodzka A, Naliwajko SK, Markiewicz-Żukowska R, Socha K (2021) Natural and conventional cosmetics—mercury exposure assessment. Molecules 26(13):4088. https://doi.org/10.3390/molecules26134088

Qureshi A (2021) Mercury in skin-care products in India and consumer exposure risks. Regul Toxicol Pharmacol 121:104870. https://doi.org/10.1016/j.yrtph.2021.104870

Randall PM, Chattopadhyay S (2013) Mercury contaminated sediment sites—an evaluation of remedial options. Environ Res 125:131–149

Rhee SW (2015) Control of mercury emissions: policies, technologies, and future trends. Energy Emission Control Technol 2016(4):1–15

Rodrigues PDA, de Pinho JV, Ramos-Filho AM, Neves GL, Conte-Junior CA (2023) Mercury contamination in seafood from an aquatic environment impacted by anthropic activity: seasonality and human health risk. Environ Sci Pollut Res 1–15. https://doi.org/10.1007/s11356-023-28435-7

Rumayor M, Lopez-Anton MA, Diaz-Somoano M, Maroto-Valer MM, Richard JH, Biester H et al (2016) A comparison of devices using thermal desorption for mercury speciation in solids. Talanta 150:272–277
Sapre S, Deshmukh R, Gontia-Mishra I, Tiwari S (2019) Problem of mercury toxicity in crop plants: can plant growth promoting microbes (PGPM) be an effective solution? In: Maheshwari D, Dheeman S (eds) Field crops: sustainable management by PGPR. Springer International Publishing, Cham, pp 253–278
Schwartz M, Smits K, Phelan T (2023) Quantifying mercury use in artisanal and small-scale gold mining for the Minamata convention on Mercury's national action plans: approaches and policy implications. Environ Sci Pol 141:1–10
Ser PH, Watanabe C (2012) Fish advisories in the USA and Japan: risk communication and public awareness of a common idea with different backgrounds. Asia Pac J Clin Nutr 21(4):487–494
Singh AD, Khanna K, Kour J, Dhiman S, Bhardwaj T, Devi K, Sharma N, Kumar P, Kapoor N, Sharma P, Arora P (2023) Critical review on biogeochemical dynamics of mercury (Hg) and its abatement strategies. Chemosphere 319:137917. https://doi.org/10.1016/j.chemosphere.2023.137917
Skalny AV, Aschner M, Sekacheva MI, Santamaria A, Barbosa F, Ferrer B, Aaseth J, Paoliello MMB, Rocha JBT, Tinkov AA (2022) Mercury and cancer: where are we now after two decades of research? Food Chem Toxicol 164:13001. https://doi.org/10.1016/j.fct.2022.113001
Soares TA, Souza-Kasprzyk J, de Assis Guilherme Padilha J, Convey P, Costa ES, Torres JPM (2023) Ornithogenic mercury input to soils of Admiralty Bay, King George Island, Antarctica. Pol Biol 1–11. https://doi.org/10.1007/s00300-023-03162-4
Soares LC, Preto O (2020) From "Blue Pills" to the Minamata convention: mercury, a singular metal. Bull Hist Chem 45(2):67–128
Streets DG, Horowitz HM, Lu Z, Levin L, Thackray CP, Sunderland EM (2019) Five hundred years of anthropogenic mercury: spatial and temporal release profiles. Environ Res Lett 14(8):084004. https://doi.org/10.1088/1748-9326/ab281f/meta
Subires-Munoz JD, Garcia-Rubio A, Vereda-Alonso C, Gomez-Lahoz C, Rodriguez-Maroto JM, Garcia-Herruzo F et al (2011) Feasibility study of the use of different extractant agents in the remediation of a mercury contaminated soil from Almaden. Sep Purif Technol 79:151–156
Tartu S, Angelier F, Wingfield JC, Bustamante P, Labadie P, Budzinski H, Weimerskirch H, Bustnes JO, Chastel O (2015) Corticosterone, prolactin and egg neglect behavior in relation to mercury and legacy POPs in a long-lived Antarctic bird. Sci Total Environ 505:180–188
Thepanondh S, Tunlathorntham V (2020) Appropriate scenarios for mercury emission control from coal-fired power plant in Thailand: emissions and ambient concentrations analysis. Heliyon 6(6). https://doi.org/10.1016/j.heliyon.2020.e04197
Tibau AV, Grube BD (2019) Mercury contamination from dental amalgam. J Health Pollut 9(22):190612. https://doi.org/10.5696/2156-9614-9.22.190612
Todd BD, Willson JD, Bergeron CM, Hopkins WA (2012) Do effects of mercury in larval amphibians persist after metamorphosis? Ecotoxicol 21:87–95
UNEP (2023) Toxic mercury, a significant threat to marine wildlife. Retrieved from: https://mercuryconvention.org/en/news/toxic-mercury-significant-threat-marine-wildlife#:~:text=Marine%20mammals%20such%20as%20dolphins,behavioral%20changes%2C%20and%20even%20death. Accessed on 30 July 2023
Wang H, Matsushita MT (2021) Heavy metals and adult neurogenesis. Curr Opin Toxicol 26:14–21
Wang Y, Li L, Yao C, Tian X, Wu Y, Xie Q, Wang D (2021) Mercury in human hair and its implications for health investigation. Curr Opin Environ Sci 22:100271. https://doi.org/10.1016/j.coesh.2021.100271
Wang Y, Wu P, Zhang Y (2023b) Climate-driven changes of global marine mercury cycles in 2100. Proceed Natl Acad Sci 120(2):e2202488120. https://doi.org/10.1073/pnas.2202488120
Wang Z, Fang X (2021) Chronic mercury poisoning from daily cosmetics: case report and brief literature review. Cureus 13(11). https://doi.org/10.7759/cureus.19916

Wang X, Xie Q, Wang Y, Lü H, Fu M, Wang D, Li J (2023a) Hg bioaccumulation in the aquatic food web from tributaries of the Three Gorges Reservoir, China and potential consumption advisories. J Hazard Mat 131902. https://doi.org/10.1016/j.jhazmat.2023.131902

Whitney MC, Cristol DA (2017) Impacts of sublethal mercury exposure on birds: a detailed review. Rev Environ Cont Toxicol 244:113–163

Wilman B, Staniszewska M, Bełdowska M (2023) Is the inhalation influence on the level of mercury and PAHs in the lungs of the baltic grey seal (*Halichoerus grypus* grypus)? Environ Pollut 320:121083. https://doi.org/10.1016/j.envpol.2023.121083

Wilson E, Lafferty JS, Thiboldeaux R, Tomasallo C, Grajewski B, Wozniak R, Meiman J (2018) Occupational mercury exposure at a fluorescent lamp recycling facility—Wisconsin, 2017. Morb Mort Week Rep 67(27):763–766

Wolf SE, Swaddle JP, Cristol DA, Buchser WJ (2017) Methylmercury exposure reduces the auditory brainstem response of zebra finches (*Taeniopygia guttata*). J Assoc Res Otolaryngol 18:569–579

Xu J, Bravo AG, Lagerkvist A, Bertilsson S, Sjöblom R, Kumpiene J (2015) Sources and remediation techniques for mercury contaminated soil. Environ Int 74:42–53

Yaginuma-Sakurai K, Murata K, Iwai-Shimada M, Nakai K, Kurokawa N, Tatsuta N, Satoh H (2012) Hair-to-blood ratio and biological half-life of mercury: experimental study of methylmercury exposure through fish consumption in humans. J Toxicol Sci 37(1):123–130

Ye BJ, Kim BG, Jeon MJ, Kim SY, Kim HC, Jang TW, Chae HJ, Choi WJ, Ha MN, Hong YS (2016) Evaluation of mercury exposure level, clinical diagnosis and treatment for mercury intoxication. J Occup Environ Med 28(1):1–8

Yu MS, Eng ML, Williams TD, Basu N, Elliott JE (2016) Acute embryotoxic effects but no longterm reproductive effects of in ovo methylmercury exposure in zebra finches (*Taeniopygia guttata*). Environ Toxicol Chem 35:1534–1540

Yuan CS, Chiang KC, Yen PH, Ceng JH, Lee CE, Du IC, Soong KY, Jeng MS (2023) Long-range transport of atmospheric speciated mercury from the eastern waters of Taiwan Island to northern South China Sea. Environ Pollut 318:120899. https://doi.org/10.1016/j.envpol.2022.120899

Zhang XY, Wang QC, Zhang SQ, Sun XJ, Zhang ZS (2009) Stabilization/solidification (S/S) of mercury-contaminated hazardous wastes using thiol-functionalized zeolite and Portland cement. J Hazard Mat 168:1575–1580

Zhang QF, Li YW, Liu ZH, Chen QL (2016) Reproductive toxicity of inorganic mercury exposure in adult zebrafish: histological damage, oxidative stress, and alterations of sex hormone and gene expression in the hypothalamic-pituitary-gonadal axis. Aquat Toxicol 177:417–424

Zhang M, Yang G, Liu S, Yu J, Li H, Zhang L, Chen Y, Guo R, Wu T (2022) MoS2 quantum dots based MoS2/HKUST-1 composites for the highly efficient catalytic oxidation of elementary mercury. J Environ Sci 116:163–174

Zhao Z, Yang J, Zhao M, Zhu J (2023) Mercury waste management in China: implications on the implementation of the Minamata Convention on Mercury. J Mat Cycl Waste Manag 1–15. https://doi.org/10.1007/s10163-023-01742-4

Zheng NA, Wang S, Dong WU, Hua X, Li Y, Song X, Chu Q, Hou S, Li Y (2019) The toxicological effects of mercury exposure in marine fish. Bull Environ Contam Toxicol 102:714–720

Zhou ZS, Wang SJ, Yang ZM (2008) Biological detection and analysis of mercury toxicity to alfalfa (*Medicago sativa*) plants. Chemosphere 70:1500–1509

Zhou J, Obrist D, Dastoor A, Jiskra M, Ryjkov A (2021) Vegetation uptake of mercury and impacts on global cycling. Nat Rev Earth Environ 2(4):269–284

Chapter 2
Mercury Contamination in Food—An Overview

Shihab Uddin, Sumona Khanom, and Md. Rafiqul Islam

2.1 Introduction

Mercury (Hg) is one of the naturally occurring substances with the greatest potential for toxicity and, even in low quantities, has no known biological function in humans or plants. Since Hg has peculiar chemical characteristics, once it enters ecosystems, it can bioaccumulate and biomagnify in food chains (Evers 2018). This is the rationale behind the World Health Organization (WHO)'s classification of Hg as a factor in global public health concerns and the extensive study and monitoring of its toxicity and environmental cycle (WHO 2007). Mercury is also known as a human carcinogen that threatens ecological stability and individual safety (Yang et al. 2020). Mercury comes into the environment after being released from natural sources and human activities. This Hg may accumulate inside living things in the food chain as a result of physical, metabolic, and biological activities as well as activities by humans (Gupta et al. 2021). Mercury permeates into the soil, the air, and water sources anywhere it can be ingested or utilized by crops or plants (Kharazi et al. 2021; Proshad et al. 2019). The fact that Hg is transmitted from agricultural soils to plants and accumulates in edible parts of plants during intensive farming is particularly alarming, even though Hg bio-availability in soils is normally low due to binding or adsorbing to soil solids (Ha et al. 2017). When higher consumers in a food chain ingest Hg-contaminated edible plant parts, they subsequently bioaccumulate and become biomagnified. In nature, Hg is continually cycling through every level of the food chain because, when it does reach the top, it is difficult to get rid of. A number of hyperaccumulating plants

S. Uddin (✉) · Md. R. Islam
Department of Soil Science, Faculty of Agriculture, Bangladesh Agricultural University, Mymensingh 2202, Bangladesh
e-mail: shihab.ss@bau.edu.bd

S. Khanom
Department of Plant Pathology, Faculty of Agriculture, Bangladesh Agricultural University, Mymensingh 2202, Bangladesh

N. Kumar (ed.), *Mercury Toxicity*, Environmental Science and Engineering,
https://doi.org/10.1007/978-981-99-7719-2_2

supply nourishment for both people and animals. As a result, Hg may persist in the ecosystem for a long time and have a variety of negative impacts. This is due to the fact that after upper consumers pass away, Hg cycles from the soil to humans, through plants, and finally back into the soil. If lifeforms ingest foods containing Hg, they could be subjected to potentially hazardous health issues (Chen et al. 2021).

Mercury contamination in our food supply presents a critical challenge to public health and environmental sustainability. Living beings mainly consume methylmercury (Me-Hg), the most toxic form of Hg, from food. Up to 90% of the time, various foods can be considered sources of the majority of organic Hg compounds. According to Mata and Gjyli (2016), various foods and beverages contained a Hg concentration of 1–50 $\mu g\ kg^{-1}$. Mercury contamination of our food supply can have serious negative health effects, especially for vulnerable groups like young children, pregnant women, and babies. Therefore, it is essential to comprehend the sources and pathways of Hg in food in order to make wise dietary decisions and minimize exposure. Oceans or other water bodies are the largest reservoirs of Hg in the environment. As a result, sea foods and fish are the most significant food sources of Hg for living beings. In addition, foods like cereals, vegetables, fruits, etc. grown in Hg-contaminated soil can also be a source of Hg exposure for humans. Consumption of these foods may create a lot of health concerns, as described in detail in this chapter. Therefore, people should have the knowledge to classify food categories with varying Hg concentrations, the impact of food processing techniques on Hg levels as they impact Hg levels in food, and the importance of adopting measures to mitigate contamination. All these issues have been discussed in this chapter. Moreover, this chapter acknowledges the significance of monitoring and regulation in ensuring food safety, reflecting on efforts by governmental and international entities to control Hg levels. The challenges of enforcing regulations and the necessity of informed health risk assessments related to Hg exposure from food are also discussed. This exploration aims to foster understanding, promote awareness, and contribute to safer dietary choices in the context of Hg contamination.

2.2 How Does Mercury Come in Food?

The precise processes by which Hg enters the food chain are still poorly understood and can differ among ecosystems. There are many natural and man-made processes by which Hg enters the food chain (Fig. 2.1). It's essential to comprehend these pathways to understand how Hg contaminates our food supply. The primary natural sources of Hg include volcanic eruptions, weathering of rocks, soil erosion, etc. Mercury is released into the sky during volcanic eruptions, where it eventually finds its way into the ground and water, resulting in the entry of Hg into terrestrial and aquatic foods. Natural weathering processes progressively release the Hg that is found in rocks and minerals into the environment, which ultimately enters terrestrial foods like cereals, vegetables, fruits, etc. Mercury is transported into rivers, lakes, and seas because of the erosion of Hg-rich soils and sediments. This Hg can then accumulate in aquatic foods like fish. Similarly, significant amounts of Hg are released into the atmosphere by human activities such as cement production, the burning of fossil fuels, coal-fired power stations, and waste incineration. After being discharged into the atmosphere, Hg can travel a great distance before settling on land and water surfaces through deposition or precipitation. This ultimately leads to the entry of Hg into terrestrial and aquatic foods. Mercury pollution in water bodies through gold extraction procedures employing Hg, like artisanal and small-scale gold mining, results in Hg assimilation in aquatic organisms and foods. In addition, the use of specific pesticides and fertilizers that contain Hg increases the amount of Hg in the soil and water, which eventually bioaccumulates in foods. Furthermore, when products like batteries and electronic waste that contain Hg are disposed of improperly, Hg is released into the environment. After being deposited, this Hg can be absorbed by plants, which can later be devoured by animals and people.

Mercury can be absorbed by plants from the soil, and when herbivores eat these plants, they can get contaminated with Hg. When carnivores eat herbivores that contain Hg, they in turn ingest more Hg. Mercury undergoes a process known as biomagnification in aquatic environments. At the apex of the food chain, predatory fish eat Hg-containing fish. As a result, higher trophic levels have higher Hg concentrations. A few microorganisms are crucial in the early stages. Inorganic Hg is ingested by bacteria that break down sulfate (SO_4^{2-}) in the environment, where it undergoes metabolic transformation into methylmercury (Me-Hg). Because Me-Hg is more poisonous and takes longer for organisms to remove, the conversion of inorganic Hg to Me-Hg is crucial. The next level of the food chain may eat these Me-Hg-containing bacteria, or the bacteria may excrete the Hg into the water, where it can swiftly bind to plankton, which is likewise eaten by the subsequent level of the food chain. Animals ingest increasing quantities of Hg at each subsequent rung of the food chain because they collect Me-Hg more quickly than they can remove it. Thus, Me-Hg in the environment can easily build up potentially hazardous amounts in fish, fish-eating species, and humans. Mercury biomagnification can have harmful consequences for consumers at the top of these aquatic food chains, even at extremely low atmospheric deposition rates at sites far from point sources. Understanding these routes enables people and decision-makers to take action to prevent Hg contamination in the environment and limit dietary Hg exposure. In order to address the issue of Hg in our food supply, regulatory actions, sustainable practices, and public awareness are crucial.

2.3 Navigating Mercury Risk in Our Diet: Identifying Foods with Higher Mercury Content

To protect our health and the welfare of our family, it is crucial that we manage the Hg risk in our diet. Strong neurotoxin Hg has negative effects on the nervous system, especially in high doses. Making informed dietary decisions and lowering exposure

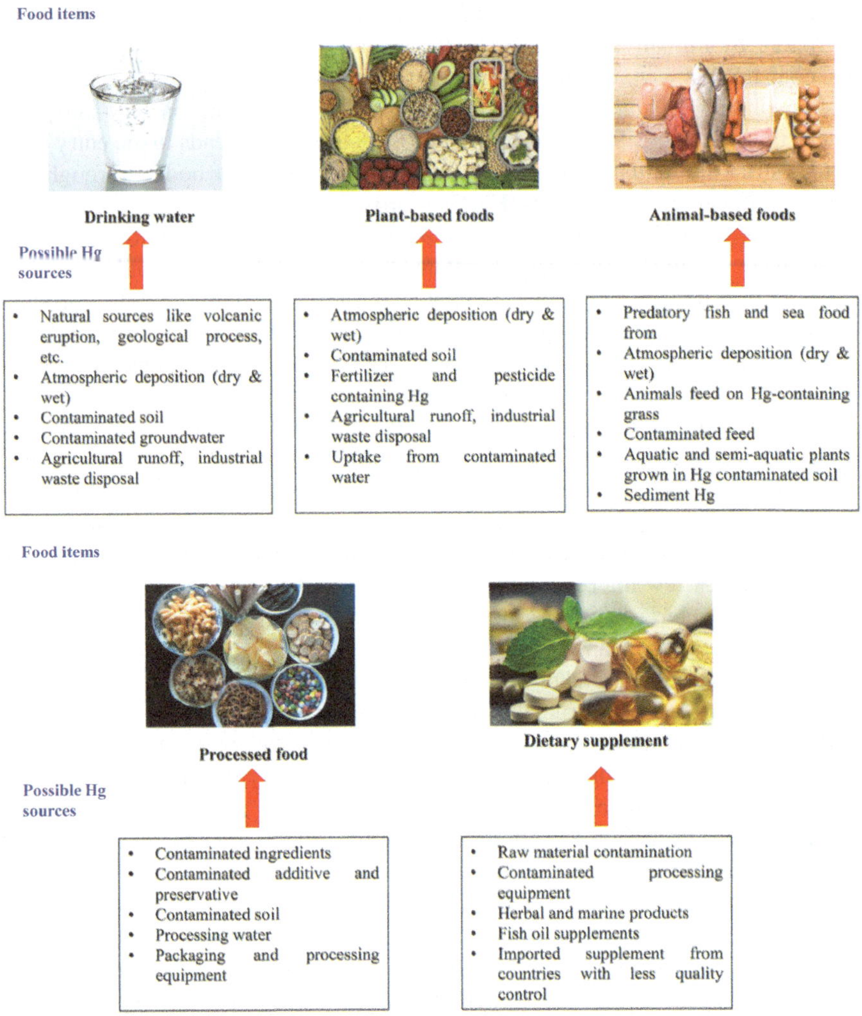

Fig. 2.1 Sources of Hg in different food items

to this dangerous metal requires knowledge about foods that contain more Hg. Fish and seafood are among the most significant dietary sources of Hg exposure. Seafood contains important nutrients like omega-3 fatty acids, but because of their place in the aquatic food chain, some fish species, especially large predatory ones, tend to acquire more Hg. It's crucial to exercise caution while selecting seafood and to refer to regional fish advisories published by health and environmental organizations if we want to keep ourselves safe. When buying seafood, we may make educated judgments if we are aware of the fish species in our area that have higher Hg levels.

Limiting the consumption of fish with high Hg content, such as swordfish, sharks, king mackerel, and tilefish, is advised to reduce Hg exposure. Choose lower-Hg alternatives instead, which are generally seen as safer selections. Examples of these include salmon, shrimp, canned light tuna, and tilapia. By including a variety of foods from several food groups in our diet, we can lessen our dependency on particular foods that can contain Hg. It is essential to understand that Hg is not just found in seafood. Foods from the earthly world, like grains, fruits, and vegetables, also contain it. Although these foods usually contain less Hg than some seafood, it is vital to be aware of their potential impact on the total amount of Hg exposure. A balanced diet that includes a variety of fruits, vegetables, whole grains, and other healthy foods can help reduce exposure from terrestrial sources. If anyone takes nutritional supplements that contain fish or animal oil, he or she needs to think about finding out how much Hg is in them. To reduce the risk of exposure to supplements, it makes sense to select renowned companies that put safety and quality first. It is advised to always check with a doctor before beginning any dietary supplements, particularly if anyone is pregnant or has underlying medical issues. Additionally, consuming fewer items that have undergone extensive processing and putting more of an emphasis on whole, unprocessed foods can help lower the risk of exposure to Hg from tainted goods. As Hg tends to concentrate in fatty tissues, thoroughly cleaning and removing skin and fat from fish before cooking can also help to lower Hg exposure. We may reduce exposure and put the health and wellbeing of us and our loved ones first by implementing these practices and remaining aware of the risk of Hg in our diet. In addition, we should seek personalized medical advice from professionals, particularly if we have particular health issues or dietary limitations. By taking these steps, we may make dietary decisions that are safer and safeguard ourselves from any potential risks associated with food that may be contaminated with Hg. A dietary chart with varying Hg concentrations is shown in Fig. 2.2.

2.4 Mercury in Seafood

For a very long time, seafood—including different types of fish, molluscs, crabs, and echinoderms—served as nutritious foods and gained popularity because of their deliciousness in countries near the oceans or sea (Olatunde and Benjakul 2018). The potential health effects of exposure to Hg in seafood make Hg a major concern. Even though fish and other seafood are great sources of protein, omega-3 fatty acids, and other vital nutrients, some species can develop higher concentrations of Hg in their organs. Through both natural and man-made processes, Hg enters aquatic ecosystems. As it goes up the food chain, it bioaccumulates and reaches higher quantities in larger predatory fish. Methylmercury, a very poisonous chemical molecule, is the type of Hg that is most frequently found in seafood. When inorganic Hg that has been released into the environment due to anthropogenic and natural causes is methylated by bacteria in water bodies, Me-Hg is the result. Small aquatic species like plankton absorb this Me-Hg afterward, and larger fish and marine mammals eat this plankton.

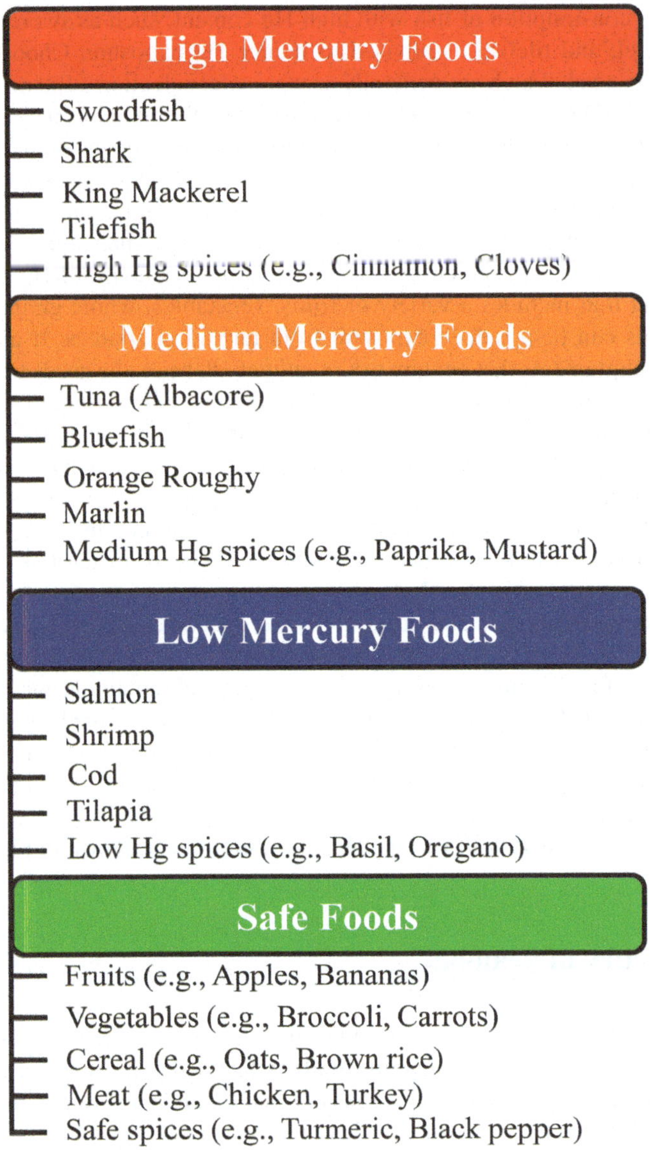

Fig. 2.2 Foods with varying Hg contents—a guide to navigating mercury risk

The level of Me-Hg rises as these species are consumed by even larger predators. Fish with longer lifespans and positions higher in the food chain tend to acquire more Hg. Large predatory fish with greater Hg concentrations include swordfish, shark, king mackerel, and tilefish (Tomada and Tomada 2023; Groth III 2010). Additionally, some tuna species, especially the bigger ones like bluefin tuna, can have high

Hg concentrations. High-Hg seafood consumption can have negative health effects, especially for some groups. Methylmercury largely impacts the nervous system, and long-term exposure, especially during pregnancy and early infancy, can have a negative impact on an infant's or young child's neurological development. To prevent potential injury to the growing fetus, pregnant women are specifically recommended to be vigilant about their seafood diet. It's important to remember nonetheless that not all seafood contains a lot of Hg. Many types of fish and seafood can be safely consumed in a balanced diet because they have a lower Hg content. Salmon, shrimp, canned light tuna, sardines, and tilapia are some examples of low-Hg seafood (Pratt 2018; Groth III 2010). These solutions provide vital nutrients while providing few hazards from Hg exposure. Following local fish advisories, which offer advice on safe seafood selections based on regional environmental monitoring, is advised to reduce Hg exposure from seafood. Diversifying the kinds of seafood eaten and choosing reduced Hg species can also help minimize exposure overall. The promotion of safe eating habits and raising public knowledge about the presence of Hg in seafood are major tasks for regulatory agencies, health groups, and consumer awareness campaigns. People can benefit from seafood while reducing potential health concerns by making educated decisions and being aware of the Hg content in seafood.

Mercury content in seafood may vary from species to species and country to country. The levels of Hg in different seafood have been reported in numerous studies. Some examples of Hg content in different seafoods across the world are presented in Table 2.1. The Hg content in various seafoods ranged from 0.526 to 2.986 ppm in the USA (Tweedy et al. 2022). In a study, Venturieri et al. (2017) reported <0.1–2.64 ppm Hg in different fish species in Brazil, while Vezzone et al. (2023) reported much lower Hg in other fish species. Shao and his research team reported 0.010–0.680 ppm Hg in different fish species collected from local markets in China in 2011 and 2012. Several authors reported 0.044–4.540 ppm Hg in different fish species in Indonesia (Table 2.1). According to the various reports, the Hg concentration in different fish species ranged from 0.020–0.470 ppm in Iran, 0.700–1.670 ppm in Italy, 0.030–0.560 ppm in Japan, 0.034–0.271 ppm in Malaysia, 0.050–0.780 ppm in Portugal, 0.010–0.420 ppm in Qatar, and 0.045–0.700 ppm in South Africa, respectively (Table 2.1). These results highlight the variation in Hg concentration among diverse seafood and fish species from different nations. Consumers must be aware of these changes and adhere to national or municipal fish warnings in order to make educated decisions about consuming seafood, particularly for expectant mothers and young children. By being aware of the Hg content of seafood, people may find a balance between reaping the nutritional advantages of fish and reducing any possible health hazards brought on by Hg exposure.

Table 2.1 Mercury content in seafish reported in different countries

Country	Fish	Study area	Hg content (mg kg^{-1})	References
Brazil	Lebranche mullet, Pearl cichlid	Estuarine survey	0.0065, 0.0580	Vezzone et al. (2023)
	Anjumara, Redeye piranha, Bicuda, Black tailed payara,	Field survey	<0.1–2.64	Venturieri et al. (2017)
China	Bighead carp, Crucian carp, Grass carp, Northern snakehead, Oriental weatherfish, Mud skipper, Yellowhead catfish	Market survey	0.010–0.680	Shao et al. (2012)
	Grass carp, Bighead carp, Crucian carp	Market survey	0.013–0.045	Shao (2011)
Indonesia	Whiting fish	Market survey	0.310–4.540	Cahyani et al. (2017)
	Fresh and saltwater fishes		0.315–0.323	Suratno et al. (2017)
	Amazon sailfin catfish		2.826–4.333	Aksari et al. (2015)
	Mantis shrimp	Sea survey	0.040–0.060	Candra et al. (2019)
	Tank goby	Market survey	Below detection-0.044	Sulistiono et al. (2018)
Iran	Sea food (Canned tuna fish)	Market survey	0.020–0.470	Andayesh et al. (2015)
	Chacunda gizzard shad, Largescale tonguesole,	Catch from sea	0.060–0.840	Keshavarzi et al. (2018)
Italy	Long nose skate, Thornback ray, Winter skate, Starry ray, Blue whiting, Striped mullet	Catch from sea	0.700–1.670	Storelli et al. (2003)
Japan	Red snow crabs	Catch from sea	0.03–0.56	Kakimoto et al. (2019)
Malaysia	Scad, Mackerel, Tuna, Prawn/Shrimp, Catfish, Snapper, Stingray, Bream		0.034–0.271	Ahmad et al. (2022)

(continued)

Table 2.1 (continued)

Country	Fish	Study area	Hg content (mg kg^{-1})	References
	Longtail tuna, Grouper, Scad, Bream, Mackerel	Sea survey	0.070–2.340 (Dry wt.)	Zulkipli et al. (2023)
Portugal	Codfish, Hake, Octopus, Horse mackerel, Sardine	Catch from sea	0.050–0.780	Rodrigues et al. (2023a, b)
Qatar	Orange-spotted grouper, Marbled spinefoot, narrow-barred Spanish mackerel, Emperor fish, mackerel tuna	Market survey	0.010–0.420	Al-Sulaiti et al. (2023)
South Africa	Mozambique tilapia, African sharptooth catfish, Brown squeaker, Bulldog, Sand Diver Lizardfish, African tetras	Catch from river and lake	0.045–0.700	van Rooyen et al. (2023)
USA	Stoneroller, Darter, Sunfish, Catfish, Gar, Smallmouth Bass	Market survey	0.526–2.986 (dry weight basis)	Tweedy et al. (2022)

2.5 Mercury in Terrestrial Food

An important part of environmental and food safety concerns is the presence of Hg in terrestrial food sources. Cereals, vegetables, fruits, and spices are only a few examples of the enormous variety of goods that fall under the category of "terrestrial food," which are staples in people's diets all around the world. Through several causes, including air deposition, soil pollution, and agricultural practices, these foods can get contaminated with Hg. The composition of the soil, how it is used, and how close it is to pollution sources are all factors that affect the amount of Hg in terrestrial food. Understanding the quantity and effects of Hg pollution in terrestrial food is essential for developing sustainable farming practices and protecting the public's health because consumers depend on these staples for nutrition. Mercury content in different terrestrial foods is discussed in this section.

2.5.1 Mercury in Cereals

Mercury contamination in cereal goods brings up serious issues with food safety and human health. For many populations, cereals are a staple food because they supply vital minerals and energy. However, these grains can also act as transporters for Hg, which can enter cereals via several different routes. The Hg level of cereals can be attributed to a variety of sources, including soil, water, and air deposition, industrial emissions, and agricultural methods. As a result, regular intake of cereal goods tainted with Hg could potentially have a negative impact on health, especially for vulnerable individuals. Mercury content in different cereal grains has been reported in different counties. These reports primarily focus on the grains collected from fields or markets close to the Hg sources. A summary table of the grain Hg content of several cereals in different countries is presented in Table 2.2. The Hg content in rice ranged from 0.00055 to 294.0 ppm across the world, and each result was observed in China (Table 2.2). The Hg content in maize ranged from 0.00072 ppm in China to 46.10 ppm in Uganda. Across the world, the reported Hg content in wheat, barley, and millet was 0.02, 0.06, and 0.10 ppm, respectively (Table 2.2). To guarantee that cereals continue to be a healthy and safe part of our diets, addressing this issue involves extensive monitoring, risk assessment, and regulatory actions.

2.5.2 *Mercury in Vegetables*

Vegetables are an indispensable source of nutrients, antioxidants, and metabolites for people globally, and they serve as buffers for the acidic substances produced during digestion. However, vegetables absorb both beneficial and toxic elements from the soil. This results in the bioaccumulation of toxic substances in the human body upon consumption. Leafy vegetables are also a significant part of the daily diet of people all over the world due to changes in food consumption patterns in recent times. However, leafy vegetables can also uptake and accumulate Hg more rapidly than other types of upland crops; thus, it has raised the danger of Hg exposure to humans (Dziubanek et al. 2015). Therefore, one of the most practical and effective methods to reduce Hg risk in soils with high Hg contamination would be to choose green vegetable species with low mercury accumulation (Wai et al. 2017; Ghasemidehkordi et al. 2018a, b). The soil properties such as pH, soil organic matter, cation exchange capacity, and soil texture (Hong et al. 2015; Jia et al. 2018), the Hg concentration (Hu et al. 2013), and the species of vegetables (Liu et al. 2013) all play a role in the absorption of soil Hg in vegetables. Table 2.3 provides data on the Hg content of different types of vegetables reported in various countries. The Hg content in vegetables is significantly different depending on the type and country. In Bangladesh, no Hg was present in vegetables, while the highest concentration of 16.45 ppm was observed in Slovenia (Table 2.3). The food safety authorities in each country play a critical role in ensuring the safety of vegetables available on the market.

2.5.3 *Mercury in Fruits*

Like vegetables, Hg contamination in fruits is also a problem for both environmental and food safety. Fruits, known for their high nutritional content and widespread consumption, can unintentionally accumulate Hg, which could be dangerous for human health. Fruits can become contaminated with Hg from several sources, such

Table 2.2 Mercury content in cereal grains reported in different countries

Country	Cereal	Study area	Hg content (mg kg^{-1})	References
Bangladesh	Rice	Field survey	0.42–14.4	Wang et al. (2020)
Cambodia	Rice	Field survey	0.006–0.012	Cheng et al. (2013)
Canada	Rice	Field survey	0.7–9.3	Lin et al. (2019)
China	Rice, Maize	Field survey	0.00055, 0.00072	Li et al. (2017a)
	Rice	Field survey	0.0049–0.215	Feng et al. (2008)
	Rice	Field survey	0.1–0.27	Li et al. (2014)
	Rice	Field survey	0.009–0.55	Qui et al. (2006)
	Maize	Field survey	< 0.01	Xu et al. (2023)
Chashula, China	Rice	Field survey near Hg mining	50.78–85.69	Meng et al. (2014)
Fankou, China	Rice	Field survey near Pb/Zn mining	0.587–5.524	Meng et al. (2014)
Guizhou, China	Rice	Field survey near Hg mining	0.0086–0.504	Zhang et al. (2010)
	Rice	Field survey near Hg mining	0.072–0.190	Rothenberg et al. (2012)
Shaanxi, China	Rice	Field survey near Hg mining	0.0058–0.643	Ao et al. (2020)
Wanshan, China	Rice	Field survey near Hg mining	7.9–316.9	Yin et al. (2018)
	Rice	Field survey near Hg mining	55.5–95.5	Li et al. (2017b)
	Rice	Field survey near Hg mining	27.9–294.89	Meng et al. (2014)
	Rice	Field survey near Hg mining	0.0041–0.034	Pang et al. (2018)
	Rice	Field survey near Hg mining	0.0424	Li et al. (2015)
Xinhuang, China	Rice	Field survey near Hg mining	11.0–58.0	Li et al. (2013)

(continued)

Table 2.2 (continued)

Country	Cereal	Study area	Hg content (mg kg^{-1})	References
Yanwuping, China	Rice	Field survey near Hg mining	0.010–0.045	Qiu et al. (2013)
Zhejiang, China	Rice	Field survey	2.56–13.20	Wu et al. (2022)
France	Miscellaneous	Market survey	0.01	Leblanc et al. (2005)
	Rice	Market survey	0.01	Leblanc et al. (2005)
India	Barley	Market survey	0.06	Singh and Garg (2006)
	Millet	Market survey	0.10	Singh and Garg (2006)
Indonesia	Rice	Field survey	0.009–0.115	Novirsa et al. (2020)
Nepal	Rice	Field survey	0.006–0.158	Wang et al. (2021)
Pakistan	Rice	Field survey	4.51	Aslam et al. (2020)
Portugal	Rice, Oat, Wheat, Corn, Rye	Market survey	0.00022–0.001	Rubio et al. (2023)
Saudi Arabia	Rice	Market survey	0.02	Ali and Al-Qahtani (2012)
	Wheat	Market survey	0.02	Ali and Al-Qahtani (2012)
Spain	Wheat	Field survey	<0.01	Hernández-Martínez and Navarro-Blasco (2012)
Tanzania	Rice	Case study	0.075–0.159	Sanga et al. (2023)
Uganda	Rice	Field survey	0.078	Ssenku et al. (2023)
	Maize	Field survey	0.004–46.10	Ssenku et al. (2023)
United Kingdom	Miscellaneous	Market survey	0.02	Rose et al. (2010)

as soil, water, and atmospheric deposition. The amount of Hg in fruits can vary depending on different variables, including industrial emissions, agricultural techniques, and proximity to pollution sources. The possibility of Hg direct exposure is a significant worry because fruits are frequently ingested unprocessed and fresh. To protect consumer health, reducing Hg contamination in fruits requires vigilant monitoring, risk analysis, and regulatory interventions. The Hg content of different fruit items reported in different countries is presented in Table 2.4. The Hg concentration ranged from below detection to 12.20 ppm in China. However, in most cases, the value is below the permissible limit of 0.1 ppm set by WHO (FAO/WHO 2016).

Table 2.3 Mercury content in vegetables reported in different countries

Country	Vegetable	Study area	Hg content (mg kg^{-1})	References
Bangladesh	Cabbage	Field survey	<0.03	Tasrina et al. (2015)
	Turnip	Market survey	Not detectable	Linkon et al. (2015)
China	Tomato, Eggplant, Cucumber, Loofah, Beans, and Cucurbita pepo	Market survey	Below detection – 9.11	Zhang et al. (2021)
	Leek, Scallion, Celery, Cabbage, romaine lettuce, Indian lettuce, and Fennel seedling		Below detection – 13.20	
	Ginger, Potato, Yam, and Onion		Below detection – 7.36	
	Lettuce, Amaranth, Water spinach	Field survey	0.00028–0.00085	Li et al. (2017a, b)
	Tomato, Eggplant, Pepper, Cucumber, Cowpea	Field survey	0.00043–0.00093	Li et al. (2017a, b)
	Chinese white cabbage, Xuelihong, Chinese radish, Spinach, Lettuce	Field survey	0.0032–0.0048	Shao et al. (2012)
	Spinach, Tung choy, Leek, Fennel, Coriander, Chinese flowering cabbage, Wuta-tsai, Pakchoi, Chicory, Crown daisy, Lettuce	Field survey	0.0142–0.0338	Yang et al. (2020)
Croatia	Potato	Market survey	0.01–0.02	Stančić et al. (2016)
Czech Republic	Spinach, Lettuce, Radish, Carrot, Beetroot, Pea	Field survey	0.01–5.8	Pelcová et al. (2021)
India	Spinach, Taro root leaves, Onion, Taro root, Potato, Bean, Pumpkin		0.2–1.28	Sharma and Bisla (2022)
	Cabbage, Carrot, Garlic, Ginger, Potato, Radish, Onion, Sugarcane, Sorghum, Fenugreek, Brinjal, Orka, Bean, Pea, Tomato	Field survey	0.001–0.356	Mawari et al. (2022)
Iran	Leek, Onion, Tarragon, Coriander, Radish, Spinach	Field survey	0.03–0.05	Ghasemidehkordi et al. (2018a, b)
Italy	Pepper, Lettuce, Turnip, Brinjal	Field survey	0.005–0.010	Melai et al. (2018)

(continued)

Table 2.3 (continued)

Country	Vegetable	Study area	Hg content (mg kg^{-1})	References
Nigeria	Pumpkin leaf, Bitter leaf, Water leaf, Uziza leaf, Oha leaf	Market survey	0.00–0.45	Ezeilo et al. (2020)
Saudi Arabia	Potato, Spinach, Turnip, Carrot	Farm survey	0.01–0.03	Ali and Al-Qahtani (2012)
Slovenia	Onion, Parsley, Lettuce, Cabbage, Pumpkin, Tomato, Carrot	Field survey	0.003–16.150	Falnoga et al. (2003)
Spain	Lettuce	Field survey	0.0003–0.00167	Margenat et al. (2018)
	Carrot, Cauliflower, Potato	Market survey	<0.01–0.01	Martorell et al. (2011)
Tanzania	Cassava, Chinese cabbage, Potato, Pumpkin	Case study	0.0556–0.3439	Sanga et al. (2023)
Turkey	Carrot, Potato, Spinach	Market survey	<0.00032	İslamoğlu et al. (2021)
Uganda	Amaranthus, Cabbage, Taro, Sweet potato, Cassava	Field survey	0.04–133.18	Ssenku et al. (2023)

2.5.4 Mercury in Spice

Spices are an essential component of culinary culture around the world, adding flavor and perfume to our food. The evidence points to relatively low Hg levels and no substantial health risks for consumers of spices. Spices' varying Hg levels can be attributable to things like soil quality, farming methods, and exposure to the environment. In most countries, spices are usually imported from other countries. Each nation's food safety authorities are essential in ensuring the security of the spices sold on the market and during importing and exporting. There are only a few reports on Hg content in spices in Italy and Tunisia. The Hg content in different spices ranged from 0.002 to 0.851 ppm in Italy (Table 2.5).

Other food items like pulse, nut, oil, etc., can also contain Hg to some extent. However, these values are not so high to affect human health. However, we should be more cautious about the Hg content in food to safeguard our health.

2.6 Mercury in Meat and Meat Products

When analyzing the total amount of Hg in our meals, it is important to consider the Hg found in meat and meat products. Although meat is not the main source of Hg exposure, several variables make it more likely to be found in foods made from animals.

Table 2.4 Mercury content in fruits reported in different countries

Country	Fruit	Study area	Hg content (mg kg^{-1})	References
Turkey	Apple	Market survey	<0.00033	İslamoğlu et al. (2021)
Chile	Miscellaneous	Market survey	<0.01	Muñoz et al. (2005)
China	Strawberry, Orange, Apple, Peach, Watermelon, Banana, and Cherry	Market survey	Below detection – 12.20	Zhang et al. (2021)
Greece	Apple	Field survey	0.001–0.008	Skordas et al. (2013)
India	Jackfruit, Blackberry, Mango, Pears	Market survey	0.45–1.12	Sharma and Bisla (2022)
	Apple, orange, papaya, tamarind	Field survey	0.001–0.147	Mawari et al. (2022)
Italy	Strawberry	Field survey	0.005	Melai et al. (2018)
Nigeria	Apple, watermelon, Pawpaw, banana, Guava, Bush mango	Market survey	0.00–0.47	Ezeilo et al. (2020)
	Watermelon	From street vendor	Below detection	Ogunkunle et al. (2014)
Slovak Republic	Blackberry, Raspberry	Polluted air	≤0.01	Vollmannova et al. (2015)
Spain	Apple, Orange, Pear, Strawberry	Market survey	≤0.01	Martorell et al. (2011)
Uganda	Banana	Field survey	0.021	Ssenku et al. (2023)

Table 2.5 Mercury content in spice reported in different countries

Country	Spice	Study area	Hg content (mg kg^{-1})	References
Italy	Black pepper, Caraway, Coriander, Laurent, Mint, Fennel	Market survey	0.061–0.100	Potortì et al. (2020)
	Cinnamon, Curcuma, Ginger	Markey survey	0.002–0.851	Bua et al. (2016)
Tunisia	Black pepper, Caraway, Coriander, Laurent, Mint, Fennel		0.089–0.311	Potortì et al. (2020)

The environment can expose livestock to Hg, notably through contaminated water sources or high-Hg soil. Additionally, the diet of the animals is important since feed tainted with Hg can increase the amount of Hg in the animals' tissues. In addition, several animal species may bioaccumulate Hg, particularly if they consume other Hg-containing organisms. Although the amounts of Hg in meat and meat products are often lower than in some fish, they should nevertheless be considered, especially for populations with dietary restrictions or unique health concerns. Farmers and food producers can take steps to lessen potential Hg exposure in cattle, and regulatory bodies monitor Hg levels in meat products to assure consumer safety. Making educated food decisions and reducing exposure to Hg can both be facilitated by being aware of potential sources of Hg in our diets and keeping a balanced, diverse diet. There are a very limited number of reports on Hg content in animal meat or meat products. In India, Mathaiyan et al. (2021) reported 0.1773 ppm Hg in different parts of the broiler. Mesinger and Ocieczek (2021) reported <0.001 ppm Hg in game meat in Poland. In another study in Poland, Nawrocka et al. (2020) reported 0.006–0.0164 ppm Hg in game animals in a survey from 2009 to 2018.

2.7 Mercury in Dietary Supplements Containing Fish and Animal Oil

Throughout the world, the number of people utilizing nutritional supplements has been rising in recent years. Dietary supplements make up for nutrient-deficiency shortages of various substances. Vegetable and fish oils can be used as dietary supplements, and both have health-promoting properties. Monounsaturated fatty acids, essential to maintaining the proper ratios between "bad" and "good" cholesterol, can be found in vegetable oils, which are a valuable source of these nutrients. Inflammation of the skin, atherosclerosis, diabetes, and cancer are all treated with them as well as prevented. They support keeping the skin looking young and the body's purported vigor (Lammari et al. 2021; Sarkar et al. 2017). However, they can have a negative impact (Bojarowicz and Dźwigulska 2012), but when administered properly, they're capable of a health-promoting effect. By interfering with metabolism, excretion, or absorption, dietary supplements can interact with medications and alter their effects when taken in an improper quantity. Concerns about Hg in dietary supplements that contain fish and animal oil are valid, particularly for people who depend on these supplements to meet their nutritional needs. Fish and animal oil supplements, such as omega-3 supplements or fish oil capsules, are well-liked because of their prospective health advantages, which include enhancing heart health and lowering inflammation. However, due to the fish or animal sources employed in their manufacture, some of these supplements may contain amounts of Hg.

The use of chemicals, exhaust emissions, and industrial activity all contribute to environmental contamination, which has an impact on the overall quality of raw materials from fish and vegetables. The succeeding steps of dietary supplement manufacture are equally susceptible to contamination with toxic metals like Hg. Very few researchers reported the presence of Hg in dietary supplements containing vegetable and animal oils. According to Smutna et al. (2009), the total Hg level in fish oils varied from 0.013 to 2.03 g kg^{-1}. In studies carried out in Poland by Brodziak-Dopiera et al. (2023), the total Hg content of cod liver oil, shark liver oil, and vegetable oil varied from 0.023 to 0.207 g kg^{-1}. Krygier et al. (2000) demonstrated in another Polish investigation that the average Hg content of cold-pressed rapeseed oil was 0.82 g kg^{-1}. In experiments carried out in New Zealand by Rucklidge and Shaw (2020), assessments of Hg in preparations containing fish oil did not reveal concentrations beyond the detection limit of the equipment (LOD = 10 g kg^{-1}). On the other hand, the study of fish oil samples performed in the USA by Foran et al. (2003) revealed mercury contents ranging from 6 to 12 g kg^{-1}. Based on the reports, it is a matter of concern that dietary supplements might also be harmful to the body if Hg or other heavy metals are present. More research should be conducted worldwide to explore the actual scenario of the existence of Hg in such a trending food.

It is essential to guarantee the products' quality and safety to reduce the danger of Hg exposure from dietary supplements. Reputable supplement producers use stringent quality control procedures to screen for pollutants like Hg. They obtain their ingredients from trustworthy vendors who follow safety regulations. In some nations, the amount of Hg in dietary supplements is also restricted by rules and regulations. When buying nutritional supplements, customers should exercise caution and stick with well-known brands that put quality and safety first. Making educated decisions concerning the Hg content of accessories can be facilitated by reading product labels and speaking with healthcare specialists. It is crucial to speak with healthcare professionals before using any dietary supplements containing fish or animal oil for susceptible populations, such as pregnant women, nursing mothers, and small children. Medical specialists can offer individualized advice and suggestions depending on a patient's health requirements and any potential concerns from exposure to Hg. Overall, even though fish and animal oil supplements can be helpful for many people, choosing high-quality products and being aware of potential Hg concentrations will help assure the safety and effectiveness of these dietary supplements. Individuals can take advantage of these supplements' potential health advantages while lowering their risk of exposure to Hg by making wise decisions.

2.8 Mercury in Baby Food

Since babies and young children are especially susceptible to the negative effects of this neurotoxic metal, the presence of Hg in baby food causes tremendous anxiety among parents and other caregivers. There are several ways that Hg can enter infant food, and when it does, it can have harmful consequences for a child's nervous system development. When certain baby food items are made, contaminated materials like fish or seafood are used, which is one of the main sources of Hg in infant food. Since fish is a common allergy for young children, some baby food producers use fish or fish oil to supply important nutrients like omega-3 fatty acids. There is a chance of exposure if the fish used to make these baby foods have high Hg levels. Additionally, environmental contamination may contribute to the amount of Hg in baby food. Mercury can be absorbed by crops and end up in infant food items if it is present in the soil or water used to grow fruits, vegetables, or cereals. Furthermore, Hg can enter the environment because of industrial emissions and waste management, resulting in Hg-contaminated substances used in food production. Manufacturers of infant food and regulatory bodies must place a high priority on product safety to address this issue. To identify and restrict Hg levels in baby food, stringent testing and quality control procedures should be in place. Guidelines and regulations are also established by governments and international organizations to guarantee the security of baby food products and shield young children from Hg exposure. In order to guarantee the safety of the food they provide to newborns and young children, parents and other caregivers play a crucial role. The danger of Hg exposure from infant food can be reduced by reading product labels, choosing companies with transparent safety testing procedures, and adhering to age-appropriate feeding recommendations. The safest and healthiest solutions for infants can be chosen by consulting with pediatricians and other medical professionals who can offer helpful advice. Parents can safeguard their children from potential health hazards linked to mercury in infant food by being knowledgeable and proactive while also promoting their general well-being and growth. Mercury concentrations have been reported in baby foods in many countries. Martins et al. (2013) assessed the Hg content of commercial baby foods available on the market in Portugal. They reported 0.5 μg Hg kg^{-1} cereal-based baby food and a maximum of 19.56 μg Hg kg^{-1} fish containing baby food. In the USA, Parker et al. (2022) reported 1.5–20.0 μg Hg kg^{-1} of commercially available baby food. In another study, Gray (2023) conducted a non-targeted survey of the presence of heavy metals in ready-to-eat baby foods in the US market and reported 0.72 μg Hg kg^{-1}. The USA government's Baby Food Safety Act sets a recommended initial action level of 2 μg Hg kg^{-1} for baby food and cereal (Bair 2022). Silva et al. (2023) determined total Hg content in Spanish baby food samples and reported 0.57–41.9 μg Hg kg^{-1} sample. In another study, Henríquez-Hernández et al. (2023) reported 28.1 μg Hg kg^{-1} ready-to-eat baby purees in the Spanish market. Lodhi et al. (2021) evaluated the Hg content in infant formula milk and cereal samples available in Pakistan market and observed a concentration of 0.95–6.58 μg total Hg kg^{-1} sample. In most of the cases, the concentration exceeded the recommended value set by WHO.

A recognized consumer protection agency looked at the Hg content of several baby food products on the market in 2022. The purpose of the experiment was to evaluate any possible dangers that regular consumption of these goods could offer to newborns and young children. The study includes evaluating a wide range of infant food products from various brands and producers, including purees, cereals, and snacks. The consumer advocacy group investigated and discovered that some infant food products had Hg levels that could be detected. It was discovered that certain goods' fish-derived constituents were the main source of Hg in infant food. These goods made the claim that they contained vital nutrients, including omega-3 fatty acids, which are good for a child's growth. The analysis, however, showed that the fish utilized in these items had high amounts of Hg while being a healthy source of nutrients. For instance, a well-known fish-based baby food puree made the claim that it was high in omega-3 fatty acids. Testing by the Watchdog group showed that the puree did indeed contain high quantities of omega-3 fatty acids, indicating the presence of components derived from fish. Further examination, however, revealed that the puree also contained minute quantities of Hg because of the fish used in its manufacture (Henríquez-Hernández et al. 2023). Despite the Hg levels being very low, they raised questions about how long infants who frequently consume these items would be exposed to them. The analysis also revealed that several fruits- and vegetable-based baby food products had low Hg levels. This was attributed to the stringent quality control methods that some manufacturers had put in place and to the cautious sourcing of materials to reduce environmental pollutants like Hg. In reaction to the findings, the consumer advocacy group shared its findings with the appropriate regulatory bodies and urged them to review the safety requirements and laws pertaining to Hg in baby food. The group also educated parents and caregivers on the value of reading product labels, selecting baby food from reputable manufacturers with open safety testing procedures, and consulting with medical specialists on infant feeding safety. Some infant food producers improved their testing and safety practices because of the inquiry to make sure that their products contained the least amount of Hg feasible. To safeguard infants from potential Hg exposure, the regulatory authorities also carried out their own evaluations, which resulted in revisions to the safety rules for the preparation of baby food. Thus, to protect the health and growth of infants, it is crucial to continuously monitor, test, and regulate baby food items. It also highlights the necessity for consumers, government agencies, and suppliers of baby food to work together to address worries about Hg in baby food and protect the welfare of the most vulnerable people in our society.

2.9 Food Processing and Mercury Level

Depending on the precise processing techniques employed, food processing can affect the amounts of Hg in food, either by reducing or raising the Hg content. Depending on the type of food and the processing methods used, the effect of food processing on Hg levels can change (Fig. 2.3).

2.9.1 Reduction of Mercury Levels

To preserve food safety and shield consumers from any health concerns linked to Hg consumption, food Hg levels must be reduced. The following techniques can be used to lower the amount of Hg in processed food.

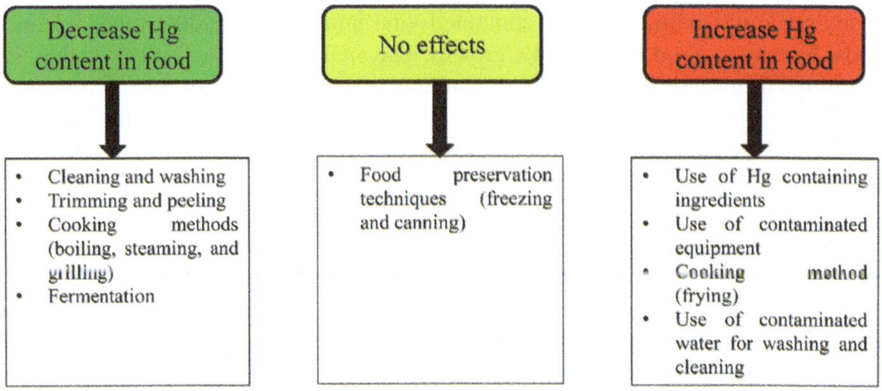

Fig. 2.3 Effect of food processing on mercury content in food

2.9.2 Cleaning and Washing

Fruits, vegetables, and cereals should be well-cleaned and washed before eating to eliminate some Hg from the environment and reduce surface contamination. For instance, washing produce under running water can help get rid of dust, grime, and perhaps Hg-containing residues, reducing the overall Hg levels in the final product. Washing with tap water, 5% and 10% acetic acid, and sodium chloride reduced Hg content by 4–23% in different vegetables (Sattar et al. 2013).

2.9.3 Trimming and Peeling

Because Hg tends to accumulate in fatty tissues, trimming the skin and fat from some fish and meat can help limit exposure. The amount of Hg in processed food can be reduced by removing the skin and visible fat, making it safer to eat.

2.9.4 Cooking Methods

By using different methods during cooking, including boiling, steaming, and grilling, the amount of Hg in food can be reduced (Costa et al. 2022; Ouédraogo and Amyot 2011). The Hg content of the fish may be further decreased by part of the Hg that may leak out into the cooking water when the fish is cooked. This works especially well for fatty fish because cooking can eliminate a considerable amount of Hg.

2.9.5 Fermentation

Mercury levels can be reduced by using fermentation techniques that are utilized in some traditional food preparations (Nie et al. 2022). For instance, fish that has been fermented over time may have less mercury in it. This is due to the fact that some microbes engaged in fermentation can bind to mercury, causing it to be removed or changed into less harmful forms.

2.9.6 Food Preservation Techniques

The amounts of Hg in food are not greatly changed by some methods of food preservation, such as freezing and canning. Making sure that the raw materials used in the processing are secure and low in Hg is crucial since properly preserved food items retain their original raw materials' Hg level.

2.9.7 Potential for Increase in Mercury Levels

While certain types of food processing can aid in lowering Hg levels, there are other circumstances in which it may actually raise Hg levels in food. The following are some elements that may cause Hg levels in processed foods to rise:

2.9.8 Frying

Foods that have been fried may contain more Hg than they originally did, especially if the oil has been used numerous times (Kalogeropoulos et al. 2012). As the frying process can concentrate the Hg present in the fish, this is especially worrying for goods that contain fried fish or shellfish.

2.9.9 Mercury-Containing Ingredients

When food is processed, the inclusion of specific Hg-tainted substances might raise the amount of Hg in the finished product. Mercury can be added to food during processing, for instance, if fish-derived substances or animal products with high Hg levels are used in infant food.

2.9.10 Contamination During Processing

The levels of Hg in food might rise if the equipment or facilities used to process food are polluted with Hg. To avoid this contamination and guarantee the safety of the food, proper hygiene and sanitation practices are important.

Overall, while food processing can have some influence on Hg levels, the presence of Hg in the raw ingredients used is the main factor determining the amount of Hg in food. Food processors must use ingredients from reliable vendors and follow food safety laws to guarantee food safety. Choosing goods from reliable brands and eating a varied and balanced diet that includes a selection of foods with lower Hg levels are additional ways for consumers to decrease their exposure to Hg in processed meals.

2.10 Monitoring and Regulation of Mercury in Food

Assuring food safety and preserving public health depend on monitoring and controlling the amount of Hg in food. Multiple governments and international organizations have devised thorough monitoring and regulation procedures to restrict Hg levels in food products due to the potential health hazards linked to Hg exposure. Effective monitoring programs are required to track Hg levels in various environmental media and food products, considering the possible health and environmental concerns associated with Hg pollution. This information is crucial for risk assessment and well-informed decision-making. Some examples of monitoring programs adopted in different countries are as follows:

2.10.1 Surveillance Programs

Countries develop surveillance programs to track Hg levels in food products. These programs methodically collect and analyze samples from a variety of categories, including processed foods, seafood, meat, vegetables, and fruits. These programs analyze Hg contamination levels and associated health risks to consumers through laboratory analysis and risk assessment. Transparent dissemination of information increases public awareness and equips customers with the knowledge they need to make knowledgeable dietary decisions. If safety limits are exceeded, regulatory action can be taken to protect public health and the security of the food supply. To reduce potential health effects, these surveillance initiatives are essential for proactively controlling Hg exposure risks and supporting risk-based decision-making.

2.10.2 Testing and Analysis

To ensure the precise detection of Hg levels in food samples, testing and analysis are essential elements of food safety measures put in place by food safety organizations and laboratories. Inductively coupled plasma techniques, mass spectrometry, and other sophisticated testing techniques are used to quantify Hg concentration precisely and sensitively. These investigations aid in pinpointing probable environmental or industrial food processing-related causes of Hg contamination in food products. Food safety organizations can evaluate compliance and take appropriate action if Hg levels exceed safety thresholds by comparing test findings to regulatory limits and standards that have been established. Monitoring the food supply chain, preserving public health, and guaranteeing that consumers have access to safe and low-Hg food options all need the use of testing and analysis as essential instruments.

2.10.3 Risk Assessment

Undoubtedly, risk assessment is a crucial stage in the process of analyzing the information gathered from programs that monitor the levels of Hg in various food products. Experts perform a detailed examination of the potential health concerns posed by the detected Hg levels in various food products as part of the risk assessment process. This evaluation considers variables including intake volume and frequency, demographic group vulnerability, and Hg toxicity. The main objective of risk assessment is to ascertain whether the amounts of Hg in particular foods are safe for human consumption or if they may be harmful to health. Regulatory agencies can set suitable regulatory limits or maximum permissible levels for Hg in food products based on the results of the risk assessment. These restrictions are meant to safeguard the general public's health by guaranteeing that dietary Hg exposure stays below levels deemed safe. Regulatory authorities can establish regulatory limits and create guidelines to reduce Hg exposure from food intake by undertaking risk assessments. This procedure assures the protection of public health while also empowering consumers to make knowledgeable dietary decisions and lowering potential health risks related to Hg contamination in food.

Regulations are essential for controlling Hg contamination and guaranteeing the protection of the environment and public health. Regulatory bodies set maximum permitted levels for Hg in various products and surroundings to limit emissions, avoid contamination, and encourage safe practices in the manufacturing of food and other items. Regulations also make it possible to set up monitoring programs that track Hg levels, aid in risk assessment, and support wise policy choices. Regulations are also the driving force behind public awareness and education campaigns, giving people the capacity to make educated decisions that will lower exposure to Hg and promote a safer, more sustainable future. Some regulatory functions are as follows:

Table 2.6 Acceptable levels for intake of mercury

Authority	Inorganic Hg (μg/kg body weight/week)	Me-Hg (μg/kg body weight/week)
European Food Safety Authority (ESFA CONTAM Panel, 2012)	4	1.3
Joint FAO/WHO Expert Committee on Food Additives (JECFA, 2011)	4*	
Joint FAO/WHO Expert Committee on Food Additives (JECFA, 2007)		1.6*
US EPA Reference Dose (adopted 2001) (US EPA, 2001)		0.7
Canada (Adopted 1997) (Health Canada, 2007)		1.4
Japan (Adopted 2005) (WHO/UNEP, 2008)		2.0
Netherlands (Adopted 2000) (WHO/UNEP, 2008)		0.7

* Preliminary tolerable weekly intake value

2.10.4 Setting Maximum Limits

The term "setting maximum limits" refers to the determination of legal thresholds or maximum permissible levels of Hg in a variety of goods, environmental media (such as air, water, and soil), and food products. These limitations are based on scientific evaluations of the toxicity of Hg and its possible effects on human health and the environment. To manage Hg emissions and releases, avoid Hg pollution, and safeguard public health and the environment from the harmful consequences of Hg exposure, regulatory authorities, such as governmental agencies and international organizations, apply these limits. To reduce Hg pollution, advance sustainable practices, and ensure the security of consumers and ecosystems, maximum limits for Hg in products and surroundings are essential. To maintain adherence to these limitations and reduce the dangers connected to Hg poisoning, regular monitoring and enforcement are crucial. The acceptable Hg intake level set by different authorities is presented in Table 2.6, and the maximum allowable Hg concentration in different environmental components and food items is presented in Table 2.7.

2.10.5 Labeling Requirements

Regulatory bodies establish labeling rules for the presence of Hg in food products to increase consumer awareness and facilitate informed decision-making. Food

Table 2.7 Maximum allowable limits of mercury in different environmental component and food items

Component/food item	Maximum allowable limit (ppm)	Regulatory authority	References
Soil	1.0	FAO/WHO	Ajani et al. (2022)
Drinking water	0.002	EPA	Griffiths et al. (2012)
	0.001	WHO	WHO (2022)
Air	50 μg Hg vapor m^{-3}	OSHA	Dudeja et al. (2023)
Cereals (barley, maize, oats, rice, rye, wheat, etc.)	0.01	European Union	EU (2018)
	0.10	WHO/FAO	Hussain et al. (2019)
Fish	0.23	FDA/EPA	FDA/EPA (2022)
Vegetables	0.10	WHO/FAO	WHO/FAO (2016)
Fruits	0.10	WHO/FAO	Hussain et al. (2019)
Oil	0.10	WHO/FAO	WHO/FAO (2016)
Pulse and spices	0.10	WHO/FAO	Hussain et al. (2019)

producers provide consumers with the power to make decisions that are in line with their dietary and health preferences by accurately and transparently disclosing information about Hg levels on food labels. This is crucial for vulnerable groups, like young children and pregnant women, who may be more prone to the negative consequences of Hg exposure. Since people can avoid items with high Hg levels and choose safer alternatives, the inclusion of Hg information on food labels adds to efforts to preserve public health.

2.10.6 Import and Export Regulations

Government-imposed restrictions that control the international commerce of Hg and goods containing Hg are known as import and export regulations for Hg. In order to avoid environmental damage, safeguard public health, and encourage responsible management of the substance, these regulations are designed to regulate and track the transportation of Hg. According to export laws, exporters must obtain licenses and abide by international treaties like the Minamata Convention, which aims to gradually phase out Hg mining and trading. To reduce exposure and potential injury, import regulations in receiving nations may place limitations on Hg imports, including bans on specific Hg-containing products. In accordance with international efforts to minimize Hg emissions, safeguard ecosystems, and promote human health, these regulations make sure that the trade in Hg is carried out safely and ethically.

2.10.7 Good Agricultural and Manufacturing Practices

Governments must strictly adhere to Good Manufacturing Practices (GMP) and Good Agricultural Practices (GAP) to reduce the danger of Hg contamination in agricultural and food processing operations. GAP emphasizes encouraging the use of Hg-free agricultural chemicals, using integrated pest management techniques, guaranteeing the management of clean water and soil, and upholding traceability and record-keeping. On the other hand, GMP places a strong emphasis on confirming the sources of raw materials, using Hg-free processing aids, adopting stringent cleaning and sanitation measures, performing routine product testing, and educating employees on safety protocols. Industries seek to maintain customer safety, product quality, and environmental protection while reducing the possible dangers related to Hg pollution by following these measures.

2.10.8 Public Awareness and Education

Initiatives to raise public awareness and educate the public are essential parts of regulatory authorities' efforts to regulate Hg content in food. These initiatives seek to inform and educate the general public, key players in the food business, and medical experts about the possible dangers of Hg exposure and healthy eating habits. These efforts enable people to make knowledgeable decisions about their dietary choices by increasing awareness about the sources of Hg in food, its health impacts, and methods to decrease exposure. Furthermore, instruction directed at those involved in the food business serves to encourage adherence to mercury rules and the implementation of best practices to reduce Hg contamination in food production and processing. Healthcare professionals are also taught on the health dangers associated with Hg, allowing them to provide vulnerable populations—such as pregnant women and young children—who may be more susceptible to negative Hg exposure effects—with the guidance and counsel they need. In general, efforts to raise public awareness and educate the public are essential to promoting public health and supporting the efficient application of Hg laws.

Countries can effectively manage the amounts of Hg in food, safeguard public health, and lower the risk of Hg-related health problems by establishing strict monitoring and regulatory procedures. Global efforts to guarantee the safety of food items and lower Hg exposure for the populace are further enhanced by ongoing research and collaboration between countries and international organizations.

2.11 Health Risk Assessment

Health risk assessment is a methodical procedure used to assess potential health risks and quantify the dangers brought on by exposure to particular materials, elements of the environment, or activities. Health risk assessment in the context of Hg in food tries to ascertain the potential negative health effects that may arise from consuming food items with various levels of Hg. This assessment is done by following several key steps.

2.11.1 Key Steps in Health Risk Assessment

Key steps refer to the fundamental or necessary steps that are normally taken when completing a procedure for assessing health risks. These crucial stages are necessary to guarantee a methodical and comprehensive assessment of potential health hazards linked to exposure to Hg. The essential steps comprise:

2.11.1.1 Hazard Identification

This step entails compiling scientific evidence from epidemiological and toxicological investigations to recognize and comprehend the health risks related to Hg exposure. It involves figuring out what kinds of negative health outcomes, such as neurological, developmental, or other effects, Hg can have on the body.

2.11.1.2 Exposure Assessment

The main goal of exposure assessment is to determine the level of exposure to Hg that people or communities receive from their diet. It considers elements including the population's dietary habits, the patterns of how various food items are consumed, and the levels of Hg present in those foods.

2.11.1.3 Dose–Response Assessment

The association between the quantity of Hg consumed (dose) and the likelihood and seriousness of unfavorable health effects (response) is established through dose–response evaluation. This process aids in comprehending the potential dangers connected to various Hg exposure levels.

2.11.1.4 Risk Characterization

To calculate the possible health hazards posed by consuming food containing Hg, risk characterization integrates the results from hazard identification, exposure assessment, and dose–response evaluation. Under various exposure situations, it offers an empirical assessment of the likelihood that health impacts may manifest in a particular population group.

2.11.1.5 Risk Communication

It is essential to effectively communicate the results to the appropriate stakeholders, decision-makers, and the general public. It entails resolving any doubts or constraints in the assessment as well as clearly and simply presenting the results.

2.11.1.6 Risk Management

To control or reduce the identified health hazards, risk-management techniques are established based on the results of the risk assessment. To decrease exposure and safeguard the public's health, this may entail establishing rules, recommendations, or other initiatives.

2.11.1.7 Monitoring and Review

As new data and information become available, health risk assessments should be reviewed and updated often. In order to make sure that the risk-management measures are successful and safeguard public health, continual monitoring is also necessary.

2.11.2 Measurement of Health Risk Assessment

Estimated dietary intakes (EDI), target hazard quotients (THQ), and hazard indexes (HI) are frequently used in describing health risk assessment (Rodrigues et al. 2023a, b; Ritonga et al. 2022; Mehouel et al. 2019). These terms are usually calculated using different formulas.

Estimated dietary intake (EDI)

The EDI can be calculated by the following equation (Antoine et al. 2017; Ju et al. 2017):

$$\mathrm{EDI} = \frac{\text{Concentration of Hg in food item } (\mu g/g) \times \text{daily consumption rate (g/day)}}{\text{Mean body weight of the population (kg)}}$$

This is typically calculated to see whether the daily or monthly Hg concentrations ingested are higher than the allowed intake limit, set by the Joint Food and Agriculture Organization (FAO)/World Health Organization (WHO) Expert Committee on Food Additives (FAO/WHO 2007). In the case of fish, it is assumed that 100% of the Hg is in Me-Hg form and full Me-Hg is absorbed in the gastrointestinal tract of the human body (Vasconcellos et al. 2021). The United States Environmental Protection Agency states that the Reference Dose (RfD), also known as safe daily consumption, is equal to 0.1 g Hg (Kg body weight)$^{-1}$ day^{-1}. The FAO/WHO standards of 0.23 g Hg (Kg body weight)$^{-1}$ day^{-1} for women of reproductive age and for children and 0.45 g Hg (Kg body weight)$^{-1}$ day^{-1} for adults generally serve the same purpose (FAO/WHO 2003; US/EPA 2000). Some other researchers also used the following formula to assess the risk based on average lifetime of a particular population (Ritonga et al. 2022; US/EPA 2000).

$$\text{EDI} = \frac{\begin{array}{c}\text{Hg conc. in food} \times \text{Daily intake rate} \times \\ \text{Exposure frequency} \times \text{Exposure duration}\end{array}}{\text{Average body weight} \times \text{Average exposure time}}$$

In the above equation, Hg conc. means the average concentration of Hg in the specific food item (μg g^{-1}), daily intake rate is the amount of that food (g) that a single person ingested every day, exposure frequency is assumed to be 365 per year (at the rate of 1 meal per day), exposure duration is assumed to be the average lifetime of a particular population (for example, 73.57 years for Bangladesh (Macrotrends 2023)), average body weight is the mean body weight of the population, and average exposure time indicates the average lifetime (73.57) years for non-carcinogenic substance, which is multiplied by 365 days.

Target hazard quotients (THQ)

The United States Environmental Protection Agency developed the target hazard quotient (THQ), a complicated measure that is frequently used to evaluate the likelihood of non-carcinogenic hazards connected with prolonged exposure to contaminants, such as Hg from foods like fish and water. The THQ can be estimated following the equation (Rodrigues et al. 2023a, b; Qing et al. 2022; Mehouel et al. 2019; Barone et al. 2015):

$$\text{THQ} = \frac{\begin{array}{c}\text{Exposure frequency} \times \text{Exposure duration} \times \\ \text{Food intake rate} \times \text{Hg conc. in food}\end{array}}{\text{Oral reference dose} \times \text{Average body weight} \times \text{Average exposure time}}$$

For the above equation, all the parameters are like EDI calculations. Where exposure frequency is assumed to be 365 days per year, exposure duration is the average life expectancy of a population, food intake rate is a measure of grams per day, Hg is measured in mg per kg of food, oral reference dose is assumed to be 0.3 and 0.1 μg per kg of body weight per day for Hg and Me-Hg (US/EPA 2017), average body weight in kg, and average exposure time is 365 days multiplied by average life expectancy (years). A THQ of less than 1 indicates that the exposure is smaller than the reference dosage and is therefore unlikely to have long-term adverse effects on health. Otherwise, there is a chance that present exposure will cause developmental neurologic problems in newborns (US/EPA 2014).

Hazard index (HI)

This represents the total of the hazard quotients for chemicals that have an impact on a given organ or group of organ systems. Typically, hazard quotients for contaminants that have negative effects due to the same hazardous mechanism should be combined (USEPA 2017). Like the hazard quotient, overall doses below 1 determined by hazard quotients are typically regarded as tolerable and are unlikely to have any long-term harmful impacts on health. The hazard index can be calculated from the THQs using the following equation (Mehouel et al. 2019; Núñez et al. 2018):

$$\mathrm{HI} = \mathrm{THQ(Hg)} + \mathrm{THQ(Me - Hg)}$$

2.11.3 Application in Food Safety

The evaluation of health risks is a crucial component of policies and regulations pertaining to food safety. Health risk evaluations are essential in defining regulatory limits for Hg in food, identifying populations at risk, and assessing interventions. The regulatory agencies can set the maximum allowed levels of Hg in food products based on the results of health risk evaluations. These restrictions guarantee that the populace's food is secure and does not pose unacceptable health risks. Health risk evaluations aid in identifying populations who may be more vulnerable to the negative consequences of Hg exposure, such as young children, pregnant women, and those who are nursing. To safeguard these susceptible populations, specific dietary advice or recommendations can be made. Policymakers can measure the efficacy of interventions targeted at lowering Hg exposure through health risk assessments. These measures might encourage safe fishing methods, put food safety programs in place, or promote dietary variety. Governments and regulatory organizations can protect the public's health and lessen the possible health risks linked to exposure to Hg in food and other environmental factors by using health risk assessment procedures. It is an essential instrument for guaranteeing food safety and defending public health.

References

Ahmad NI, Mahiyuddin WRW, Azmi WNFW, Azlee RSR, Shaharudin R, Sulaiman LH (2022) Exposure assessment of methyl mercury from consumption of fish and seafood in Peninsular Malaysia. Environ Sci Pollut Res 29(17):24816–24832

Ajani MB, Maleka PP, Penabei S, Usman IT (2022) Health risk assessment of heavy metals concentration from soil; a case study of the Mayo-Dallah in Southern area of Chad. J Rad Res Appl Sci 15(1):130–138

Aksari YD, Perwitasari D, Butet NA (2015) Concentration of heavy metals (Cd, Hg, and Pb) of Amazon sailfin catfish, *Pterygoplichthys pardalis* (Castelnau, 1855) in Ciliwung River West Java. J Iktiolog Indonesia 15(3):257–266

Ali MH, Al-Qahtani KM (2012) Assessment of some heavy metals in vegetables, cereals and fruits in Saudi Arabian markets. Egypt J Aquat Res 38(1):31–37. https://doi.org/10.1016/j.ejar.2012.08.002

Al-Sulaiti MM, Al-Ghouti MA, Ramadan GA, Soubra L (2023) Health risk assessment of methyl mercury from fish consumption in a sample of adult Qatari residents. Environ Mon Assess 195(5):617. https://doi.org/10.1007/s10661-023-11194-w

Andayesh S, Hadiani MR, Mousavi Z, Shoeibi S (2015) Lead, cadmium, arsenic and mercury in canned tuna fish marketed in Tehran, Iran. Food Addit Contam B 8:93–98

Antoine JMR, Fung LAH, Grant CN (2017) Assessment of the potential health risks associated with the aluminium, arsenic, cadmium and lead content in selected fruits and vegetables grown in Jamaica. Toxicol Rep 4:81–187

Ao M, Xu X, Wu Y, Zhang C, Meng B, Shang L, Liang L, Qiu R, Wang S, Qian X, Zhao L, Qiu G (2020) Newly deposited atmospheric mercury in a simulated rice ecosystem in an active mercury mining region: High loading, accumulation, and availability. Chemosphere 238:124630. https://doi.org/10.1016/j.chemosphere.2019.124630

Aslam MW, Ali W, Meng B, Abrar MM, Lu B, Qin C, Zhao L, Feng X (2020) Mercury contamination status of rice cropping system in Pakistan and associated health risks. Environ Pollut 263:114625. https://doi.org/10.1016/j.envpol.2020.114625

Bair EC (2022) A narrative review of toxic heavy metal content of infant and Toddler Foods and evaluation of United States policy. Front Nutr 9:919913. https://doi.org/10.3389/fnut.2022.919913/full

Barone G, Storelli A, Garofalo R, Busco VP, Quaglia NC, Centrone G, Storelli MM (2015) Assessment of mercury and cadmium via seafood consumption in Italy: estimated dietary intake (EWI) and target hazard quotient (THQ). Food Addit Contam Part A 32(8):1277–1286

Bojarowicz H, Dźwigulska P (2012) Suplementy diety. Część III. Interakcje suplementów diety z lekami. Hyg Pub Health 47:442–447

Brodziak-Dopierała B, Fischer A, Chrzanowska M, Ahnert B (2023) Mercury exposure from the consumption of dietary supplements containing vegetable, cod liver, and shark liver oils. Int J Environ Res Pub Health 20(3):2129. https://doi.org/10.3390/ijerph20032129

Bua DG, Annuario G, Albergamo A et al (2016) Heavy metals in aromatic spices by inductively coupled plasma-mass spectrometry. Food Addit Contam Part B 9:210–216

Cahyani N, Lumban, Batu DTF, Sulistiono S (2017) Heavy metal contain Pb, Hg, Cd and Cu in whiting fish (*Sillago sihama*) muscle in estuary of Donan River, Cilacap, Central Java. J.I Pengol. Hasil Perikan Indonesia 19(3):267. https://doi.org/10.17844/jphpi.v19i3.15090

Candra YA, Syaifullah M, Irawan B, Putranto TWC, Hidayati D, Soegianto A (2019) Concentrations of metals in mantis shrimp Harpiosquilla harpax(de Haan, 1844) collected from the eastern region of Java Sea Indonesia, and potential risks to human health. Reg Stud Mar Sci 26:100507. https://doi.org/10.1016/j.rsma.2019.100507

Chen Z, Muhammad I, Zhang Y, Hu W, Lu Q, Wang W et al (2021) Transfer of heavy metals in fruits and vegetables grown in greenhouse cultivation systems and their health risks in Northwest China. Sci Total Environ 766:142663. https://doi.org/10.1016/j.scitotenv.2020.142663

Cheng Z, Wang HS, Du J, Sthiannopkao S, Xing GH, Kim KW, Yasin MS, Hashim JH, Wong MH (2013) Dietary exposure and risk assessment of mercury via total diet study in Cambodia. Chemosphere 92:143–149

Costa F, Mieiro CL, Pereira ME, Coelho JP (2022) Mercury bioaccessibility in fish and seafood: effect of method, cooking and trophic level on consumption risk assessment. Mar Pollut Bull 179:113736. https://doi.org/10.1016/j.marpolbul.2022.113736

Dudeja P, Dudeja K, Grover S, Singh H, Jabin Z (2023) Pathway to mercury-free dentistry: an insight into past, present, and future. Eur Oral Res 57(2):60–67

Dziubanek G, Piekut A, Rusin M, Baranowska R, Hajok I (2015) Contamination of food crops grown on soils with elevated heavy metals content. Ecotoxicol Environ Saf 118:183–189

EFSA CONTAM Panel (2012) Scientific opinion on mineral oils hydrocarbons in food. EFSA J 10(6):2704. https://doi.org/10.2903/j.efsa.2012.2704. Available online: www.efsa.europa.eu/efsajournal

EU (2018) Commission regulation (EU) 2018/73 of 16 January 2018 amending Annexes II and III to Regulation (EC) No 396/2005 of the European Parliament and of the Council as regards maximum residue levels for mercury compounds in or on certain products (Text with EEA relevance). Off J Eur Union L 138:8–20

Evers D (2018) OBSOLETE: Mercury in higher biota. Biol Eff Ref Mod Earth Syst Environ Sci. https://doi.org/10.1016/B978-0-12-409548-9.09985-1

Ezeilo CA, Okonkwo SI, Chibuzor C, Onuorah ALC, Ugwunnadi NE (2020) Determination of heavy metals in some fruits and vegetables from selected market's in Anambra state. ACTA Sci Nutr Heal 4:163–171

Falnoga I, Jereb V, Smrkolj P (2003) Hg and Se in foodstuffs grown near a Hg mining area. J Phys IV France 107:447–450

FAO/WHO (2003) Evaluation of certain food additives and contaminants: sixty-first report of the joint FAO/WHO expert committee on food additives. In: Proceedings of the joint FAO/WHO expert committee on food additives (JECFA), Rome, Italy, pp 10–19

FAO/WHO (2007) Meeting (68th: 2007: Geneva, Switzerland), world health organization & food and agriculture organization of the United Nations. Evaluation of certain food additives and contaminants: sixty-eighth report of the joint FAO/WHO expert committee on food additives. World Health Organization, pp. 225. Available in: https://apps.who.int/iris/handle/10665/43870. Accessed on 6 Aug 2023

FDA/EPA (2022) Technical information on development of FDA/EPA advice about eating fish for those who might become or are pregnant or breastfeeding and children ages 1–11 years. https://www.fda.gov/food/environmental-contaminants-food/technical-information-development-fdaepa-advice-about-eating-fish-those-who-might-become-or-are#:~:text=Any%20fish%20with%20an%20average,week%20%3D%200.23%20%C2%B5g%2Fg. Accessed on 6 Aug 2023

Foran SE, Flood JG, Lewandrowski KB (2003) Measurement of mercury levels in concentrated over-the-counter fish oil preparations: is fish oil healthier than fish? Arch Pathol Lab Med 127:1603–1605

Ghasemidehkordi B, Malekirad AA, Nazem H, Fazilati M, Salavati H, Shariatifar N, Rezaei M, Fakhri Y, Mousavi Khaneghah A (2018a) Concentration of lead and mercury in collected vegetables and herbs from Markazi province, Iran: a non-carcinogenic risk assessment. Food Chem Toxicol 113:204–210

Ghasemidehkordi B, Malekirad AA, Nazem H, Fazilati M, Salavati H, Shariatifar N, Rezaei M, Fakhri Y, Khaneghah AM (2018b) Concentration of lead and mercury in collected vegetables and herbs from Markazi province, Iran: a non-carcinogenic risk assessment. Food Chem Toxicol 113:204–210

Gray PJ (2023) A survey of toxic elements in ready to eat baby foods in the US market 2021. Food Addit Contam Part B 16(2):79–85

Griffiths C, Klemick H, Massey M, Moore C, Newbold S, Simpson D, Walsh P, Wheeler W (2012) US Environmental Protection Agency valuation of surface water quality improvements. Rev Environ Econ Pol 6(1). https://doi.org/10.1093/reep/rer025?journalCode=reep

Groth E III (2010) Ranking the contributions of commercial fish and shellfish varieties to mercury exposure in the United States: implications for risk communication. Environ Res 110(3):226–236
Gupta N, Yadav KK, Kumar V, Krishnan S, Kumar S, Nejad ZD et al (2021) Evaluating heavy metals contamination in soil and vegetables in the region of North India: levels, transfer and potential human health risk analysis. Environ Toxicol Pharmacol 82:103563. https://doi.org/10.1016/j.etap.2020.103563
Ha E, Basu N, Bose-O'Reilly S, Dórea JG, McSorley E, Sakamoto M, Chan HM (2017) Current progress on understanding the impact of mercury on human health. Environ Res 152:419–433. https://doi.org/10.1016/j.envres.2016.06.042
Health Canada (2007) Human health risk assessment of mercury in fish and health benefits of fish consumption. Health Canada, Ottawa, Canada. https://www.canada.ca/content/dam/hc-sc/migration/hc-sc/fn-an/alt_formats/hpfb-dgpsa/pdf/nutrition/merc_fish_poisson-eng.pdf . Accessed on 6 Aug 2023
Hernández-Martínez R, Navarro-Blasco I (2012) Estimation of dietary intake and content of lead and cadmium in infant cereals marketed in Spain. Food Control 26:6–14. https://doi.org/10.1016/j.foodcont.2011.12.024
Henríquez-Hernández LA, Acosta-Dacal AC, Boada LD, Zumbado M, Serra-Majem L, Luzardo OP (2023) Concentration of essential, toxic, and rare earth elements in ready-to-eat baby purees from the Spanish market. Nutrients 15(14):3251. https://doi.org/10.3390/nu15143251
Hong ZC, Guo YL, Wang G (2015) Mercury pollution and its influencing factors of vegetable soils in Fujian Province, China. J Agr Res Environ 32(2):198–203 (in Chinese)
Hu J, Wu F, Wu S, Cao Z, Lin X, Wong MH (2013) Bioaccessibility, dietary exposure and human risk assessment of heavy metals from market vegetables in Hong Kong revealed with an in vitro gastrointestinal model. Chemosphere 91(4):455–461. https://doi.org/10.1016/j.chemosphere.2012.11.066
Hussain S, Rengel Z, Qaswar M, Amir M, Zafar-ul-Hye M (2019) Arsenic and heavy metal (cadmium, lead, mercury and nickel) contamination in plant-based foods. Plant Hum Health: Phytochem Mol Asp 2:447–490
İslamoğlu AH, Kahvecioğlu T, Bönce G, Gedik E, GÜNEŞ F (2021) Determination of heavy metals in some fruits, vegetables and fish by ICP-MS. Eur J Food Sci Technol 5(1):67–76
JECFA (2007) Evaluation of certain food additives and contaminants.: sixty-seventh report of the joint FAO/WHO expert committee on food additives. WHO Technical Report Series, vol 940. World Health Organization, Geneva Switzerland
JECFA (2011) Evaluation of certain contaminants in food.: seventy-second report of the joint FAO/WHO expert committee on food additives. WHO Technical Report Series, vol 959. World Health Organization, Geneva Switzerland
Jia Q, Zhu X, Hao Y, Yang Z, Wang Q, Fu H et al (2018) Mercury in soil, vegetable and human hair in a typical mining area in China: implication for human exposure. J Environ Sci 68(6):73–82
Ju YR, Chen CW, Chen CF, Chuang XY, Dong CD (2017) Assessment of heavy metals in aquaculture fishes collected from the southwest coast of Taiwan and human consumption risk. Int Biodeterior Biodegrad 124:314–325
Kakimoto S, Yoshimitsu M, Akutsu K, Kiyota K, Fujiwara T, Watanabe T, Kajimura K, Yamano T (2019) Concentrations of total mercury and methylmercury in red snow crabs (*Chionoecetes japonicus*) caught off the coast of Japan. Mar Pollut Bull 145:1–4. https://doi.org/10.1016/j.marpolbul.2019.04.062
Kalogeropoulos N, Karavoltsos S, Sakellari A, Avramidou S, Dassenakis M, Scoullos M (2012) Heavy metals in raw, fried and grilled Mediterranean finfish and shellfish. Food Chem Toxicol 50(10):3702–3708
Keshavarzi B, Hassanaghaei M, Moore F, Mehr MR, Soltanian S, Lahijanzadeh AR, Sorooshian A (2018) Heavy metal contamination and health risk assessment in three commercial fish species in the Persian Gulf. Mar Pollut Bull 129:245–252

Kharazi A, Leili M, Khazaei M, Alikhani MY, Shokoohi R (2021) Human health risk assessment of heavy metals in agricultural soil and food crops in Hamadan, Iran. J Food Comp Anal 100:103890. https://doi.org/10.1016/j.jfca.2021.103890

Krygier K, Wroniak M, Wódka M, Grzéskiewicz S, Obiedzí nski S (2000) Badania wpływu czasu tłoczenia na jakość oleju rzepakowego tłoczonego na zimno. Z˙ ywność Nauka Technologia Jakość 7:39–48

Lammari N, Louaer O, Meniai AH, Fessi H, Elaissari A (2021) Plant oils: from chemical composition to encapsulated form use. Int J Pharm 601:120538. https://doi.org/10.1016/j.ijpharm.2021.120538

Leblanc JC, Guérin T, Noël L, Calamassi-Tran G, Volatier JL, Verger P (2005) Dietary exposure estimates of 18 elements from the 1st French Total Diet Study. Food Addit Contam 22(7):624–641. https://doi.org/10.1080/02652030500135367

Li B, Shi JB, Wang X, Meng M, Huang L, Qi XL, He B, Ye ZH (2013) Variations and constancy of mercury and methylmercury accumulation in rice grown at contaminated paddy field sites in three Provinces of China. Environ Pollut 181:91–97

Li P, Feng X, Chan HM, Zhang X, Du B (2015) Human Body burden and dietary methylmercury intake: the relationship in a rice-consuming population. Environ Sci Technol 49:9682–9689. https://doi.org/10.1021/acs.est.5b00195

Li R, Wu H, Ding J, Fu W, Gan L, Li Y (2017a) Mercury pollution in vegetables, grains and soils from areas surrounding coal-fired power plants. Sci Rep 7(1):46545

Li P, Du B, Maurice L, Laffont L, Lagane C, Point D, Sonke JE, Yin R, Lin CJ, Feng X (2017b) Mercury isotope signatures of methylmercury in rice samples from the Wanshan mercury mining area, China: environmental implications. Environ Sci Technol 51:12321–12328

Lin H, Santa-Rios A, Barst BD, Basu N, Bayen S (2019) Occurrence and bioaccessibility of mercury in commercial rice samples in Montreal (Canada). Food Chem Toxicol 126:72–78. https://doi.org/10.1016/j.fct.2019.02.006

Linkon KMMR, Satter MA, Jabin SA, Abedin N, Islam MF, Lisa LA, Paul DK (2015) Mineral and heavy metal contents of some vegetable available in local market of Dhaka city in Bangladesh. IOSR J Environ Sci Toxicol Food Technol 9:01–06. https://doi.org/10.9790/2402-09510106

Liu XM, Song QJ, Tang Y, Li WL, Xu JM, Wu JJ, Wang F, Brookes PC (2013) Human health risk assessment of heavy metals in soil-vegetable system: a multi-medium analysis. Sci Total Environ 463–464:530–540

Lodhi RS, Kazi TG, Talpur FN, Afridi HI, Baig JA, Rajput AA (2021) Evaluation of the methyl and inorganic mercury in infant formula milk and cereals samples: estimated risk assessment in children under 2.0 years. ACS Food Sci Technol 1(10):1757–1763

Macrotrends (2023) Bangladesh life expectancy 1950–2023. Retrieved from: https://www.macrotrends.net/countries/BGD/bangladesh/life-expectancy. Accessed on 6 Aug 2023

Margenat A, Matamoros V, Díez S et al (2018) Occurrence and bioaccumulation of chemical contaminants in lettuce grown in peri-urban horticulture. Sci Total Environ 637–638:1166–1174

Martins C, Vasco E, Paixão E, Alvito P (2013) Total mercury in infant food, occurrence and exposure assessment in Portugal. Food Addit Contam: Part B 6(3):151–157

Martorell I, Perelló G, Martí-Cid R et al (2011) Human exposure to arsenic, cadmium, mercury, and lead from foods in Catalonia, Spain: temporal trend. Biol Trace Elem Res 142:309–322

Mathaiyan M, Natarajan A, Rajarathinam X, Rajeshkumar S (2021) Assessment of Pb, Cd, As and Hg concentration in edible parts of broiler in major metropolitan cities of Tamil Nadu, India. Toxicol Rep 8:668–675

Matta G, Gjyli L (2016) Mercury, lead and arsenic: impact on environment and human health. J Chem Pharm Sci 9(2):718–725

Mawari G, Kumar N, Sarkar S, Daga MK, Singh MM, Joshi TK, Khan NA (2022) Heavy metal accumulation in fruits and vegetables and human health risk assessment: findings from Maharashtra, India. Environ Health Insights 16:11786302221119152. https://doi.org/10.1177/11786302221119151

Mehouel F, Bouayad L, Berber A, Van Hauteghem I, Van de Wiele M (2019) Risk assessment of mercury and methyl mercury intake via sardine and swordfish consumption in Algeria. J Hell Vet Med Soc 70(3):1679–1686

Melai V, Giovannini A, Chiumiento F et al (2018) Occurrence of metals in vegetables and fruits from areas near landfill in Southern Italy and implications for human exposure. Int J Food Cont 5:8. https://doi.org/10.1186/s40550-018-0070-5

Meng M, Li B, Shao JJ, Wang T, He B, Shi JB, Ye ZH, Jiang GB (2014) Accumulation of total mercury and methylmercury in rice plants collected from different mining areas in China. Environ Pollut 184:179–186

Mesinger D, Ocieczek A (2021) Risk assessment of wild game meat intake in the context of the prospective development of the venison market in Poland. Polish J Environ Stud 30(2):1307–1315

Muñoz O, Bastias JM, Araya M et al (2005) Estimation of the dietary intake of cadmium, lead, mercury, and arsenic by the population of Santiago (Chile) using a Total Diet Study. Food Chem Toxicol 43:1647–1655

Nawrocka A, Durkalec M, Szkoda J, Filipek A, Kmiecik M, Żmudzki J, Posyniak A (2020) Total mercury levels in the muscle and liver of livestock and game animals in Poland, 2009–2018. Chemosphere 258:127311. https://doi.org/10.1016/j.chemosphere.2020.127311

Nie J, Fu X, Wang L, Xu J, Gao X (2022) A systematic review of fermented Saccharina japonica: fermentation conditions, metabolites, potential health benefits and mechanisms. Trends Food Sci Technol 123:15–27

Novirsa R, Dinh QP, Jeong H, Addai-Arhin S, Nugraha WC, Hirota N, Wispriyono B, Ishibashi Y, Arizono K (2020) The dietary intake of mercury from rice and human health risk in artisanal small-scale gold mining area, Indonesia. Fundam Toxicol Sci 7(5):215–225

Núñez R, García MA, Alonso J, Melgar MJ (2018) Arsenic, cadmium and lead in fresh and processed tuna marketed in Galicia (NW Spain): risk assessment of dietary exposure. Sci Total Environ 627:322–331

Ogunkunle ATJ, Bello OS, Ojofeitimi OS (2014) Determination of heavy metal contamination of street-vended fruits and vegetables in Lagos state, Nigeria. Int Food Res J 21:2115–2120

Olatunde OO, Benjakul S (2018) Nonthermal processes for shelf-life extension of seafoods: a revisit. Comp Rev Food Sci Food Saf 17(4):892–904

Ouédraogo O, Amyot M (2011) Effects of various cooking methods and food components on bioaccessibility of mercury from fish. Environ Res 111(8):1064–1069

Pang J, Han J, Fan X, Li C, Dong X, Liang L, Chen Z (2018) Mercury speciation, bioavailability and risk assessment on soil–rice systems from a watershed impacted by abandoned Hg mine-waste tailings. Acta Geochim 38:391–403. https://doi.org/10.1007/s11631-018-0305-4

Parker GH, Gillie CE, Miller JV, Badger DE, Kreider ML (2022) Human health risk assessment of arsenic, cadmium, lead, and mercury ingestion from baby foods. Toxicol Rep 9:238–249

Pelcová P, Ridošková A, Hrachovinová J, Grmela J (2021) Evaluation of mercury bioavailability to vegetables in the vicinity of cinnabar mine. Environ Pollut 283:117092. https://doi.org/10.1016/j.envpol.2021.117092

Potortì AG, Bua GD, Lo Turco V et al (2020) Major, minor and trace element concentrations in spices and aromatic herbs from Sicily (Italy) and Mahdia (Tunisia) by ICP-MS and multivariate analysis. Food Chem 313:126094. https://doi.org/10.1016/j.foodchem.2019.126094

Pratt SG (2018) Food for thought: how to steer patients toward healthier habits: food may act as medicine to promote wellness and minimize chronic disease, says the ophthalmologist who coined the word "superfood." Rev Optom 155(9):S14–S14

Proshad R, Kormoker T, Islam MS, Chandra K (2019) Potential health risk of heavy metals via consumption of rice and vegetables grown in the industrial areas of Bangladesh. Hum Ecol Risk Assess 24(4):921–943

Qing Y, Li Y, Yang J, Li S, Gu K, Bao Y, Zhan Y, He K, Wang X, Li Y (2022) Risk assessment of mercury through dietary exposure in China. Environ Pollut 312:120026. https://doi.org/10.1016/j.envpol.2022.120026

Qiu G, Feng X, Meng B, Zhang C, Gu C, Du B, Lin Y (2013) Environmental geochemistry of an abandoned mercury mine in Yanwuping, Guizhou Province, China. Environ Res 125:124–130

Ritonga IR, Bureekul S, Sompongchaiyakul P (2022) Risk assessment of total mercury (T-Hg) in commercial seafood marketed in Bangkok, Thailand. Appl Environ Res 44(1). https://doi.org/10.35762/AER.2021.44.1.10

Rodrigues ET, Coelho JP, Pereira E, Pardal MA (2023a) Are mercury levels in fishery products appropriate to ensure low risk to high fish-consumption populations? Mar Pollut Bull 186:114464. https://doi.org/10.1016/j.marpolbul.2022.114464

Rodrigues PDA, de Pinho JV, Ramos-Filho AM, Neves GL, Conte-Junior CA (2023b) Mercury contamination in seafood from an aquatic environment impacted by anthropic activity: seasonality and human health risk. Environ Sci Pollut Res 30:85390–85404

Rose M, Baxter M, Brereton N, Baskaran C (2010) Dietary exposure to metals and other elements in the 2006 UK Total Diet Study and some trends over the last 30 years. Food Addit Contam 27(10):1380–1404. https://doi.org/10.1080/19440049.2010.496794

Rothenberg SE, Feng X, Zhou W, Tu M, Jin B, You J (2012) Environment and genotype controls on mercury accumulation in rice (*Oryza sativa* L.) cultivated along a contamination gradient in Guizhou, China. Sci Total Environ 426:272–280

Rubio C, Gutiérrez ÁJ, Hardisson A, Martín V, Revert C, Pestana Fernandes PJ, Horta Lopes DJ, Paz-Montelongo S (2023) Dietary exposure to toxic metals (Cd, Pb and Hg) from cereals marketed in madeira and the azores. Biol Trace Element Res 1–10

Rucklidge JJ, Shaw IC (2020) Are over-the-counter fish oil supplements safe, effective and accurate with labelling? Analysis of 10 New Zealand fish oil supplements. N Z Med J 133:52–62

Sanga TR, Maseka KK, Ponraj M, Tungaraza C, Mwakalapa EB (2023) Accumulation and distribution of mercury in agricultural soils, food crops and associated health risks: a case study of Shenda gold mine-Geita Tanzania. Environ Chall 11:100697. https://doi.org/10.1016/j.envc.2023.100697

Sarkar R, Podder I, Gokhale N, Jagadeesan S, Garg VK (2017) Use of vegetable oils in dermatology: an overview. Int J Dermatol 56:1080–1086

Sattar MU, Khan MA, Khalil AA, Amir RM (2013) Mitigation of heavy metals in vegetables through washing with house hold chemicals. Int J Agric Sci Res 3(5):1–11

Shao DD, Wu SC, Liang P, Kang Y, Fu WJ, Zhao KL, Cao ZH, Wong MH (2012) A human health risk assessment of mercury species in soil and food around compact fluorescent lamp factories in Zhejiang Province, PR China. J Hazard Mat 221:28–34

Shao DD (2011) Human risk assessment based on mercury contamination in food and environmental matrix at two regions in Guangdong and Zhejiang Provinces. PhD thesis, Hong Kong Baptist University

Sharma D, Bisla G (2022) Assessment of heavy metals in fruits and vegetables collected from Bareilly Local Market, Uttar Pradesh State, India. Int J Res Appl Sci Eng Technol 10:501–509

Singh V, Garg AN (2006) Availability of essential trace elements in Indian cereals, vegetables and spices using INAA and the contribution of spices to daily dietary intake. Food Chem 94(1):81–89. https://doi.org/10.1016/j.foodchem.2004.10.053

Silva MJD, Paim APS, da Silva IJ, Pimentel MF, Cervera ML, Guardia MDL (2023) Determination of total mercury in Spanish samples of baby food, fast food, and daily meal. J Brazil Chem Soc 34:517–526

Skordas K, Papastergios G, Filippidis A (2013) Major and trace element contents in apples from a cultivated area of central Greece. Environ Monit Assess 185:8465–8471

Smutna M, Kruzikova K, Marsalek P, Kopriva V, Svobodova Z (2009) Fish oil and cod liver as safe and healthy food supplements. Neuroendocrinol Lett 30:156–162

Ssenku JE, Naziriwo B, Kutesakwe J, Mustafa AS, Kayeera D, Tebandeke E (2023) Mercury accumulation in food crops and phytoremediation potential of wild plants thriving in artisanal and small-scale gold mining areas in Uganda. Pollutants 3(2):181–196

Stančić Z, Vujević D, Gomaz A, Bogdan S, Vincek D (2016) Detection of heavy metals in common vegetables at Varaždin City Market, Croatia. Arhiv za higijenu rada i toksikologiju 67(4):340–350. https://doi.org/10.1515/aiht-2016-67-2823
Storelli MM, Stuffler RG, Storelli A, Marcotrigiano GO (2003) Total mercury and methylmercury content in edible fish from the Mediterranean Sea. J Food Prot 66(2):300–303
Sulistiono S, Irawati Y, Batu DTFL (2018) Heavy metal content of tank goby (*Glossogobius giuris*) in East Segara Anakan, Cilacap, Central Java, Indonesia. J Pengolah Has Perikan Indo 21(3):423–432
Suratno S, Cordova MR, Arinda S (2017) Mercury content in commercial fishes of Bantul areas and Yogyakarta. Oseanologi Dan Limnologi Di Indonesia 2(1):15–23
Tasrina RC, Rowshon AAMR, Mustafizur AMR, Rafiqul I, Ali MP (2015) Heavy metals contamination in vegetables and its growing soil. J Environ Ana Chem 2:142. https://doi.org/10.4172/2380-2391.1000142
Tomada I, Tomada N (2023) Mediterranean diet and male fertility. Endocrines 4(2):394–406
Tweedy BN, Sansom B, Vaughn CC (2022) Mercury contamination of fish and their prey across a riverine food web. The Am Midl Nat 188(2):234–243
US/EPA (2000) Reference dose for methylmercury; U.S. Environmental Protection Agency, Washington, DC, USA
US/EPA (2001) Water quality criterion for the protection of human health: methylmercury. Office of Water, Washington (DC). Report no. EPA-823-R-01–001
US/EPA (2014) United States environmental protection agency. Risk-Based Concentrations Table
US/EPA (2017) Human health risk assessment. Retrieved from: https://www.epa.gov/risk/human-health-risk-assessment. Accessed on 6 Aug 2023
van Rooyen D, Erasmus JH, Gerber R, Nachev M, Sures B, Wepener V, Smit NJ (2023) Bioaccumulation and trophic transfer of total mercury through the aquatic food webs of an African sub-tropical wetland system. Sci Total Environ 889:164210
Vasconcellos ACSD, Hallwass G, Bezerra JG, Aciole ANS, Meneses HNDM, Lima MDO, Jesus IMD, Hacon SDS, Basta PC (2021) Health risk assessment of mercury exposure from fish consumption in Munduruku indigenous communities in the Brazilian Amazon. Int J Environ Res Pub Health 18(15):7940. https://doi.org/10.3390/ijerph18157940
Venturieri R, Oliveira-da-Costa M, Gama C, Jaster CB (2017) Mercury contamination within protected areas in the Brazilian northern Amazon-Amapá State. Am J Environ Sci 13(1):11–21
Vezzone M, Cesar R, Serrano A, Lourenço R, Rodrigues AP, Castilhos Z, dos Anjos RM, Polivanov H (2023) Mercury contamination in sediments and fish from an urban tropical estuary: ecological and human health risks. Water, Air, Soil Pollut 234(2):72. https://doi.org/10.1007/s11270-023-06089-y
Vollmannova A, Zupka S, Bajcan D et al. (2015).Dangerous heavy metals in soil and small forest fruit as a result of old environmental loads. In: Proceedings of the 14th international conference on environmental science and technology, pp 3–5
Wai K, Dai J, Yu P, Zhou X, Wong C (2017) Public health risk of mercury in China through consumption of vegetables, a modelling study. Environ Res 159:152–157
Wang L, Han J, Katuwal HB, Xia P, Xu X, Feng X, Qiu G (2021) Occurrence of total mercury and methylmercury in rice: exposure and health implications in Nepal. Ecotoxicol Environ Saf 228:113019. https://doi.org/10.1016/j.ecoenv.2021.113019
WHO (2007) Exposure to mercury: a major health concern. World health organization, Geneva, Switzerland 1
WHO (2022) Guidelines for drinking-water quality: recommendations, 4th edn. World Health Organization. https://www.who.int/publications/i/item/9789240045064
WHO/FAO (2016) General standard for contaminants and toxins in food and feed. Food and Agriculture Organization, World Health Organization, Rome, Geneva
WHO/UNEP (2008) Technical background report to the global atmospheric mercury assessment. Artic Monitoring and Assessment Programme/UNEP Chemicals Branch, pp 159

Wu Z, Li P, Feng X (2022) Assessing the factors impacting the bioaccessibility of mercury (Hg) in rice consumption by an in-vitro method. J Environ Sci (China) 119:119–129

Xu J, Zhang Q, Wang S, Nan Z, Long S, Wu Y, Dong S (2023) Bioavailability, transfer, toxicological effects, and contamination assessment of arsenic and mercury in soil-corn systems. Environ Sci Pollut Res 30(4):10063–10078

Yang B, Gao Y, Zhang C, Zheng X, Li B (2020) Mercury accumulation and transformation of main leaf vegetable crops in Cambosol and Ferrosol soil in China. Environ Sci Pollut Res 27:391–398

Yin D, He T, Yin R, Zeng L (2018) Effects of soil properties on production and bioaccumulation of methylmercury in rice paddies at a mercury mining area, China. J Environ Sci (China) 68:194–205

Zhang H, Feng X, Larssen T, Qiu G, Vogt RD (2010) In inland China, rice, rather than fish, is the major pathway for methylmercury exposure. Environ Health Perspect 118:1183–1188

Zhang T, Zhang Y, Li W, Wang L, Jiao Y, Wang Y, Jiang D, Gao X (2021) Occurrence and dietary exposure of heavy metals in marketed vegetables and fruits of Shandong Province, China. Food Sci Nutr 9(9):5166–5173

Zulkipli SZ, Tan CP, Seah YG, Liew HJ, Sung YY, Ando M, Wang M, Liang Y, McMinn A, Mok WJ (2023) Assessment of mercury contamination and food composition in commercially important marine fishes in the southern South China Sea. Reg Stud Mar Sci 58:102795. https://doi.org/10.1016/j.rsma.2022.102795

Chapter 3
Research Trend on Mercury (Hg) Contamination of Water Resources: A Bibliometric Review

Gyanajeet Yumnam, Yumnam Gyanendra, and Wazir Alam

3.1 Introduction

Mercury (Hg) is a toxic heavy metal that has been regarded as one of the "ten leading chemicals of concern" that poses significant human health risks and environmental concern (WHO 2017). According to the United Nations Environment Programme (UNEP) report (UNEP 2018), it is estimated that global Hg-emissions to air from anthropogenic sources in 2015 were about 2220 tons. Among the anthropogenic sources, stationary combustion of fossil fuels accounts for 24% of the estimated emissions, primarily from burning of coal (21%) (UNEP 2018). The majority of Hg-contamination comes from mining, agriculture, and industry, with most in the form of inorganic Hg (Hg (II)) (Rumayor et al. 2017; Bindler et al. 2012). The presence of Hg in water is a major concern due to its potential impacts bioaccumulation risk on human health and the environment. Consumption of contaminated water is one of the major pathways through which mercury enters the human body. Similarly, fish and other aquatic organisms are particularly vulnerable to mercury contamination (Parang and Esmaeilbeigi 2022; Perelonia et al. 2021). Consumption of Hg-contaminated water, directly or through contaminated fish and seafood, it can lead to serious human health risks. Generally, methylmercury gets bioaccumulated in the tissues of higher organisms in food chain as they feed on smaller organisms (Al-Sulaiti et al. 2022). This bioaccumulation and biomagnification process can lead to higher Hg-concentration in fish and seafood than in the surrounding water (Qiu and Wang 2016). Consequently, eating contaminated fish is a significant source of mercury exposure to

G. Yumnam
Department of Library and Information Science, Manipur University, Canchipur, India

Y. Gyanendra · W. Alam (✉)
Department of Forestry and Environmental Science, Manipur University, Canchipur, India
e-mail: wazir@manipuruniv.ac.in

N. Kumar (ed.), *Mercury Toxicity*, Environmental Science and Engineering,
https://doi.org/10.1007/978-981-99-7719-2_3

humans, leading to widespread advisories on fish consumption, especially for pregnant women and young children (Chen and Dong 2022; Castaño et al. 2015). Studies have shown that prolonged exposure to low levels Hg-contaminated water can lead to various health issues related to neurological symptoms such as tremors, kidneys failure, weak immune system, cardiovascular disorders, memory loss, and cognitive impairments (Mitra et al. 2022; Azar et al. 2021; Clarkson et al. 2003). In severe cases, high-level exposure to mercury can result in Minamata disease, a neurological disorder characterized by numbness, muscle weakness, and incoordination (Yang et al. 2020; Harada 1995).

The effects of mercury contamination on ecosystems are far-reaching as it can disrupt the balance of aquatic ecosystems by impairing the reproduction and growth of aquatic plants and animals. Some species may be highly sensitive to mercury, while others may accumulate high levels without being significantly affected. This imbalance can lead to loss of biodiversity and sensitive species, disrupting the overall functioning of the ecosystem. Various regulations and guidelines have been established to protect human health to limit Hg-concentration in drinking water. For instance, the World Health Organization (WHO) has set a provisional guideline value of 6 μg per liter (μg/L) for total mercury in drinking water (WHO 2005). Similarly, the United States Environmental Protection Agency (USEPA) has established a maximum contaminant level (MCL) of 2 μg/L for Hg in drinking water (EPA 2020). International agreements like the Minamata Convention on Mercury aim to control and reduce mercury emissions from various industrial sectors (UNEP 2018). Additionally, technologies for wastewater treatment and control measures in industries that release mercury have been developed to minimize Hg-contamination in water bodies (García et al. 2018). Efforts have also been made to reduce Hg-pollution at the source by various control measures (EPA 2020).

In the past 30 years, the field of Hg-contamination in water has witnessed a significant rise in research publications. Scholars have increasingly relied on bibliometric studies to gain a comprehensive understanding of the current state of research and to identify emerging trends within the field (Gyanajeet and Ibohal 2023; Gyanendra et al. 2022; Shi and Yin 2021). These studies have employed systematic and quantitative analyses of scientific literature to enable researchers to assess publication volume, track the field's growth, and explore statistical and qualitative patterns and themes prevalent in mercury contamination in water research from 1989 to 2021. A set of research questions (RQ) has been formulated to facilitate this bibliometric study, and subsequent efforts have been made to address these questions comprehensively.

RQ1. How has the research on mercury contamination of water evolved in terms of publication output and citation trends?

RQ2. Which scientific journals have published the most articles on mercury contamination of water, and what is their impact factor distribution?

RQ3. What are the leading countries, institutions and their collaborations patterns, and prolific authors to the literature on mercury in water?

RQ4. What are the most cited research papers in the field of Hg-contamination of water?

RQ5. What are the key trends and subject research areas in research publications on mercury contamination in water bodies over the past decade?

3.2 Materials and Methods

3.2.1 Data Sources

The study utilized scientific data retrieved from the Clarivate Analytics Web of Science (WoS) Core Collection database, which includes the Science Citation Index-Expanded (SCI-E) and the Social Sciences Citation Index (SSCI). This database is renowned as one of the largest repositories of abstracts and citations for scientific content, and it is widely used in bibliometric studies (Sharifi et al. 2020). The database allows the download of full citation records along with cited references. To search for document titles related to Mercury in Water, an advanced search strategy was employed using the Boolean Logical Operator "OR," i.e., Title = ("Mercury" Or "Hg" AND "Water"). The data retrieval was conducted on April 12, 2023, to ensure consistency, as the Web of Science database is regularly updated.

3.2.2 Exclusion and Inclusion Criteria

The initial outcome from the database consisted of 1737 documents. After examining all the documents, 8 irrelevant document types were identified and excluded from the analysis (refer to Fig. 3.1). Following the application of exclusion and inclusion criteria, a total of 1729 documents were included for further review analysis that includes documents comprising of mainly three types: articles (1614), reviews (25), and proceeding papers (90). These document types were chosen as they were expected to provide more comprehensive data. The study period was defined as 1989 to 2021, starting from the first article indexed in WoS in 1989. Articles authored by individuals from England, Northern Ireland, Scotland, and Wales were grouped under the United Kingdom (UK) heading. To gather information on the journal impact factor (IF) and quartiles (Q) of the journals, data from the Journal Citation Report of 2022, published by Clarivate Analytics Web of Science, were utilized.

3.2.3 Analysis Tool

The comprehensive analysis was carried out utilizing the retrieved data (1729) by means of statistical tools and bibliometric software. Calculation packages utilized included Microsoft Excel 2019 for data filtration, cleaning, and statistical analysis. Biblioshiny (v4.1.2) (Aria and Cuccurullo 2017) and BibExcel (v 2017) (Persson

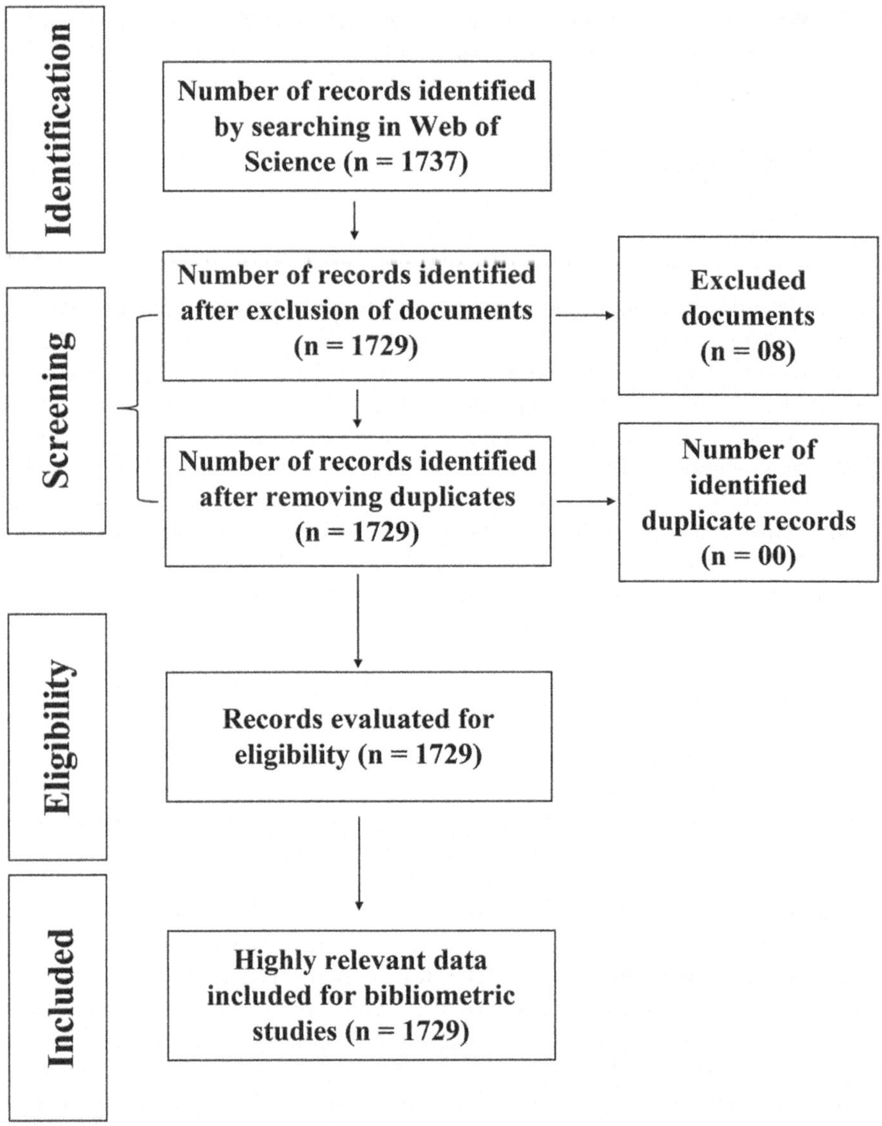

Fig. 3.1 Four phase for data extraction, screening and filtration process

et al. 2009) aided bibliometric analysis, generating citation analysis, collaborations network, and thematic maps. VOS viewer (v1.6.18) (van Eck and Waltman 2009) facilitated mapping and data visualization. These open-source educational software tools are widely employed for academic and research purposes. Data was exported from the database in plain text (.txt) and converted to comma-separated values (.csv) for analysis.

3.3 Results

3.3.1 Annual Research Growth and Trends of Citations

The presentation of the annual growth of publications and citation trends from 1989 to 2022 were measured in the number of papers published annually (Table 3.1). The findings reveal a significant increase in the number of publications, from 23 papers in 1989 to 102 in 2022. Notably, the highest growth occurred in 2022, with 102 publications, followed by 98 papers in 2020 and 95 in 2019. Over the entire period under consideration i.e., from 1989 to 2022, the research area experienced an annual growth rate of 4.66% in the number of papers published each year, indicating an exponential growth in scientific literature. Regarding total citations (TC), the research area gained 52,368 citations from 1,729 publications. The year 2014 witnessed the highest number of citations, with a total of 3,741, followed by 2009 (TC = 3,704) and 2012 (TC = 2,396). These analyses demonstrate that both publications and citations have experienced exponential growth, with citations outpacing publications, which is highly appreciable for the emerging research on mercury. Notably, the most recently published papers tend to have lower citation scores, as it takes at least two years to gather substantial data for meaningful analysis (Fig. 3.2).

Table 3.1 Annual publication and citation structure between 1989 and 2022

Year	TP	TC	C/P	h-index	h-core	Year	TP	TC	C/P	h-index	h-core
1989	23	462	20.09	11	425	2006	39	1592	40.82	20	1417
1990	36	1203	33.42	14	1112	2007	39	2095	53.72	22	1927
1991	33	1194	36.18	15	1074	2008	43	1160	26.98	21	943
1992	30	1786	59.53	16	1703	2009	66	3704	56.12	32	3243
1993	27	1231	45.59	14	1170	2010	47	2140	45.53	23	1882
1994	25	927	37.08	16	895	2011	59	1910	32.37	25	1475
1995	40	1231	30.78	18	1034	2012	72	2396	33.28	26	1759
1996	32	756	23.63	15	683	2013	71	1617	22.77	23	1161
1997	30	919	30.63	15	848	2014	81	3741	46.19	32	3085
1998	24	907	37.79	13	845	2015	73	2198	30.11	25	1651
1999	27	1034	38.30	16	975	2016	72	1760	24.44	25	1228
2000	38	1857	48.87	24	1733	2017	73	2302	31.53	28	1713
2001	27	977	36.19	16	903	2018	94	2252	23.96	29	1534
2002	25	711	28.44	12	639	2019	95	1904	20.04	24	1303
2003	33	1127	34.15	20	1034	2020	98	1668	17.02	21	1018
2004	32	1200	37.50	20	1077	2021	89	728	8.18	14	352
2005	29	1386	47.79	17	1302	2022	102	293	2.87	8	127

* *TP* = Total number of publications; *TC* = Total citations; *C/P* = Average citations per publication; *h-core* = Citation sum within the h-core

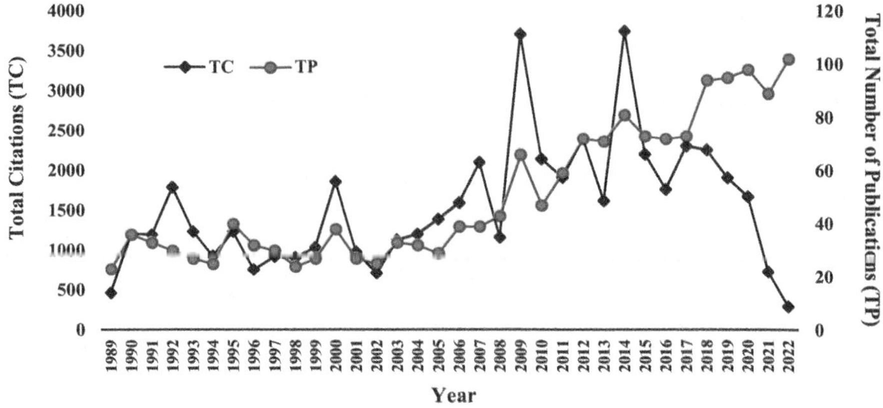

Fig. 3.2 Publications and citation pattern on mercury in water from 1989 to 2022

3.3.2 Journal Performance

A total of 1731 research publications focused on Hg-contamination of water which were sourced from 539 journals. The top 20 most productive journals contributed 33.16% of the publications (Table 3.2). Leading the pack was the journal "Science of the Total Environment," which began publishing in this research area in 1990 and produced 57 publications (TP). It was closely followed by "Water Air and Soil Pollution" and "Journal of Hazardous Materials," with 55 and 43 publications, respectively. Interestingly, the "Journal of Hazardous Materials" garnered the highest number of citations (TC), reaching 3413 from its publications, and boasted an impressive h-index of 30, along with an impact factor of 14.255. "Environmental Science & Technology" and "Analytica Chimica Acta" followed with TCs of 2968 and 2382, respectively. In terms of average citations per publication (C/P), "Environmental Science & Technology" stood out with an outstanding C/P of 80.22, indicating the significant impact of the papers published in this journal. Analyzing quartiles, thirteen of the top 20 journals fell within the first quartile (Q1), while three were in the second (Q2) and third quartile (Q3), and only one was in the fourth quartile (Q4). This suggests that most of the top journals in this research area are highly quality, influential, and hold substantial sway within the scientific community.

3.3.3 Prolific Authors

An extensive review of research publications on mercury in water from 1989 to 2022 reveals the active involvement of 5936 authors in this field. Table 3.3 highlights the top 15 authors who have made significant contributions to the field of bioremediation, along with their citation impacts. Eduarda C Pereira, affiliated with the University of

Table 3.2 Top 20 prolific journals published on mercury in water from 1989 to 2022

Journal	TP	TC	C/P	h-index	IF	Q	PY-Start
Science of the Total Environment	57	1952	34.25	28	10.753	Q1	1990
Water Air and Soil Pollution	55	1942	35.31	23	2.984	Q3	1989
Journal of Hazardous Materials	43	3413	79.37	30	14.255	Q1	2007
Analytica Chimica Acta	37	2382	64.38	25	6.911	Q1	1990
Environmental Science & Technology	37	2968	80.22	24	11.357	Q1	1990
Talanta	35	1554	44.40	25	6.556	Q1	1990
Chemosphere	28	913	32.61	17	8.943	Q1	1995
Microchemical Journal	26	704	27.08	17	5.304	Q1	1998
Environmental Science and Pollution Research	26	383	14.73	9	5.19	Q2	2012
Journal of Electroanalytical Chemistry	24	414	17.25	13	4.598	Q2	1989
Journal of Analytical Atomic Spectrometry	23	1110	48.26	19	4.351	Q1	1993
Sensors and Actuators B-Chemical	23	813	35.35	17	9.221	Q1	2001
RSC Advances	23	391	17.00	14	4.036	Q2	2014
International J of Environmental Analytical Chemistry	22	406	18.45	11	2.731	Q3	1994
Analytical Methods	21	316	15.05	11	3.532	Q1	2010
Analytical Sciences	21	314	14.95	11	1.967	Q4	1991
Analyst	19	579	30.47	14	5.227	Q1	1996
Environmental Pollution	19	569	29.95	14	9.988	Q1	1990
Bulletin of Environmental Contamination and Toxicology	18	209	11.61	9	2.807	Q3	1989
Spectrochimica Acta Part A-Molecular and Biomolecular Spectroscopy	17	334	19.65	11	4.831	Q1	2008

* *TP* = Total number of publications; *TC* = Total citations; *C/P* = Average citations per publication; *h-core* = Citation sum within the h-core; *PY-Start* = Publications Year-Start; *IF* = Impact Factor; *Q* = Quartiles

Aveiro in Portugal, emerges as the most prolific author, having an impressive tally of 17 research publications (TP). Notably, Eduarda C Pereira also possesses the highest h-index of 11 and h-core value of 361, indicating the significance and impact of her work. Close behind, Xinbin Feng from the Chinese Academy of Sciences in China has contributed 15 publications, while Shaofeng Wang from the Dalian University of Technology in China and Yue Zhang from the University of Southampton in the UK have each published 11 papers. Remarkably, Xinbin Feng holds the record for the highest number of citations, with a total count (TC) of 577, followed closely by David P. Krabbenhoft (TC = 568) and Margareta Horvat (TC = 547). Researchers and scholars in the field can greatly benefit from keeping track of these influential authors, as their work promises remarkable and influential publications within the domain of mercury in water research.

The field of research being discussed holds significant promise and is undergoing rapid development. It continues attracting the attention of numerous authors actively contributing to its advancement. In order to foster potential collaborations, we have visually depicted the top ten authors in Fig. 3.3. This visual representation showcases their paper production and the total citations they have received from 1989 to 2022. The graph utilizes blue bubble circles to represent the number of publications and darker circles to represent the number of citations received annually by each author. The horizontal line in the graph signifies an author's timeline, providing a clear depiction of their career progression. The size of the bubbles corresponds to the number of publications, with larger bubbles indicating higher publication rates in specific years. On the other hand, the darkness of the bubbles indicates the number of citations, with darker bubbles representing a higher number of citations received each year. This graph serves as a valuable tool for researchers and institutions as it allows them to identify an author's productivity pattern over time swiftly. It facilitates the evaluation of an author's research output and enables tracking of their scholarly impact. Additionally, it aids in identifying productive periods in an author's career and monitoring changes in their research focus.

3.3.4 Contribution of Country and Collaboration

Between 1989 and 2022, a total of 84 countries actively contributed to global research publications on mercury in water. Table 3.4 overviews the top twenty countries based on their total publications (TP). China is Leading the pack, which produced a remarkable 721 research outputs, followed by the USA with 577 and India with 256. This highlights the significant emphasis these three countries' researchers have on water and mercury-related areas of study. Following closely behind are Iran, Canada, Japan, France, Spain, Brazil, and Portugal, with over 100 publications. The high publication output from these countries can be attributed to their robust economic power and substantial investment in research, development, and innovation. Seven of the top twenty countries belong to the major industrialized G7 nations: The USA, the UK, Germany, France, Italy, Japan, and Canada. Also, five of

Table 3.3 Top 20 prolific authors on mercury in water research

Authors	Affiliations	TP	TC	C/P	h-index	h-core	PY-start
Eduarda C Pereira	University of Aveiro, Portugal	17	389	22.88	11	361	2006
Xinbin Feng	Chinese Academy of Sciences, China	15	577	38.47	10	536	2001
Shaofeng Wang	Dalian University of Technology, China	11	402	36.55	9	394	2004
Yue Zhang	University of Southampton, UK	11	530	48.18	6	520	2006
David Amouroux	University of Pau and the Adour Region, France	10	367	36.70	9	365	2002
Armando Duarte	University of Aveiro, Portugal	10	245	24.50	8	238	2003
David P. Krabbenhoft	United States Geological Survey, USA	10	568	56.80	10	568	1994
Margareta Horvat	Jožef Stefan Institute, Slovenia	9	547	60.78	9	547	1993
Claudia B. Lopes	University of Aveiro, Portugal	9	220	24.44	7	211	2013
Dingyong Wang	Southwest University, China	9	121	13.44	6	115	2011
Xiao Wang	Xiamen University, China	9	216	24.00	6	214	2009
Robert de Levie	Bowdoin College, USA	8	181	22.63	7	178	1991
Francis Ribeyre	Inst Polytech Bordeaux, France	8	201	25.13	6	190	1991
Rodolfo G Wuilloud	National University of Cuyo, Argentina	8	370	46.25	8	370	2001
Andrea G Bravo	Spanish National Research Council, Spain	7	327	46.71	6	322	2011

* *TP* = Total number of publications; *TC* = Total citations; *C/P* = Average citations per publication; *h-core* = Citation sum within the h-core

Fig. 3.3 Temporal evolution of Top 10 Authors with their research production (1989–2022) (*Source* Biblioshiny)

the BRICS countries-Brazil, Russia, India, China, and South Africa-made the list. These countries' advanced scientific capabilities and technological advancements, combined with early recognition of pollution issues in industrialized nations, have fueled extensive research.

On the other hand, the USA received a maximum citation of 25,165 than China, which received 19,158 citations, indicating the need for further efforts of China to enhance academic influence and also got the highest C/P of 43.61. The h-index measures an article's quality and can provide insights into a country's influence on mercury in water. A higher h-index indicates a more significant influence (Prathap, 2010). In this context, the USA holds the highest h-index (79), followed by China (70), India (42), France (878), and Canada (862). On the other hand, Russia's h-index is considerably lower at 11, suggesting that the research findings and outcomes related to mercury in water from Russia may be comparatively less innovative and significant. It is important to note that evaluating a country's influence solely based on publication output would be incomplete, considering the high productivity but relatively low h-index.

International collaboration is vital in disseminating knowledge and exchanging ideas among experts (Chen et al. 2020). The collaborative relationships between countries, represented visually in Fig. 3.4, demonstrate the significance of such collaborations. The thickness of the curved lines in the figure signifies the strength of collaboration (Tan et al. 2021). Notably, a robust research collaboration in water and Hg-related research has emerged between China and the USA. This collaboration encompasses developed and developing countries, driven by their shared development levels and similar environmental challenges. Among these collaborations, the highest frequency of research publications is observed between China and the USA in 22 publications, followed by USA-Canada (13), Egypt-Saudi Arabia (10), and France-Switzerland (7). These findings provide valuable insights to foster further

Table 3.4 Leading countries that have published related to mercury in water

Country	Continent	TP	TC	C/P	h-index	h-core
China	Asia	721	19,158	26.57	70	8294
USA	North America	577	25,165	43.61	79	14,259
India	Asia	256	5668	22.14	42	3004
Iran	Asia	166	4366	26.30	38	2589
Canada	North America	157	5346	34.05	39	3651
Japan	Asia	111	3085	27.79	33	2142
France	Europe	108	4151	38.44	37	3188
Spain	Europe	103	3908	37.94	38	2912
Brazil	South America	103	2687	26.09	28	1889
Portugal	Europe	102	2185	21.42	31	1497
Sweden	Europe	77	2242	29.12	28	1665
Turkey	Asia	72	1640	22.78	22	1240
Italy	Europe	71	1493	21.03	24	1122
Germany	Europe	71	2019	28.44	22	1453
Egypt	Asia	69	2112	30.61	26	1615
South Korea	Europe	65	2016	31.02	24	1587
UK	Europe	54	1791	33.17	22	1458
Poland	Europe	54	1254	23.22	19	957
Russia	Asia	53	367	6.92	11	276
Saudi Arabia	Asia	50	1066	21.32	18	856

TP = total number of publications; *TC* = total citations; *C/P* = average citations per publication; *h-core* = citation sum within the h-core

collaborations between countries and identify potential research partners, ultimately advancing knowledge in the research area.

3.3.5 *Performance of Institutions*

According to the authors' affiliation information, a total of 1806 institutions contributed to research on mercury in water. The institutions were ranked based on the number of publications they produced, as analyzed in the country's performance. China and the USA had six representatives among the top 20 productive institutes, while Iran had two. Additionally, one institute each was from India, Portugal, Russia, Malaysia, Switzerland, and Spain. Specifically, the Chinese Academy of Science (CAS) China had the highest number of publications, totaling 43. The Chinese Academy of Sciences (CAS) holds the distinction of being China's largest and most prestigious research institution, boasting over 100 affiliated research institutes. The

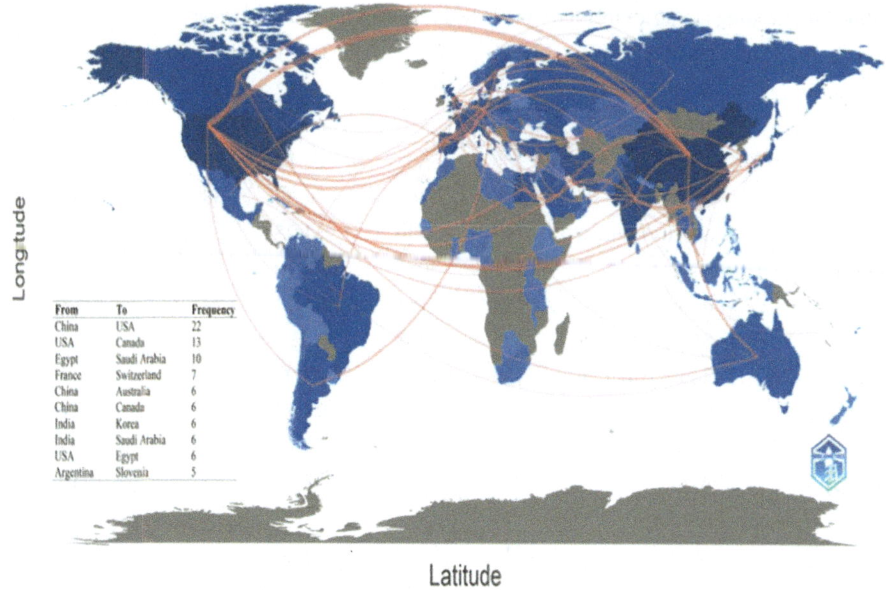

Fig. 3.4 Collaboration of countries on the research publication of mercury in water

publications from all these research institutions were considered part of the overall CAS output, contributing significantly to its top-ranking status and followed by the US Geological Survey from the USA and the University Aveiro from Portugal, with 21 and 20 publications, respectively (Table 3.5). However, other renowned institutions and research agencies worldwide, such as the Islamic Azad University, the US Environmental Protection Agency, Wuhan University, and the Indian Institute of Technologies, also appeared as affiliations in this type of research. Regarding citation counts, research publications from the Chinese Academy of Science researchers received the highest number of citations (TC) with 1565, followed by the US Geological Survey with 989 citations, and the Islamic Azad University with 509 citations.

Figure 3.5 illustrates the extensive collaborative network focused on mercury in water in the research area. A total of 1806 institutions have been involved in this research, with 519 institutions identified as having the highest level of collaboration. The connections depicted in the figure represent the collaborative relationships between these institutions. Each node in the figure corresponds to a distinct institution, and the thickness of the connecting lines reflects the degree of collaboration and the number of partnerships established between institutions. The weight of a link, indicated by the line thickness, signifies the quantity of co-authored papers by researchers from different institutions. Consequently, the thicker the line, the stronger the collaborative bond between the institutions involved.

Table 3.5 Leading institutions on mercury in water research publications

Institutions	Country	TP	TC	C/P	h-core	h-index
Chinese Academy of Science	China	43	1565	36.40	1339	23
US Geological Survey	USA	21	989	47.10	937	15
University Aveiro	Portugal	20	423	21.15	382	12
Russian Academy of Science	Russia	19	85	4.25	73	6
Islamic Azad University	Iran	18	509	28.28	451	9
Oak Ridge National Laboratory	USA	10	273	27.30	268	9
Southwest University	China	10	143	14.30	137	7
US Environmental Protection Agency	USA	10	186	18.60	176	7
University of Tehran	Iran	9	268	29.78	267	8
University of Minnesota	USA	9	343	38.11	338	7
Wuhan University	China	9	402	44.67	395	7
Universiti Malaya	Malaysia	9	266	29.56	258	6
Nankai University	China	8	382	47.75	375	7
Jilin University	China	8	239	29.88	237	7
Indian Institute of Technologies	India	8	342	42.75	334	6
University of Geneva	Switzerland	8	287	35.88	279	6
University of Nevada	USA	7	257	36.71	257	7
University of Oviedo	Spain	7	321	45.86	321	7
South China University of Technology	China	7	335	47.86	331	6
University of Florida	USA	6	248	41.33	248	6

* *TP* = Total number of publications; *TC* = Total citations; *C/P* = Average citations per publication; *h-core* = Citation sum within the h-core

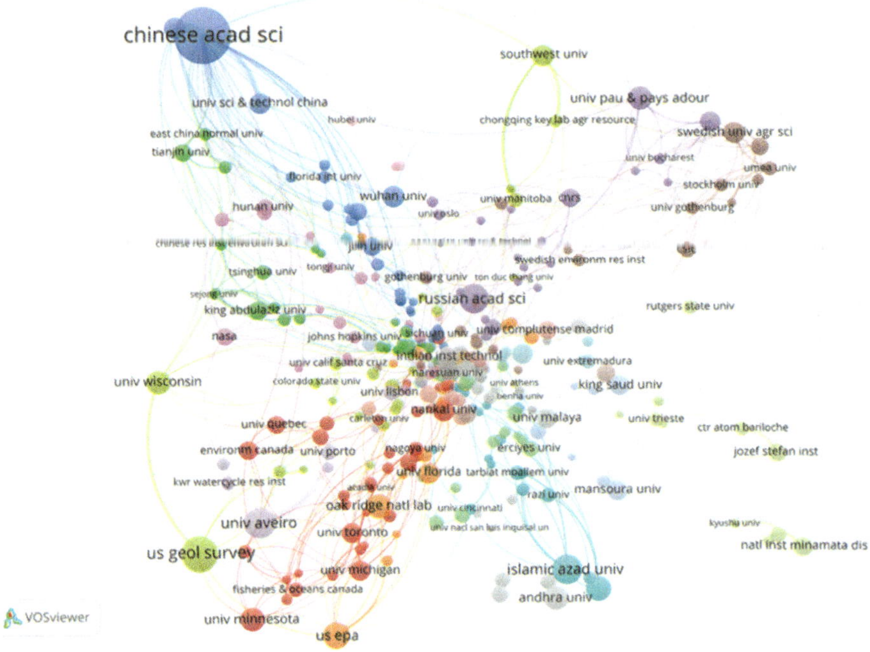

Fig. 3.5 Illustration of the extensive collaborative network focused on mercury in water in the research area

3.3.6 Most Frequently Cited Papers

Table 3.6 provides an intriguing and comprehensive overview of the most influential and frequently referenced research papers in the domain of mercury in water, spanning the years 1989 to 2022. These papers have been ranked based on various key metrics such as total citations (TC), total citations per year (TCPY), and normalized total citations (NTC), highlighting their significant impact on the field. Notably, the publication titled "*Sulfate Stimulation of Mercury Methylation in Fresh-Water Sediments*," authored by Gilmour et al. and published in the esteemed journal "Environmental Science & Technology" in 1992, emerges as the most highly cited article with an impressive TC of 819, a TCPY of 25.59, and an NTC of 13.76. The second position is secured by a collaborative review paper authored by Juhasz and Naidu, titled "*EDTA functionalized magnetic graphene oxide for removal of Pb(II), Hg(II), and Cu(II) in water treatment: Adsorption mechanism and separation property*," published in 2000 in the journal "Chemical Engineering Journal," boasting a TC of 473. Among the top 10 highly cited papers, one stands out as a review paper emphasizing the significance of comprehensive analyses and synthesis of existing knowledge in propelling the progress of mercury in water research.

Table 3.6 Top ten highly cited research literature on mercury in water

Rank	TC	Title of the Paper	Authors	Journal	Year	DT	TCPY	NTC
1	819	Sulfate Stimulation of Mercury Methylation in Fresh-Water Sediments	Gilmour et al.	Environmental Science & Technology	1992	Article	25.59	13.76
2	473	EDTA functionalized magnetic graphene oxide for removal of Pb(II), Hg(II) and Cu(II) in water treatment: Adsorption mechanism and separation property	Cui et al.	Chemical Engineering Journal	2015	Article	52.56	15.71
3	422	Synthesis of amino functionalized magnetic graphene's composite material and its application to remove Cr(VI), Pb(II), Hg(II), Cd(II) and Ni(II) from contaminated water	Guo et al.	Journal of Hazardous Materials	2014	Article	42.20	9.14
4	401	Turn-On and Ratiometric Mercury Sensing in Water with a Red-Emitting Probe	Nolan & Lippard	Journal of the American Chemical Society	2007	Article	23.59	7.46
5	394	Hg(II) removal from water by chitosan and chitosan derivatives: A review	Miretzky & Cirelli	Journal of Hazardous Materials	2009	Review	26.27	7.02
6	384	Regenerable DNA-Functionalized Hydrogels for Ultrasensitive, Instrument-Free Mercury(II) Detection and Removal in Water	Dave et al.	Journal of the American Chemical Society	2010	Article	27.43	8.43

(continued)

Table 3.6 (continued)

Rank	TC	Title of the Paper	Authors	Journal	Year	DT	TCPY	NTC
7	384	Comparison of distillation with other current isolation methods for the determination of methyl mercury compounds in low-level environmental samples: Part II. Water	Horvat et al.	Analytica Chimica Acta	1993	Article	12.39	8.42
8	350	A Bright and Specific Fluorescent Sensor for Mercury in Water, Cells, and Tissue	Yoon et al.	Angewandte Chemie	2007	Article	20.59	6.52
9	345	Mercury removal from water using activated carbons derived from organic sewage sludge	Zhang et al.	Water Research	2005	Article	18.16	7.22
10	331	Facile Synthesis of Water-Soluble Fluorescent Silver Nanoclusters and Hg^{II} Sensing	Adhikari & Banerjee	Chemistry of Materials	2010	Article	23.64	7.27

* *TC* = Total Citations; *DT* = Document Types; *TCPY* = Total Citations per Year; *NTC* = Normalized Total Citations

3.3.7 Subject WoS Categories

The entire collection of publications has been extensively analyzed and categorized into 50 subject research categories based on the Web of Science. This categorization is depicted in Fig. 3.6, providing a visual representation of the distribution of publications across these various fields of study. Among the numerous subject research areas, it is noteworthy that the field of "Environmental Sciences" emerged as the most popular category. This particular area of research contributed substantially to the collection, with a total of 538 publications, accounting for approximately 19% of the entire corpus. This finding indicates the significance and prevalence of environmental science-related studies within the publication dataset. Following closely behind, the subject research category labeled "Chemical Analytical" also demonstrated considerable prominence. It comprised 403 publications, constituting around 14% of the total collection. A substantial number of publications in this field suggests a strong focus on analytical chemistry research within the dataset. Furthermore, the

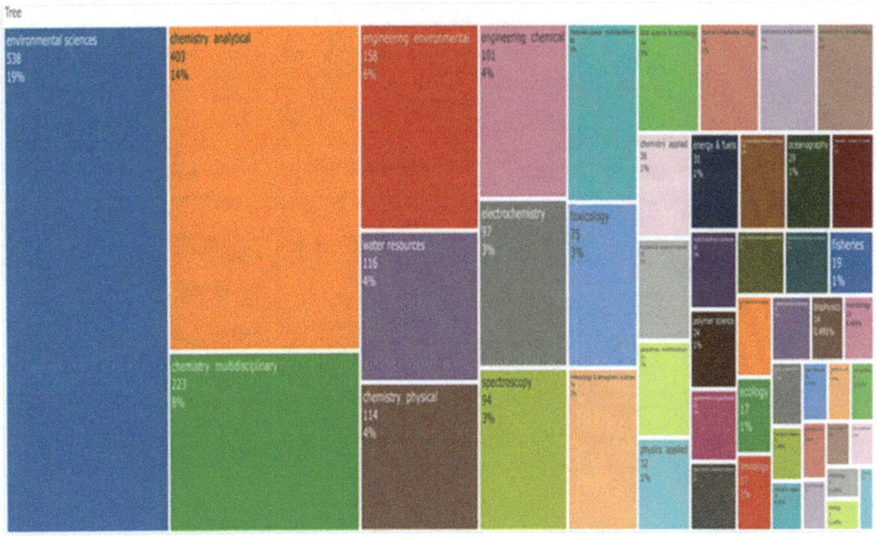

Fig. 3.6 TreeMap showing the subject-wise categories in which mercury in water were published

subject research category "Chemistry Multidisciplinary" stood out with 223 publications, making up approximately 8% of the overall collection. This classification suggests that the publications within this category covered diverse aspects of chemistry, crossing multiple sub-disciplines and addressing interdisciplinary approaches. Lastly, the subject research category "Engineering Environmental" accounted for 158 publications, comprising around 6% of the total.

3.4 Keywords Occurrence Analysis

Figure 3.7 demonstrates the co-occurrence network analysis of keywords, comprising a total of 3483 keywords. To participate in the analysis, a minimum frequency of occurrence of 3 was set, resulting in 308 keywords meeting the requirements. Ultimately, 306 keywords appeared in the network map. The size of each circle in the map represents the frequency of the corresponding keyword occurrence. Consequently, larger circles indicate higher frequencies.

Among the keywords, "Mercury" occupied the largest circle, appearing with a frequency of 446, making it a prominent and highly discussed term in the field. Its total link strength amounted to 786. Following closely behind, "Adsorption" emerged as the second high-frequency keyword, occurring 82 times with a total link strength of 174. "Methylmercury" ranked third with a frequency of 60 and a total link strength of 131. The research conducted by domestic and foreign scholars on mercury in water environments predominantly focused on several key areas, namely, adsorption,

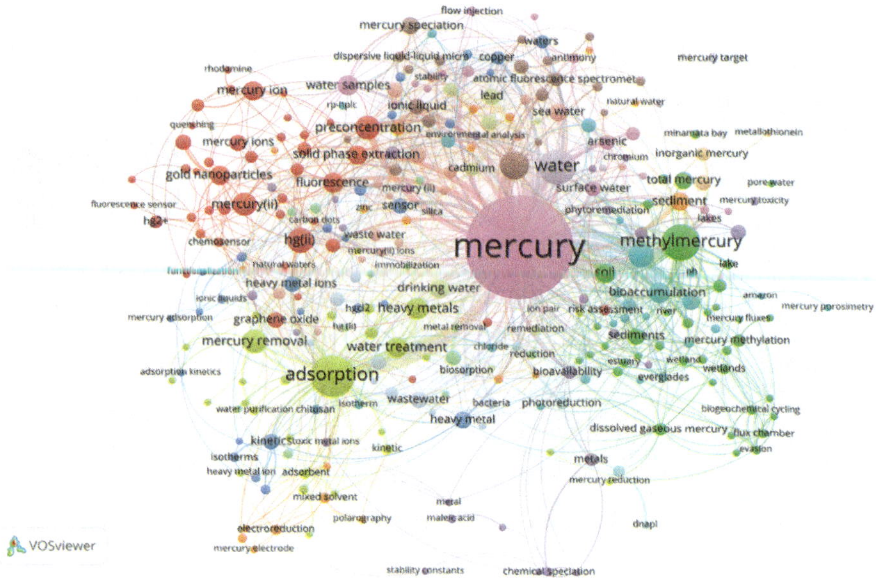

Fig. 3.7 Authors keywords network visualization in water were published

speciation, fish, pre-concentration, water treatment, and mercury removal. These keywords can be classified into seven clusters, each represented by a distinct group of colors.

3.5 Limitation

One of the notable constraints of this article pertained to its reliance on a solitary database for citation review, potentially overlooking the impact of other databases on citation counts and subsequent results. Furthermore, it is worth acknowledging that articles published in earlier years tend to accumulate more citations. In contrast, studies published within the past 1 or 2 years may not have garnered substantial citation numbers yet. Lastly, the study's keyword search methodology might have inadvertently omitted relevant articles, thus limiting the scope of analysis. These considerations highlight the need for a more comprehensive approach to citation analysis and database selection in future research.

3.6 Discussion and Conclusion

This review study represents a pioneering global bibliometric analysis focusing on Hg- contamination in water, making it the first of its kind. Spanning the extensive period from 1989 to 2022, this comprehensive analysis examined a substantial collection of 1729 pertinent research papers sourced from the esteemed WoS-database. The findings obtained from this extensive examination divulged a striking and consistent increase in the number of publications, demonstrating an impressive annual growth rate. Among the various types of publications, research articles dominated the scholarly landscape. This preference for research articles can be attributed to their unique ability to generate substantial impact, garner attention, and maintain longevity within the scientific community. Consequently, this finding implies that the production of scholarly articles in this field will continue to grow exponentially in the foreseeable future. The countries that stand out prominently for their significant contributions to mercury contamination in water, in terms of depth, content, and direction, are China, the USA, and India. The crucial outcome of this study lies in its facilitation of mapping the knowledge domain and intellectual landscape of this field, thereby enabling the identification of emerging research themes, significant articles, and prevailing trends. Furthermore, analyses of authors and institutions shed light on the dynamics of research collaborations within the scientific community. The increasing recognition of the scientific community's importance of mercury contamination research indicates this issue's growing significance. The implications of this study hold paramount importance for academicians and policymakers alike, as they align with the pursuit of sustainable growth by establishing an integrated framework that considers both economic and environmental progress. Hg-contamination in water poses a significant environmental and public health concern. It can potentially harm human health, particularly the neurological development of children, and disrupt aquatic ecosystems. Thus, monitoring, regulation, and pollution prevention efforts are crucial for minimizing Hg-contamination and safeguarding the health of humans and the environment.

Raising awareness about the dangers of mercury and promoting sustainable practices are essential steps toward mitigating its impact and creating a safer and healthier future. Efforts to address Hg-contamination in water bodies involve a combination of monitoring, regulation, and pollution prevention measures. This encompasses the implementation of cleaner production technologies, the promotion of energy efficiency, and the reduction in the use of mercury in industrial processes. Furthermore, wastewater treatment plants play a pivotal role in removing mercury from effluent before it is discharged into water bodies, thereby preventing its entry into the environment. In summary, this study marks a significant milestone in understanding the global landscape of mercury contamination in water through its pioneering bibliometric analysis. The findings highlight the increasing significance of this issue and emphasize the need for continuous research, monitoring, and regulation to minimize the adverse effects on human and environmental health. By promoting awareness and adopting sustainable practices, we can collectively mitigate the impact of Hg-contamination and work toward a safer and healthier future for all.

References

Adhikari B, Banerjee A (2010) Facile synthesis of water-soluble fluorescent silver nanoclusters and HgII sensing. Chem Mater 22(15):4364–4371. https://doi.org/10.1021/cm1001253

Al-Sulaiti MM, Soubra L, Al-Ghouti MA (2022) The causes and effects of mercury and methylmercury contamination in the marine environment: a review. Curr Pollut Rep 8(3):249–272. https://doi.org/10.1007/s40726-022-00226-7

Aria M, Cuccurullo C (2017) Bibliometrix: an R-tool for comprehensive science mapping analysis. J Informetr 11:959–975. https://doi.org/10.1016/j.joi.2017.08.007

Azar J, Yousef MH, El-Fawal HA, Abdelnaser A (2021) Mercury and Alzheimer's disease: a look at the links and evidence. Metab Brain Dis 36:361–374. https://doi.org/10.1007/s11011-020-00649-5

Bindler R, Yu R, Hansson S, Claßen N, Karlsson J (2012) Mining, metallurgy and the historical origin of mercury pollution in lakes and watercourses in Central Sweden. Environ Sci Technol 46(15):7984–7991. https://doi.org/10.1021/es300789q

Castaño A, Cutanda F, Esteban M, Pärt P, Navarro C, Gómez S, Posada M (2015) Fish consumption patterns and hair mercury levels in children and their mothers in 17 EU countries. Environ Res 141:58–68. https://doi.org/10.1016/j.envres.2014.10.029

Chen B, Dong S (2022) Mercury contamination in fish and its effects on the health of pregnant women and their fetuses, and guidance for fish consumption—a narrative review. Int J Environ Res Public Health 19(23):15929. https://doi.org/10.3390/ijerph192315929

Chen W, Geng Y, Zhong S, Zhuang M, Pan H (2020) A bibliometric analysis of ecosystem services evaluation from 1997 to 2016. Environ Sci Pollut Res 27:23503–23513. https://doi.org/10.1007/s11356-020-08760-x

Clarkson TW, Magos L, Myers GJ (2003) The toxicology of mercury—current exposures and clinical manifestations. N Engl J Med 349(18):1731–1737

Cui L, Wang Y, Gao L, Hu L, Yan L, Wei Q, Du B (2015) EDTA functionalized magnetic graphene oxide for removal of Pb(II), Hg(II), and Cu(II) in water treatment: adsorption mechanism and separation property. Chem Eng J 281:1–10. https://doi.org/10.1016/j.cej.2015.06.043

Dave N, Chan MY, Huang PJJ, Smith BD, Liu J (2010) Regenerable DNA-functionalized hydrogels for ultrasensitive, instrument-free mercury(II) detection and removal in water. J Am Chem Soc 132(36):12668–12673. https://doi.org/10.1021/ja106098j

García LO, García PL, Bartlett JR (2018) A review of mercury emission sources, control technologies, and regulations in the cement industry. Environ Sci Pollut Res 25(4):3137–3148

Gilmour CC, Henry EA, Mitchell R (1992) Sulfate stimulation of mercury methylation in freshwater sediments. Environ Sci Technol 26(11):2281–2287. https://doi.org/10.1021/es00035a029

Gyanajeet Y, Ibohal SC (2023) A scientometric visualization of 32 years of leukemia research in India. Sci Technol Libr. https://doi.org/10.1080/0194262X.2023.2181279

Gyanendra Y, Yumnam G, Alam W, Ibohal Singh C (2022) A bibliometric analysis and assessment of scientific studies trend on groundwater research in India during 1989–2020. Arab J Geosci 15. https://doi.org/10.1007/s12517-022-10707-0

Harada M (1995) Minamata disease: methylmercury poisoning in Japan caused by environmental pollution. Crit Rev Toxicol 25(1):1–24

Horvat M, Bloom NS, Liang L (1993) Comparison of distillation with other current isolation methods for the determination of methyl mercury compounds in low level environmental samples. Part 1. Sediments. Anal Chim Acta 281(1):135–135. https://doi.org/10.1016/0003-2670(93)80364-Q

Miretzky P, Cirelli AF (2009) Hg (II) removal from water by chitosan and chitosan derivatives: a review. J Hazard Mater 167(1–3):10–23. https://doi.org/10.1016/j.jhazmat.2009.01.060

Mitra S, Chakraborty AJ, Tareq AM, Emran TB, Nainu F, Khusro A, Simal-Gandara J (2022) Impact of heavy metals on the environment and human health: Novel therapeutic insights to counter the toxicity. J King Saud Univ Sci 34(3):101865. https://doi.org/10.1016/j.jksus.2022.101865

Nolan EM, Lippard SJ (2007) Turn-on and Ratiometric mercury sensing in water with a red-emitting probe. J Am Chem Soc 129(18):5910–5918. https://doi.org/10.1021/ja068879r

Parang H, Esmacilbcigi M (2022) Total mercury concentration in the muscle of four mostly consumed fish and associated human health risks for fishermen and non-fishermen families in the Anzali Wetland, Southern Caspian Sea. Reg Stud Mar Sci 52:102270. https://doi.org/10.1016/j.rsma.2022.102270

Perelonia KBS, Benitez KCD, Banicod RJS, Tadifa GC, Cambia FD, Montojo UM (2021) Validation of an analytical method for the determination of cadmium, lead and mercury in fish and fishery resources by graphite furnace and cold vapor atomic absorption spectrometry. Food Control 130:108363. https://doi.org/10.1016/j.foodcont.2021.108363

Persson O, Danell R, Schneider JW (2009) How to use BibExcel for various types of bibliometric analysis. Celebrating scholarly communication studies: a Festschrift for Olle Persson at his 60th birthday 5:9–24

Prathap G (2010) Is there a place for a mock h-index? Scientometrics 84(1):153–165. https://doi.org/10.1007/s11192-009-0066-2

Qiu YW, Wang WX (2016) Comparison of mercury bioaccumulation between wild and mariculture food chains from a subtropical bay of Southern China. Environ Geochem Health 38:39–49. https://doi.org/10.1007/s10653-015-9677-0

Rumayor M, Gallego JR, Rodríguez-Valdés E, Díaz-Somoano M (2017) An assessment of the environmental fate of mercury species in highly polluted brownfields by means of thermal desorption. J Hazard Mater 325:1–7. https://doi.org/10.1016/j.jhazmat.2016.11.068

Sharifi A, Simangan D, Kaneko S (2020) Three decades of research on climate change and peace: a bibliometrics analysis. Sustain Sci. https://doi.org/10.1007/s11625-020-00853-3

Shi S, Yin J (2021) Global research on carbon footprint: a scientometric review. Environ Impact Asses 89:106571. https://doi.org/10.1016/j.eiar.2021.106571

Tan H, Sun J, Wenjia W, Zhu C (2021) User experience & usability of driving: a bibliometric analysis of 2000–2019. Int J Hum Comput Int 37:297–307. https://doi.org/10.1080/10447318.2020.1860516

UNEP (2018) Global mercury assessment. United Nations Environment Programme, Geneva

van Eck NJ, Waltman L (2009) Software survey: VOSviewer, a computer program for bibliometric mapping. Scientometrics 84:523–538. https://doi.org/10.1007/s11192-009-0146-3

WHO (2005) Mercury in drinking-water: background document for development of WHO guidelines for drinking-water quality. https://www.who.int/docs/default-source/wash-documents/wash-chemicals/mercury-background-document.pdf?sfvrsn=9b117325_4

WHO (2017) Ten chemicals of major health concern. World Health Organization

Yang L, Zhang Y, Wang F, Luo Z, Guo S, Strähle U (2020) Toxicity of mercury: molecular evidence. Chemosphere 245:125586. https://doi.org/10.1016/j.chemosphere.2019.125586

Yoon S, Miller EW, He Q, Do PH, Chang CJ (2007) A bright and specific fluorescent sensor for mercury in water, cells, and tissue. Angew Chem 119(35):6778–6781. https://doi.org/10.1002/anie.200701785

Zhang FS, Nriagu JO, Itoh H (2005) Mercury removal from water using activated carbons derived from organic sewage sludge. Water Res 39(2–3):389–395. https://doi.org/10.1016/j.watres.2004.09.027

Part II
Health Risks Linked to Mercury Toxicity

Chapter 4
Impact of Mercury and Its Toxicity on Health and Environment: A General Perspective

Mahua Basu

4.1 Introduction

Mercury, a universal contaminant, poses severe hazards to the health of ecological systems and biota including humans. It is persistent and undergoes long-range transport in the environment accumulating in and affecting the environmental components such as land, air, water, and biodiversity (Nriagu 1979). Hence it has been a global concern for decades. It has a complicated biogeochemical cycle with several transformations and transport processes in the environment, the understanding of which helps in better assessment of the risks of mercury contamination (Fitzgerald and Lamborg 2003; Fitzgerald et al. 2005).

Mercury, also known as quicksilver, is one of the few elements that have been known since ancient times. Cinnabar was used by the Paleolithic painters of thirty thousand years ago to decorate the caves in Spain and France. It has been used by the Egyptians, Greeks, Romans, and Chinese civilizations. A human skeleton covered in vermillion was discovered dating 5000 BCE. The use of mercury was also validated to be in the fifteenth century BCE Egyptian tomb ceremonial cup. Mercury paint was found in the Roman houses covered by the volcanic ash of Mount Vesuvius in 79 CE. The ancient Greeks and Romans used mercury in various applications, including cosmetics, medicines, and alchemy. They recognized its unique properties, such as its liquid form at room temperature and its ability to dissolve other metals. During the medieval and Renaissance periods, mercury played a significant role in alchemy, a precursor to modern chemistry. Alchemists believed that mercury held mystical properties and could be transformed into gold, leading to the pursuit of the 'Philosopher's Stone'. Aristotle is credited with the academic record of mercury during the fourth century BCE as 'fluid silver' or 'quick silver'.

M. Basu (✉)
St. Xavier's College (Autonomous), Kolkata, West Bengal, India
e-mail: mbasu@sxccal.edu

N. Kumar (ed.), *Mercury Toxicity*, Environmental Science and Engineering,
https://doi.org/10.1007/978-981-99-7719-2_4

Alchemists also explored the medicinal uses of mercury, believing it had the power to cure diseases and prolong life. However, their understanding of its toxic nature was limited at the time. The Spanish conquerors exploited larger cinnabar deposits at Huancavelica for gold extraction. In 1848 the Californian Gold Rush miners used mercury from the New Almaden Mines, California, and the USA. The eighteenth and nineteenth centuries marked the rise of the industrial revolution, during which mercury found numerous applications in various industries, including mining, metal extraction, and the production of thermometers, barometers, and electrical equipment. Mercury's unique properties, such as its high density and conductivity, made it valuable in scientific instruments and electrical applications.

4.2 Distribution

Present ubiquitously as a naturally occurring substance, it is extremely rare with an average abundance of 0.08 ppm by mass. It is a toxic, persistent, mobile contaminant (Gworek et al. 2020) found in the air, water, and soil. It is mostly produced in Spain (Almaden mine), Peru (Huancavelica), Yugoslavia, Italy (Iudrio), and the USA (coastal ranges of California). It is usually mined as mercuric sulfide (HgS) or (cinnabar ore). Seven countries, namely the USA, Spain, Yugoslavia, Italy, the former Soviet Union, China, and Mexico provide more than 90% of the global supply.

4.3 Sources of Mercury in the Environment

Both natural and anthropogenic release occurs for mercury into the land, air, and water, though the main route for environmental mercury entry is the atmospheric emission (Berg et al., 2006). The collective input from natural and anthropogenic sources in the air is estimated to be about 5000–6000t of mercury worldwide (Lamborg et al. 2002; Gray and Hines 2006). Over 99% of mercury is the gaseous elemental mercury from natural releases. Natural mercury emissions also comprise reemissions of formerly deposited atmospheric mercury by wet and dry processes from either natural or anthropogenic sources (Pacyna et al. 2006). For example, background soil and low mercury-containing substrates can reemit formerly deposited mercury (Seigneur et al. 2004; Shetty et al. 2008).

4.3.1 Natural Sources

Geological processes like volcanic activity, geothermal emission, volatilization of mercury in marine environments, and release from terrestrial environments comprise the predominant natural source of mercury in the air (Nriagu 1989; Nriagu and Becker

2003; Gray and Hines 2006). The temporal and spatial complexity of the geological processes and lack of data make the accurate estimation of mercury emissions rather difficult (Gustin et al. 2008). The global mercury emissions were estimated to range between 800 and 5800t every year with a middle range from 1800t to 3000t (Nriagu 1989; Pacyna et al. 2001; Lamborg et al. 2002; Shetty et al. 2008). Annually, volcanic activities emit between 1 and 700t, geothermal emissions around 60t, oceanic release between 899 and 2600t, and terrestrial emissions between 1000 and 3200t (Nriagu 1989; Lamborg et al. 2002; Seigneur et al. 2004).

4.3.2 Anthropogenic Sources

Historically and worldwide, ever since the start of the industrialized period, there has been an increase in the mercury burden of air by a factor between 2 and 5 (Keating, et al. 1997). Despite the global decline in anthropogenic mercury release in the last 30 years, humans are accountable for a considerable proportion of mercury input globally. Human activity contributes to 50% of the total atmospheric global emissions (i.e., 5000–6000t) of mercury (Lamborg et al. 2002; Gray and Hines 2006). Estimated global mercury emissions to range between 2000 and 2600t every year (Pacyna et al. 2006). Mercury emissions may be from point sources as well as from non-point sources. Point sources such as combustion (medical waste incinerators, sewage sludge incinerators, municipal waste combustors, industrial and residential heating), coal-fired power plants, manufacturing units, and dental amalgam account for more than 95% of the anthropogenic emissions (Keating et al. 1997). Combustion alone comprises 80% while gold mining contributes about 20% of the anthropogenic emissions. Industrial operations such as gold mining, chlor-alkali production, paper, pulp manufacturing, and others release mercurial compounds directly into the aquatic and terrestrial environments (Keating et al. 1997; Branch 2008). Non-point or diffuse sources can be landfills, sewage sludge amended fields, mining wastes, and others (Nriagu 1989). Fossil fuel combustion for electricity both in industrial and residential heating accounts for 880t–1930t, i.e., 45% of the total mercury emission worldwide. Historically, Asia remains the largest emitter contributing 65% of the global mercury emission.

4.4 Forms of Mercury in the Environment

A naturally occurring crustal element with atomic number 80, it is a silver-white, shiny, liquid metal at room temperature. It primarily exists in three forms; elemental, inorganic, and organic. A summary of the physicochemical properties and toxicological profile of different species of mercury is given in Table 4.1.

Table 4.1 Properties and toxicological profile of different species of mercury

Mercury species	Physical properties	Chemical properties	Toxicological profile
Elemental mercury Hg^0	– Appearance: Liquid at room temperature – Specific gravity:13.6 at 25 °C – Color: Silver-white – Molecular weight –200.59 – Odor: Odorless – Melting Point: –38.87 °C – Boiling Point: 356.72 °C (674.9°F) – Density: 13.534 g/cm^3 at 25 °C – Insoluble in water – Soluble in organic solvents	– Forms amalgams with metals – Reacts with halogens, sulfur, and some metals – Oxidized to divalent mercury (Hg^{2+}) in the atmosphere	– Highly toxic when inhaled as a vapor may cause pulmonary inflammation, dyspnea, chest pain, and dry cough. Pulmonary edema, respiratory failure, and even death – Acute toxicity causes renal damage, tachycardia, hypertension – The patient may have a metallic taste, salivation, dysphagia, abdominal cramps, and diarrhea – Effect on CNS includes headache, weakness, and visual disturbances – Skin absorption can also lead to contact dermatitis – Ingestion may result in gastrointestinal symptoms such as abdominal pain and vomiting
Methylmercury $Hg(CH_3)$	– Physical State: Liquid at room temperature – Molecular weight –251.1 – Color: Colorless to pale yellow – Odor: Odorless – Melting Point: –60.5 °C – Boiling Point: 83.0 °C – Density: 4.06 g/cm^3 – Soluble in organic solvents, not soluble in water	– Easily oxidized to Hg^{2+} – Forms complexes with sulfhydryl groups in proteins – Reacts with certain metals to form amalgams – Methylated by anaerobic microorganisms in sediments	– A highly toxic organic form of mercury – Bioaccumulate and biomagnifies in the food chain – Primary route of exposure is through the consumption of contaminated fish and seafood – Affects the CNS, causing developmental delays in children and neurological impairments in adults; increases cardio-vascular risks – Cross the placental barrier to affect the fetus – Clinical manifestation includes paresthesia, ataxia, tremors, spasticity, paralysis, visual and hearing loss, and speech impairment – Prenatal exposure can result in neurodevelopmental disorders in infants

(continued)

Table 4.1 (continued)

Mercury species	Physical properties	Chemical properties	Toxicological profile
Dimethylmercury $(CH_3)_2Hg$	– Physical State: colorless liquid at room temperature Molecular weight –230.66 – Color: Colorless to pale yellow – Odor: Sweet, fruity Melting Point: –43 °C – Boiling Point: 93 °C – Density: 3.1874 g/mL at 20 °C – Soluble in organic solvents, not soluble in water	– Easily oxidized to Hg^{2+} – Forms complexes with sulfhydryl groups in proteins – Reacts with certain metals to form amalgams – Methylatesthiols, amines, and alkyl halides	– Extremely toxic, even in small quantities; lethal dose 400 mg – Enters through inhalation or skin; even through latex gloves Being lipophilic, rapidly crosses the blood–brain barrier forming a methylmercury-cysteine complex, leading to severe neurological damage – Inhalation results in cough, incoordination, tremor, slurring of speech, and confusion – Ingestion or skin absorption can result in acute and delayed symptoms, including tremors, ataxia, and eventual death
Phenylmercury $C_6H_5Hg^+$	– Physical State: Solid lustrous crystalline powder at room temperature – Color: White to creme – Molecular weight – 336.742 – Odor: Odorless – Melting Point: 148 °C – Boiling Point: 291 °C – Insoluble in water, soluble in organic solvents	– Forms complexes with sulfhydryl groups in proteins – Reacts with halogens, sulfur, and other organic compounds	– Toxic organic compound, especially phenylmercuric acetate – Absorption through the skin is slow by efficiency by the GIT – Concentrates inside the RBC and also the blood–brain barrier – Inhalation causes nausea, vomiting, diarrhea, – Neurotoxic effects include confusion, emotional instability, hallucination, and loss of sensation. tremors, cognitive impairments, and damage to the central nervous system

(continued)

Table 4.1 (continued)

Mercury species	Physical properties	Chemical properties	Toxicological profile
Ethylmercury $C_2H_5Hg^+$	– Physical State: Liquid at room temperature – Color: Colorless – Odor: Odorless – Molecular weight: 229.65 – Soluble in organic solvents, insoluble in water	– Easily oxidized to Hg^{2+} – Forms compounds with chloride, thiolate, or another organic group – Reacts with certain metals to form amalgams	– Cross the blood–brain barrier and placental barrier – Used as a preservative in some vaccines – Considered less toxic than methylmercury, but the potential risk to neurodevelopment in infants – Rapidly metabolized and excreted from the body
Mercurous salt Hg_2Cl_2	– Appearance: white solid – Molecular weight: 472.09 – Odor: odorless Solubility: practically insoluble in water Density: 7.15 gm/cc	– Can oxidize to the mercuric state via the hydrogen peroxide pathway	– Enters through inhalation, dermal, and oral routes – Tightness and chest pain, cough, breathing difficulties; – Dermal exposure causes eye and skin irritation – Chronic exposure affects the brain, liver, and kidneys, loosens teeth, loss of appetite, and skin ulcers – Calomel is associated with acrodynia or pink disease – Do not cross the blood–brain barrier
Mercuric salt $HgCl_2$	– Appearance: colorless or white solid – Molecular weight: 271.52 – Odor: odorless – Melting point: 276 °C – Boling point: 304 °C – Solubility: soluble in alcohol, acetone, ethyl acetate	– Forms amalgam with metals like aluminium	– Ulcers in stomach, mouth, and throat; corrosive effect on intestines; causing stomach pain, abdominal discomfort, and blood vomiting – Accumulate in kidneys causing renal failure – Do not cross the blood–brain barrier

4.4.1 Elemental or Metallic Mercury

Physical Properties: It is unique and the only liquid metal at STP. It has a low melting point of −38.87 °C and a boiling point of 356.72 °C. It has a high density and is a poor conductor of heat compared to other metals.

Toxicity: Being highly toxic to humans and animals, it can be absorbed through inhalation, ingestion, or skin contact. Upon inhalation, mercury can enter the bloodstream and distribute throughout the body, including the brain, where it can cause neurological damage. Prolonged exposure to mercury vapor can lead to symptoms such as tremors, memory loss, and kidney and lung damage.

Sources of Exposure: Metallic mercury finds its use in various industrial applications. It has been utilized in thermometers, barometers, electrical switches, dental amalgams, and certain chemical processes. However, many of these applications have transitioned to alternative materials and technologies to minimize mercury exposure and pollution.

Environmental Impact: Spills or improper disposal of metallic mercury can contaminate soil, water bodies, and air. Mercury can transform into more toxic forms, such as methylmercury, naturally or biologically. Methylmercury is readily absorbed by organisms and bioaccumulated in the food chain, posing a risk to both wildlife and humans.

Regulation and Precautions: Owing to its toxicity, handling metallic mercury demands extreme care. Protective measures, such as gloves, goggles, and ventilation, should be used when working with or cleaning up mercury spills. Proper disposal guidelines and methods for mercury-containing devices and waste should be followed to prevent environmental contamination.

It is important to note that the use of metallic mercury is heavily regulated or banned in many countries to mitigate the risks associated with its toxicity and environmental impact.

4.4.2 Inorganic Mercury

Inorganic mercury can exist as salts or complexes and is classified into two main categories: mercuric compounds and mercurous compounds.

Physical Properties: Mercuric compounds with a + 2 oxidation state are typically white crystalline solids. Mercurous compounds with a + 1 oxidation state are usually white or yellowish solids.

Toxicity: Mercuric chloride ($HgCl_2$) and mercuric oxide (HgO), are highly toxic and can be absorbed through inhalation, ingestion, or skin contact. They pose similar health risks as metallic mercury, affecting the nervous system, kidneys, and other organs. Mercurous compounds with a + 1 oxidation state are usually white or yellowish solids. Mercurous chloride (Hg_2Cl_2) and mercurous nitrate ($Hg_2(NO_3)_2$) are generally less toxic than mercuric compounds but can still be harmful. They can get transformed in the environment and convert into more toxic forms, such as methylmercury. The toxicity and environmental behavior of inorganic mercury compounds can vary depending on the specific compound and its chemical properties.

Sources of Exposure: Inorganic mercury compounds can be released into the environment through various sources, including industrial processes, mining, coal combustion, waste incineration, and the improper disposal of mercury-containing products.

Environmental Impact: Inorganic mercury compounds can contaminate ecosystems, leading to potential exposure to humans and wildlife, especially when converted to methylmercury by bacteria and other microorganisms.

Regulation and Precautions: Inorganic mercury compounds are regulated by various environmental and health agencies globally to minimize exposure and protect human health and the environment. Proper handling, storage, and disposal practices should be followed to prevent contamination. Occupational safety measures and personal protective equipment should be employed when working with inorganic mercury compounds to minimize the risk of exposure.

4.4.3 Organic Mercury

These compounds are a class of mercury compounds containing carbon atoms bonded to mercury, like methylmercury (CH_3Hg), dimethylmercury, phenylmercury, ethylmercury, and likewise.

Physical Properties: Methylmercury and phenylmercury exist as salts like methylmercuric chloride and phenylmercuric acetate.

Toxicity: The highly toxic methylmercury is formed through natural processes and human activities. Methylmercury can bioaccumulate in the organisms involved in the food chain, reaching high concentrations (biomagnifications) in predator fish and marine mammals. It results in adverse effects on aquatic organisms, birds, and other wildlife including humans, impacting their reproductive success, neurological functions, and overall health. It is particularly harmful to the nervous system, especially during fetal development and early childhood. Exposure to methylmercury can lead to developmental delays, cognitive impairments, and neurological disorders.

Sources of Exposure: The main source of personal exposure to organic mercury, specifically methylmercury, is by the way of consuming contaminated fish and shellfish. Industrial activities, such as coal combustion and artisanal gold mining, can also contribute to the release of inorganic mercury into the environment, where it can be converted into methylmercury by natural processes primarily through the conversion of mercury by bacteria in aquatic environments.

Environmental Impact: Methylmercury especially poses significant risks to the environment. They can contaminate water bodies and accumulate in aquatic organisms. Methylmercury biomagnifies as it moves up the food chain, resulting in higher concentrations in top predators. This process can have detrimental effects on the health and reproductive success of wildlife, including fish, birds, and mammals. It can disrupt ecosystems and contribute to biodiversity loss.

Regulation and Precautions: Due to the toxicity, there are regulations in place to limit their release into the environment and guidelines for acceptable levels in food and drinking water. Monitoring programs and advisories are implemented to inform the public about the consumption of fish from contaminated areas. Pregnant women, nursing mothers, and young kids are typically recommended to limit their intake of fish known to have elevated levels of mercury.

Remediation: Cleaning up environments contaminated with organic mercury compounds is challenging. Strategies may include reducing mercury emissions from industrial processes, implementing best practices in waste management, and remediating contaminated sites. Additionally, efforts are focused on mitigating mercury pollution at its source, such as reducing mercury use in industrial applications and promoting cleaner technologies. Properties and toxicological profile of different species of mercury are presented in Table 4.1.

4.5 Global Mercury Cycling

The transformation between different mercury species, environmental transport, and other processes is crucial in the biogeochemical cycling of mercury. Oxidation of Hg^0 and reduction of Hg^{2+} occurs during its biogeochemical cycle. Hg^0 enters the atmosphere through coal combustion, and mining and contaminates the northern hemisphere. They can form on the water droplets in the cloud and deposit on land or water, or taken up by plants during transpiration. Inter-conversion between elemental mercury and Hg^{2+} can be facilitated by photo-oxidation and photo-reduction (Cooke Andrews 2006). Elemental mercury is gaseous and more than 90% of the atmospheric mercury undergoes long-range global transport in the air (Keating et al. 1997). Hg^0 gets easily oxidized to Hg^{2+} under aerobic conditions, with higher solubility than Hg^0.

Hg^{2+} can enter the ecosystem through erosion, urban discharges, agricultural sources, mining discharges, and mercury amalgams (Cooke Andrews 2006). Hg^{2+} in atmospheric waters occurs in a dissolved state or is adsorbed in droplets onto particles. They show a trend to quickly settle on the earth's surface by wet and dry deposition (Nriagu 1979). Mercury exists as mercuric compounds in water, soil, and sediments.

They can be primarily mercuric hydroxide and methyl mercuric compounds. Secondarily present mercuric oxide plays an important role in mercury exchange between air, eater, and terrestrial environment. These compounds exist in diverse physical and chemical forms by forming complexes with chloride, sulfide, and organic ligands (Ullrich et al. 2001). Methylation and dimethylation depend on the source and environmental condition (Cooke Andrews 2006). Methylation of mercury is generally carried out by sulfate-reducing bacteria (SRB). To a lesser extent, inorganic transformations are carried out by fulvic acid and humic acid (Ravichandran 2004). Methylation is directly dependent on the bioavailability of mercury to the sulfate-reducing bacteria (SRB), bacterial activities, pH, temperature, dissolved oxygen (DO), redox activity, and electron donors and acceptors (Cooke Andrews 2006). Methylmercury constitutes a small fraction of the total mercury present; less than 10% in water and less than 3% in soil or sediment. Methylmercury bioaccumulates and reaches high concentrations in organisms through biomagnifications (Stein et al. 1996).

4.5.1 The Biochemical and Molecular Mechanism of Mercury Methylation

Recent studies exposed that methylmercury can be produced in a wide variety of environments. Earlier pieces of evidence showed mercury methylation in water was primarily in sediments under strict anaerobic conditions. Sediments and sinking particles form a complex matrix of solid phases comprising clay, quartz, metal-oxides, carbonates, sulfides, and other minerals that form a microenvironment and habitat for bacteria, archaea, algae, and others. Till date, biological mercury methylation is accepted to be catered by species having *hgcAB* gene clusters. Later several studies showed that mercury methylation can also occur in oxygen-deficient zones.

Research after 50 years of the Minamata incident suggested surface sediments and bacterial activity to be responsible for mercury methylation. The mercury methylation ability of *Pseudomonas fluorescens*, *Mycobacterium phlei*, *Escherichia coli*, *Aerobacter aerogenes*, and *Bacillus megaterium* was evaluated to validate the observation. After another ten years, sulfate-reducing bacteria (SRB) were ultimately recognized as the mercury methylator in salt marsh through inhibition of their activity with sodium molybdate along with isolation of *Desulfovibrio desulfuricans* from sediments. The fact that mercury methylation could be linked to specific bacterial metabolism was corroborated by comparable results of sulfate reduction and mercury methylation. Hence sulfate reduction was assumed to be the key metabolic pathway related to mercury methylation. Subsequent studies in 2006 and 2010 discovered the role of Fe-reducing bacteria (FeRB) on mercury methylation in ferruginous conditions and methanogens as major mercury methylators in lake periphytons respectively. Non-enzymatic mercury methylation occurs from methylcobalamin at pH 4.5 in the following way.

$$CH_3 - \text{cobalamine} + Hg^{2+} \rightarrow CH_3Hg^+ + \text{cobalamine}$$

Biochemical pathway for enzyme-mediated mercury methylation is depicted in Fig. 4.1. The discovery of a two-gene cluster, *hgcAB* concerned with C1 metabolism and the acetyl-CoA pathway brought a breakthrough in comprehending the genetic basis of biological mercury methylation. The gene *hgcA* codes for a corrinoid protein (HgcA) with a corrinoid binding domain (CBD) that faces the cytoplasm and a trans-membrane domain (TMD) that remains embedded in the cytoplasmic membrane. This is necessary for the synthesis of the folate branch of the acetyl-CoA pathway. The corrinoid binding domain of HgcA is homologous to the corrinoid binding domain of the Corrinoid Fe-S-protein (CFSP) from the reductive acetyl-CoA pathway of C-fixation. The corrinoid binding domain (CBD) of HgcA comprises a rigidly conserved cysteine (cys) that is crucial for mercury methylation in vivo. The gene *hgcA* is located immediately downstream of *hgcB*. The gene *hgcB* encodes a ferredoxin-like protein (HgcB) with 2 binding motifs that are presumed to be an electron donor to *hgcA*. HgcB with two clusters binding domain is unique in architecture having a supplementary rigidly conserved Cys residue and a pair of conserved cysteines at C-terminus. Whether both *hgcA* and *hgcB* provide methyl groups for mercury methylation is not clear. Else deletion of either of the genes fails mercury methylation in *Desulfovibrio desulfuricans* ND132 (Bravo and Cosio 2020). Cellular disposition of HgcAB protein is presented in Fig. 4.2 and genetic basis of mercury methylation is depicted in Fig. 4.3. At first, the HgcB protein provides low-potential electrons to the oxidized corrinoid cofactor of HgcA to produce a supernucleophilic Co(I) state. This Co(I)-corrinoid then receives a methyl group from methyltetrahydrofolate (CH_3-H_4-folate) to produce CH_3-Co(III)-corrinoid. The methyl-corrinoid then shifts the methyl group to the Hg(II)-substrate to produce methylmercury (CH_3-Hg^+). Subsequently, the corrinoid is in an oxidized state (Date et al. 2019). The corrinoid remains in an oxidized state after the methyl transfer. The oxidized corrinoid is reduced to Co(I) state by receiving low-potential electrons from the HgcB after each cycle. The biogeochemical cycling of mercury is presented in Fig. 4.4

4.6 Uses of Mercury

Mercury has been used in various applications throughout history due to its unique properties. Before 1900, nearly all mined mercury was utilized in gold and silver mining. Its utility diversified in the twentieth century. In 1970 mercury use in consumer products like paints, batteries, and chlor-alkali plants reached its peak. However, its usage has significantly decreased or been banned in many countries in recent years due to the recognition of its harmful effects on human health and the environment. However, post-1970, the total consumption fell owing to implemented regulations on use and release by the developed countries. Since then developing countries took the lead (Horowitz et al. 2014). Efforts are being made globally to

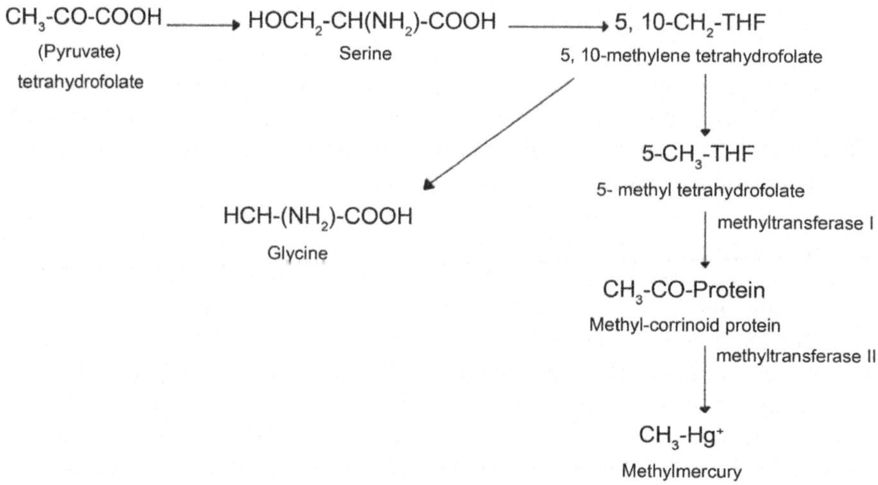

Fig. 4.1 Biochemical pathway for enzyme-mediated mercury methylation

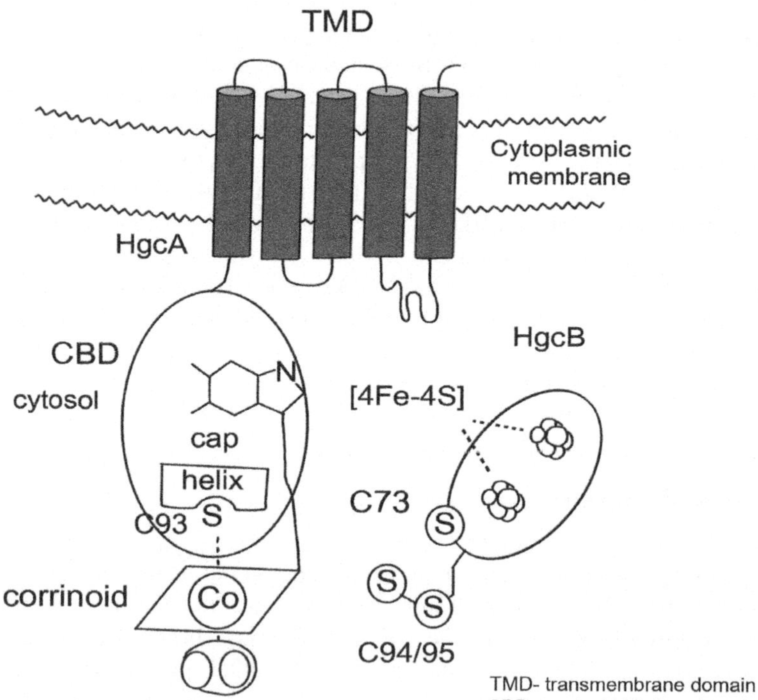

Fig. 4.2 Cellular disposition of HgcAB protein

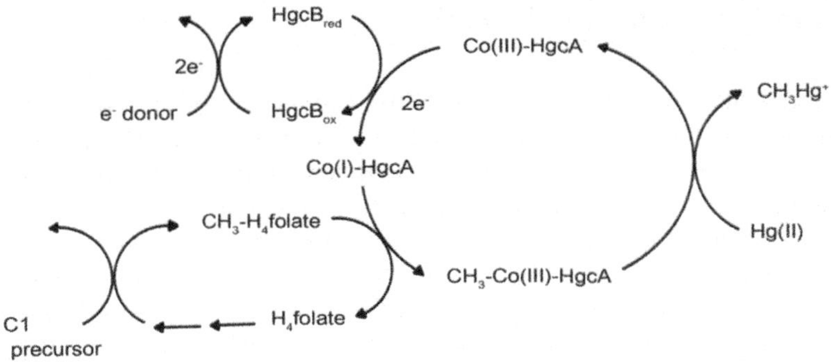

Fig. 4.3 Genetic basis of mercury methylation

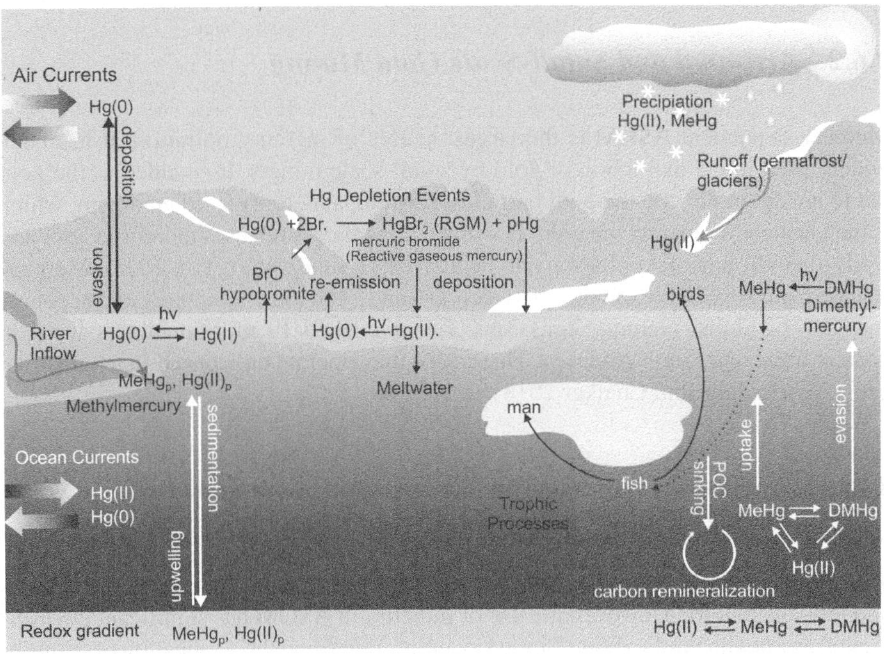

Fig. 4.4 Biogeochemical cycling of mercury

phase out or minimize mercury usage and promote safer alternatives. Some historical and current uses of mercury are as follows:

4.6.1 In Traditional Medicines

Metallic mercury is used in herbal remedies in some spiritual practices, sold under the name 'azogue' in botanicas stores, quite common in Hispanic and Haitian communities (ATSDR 1999). The element was used to treat indigestion. In certain traditional medical practices, mercury has been used in small quantities for its perceived therapeutic properties. Cinnabar has been used for two thousand years in traditional Chinese medicines (Liu et al. 2008). The Chinese used mercury in making pigments, soaps, cosmetics, and laxatives. It was also included in the list of drugs in the *Chinese Materia Medica* around 2735 BCE. Mercury amalgamated with silver was used for making '*lingas*', '*Sri yantras*', etc. by the Hindus. Ayurvedic medicines are also reported to contain mercury (Masur 2011).

4.6.2 Artisanal and Small-Scale Gold Mining

Mercury-dependent ASGM is the largest source of mercury pollution. It has been widely used in the extraction of gold by small-scale miners. It is added to the gold ore to amalgamate with the gold particles, forming a mercury-gold amalgam, which is then heated to separate the gold. However, this process releases mercury vapor and leads to environmental pollution and health risks (Nandiyanto et al. 2023). Mercury release from ASGM is estimated between 410 and 1400 tonnes annually and accounts for 37% of global mercury emissions. Around 10 to 19 million miners work in Asia, Africa, and South America. These activities operate unlicensed in an informal economy (Esdaile and Chalker 2018).

4.6.2.1 Application of Mercury in Artisanal and Small-Scale Gold Mining (ASGM)

The process, known as amalgamation, has been widely practiced due to its simplicity and low-cost nature. However, the use of mercury in ASGM has significant environmental and health implications. Some key points regarding the application of mercury in ASGM:

Ore Crushing: The gold-bearing ore is first crushed into smaller pieces to the consistency of coarse sand to facilitate the extraction process. This can be done manually using hammers or mechanically using crushers.

Grinding and Pulverizing: The crushed ore is further grounded into a fine powder through grinding and pulverizing to increase the surface area of the ore, facilitating the release of gold particles during subsequent processing steps.

Gold Amalgamation: The powdered ore is mixed with water to form slurry into motorized mills or trommels. Heavy steel or tungsten bars or balls are added to further pulverize. Liquid mercury is then added to the slurry in a rotating drum or grinding mill and mixed for several hours. Around 0.3 to 1 kg of mercury is added to 29 kg of crushed ore. Mercury has a strong affinity for gold and forms a bond with the gold particles, creating a mercury-gold amalgam.

Separation of Amalgam: Being dense, the gold-mercury amalgam is separated from the finely crushed rocks and sand by either gravity separation methods, such as panning or sluicing, or by using mercury-coated copper plates or mercury capture mats. The amalgam adheres to these surfaces, allowing for the separation of the mercury-gold mixture from the rest of the slurry.

Gold Extraction and Mercury Recovery: The amalgam is further processed to separate the gold from the mercury. This is usually done by heating the amalgam, in a process called retorting which causes the mercury to vaporize, leaving behind the gold. The mercury vapor is collected, and condensed, typically using a cooling system, to recover the mercury for reuse. This mercury distillation step is often carried out in open-air settings, and is especially dangerous, because of the exposure to mercury gas. Some mercury vapor is released into the environment.

4.6.3 Chlorine Production

Mercury cells were historically used in the production of chlorine and caustic soda through the chlor-alkali process, further used in the production of detergents, herbicides, pesticides, pharmaceuticals, plastics like PVC, and soaps (Lakshmanan and Murugesan 2014). This electrolytic process is a primary industry in the USA, Western Europe, and Japan by diaphragm cell process mercury cell process, and membrane cell process. The first two are environment unfriendly (Dufault et al. 2009). However, most industrialized countries have phased out the use of mercury cells and shifted to more eco-friendly membrane cell process technologies.

4.6.4 Electrical and Electronic Equipment

Mercury has been used in certain electrical and electronic devices, including switches, relays, and fluorescent lamps. The standard non-rechargeable battery uses a mercuric oxide cathode. Besides standard mercury batteries, other batteries also contain small amounts of mercury. In 1993, a button cell battery contained 0.00428 gm and a D cell battery had 0.02973 gm of mercury on average. Other electrical uses include fluorescent and mercury vapor lamps. The average mercury content in a 4-foot tube with a 4-year lifespan was 41.65 mg, whereas that of a compact fluorescent lamp is 5 mg. The high-intensity mercury vapor lamps are used in street lighting, photography, floodlighting, and underwater lighting. It uses less mercury per bulb in comparison to fluorescent bulbs (Jasinski 1995). There are different types of mercury lamps,

Table 4.2 Summary of various uses of mercury

	Species	Key uses
INORGANIC	Elemental Hg	Industries including chlor-alkali, Scientific, and electrical instruments, gold processing, and dentistry
	Mercurous Hg(I) (such as calomel, HgCl)	Laxative, vermifuge, teething powder
	Mercuric Hg(II) (e.g. corrosive sublimate $HgCl_2$)	Antiseptic
ORGANIC	Aryl, Alkoxy, alkyl (such as CH_3Hg^+)	Paper and pulp, fungicides (now banned), diuretics, preservatives

including low-pressure mercury lamps, high-pressure mercury lamps, and mercury-xenon lamps differing in operating characteristics and applications. However, efforts have been made to reduce the mercury content in these products and promote alternative technologies that are mercury-free. Various uses of mercury are summarized in Tables 4.2, 4.3 and 4.4.

4.6.5 Healthcare and Dental Amalgams

Mercury has historically been used in various healthcare applications due to its antimicrobial properties. In the past, certain mercury-containing compounds, such as mercuric chloride and merbromin, were used as antiseptics or disinfectants. They were applied topically to prevent or treat infections (Bharti et al. 2010). Other examples include chlormerodrin, merbaphen, and mercurophylline used as diuretics, and phenylmercury nitrate as a disinfectant. Some skin-lightening creams (Hg_2Cl_2) and soaps also contain mercury. Hg_2Cl_2 in such creams removes freckles and spots formed by excess deposition of melanin. Mercurous chloride prevents melanin formation by irreversibly inhibiting tyrosinase activity by replacing the copper cofactor. Mercury-containing thimerosal is used to preserve multidose vials of vaccines (Balali-Mood et al. 2021). The use of mercury in dentistry has re-emerged as a contentious issue in public health exposing quite a few people to mercury. Dental amalgam is a dental filling material that has been used for over 150 years to restore teeth affected by decay or damage. It is valued for its durability, strength, and resistance to wear, making it suitable for use in load-bearing areas of the mouth. It comprises a mixture of mercury, silver, tin, and copper, have been used for dental fillings. It is considered inert after initial setting and hardening. Reports show swallowing or lodging of dental amalgam debris in the periodontal tissues along with corrosion at the interface of the tooth and abrasion due to masticatory load. Whether the mercury is released in vapor or metallic ion form is still unknown (Spencer 2000). However, the use of dental amalgams containing mercury has decreased in

Table 4.3 Mercury Switches in Electrical Applications

Switch	Quantity of mercury and description	Applications	Available alternatives
Tilt switches			
Thermostats	3,000–6,000 mg A temperature-sensitive switch that controls heating or cooling systems	HVAC systems, appliances, temperature control devices	Electronic type and snap switches
Float control (in the Septic tank and Sump pumps)	Not specified A switch activated by the rise or fall of a liquid level	Sump pumps, septic tanks, water tanks, liquid level control	Magnetic dry reed switches, optic sensors, or mechanical switches
Freezer light	2,000 mg A switch that activates the light inside a freezer when the door is opened	Freezers, refrigerators, cold storage units	Mechanical switch
Washing machine (power shut off)	2,000 mg A switch that cuts off power to the washing machine when the lid or door is opened	Washing machines, laundry appliances	Mechanical switch
Silent switches (light switches before 1991)	2,600 mg Older-style light switches that use mercury to make or break electrical connections	Residential and commercial lighting systems (before 1991), lamps, light fixtures, wall switches	Mechanical switch
Thermo-electrical applications			
Accustat (a calibrated device similar to a thermometer that provides precise temperature control for dedicated applications)	~1,000 mg A calibrated device similar to a thermometer that provides precise temperature control for specialized applications	Laboratory equipment, scientific instruments, temperature-controlled environments	Digital thermostats or other precise temperature control devices
Flame sensor (mercury in capillary tube when heated vaporizes and opens gas valve or operates switch. Used for both electrical and mechanical output in residential and commercial gas ranges)	2,500 mg A switch that uses a capillary tube containing mercury, which vaporizes when heated to open a gas valve or operate a switch	Gas ranges, gas-powered appliances, flame detection systems, heating systems, industrial furnaces	Hot surface ignition system for devices or products that have electrical connections

Table 4.4 Types of Bulbs and Lamps that Contain Mercury

Bulb/lamp type	Description	Mercury content
Compact fluorescent lamp (CFL)	Energy-efficient bulbs that produce light by exciting mercury vapor were first used as overhead lighting in offices	A small amount of mercury, typically 3–5 mg
Fluorescent tube	Linear tubes that use mercury vapor to generate ultraviolet light	Varies depending on the length and wattage of the tube
High-pressure sodium vapor lamps	Operate by using a high-pressure arc discharge through a tube filled with a mixture of noble gases, sodium, and mercury. The light emitted by HPS lamps is characterized by a warm golden-yellow hue, making them commonly used in street lighting. These lamps are known for their high energy efficiency and long lifespan, making them suitable for outdoor and area lighting applications	Do contain mercury as part of their operating mechanism, but the amount is relatively low compared to other types of lamps The mercury content in HPS lamps is typically less than 10 mg (mg)
Metal halide lamp	A type of HID lamp that produces bright, white light for homes and offices applications	Contains mercury as part of the metal halide mixture
Mercury vapor lamp	Lamps that produce light by electrically exciting mercury vapor, first high-intensity discharge (HID) lamps with blue-white light, initially used as farmyard lights	Contains mercury vapor
Ultraviolet (UV) lamp	Lamps that emit ultraviolet light for various applications, such as sterilization	May contain mercury as a component or amalgam
Germicidal lamp	Specialized UV lamps used for disinfection purposes	May contain mercury amalgam or other UV-emitting gases
Neon sign	Tubes filled with low-pressure neon gas and small amounts of mercury are characteristically used in advertising; nearly all colors contain mercury except red, orange, and pink	Contains a small amount of mercury

many countries. Currently used alloy contains zinc, palladium, and indium is less corrosive and tarnishing. Resin can be coated to prevent micro-leakage (Bharti et al. 2010).

4.6.6 *Laboratory and Scientific Applications*

Mercury has been used in laboratories and scientific research. Mercurous chloride is used in medicine and acousto-optical filters. Mercury fulminate is used as a detonator in explosives. Mercury selenide, mercury telluride, mercury cadmium telluride, and mercury zinc telluride are used in semiconductors and infrared detector materials (Rustagi and Singh 2010). Though many actions have been taken to reduce the use and emissions of mercury, some uses remain. Barometers, manometers, thermometers, mercury sphygmomanometers, and strain gauges contain mercury as an integral part. Mercury is used in porosimeters and pycnometers to determine the softening point. It is also used as a catalyst in certain chemical reactions. However, alternatives to mercury-based instruments and processes are now widely available and preferred. Presently three categories of button cell batteries may contain small amounts of mercury. They are zinc-air, silver oxide, and alkaline batteries and do not pose any health risks.

4.7 Mercury and the Environment

The introduction of mercury into the environment can occur in 3 ways:

i. natural volcanic emission bringing mercury from below the earth's crust, weathering of rocks, forest fires, and soil
ii. combustion of fossil fuels, municipal and biomedical wastes
iii. reintroduction of mercury through evaporation of ocean water.

4.7.1 *Environmental Transport*

More than 90% of the atmospheric mercury is mercury vapor. After release, mercury can be transported to long distances by wind before being deposited on the earth's surface. Water-mediated transport also occurs. Deposition time can vary between 5 and 14 days or even 1 year. It can tour as much as 2,500 km in about 72 h with airborne residence time varying between six days to six years after which the mercury is redeposited but that of organic mercury is for decades. After being deposited on the ground, mercury can be carried by rain and snow-melted runoff to reach the surface waters. Elemental mercury can form salts by reacting with chlorine, sulfur, etc. Groundwater mercury is mostly from the air. In soil and water mercury is mostly

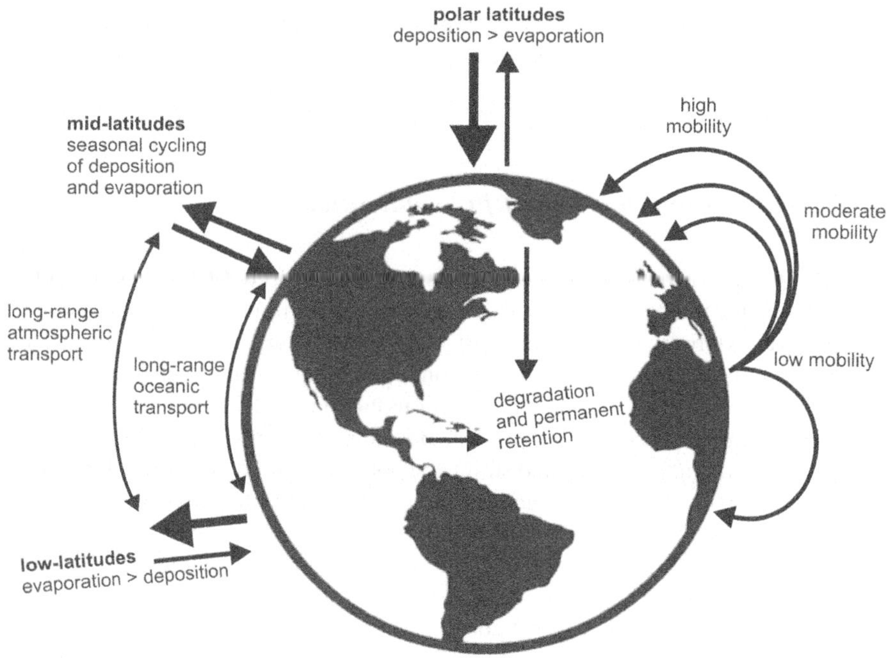

Fig. 4.5 Global transport of mercury

in monovalent and divalent states. The specific valence state is dependent on pH, the redox potential of the medium, and the strength of the ligands present. Though the sorption of mercury into the soils decreases with the increase of pH or chloride ion concentration, it binds strongly with humic materials and sesquioxides at higher pH and the peat surface layer. Both inorganic and organic mercury bind strongly to the organic particulates in water. The mercury mobilization from the sorbed particles in soil or sediment takes place through chemical or biological reduction or by microbial methylmercury conversion. Mercury is persistent, bioaccumulative, and cycles back and forth amid air, soil, and water in diverse chemical forms. The extent of bioaccumulation is species-dependent (Fisher and Organization 2003). The biogeochemical cycling of mercury is presented in Fig. 4.4 and global transport of mercury is depicted in Fig. 4.5.

4.7.2 Environmental Distribution

Mercury concentrations have increased even in pristine conditions due to an increase in atmospheric deposition. In the north-eastern Atlantic Ocean, the mercury concentrations in the feathers of fish-eating seabirds have progressively increased for the last 100 years. The increase in the proportion of dissolved oxygen content (DOC) and pH of water boosts mercury mobility to enter the food chain which is further reflected in increased levels of mercury in the fishes. Industrialization sediment deposits were found to contain 3–5 times mercury concentrations. Anthropogenic activities have turned some places into hotspots of methylmercury such as Polar Regions, paddy soils, acidic lakes in the USA, parts of Florida Everglades, coastal wetlands along the Gulf of Mexico, the Atlantic Ocean, and San Francisco Bay. The addition of sulfur, and acidification of lakes, triggers methylation and mercury mobilization in soils.

4.7.3 Environmental Transformation and Impacts

Oxido-reduction of atmospheric elemental mercury vapor may take place in the presence of dissolved ozone, H_2O_2, hypochlorites, or organoperoxy compounds. Mercury vapor may remain for 2 years in the air but may oxidize rapidly in the clouds in the presence of ozone. Mercuric hydroxide may be reduced quickly to monovalent mercury in sunlight, but mercuric sulfide in the air is quite stable (Fisher and Organization 2003).

In an aqueous environment mercury transformation is mostly the microbe-aided biological conversion into organomercury compounds, especially the sulphur-reducing anaerobic bacteria at low pH and high sedimentary mercury concentrations. Yeast like *Candida albicans* and *Saccharomyces cerevisiae* are also able to methylate mercury at low pH and even ionic mercury to elemental mercury. Similarly acidified lakes and industrial runoff favor methylation. Methylation increases due to anaerobic conditions and increased dissolved organic carbon levels (Fisher and Organization 2003).

In soil, the mercury transformation process is also dependent on the microbial and abiotic processes as in water. Methylation decreases with increasing chloride ion concentration, but chloride ion presence increases the rate of mercury release from sediments (Fisher and Organization 2003).

A very significant impact of mercury on the environment lies in its competence to build up in the organisms (bioaccumulation) and move up the trophic levels irrespective of its forms. But methylmercury is absorbed and accumulated to a larger extent than any other type of Hg. In fish tissue, most of the methylmercury binds to the protein sulfhydryl groups which result in a long half-life for elimination. Mercury concentration is lowest in small, non-predatory fishes but can magnify manifold along the food chain. Bioaccumulation is not only dependent on Hg-concentration in food but also on methylation and demethylation.

Significant disturbance to aquatic life is caused as methylmercury undergoes biomagnification. Humans are exposed to the consumption of contaminated aquatic food. Being lipophilic, these organomercury compounds can easily cross cell membranes. Hence high concentrations of mercury are found in the liver of lean species and fatty fish species. Methylmercury concentrations increase with age of fish and trophic levels. Predatory species like tuna, halibut, red fish, shark, and swordfish were reported to have high mercury concentrations (Matta and Gjyli

2016). Biomagnification of mercury through trophic levels is presented in Fig. 4.6. Maximum permissible limit for mercury in water for safe aquatic life is presented in Table 4.5 and environmental standards for mercury in potable water are explained in Table 4.6. A notable disaster linked with mercury poisoning is the Minamata disaster. Minamata was a small farming and fishing village facing the Yatsushiro Sea (Shiranui Sea) in Kumamoto Prefecture on Kyushu Island in southern Japan. The Chisso Corporation started a chemical factory in 1908 producing fertilizers that further expanded to produce other chemicals. A five-year-old girl was first reported to manifest unusual neurological symptoms in 1956. Methylmercury was a byproduct of acetaldehyde manufacture and was discharged into the sea. The official investigation by the Ministry of Health and Welfare identified the cause as methylmercury poisoning that led to Minamata disease well correlated with heavy consumption of contaminated fish and shellfish. The mercury entered the food chain, bioconcentrate, and biomagnified along the food chain. The Kumamoto University Research group was formed. Upon investigation, the staple foods of victims were found to be fish and shellfish which led them to believe in some kind of poisoning. The company's waster contained multiple heavy metals but after a thorough study, organic mercury was identified as the main cause. Unexplainable occurrences created panic in the area such as birds dropping from the sky and cats committing suicide. Inhabitants were detected with high levels of mercury in their hair samples along with symptoms like sensory disturbances, ataxia, dysarthria, muscular weakness, damage to hearing and speech, loss of peripheral vision, cough, rashes, pruritus, headache, nausea, vomiting, restlessness, and insomnia. Extreme cases like insanity, paralysis, coma, and death were also reported. The toxicity was congenital affecting the fetus with microcephaly and cerebral damage. Over 3000 people suffered from this degenerating neurological disease, and many of them died of toxicity. The Minamata disease broke out again in 1965 along the Agano river banks in Niigata Prefecture.

The Minamata disaster prompted a significant response from the Japanese government, leading to the enactment of the Minamata Disease Compensation Law in 1969. The law aimed to provide medical support, financial compensation, and relief to the victims and their families. The Minamata disaster served as a turning point in environmental regulations and pollution control measures in Japan and globally.

Iraq went through a similar case of mercury poisoning in 1971–72 through the consumption of seeds treated with organomercury. Around 6,530 people were admitted to the hospital of whom 459 died. The poisoning involved every age and is irrespective of males and females. Alkylmercury compound namely methylmercury was used as a fungicide and poisoning might have occurred due to inhalation of dust or vapor as a part of occupational exposure. Around 73,000 metric tonnes of wheat and 22,000 metric tonnes of barley were distributed, mostly treated with methylmercury compounds leaving a small proportion that is treated with phenylmercury compounds (Al-Tikriti and Al-Mufti 1976). Similar incidents have also occurred in Guatemala, Pakistan, and Ghana.

Grassy Narrows, another mercury disaster, caught public attention in 1970 when an alarming number of people reported symptoms of Minamata disease. The source

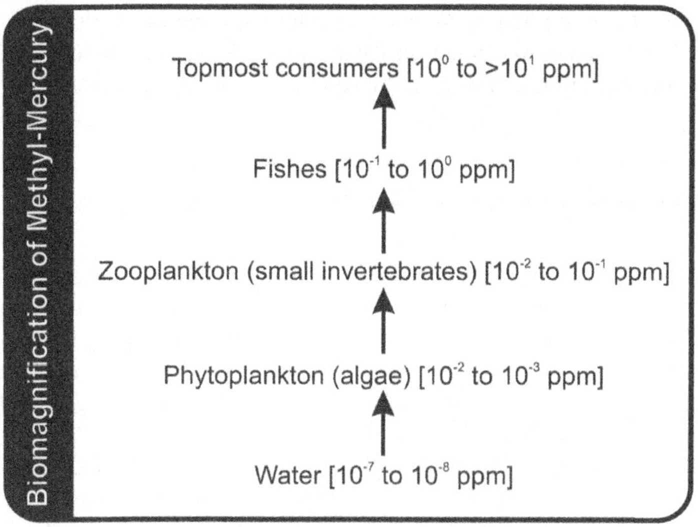

Fig. 4.6 Biomagnification of mercury through trophic levels

Table 4.5 Maximum permissible limit for mercury in water for safe aquatic life

Mercury type	Permissible limit	
Mercury (inorganic)	0.026 μg/l	0.016 μg/l
Mercury (organic)	0.004 μg/l	

Table 4.6 Environmental standards for mercury in potable water

IS:10,500 BIS (Bureau of Indian Standards)	World Health Organization (WHO)
0.001 mg/l	0.001 mg/l

was identified as the Reed Paper chemical plant that had disposed of 4–9 kg of mercury daily in the English-Wabigoon River system located in North-western Ontario between 1962 and 1970. The river system was a vital source of food and water for the communities, and the contamination led to the bioaccumulation of mercury in fish, particularly walleye, which was a staple of their diet. The mercury contamination had devastating effects on the health and well-being of the indigenous Whitedog First Nations community, downstream. Reed Paper Ltd. ceased to operate in 1975 but left a devastating economic and social impact.

Mercury contamination was also reported from Kodaikanal, Tamil Nadu that originated from the thermometer factory owned by Hindustan Unilever Limited. Improper treatment and disposal led to its release into the environment. The issue gained widespread attention, and legal action was taken against Hindustan Unilever Limited. After years of legal battles and public pressure, a settlement was reached between

HUL and the affected workers' association. The settlement included compensation for the workers and a commitment to environmental remediation and cleanup. The mercury levels are still much high in the soil than in other regions.

4.8 Impact of Mercury on Health

People are exposed to mercury mostly in workplaces where mercury is used like chlor-alkali, electrical bulb manufacturing, thermometers, and other industries. Elemental Hg crosses the placental barriers exposing the developing fetus to mercury. Infants are exposed through their mother's breast milk (Fisher and Organization 2003). The toxicity related to various forms of mercury decreases as CH_3-Hg > Hg^+ > Hg^{2+} > Hg^0 (Balali-Mood et al. 2021).

Mercury is a toxic heavy metal that can have harmful effects on the neurological, digestive, respiratory, muscular, and renal systems and cellular systems of the body. But the brain remains the primary target organ for mercury. There exist considerable differences in the toxicity of metallic mercury, inorganic mercurous and mercuric salts, and organometallic mercury compounds. Toxicity results from both acute and chronic exposure. The severity of health impacts varies with the chemical form of the mercury, quantity, and duration of exposure, route of exposure, age of the exposed person, and the health status of the exposed person (Bhawan and Nagar 2020).

When inhaled, metallic mercury vapor gets deposited in the CNS, whereas rare acute toxic effects are reported when taken orally. This is attributed to low digestive tract absorption, i.e., below 0.01% of the dose. LD_{10} in men is reported to be 1429 mg/kg orally, around 100 g for an adult of 70 kg body weight. The transcutaneous uptake is little (Bhawan and Nagar 2020) and nearly 2% enter through the lungs. Inhaled mercury vapor freely passes the alveolar membrane with approximately 100% bioavailability. Readily after absorption, while a significant amount of mercury is taken up by the RBC, some amount remains in the bloodstream. This mercury is rapidly distributed throughout the body including the CNS. Inside the RBC, liver, and CNS, the mercury gets oxidized to mercuric oxide via the catalase-peroxidase pathway (Langford and Ferner 1999). A slow H_2O_2 production rate limits the oxidation of mercury vapor (Bhawan and Nagar 2020). The reaction rate is dependent on the presence or absence of catalase inhibitors and competitive substrates like ethanol (Langford and Ferner 1999). Adding hydrogen peroxide to blood samples increases the mercury uptake by 6 times. Experiments with horseradish peroxidase, and beef liver catalase indicate the metallic elemental mercury acts as an electron donor for complex I of catalase that is produced from the first reaction of catalase with hydrogen peroxide (Magos et al. 1978). Mercury disrupts the membrane potential and disrupts intracellular calcium homeostasis. It plays a crucial role in damaging the tertiary and quaternary structure of proteins (Jaishankar et al. 2014).

Mercury is readily excreted primarily from the kidneys soon after absorption. Other routes include the GI tract, saliva, lungs, nails, hair, and sweat glands. The half-life of mercury excreted via urine is 30–60 days, while the half-life of mercury deposited in the encephalon may exceed many years. A high proportion of mercury in the cerebellar Purkinje cells and certain neurons of the spinal cord and mesencephalon is reported in rats. Patients with mercury exposure complain about a metallic taste. Coughing, shortness of breath, chest tightness, sluggishness, anxiety and uneasiness, fever, and signs of pneumonitis emerge after 3–5 h of acute exposure. With repeated exposure in sufficient quantities tremors and erethism increase coupled with vasomotor disturbances like excess perspiration and blushing. Erythism symptomizes extreme shyness, lack of self-confidence, vague fears, bad temperament, insecurity, and suicidal melancholia. Mouth ulceration, bleeding gums, and loosening of teeth are also reported. Chronic mercury poisoning mostly affects the CNS, and kidneys with tremors of the face muscles and eyelids. Slowly the limbs are affected by illegible handwriting, alphabet omissions, and ultimately entire words. The affected person even faces problems with dressing like simple tasks. With kidneys as the prime elimination route, proteinuria reflects glomerular damage.

Mercury deposits were reported in the different parts of the brain and spinal cord in animal models exposed to mercury vapor (550 $\mu g/m^3$). Such exposure in mice results in a 70% increase in metallothionein (a metal-binding protein). Chronic exposure to mercury vapor injures the kidneys. Alteration of renal tissue was observed after 45 days in Wister rats that were exposed to 1 mg/m^3 Hg every day. Deposition in the kidneys might be slow owing to γ-glutamyltranspeptidase inhibition and thus causing a rise in urinary mercury excretion. The reason for such reduction was due to the role of γ-glutamyltranspeptidase in the reabsorption of the mercury-glutathione conjugate in the glomerulus (Balali-Mood et al. 2021).

The toxicity of inorganic mercury salts varies with their solubility. Mercuric salts are more toxic as compared to mercurous salts owing to their more solubility. Inorganic salts of mercury are more hazardous due to their water solubility when taken orally in comparison to metallic mercury. The lethal dose of mercuric chloride is only 0.5gms while for metallic mercury it is 100 gm. Non-volatile poisoning by inhalation of mercury salt is uncommon. Upon adsorption mercury quickly reacts with the thiol group(s) of amino acids (cysteine for example) in the protein. The protein metallothionein formed acts as a natural chelator to prevent further damage by mercury and other transitional metals. If metallothionein protein gets saturated, other structurally correlated proteins can be damaged. Such ingested mercury salts deposit in the liver to be excreted in bile or in the kidneys to be excreted in the urine. Being lipid insoluble it does not cross the blood–brain barrier in considerable amounts. Soon after ingestion chest pain, mucosal membrane discoloration, and stomach pain are systematized. Vomiting and excessive blood diarrhea can lead to hypovolaemic shock and death. Upon survival systemic effects manifest with glossitis, ulcerative gingivitis, excessive salivation, metallic taste, teeth loosening, and kidney damage (Ogata and Aikoh 1983). Inorganic mercury is reabsorbed from the PCT of the nephron as Cys-S-Hg-S-Cys or from the basolateral membrane by organic anion transporters. Inorganic Hg is unable to cross the blood–brain -barrier (BBB) and placenta (Balali-Mood et al. 2021). The deposition of mercury salt in the PCT of the nephron secondarily causes transient polyuria, proteinuria, haematuria, anuria, and renal acidosis. Chronic poisoning of inorganic mercury is rare.

However, combined with elemental mercury, chronic poisoning of inorganic mercury affects the CNS. It gives rise to acrodynia or 'Pink disease' manifested by severe leg cramps, irritability, paraesthesia, pink extremities, and exfoliation of skin (Ogata and Aikoh 1983). Mercury vapor is possibly oxidized by the RBC through the creation of glutathione radicals, releasing nascent oxygen or peroxidase reacting with catalase complex I. Studies indicate mercury oxidation to be dependent on the rate of H_2O_2 production and on the specific activity of red call catalase (Halbach and Clarkson 1978).

$HgCl_2$-induced renal injury in rats compromises the antioxidative system of rat kidneys. After 72 h of exposure, morphological changes in the renal tissue were observed along with reduced glutathione peroxidase activity and an increase in oxidative lipids and proteins. Intestinal fluid accumulation, diarrhea, and mucosal damage were reported in rats orally exposed to $HgCl_2$ for four days. Administering 5 mg/kg of mercury reduced mRNA and downregulated aquaporins in the digestive tract which are said to be involved in harmful GI effects related to mercury uptake. Animals treated with $HgCl_2$ showed hepatotoxicity that was determined by histopathological examination and with increased levels of hepatic enzymes like alanine transaminase (ALT), aspartate transaminase (AST), and alkaline phosphatase (ALP). Such hepatotoxicity exhibited gender biases in Wister rats. Hepatic damage was suggested to be related to organic anion transporter 3 (Oat3). Low Oat3 limits mercury uptake. Male rats exhibited reduced expression of Oat3 and hence low hepatotoxicity. It seems like Oat3 transport is sex hormones-dependent (Balali-Mood et al. 2021).

The oxidation of elemental Hg in vitro by the RBC of the normal, hypocatalasemia, and acatalasemia does not bear proportional relation with catalase activity. Hg oxidation is accelerated by methemoglobin-hydrogen peroxidase compounds and glutathione peroxidase in the RBC. The in vivo oxidation of elemental Hg in the lungs and blood of the normal, hypocatalasemia, and acatalasemia, exposed to mercury is more or less proportional to the decrease in catalase activity. Catalase, cyt C, methemoglobin, lactoperoxidase, ferritin, and ferric ion possess the ability to oxidize elemental Hg in the presence of hydrogen peroxide (Ogata and Aikoh 1983). In vivo transformation of mercury is described in Fig. 4.7.

Mercury species transformation in mammalian tissues: Ingested, absorbed, or inhaled mercuric ions combine with glutathione to form G-S-Hg-S-G, and with cysteine (Cys) to form Cys-S-Hg-S-Cys conjugates. With methyl mercury, it forms CH_3Hg-S-G and CH_3Hg-S-Cys respectively. G-S-Hg-S-G and CH_3Hg-S-G are then converted to Cys-S-Hg-S-Cys and CH_3Hg-S-Cys in the presence of γ-glutamyltransferase and cysteinylglycinase.

Organomercury compounds like ethylmercury and methylmercury are more toxic than inorganic ones. They are readily absorbed from the digestive tract and later spread all through the body. CH_3-Hg binds to thiol-containing molecules like cysteine to form CH_3-Hg-Cys and readily crosses the blood–brain barrier. Ethylmercury follows similar pharmacokinetics. Since organic mercury accumulates in the hair, it can be considered as an index of mercury exposure (Balali-Mood et al. 2021). They can enter the human body via the aquatic food chain either directly or indirectly. Microorganisms convert elemental mercury to organomercury compounds

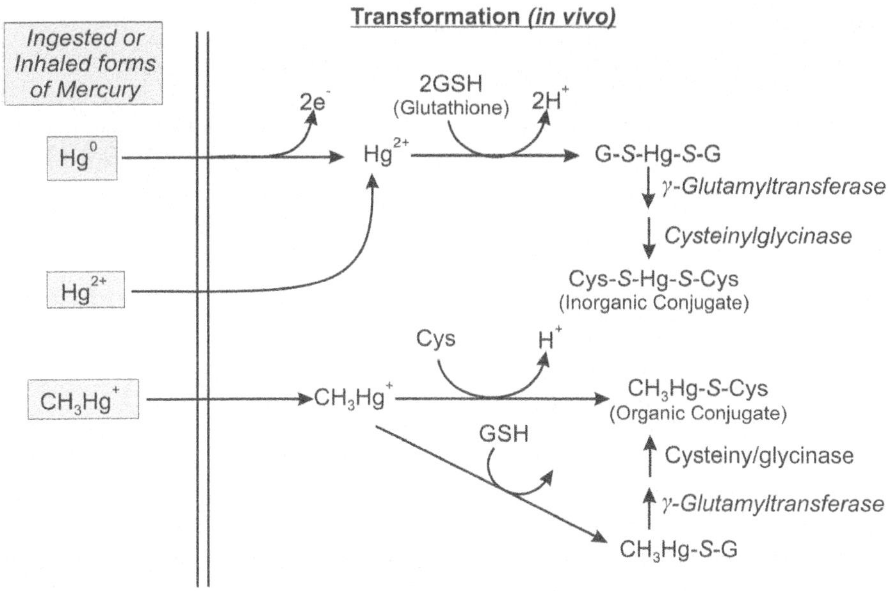

Fig. 4.7 In vivo transformation of mercury

that are consumed by the subsequent invertebrates, and vertebrates including man. Organomercury compounds are highly soluble in lipids with their absorption efficiency varying between 90 and 100%. Though a considerable part is converted to inorganic mercury, it fails to explain CNS toxicity. The cerebellum and visual cortex are specifically damaged. Organomercury compounds are at first excreted through the liver in the bile. Organomercury compounds undergo enterohepatic recirculation leading to reabsorption and uptake into RBC, where they undergo metabolism to form inorganic salts. The effects might get delayed from weeks to months following organomercury compounds exposure and mostly affect the GIT and CNS. Nausea vomiting and abdominal pain are common with high doses causing diarrhea and colitis. Sialorrhoea (hypersalivation), perioral paraesthesia (mouth numbness), and gum discoloration are also seen. CNS toxicity is exhibited by numbness of limbs, tremors, ataxia, dysarthria (speech difficulty), and constricted visual fields (Ogata and Aikoh 1983). Methylmercury causes microtubule damage, mitochondrial damage, and lipid peroxidation and leads to the accumulation of serotonin, aspartate, and glutamate-like neurotoxic molecules (Jaishankar et al. 2014). Undergoing reaction with CH_3-mercury alters cellular function by binding to the selenohydryl and sulfhydryl groups. It interferes with transcription and translation leading to ribosomal and ER disappearance as well as activity of the NK cells. It affects cellular integrity by forming free radicals. Even though the mercury sulfhydryl bond is stable, a free sulfhydryl group is formed that promotes metal mobility within the ligands (Jaishankar et al. 2014).

Table 4.7 Guidelines for chronic mercury exposure

	Component of environment	Standard	Organization/agency
1	Air (metallic)	0.2–0.3 mg/m^3	Agency for Toxic Substance and Disease Registry (ATSDR), USA
2	Air (inorganic form	1 mg/m^3	WHO
3	Drinking water and bottled water	1 mg/l^1 for total Hg and2mg/ l^1 for inorganic mercury	WHO
4		0.1 mg/kg/day 0.3 mg/kg/day 1 mg/g	EPA ATSDR FDA
5		0.3 mg/kg/day	EPA

Methyl mercury chloride was reported to induce CNS damage in rats that were administered with different doses of mercury. Such damage was shown through an amplified expression of c-fos protein in the cerebral cortex and hippocampus as crucial signal transduction pathways (Balali-Mood et al. 2021). A guideline for chronic mercury exposure is presented in Table 4.7.

4.8.1 Mercury-Induced Cellular Changes

Mercury hits the living system either by breaking H-bonds, dislodging other metallic ions, altering protein structures, or disturbing catalytic activity. This may modify the translation in a way to elicit mutagenic or carcinogenic activity and to bind to sulfhydryl and selenohydryl groups. Patients with multiple sclerosis (MS) are seen to possess higher mercury content in their cerebrospinal fluid than healthy persons (Tabatabaei et al. 2022).

Mercury can induce a range of cellular changes that can have detrimental effects on various tissues and organs in the body. It binds with the sulfhydryl and selenohydryl groups on albumin molecules present in the plasma membranes, receptors, and intracellular signal links. It induces the production of free radicals and alters cellular redox potential. A low concentration of mercury provokes intracellular signaling through phosphorylation. It prevents cytoskeleton structural development in the nerve cells. It causes an apoptotic cascade in human T lymphocytes. Patients with multiple sclerosis (MS) are seen to possess higher mercury content in their serum than normal people (Attar et al. 2012).

Mercury exposure can disrupt cellular signaling pathways, including those involved in cell growth, differentiation, and apoptosis. It can interfere with signaling molecules and receptors, leading to dysregulated cellular responses and altered cell fate. MS affects over 2.3 million people worldwide and is a chronic and progressive autoimmune disease of the CNS with hardly any effective cure and unknown etiology. Presumably, MS can be caused by excessive mitochondrial dysfunction and associated oxidative stress.

In experiments with C57BL/6 mice, mercury is found to cause behavioral, neuromuscular, and sensorimotor problems in experimental autoimmune encephalomyelitis (EAE) + Hg^{2+} treated mice in comparison to the Hg^{2+} group. Oxidative stress-related mitochondrial dysfunction is caused by the increase in reactive oxygen species (ROS) production. This may lead to mitochondrial membrane potential collapse, cyt C release, mitochondrial inflammation, and initiate signaling of mitochondrial pathway apoptosis (Kahrizi et al. 2016).

Five types of brain cells showed mercury uptake after analyzing the brain of a dead man injected with metallic mercury. They are astrocytes, scattered grey matter oligodendrocytes, cortical neurons, locus ceruleus neurons, and cerebral microvessels (Pamphlett and Kum Jew 2018). Mercury stimulates oxidative stress; triggers autoimmunity and causes DNA damage. A biopsy of a nerve revealed axonal degeneration and nerve demyelination taken from a man using ammoniated mercury ointment. Brain atrophy is seen in both mercury poisoning and MS patients. The tendency of mercury to accumulate in the basal ganglia may lead to involuntary movements, trembling, and weakness; this is also seen in patients with multiple sclerosis (MS). The similarity in pathological and physiological characteristics such as demyelination, reduced impulse conduction velocity, decreased viral immunity, damaged blood–brain barrier, and so forth probably point out mercury to be one of the etiological factors in multiple sclerosis (MS) (Siblerud and Mutter 2020).

The ability of mercury to produce oxidative stress, trigger autoimmunity, damage DNA, mitochondria, and lipid membranes and its disposition in the CNS strongly suggest its potential role in the pathogenesis of multiple sclerosis, Alzheimer's disease, amyotrophic lateral sclerosis (ALS), and glial tumors (Pamphlett and Kum Jew 2018). Both mercury toxicity and AD show enzyme dysfunction. They are BACE 1, γ-secretase, cyclooxygenase-2, cytochrome–c-oxidase, protein kinases, monoamine oxidase, nitric oxide synthetase, acetylcholine transferase, and caspases. Vitamin and mineral abnormalities of B1, B12, Vit C, Vit E, aluminium (Al), calcium (Ca), copper (Cu), iron (Fe), magnesium (Mg), selenium (Se), zinc (Zn) are seen in patients with both AD and mercury toxicity (Siblerud et al. 2019).

Neuropathy is one of the prominent pathological features in the nervous system caused by both inorganic and organic mercury exposure; with organic mercury toxicity producing far-reaching lesions. Mercury affects sensory nerves more than the motor nerves. Mercury also damages the neurons in the cerebellum, visual cortex, and spinal ganglia. Regardless of the mercury species, within 12–24 h of administration, they are identified as fine dense electron granules bound to Schwann cells but absent in axoplasm. The same after 4 days are detected inside several nerve fibers bound to Schwann cells, myelin sheaths, mitochondria, and microfilaments. Mercury denatures protein and has been shown to be involved with protein folding processes, leading to misfolding and aggregation. Lesions form in the dorsal and ventral root fibers with progression. Large vacuolated axonal spaces are formed when the axon shrinks and the axolemma detaches from the myelin sheath. The myelin sheath also detaches from the Schwann cell. Mercury accumulates in the axon through active

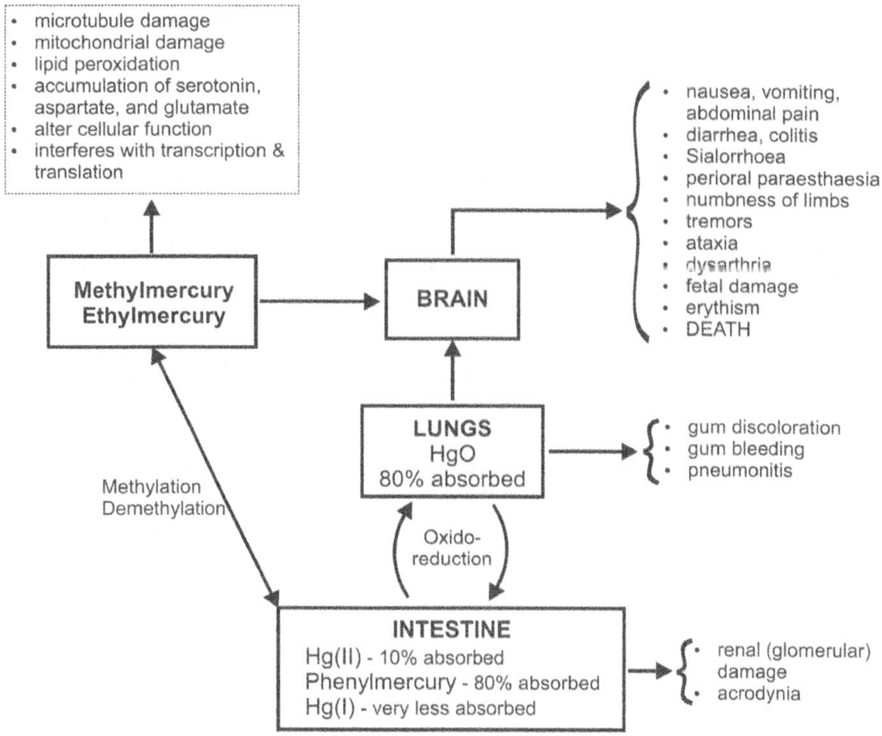

Fig. 4.8 Significant impacts of mercury on human health

transport from the nerve cell body or from the neighboring Schwann cell and myelin sheath (Chang and Hartmann 1972). A significant impact of mercury on human health is described in Fig. 4.8.

4.8.2 Mercury-Induced Chromosomal, Molecular, and Biochemical Changes

The mutagenic ability of metallic mercury is convincingly evident from the ability of inorganic mercury compounds to damage DNA in vitro. Information from in vitro studies also suggests that inorganic mercury induces clastogenic effects in somatic cells. Inorganic mercury compound administered parenterally in quite high doses, in rodents was proved to be embryotoxic and teratogenic (Fisher and Organization 2003). Mercury inhibits DNA replication in the hippocampus and correlates with genetic mutation of presenilin 1 and 2, found in Alzheimer's disease (Siblerud et al. 2019). In vitro, DNA damage in terms of polyploidy was also observed in a human fetal liver cell line (WRL-68) with $HgCl_2$ by means of comet assay (Sánchez-Alarcón et al. 2021).

Mercury exposure has been associated with various structural alterations and numerical chromosomal abnormalities such as chromosomal breakage, translocations, inversions, deletions, and aneuploidy leading to aberrations. They can disrupt normal gene expression patterns, affect cell division and growth, and increase the risk of genetic disorders and cancer. Mercury exposure has been associated with an increased risk of aneuploidy, particularly in developing embryos and fetuses. This can lead to conditions such as Down syndrome (trisomy 21) and Turner syndrome (monosomy X), among others. Mercury exposure has been linked to an increased frequency of micronuclei formation in cells. Ethylmercury and phenylmercury chloride compounds were found to induce genotoxicity through chromosomal alterations in the human HeLa S3 (human cervix epithelioid carcinoma cell line) cell cultures. Phenylmercury chloride was found to increase the sisters chromatid exchange. A study found a similar genotoxic potential for mercuric nitrate in human lymphocytes. Dimethylmercury induced both numerical and structural chromosomal aberrations in human lymphocytes. Despite being used thiomersal is reported to decrease mitotic index in human lymphocytes. Thiomersal or thimerosal sold under trade names merthiolate, mertodol, or mertogan is a mercury ((o-carboxyphenyl)thio)ethyl, sodium salt ($C_9H_9HgNaO_2S$), used as an antiseptic and antifungal agent. Methylmercury also induces chromosomal aberrations like an increase in the sisters chromatid exchange in both human lymphocytes and whole blood cultures; polyploidy induction lowered mitotic index is observed in peripheral blood lymphocytes. Experimental evidence shows methyl mercury chloride to be more genotoxic in human neuroblastoma cells than methylmercury hydroxide. Occupational and environmental exposure to mercury in several cases showed ascends in aneuploidy to be more with organic mercury compounds. Occupational exposure also showed an increase in chromosomal aberrations, micronuclei, sisters chromatid exchange, and hypoxanthine–guanine phosphoribosyltransferase mutations (Sánchez-Alarcón et al. 2021).

The predisposed targets for methylmercury are the brain cortex and cerebellum specifically the cerebellar granule cells (CGC). Being a soft electrophile it has a strong affinity for nucleophilic thiol (–SH) and selenol (–SeH) groups and this plays a predominant role in methylmercury toxicity. The high affinity for these groups on amino acids like cysteine and selenocysteine blocks catalytic functional groups of the proteins. This disrupts several intracellular functions such as the increase in reactive oxygen species (ROS). Methylmercury can also disrupt intracellular homeostasis by alterations in glutamate recycling and calcium balance.

Methylmercury preferentially accumulates in the astrocytes and inhibits glutamate uptake of astrocytes. Spontaneous glutamate release from presynaptic terminals coupled with inhibition of vesicular uptake is critically linked with methylmercury excitotoxicity. As extracellular glutamate levels increase, the *N*-methyl D-aspartate receptors are over-activated which leads to neuronal Ca^{2+} influx. This then leads to the activation of significant pathways concerned with cell death. Such effects are found to be mediated by increased production of mitochondrial and intracellular compartmental ROS. Increased Ca^{2+} influx stimulates GABA (gamma-aminobutyric acid) release in cerebellar neurons. The early stimulatory effect on excitatory synaptic transmission is due to the GABA receptor blockage in hippocampal CA1 neurons. Methylmercury also targets dopaminergic neurons and dopamine metabolism. It induces increased striatal release of dopamine (DA) and acidic dopamine catabolites homovanillic acid and 3,4-dihydroxyphenylacetic acid (DOPAC) and inhibits aldehyde dehydrogenase ALDH). Methylmercury can be compared to a neurotoxin, 1-methyl-4-phenylpyridinium (MPP(+)), in impairing DA metabolism. Methylmercury treatment also inhibits serotonin (5HT) uptake in astrocytes and is linked to decreased monoamine oxidase activity (Ke et al. 2019).

Methylmercury can induce concentration-dependent impairment of mitochondrial membrane potential. This process is probably linked with increased Ca^{2+} level that triggers ROS production and increased oxidative stress. Methylmercury also induces dose-dependent and time-dependent increases in lipid peroxidation and ROS formation. Mitochondria owing to a large number of redox centers is a potential site for reacting with metals and xenobiotics leading to the formation of ROS and reactive metabolites. One such site is the electron transport chain (ETC). The oxidative phosphorylation occurs in the inner mitochondrial membrane comprising 5 multi-subunit protein complexes (CI to CV). Methylmercury triggers changes in CII and CIII, lower respiration, ATP generation, and swelling of the mitochondrial matrix. In addition to oxidative stress and mitochondrial dysfunction, methylmercury also triggers cellular NAD + level depletion. Besides regulating energy metabolism mitochondria play a key role in regulating apoptosis. It is reported that methylmercury induces apoptosis through a classical mitochondrial pathway where the impaired mitochondrial membrane potential may induce the secretion of pro-apoptotic mediators such as cyt C and Bax activation (Ke et al. 2019).

Post-translational modification for protein biosynthesis comprises phosphorylation, glycosylation, ubiquitination, nitrosylation, methylation, acetylation, lipidation, proteolysis, and so on, phosphorylation being the most prevalent. In rats, methylmercury was reported to modulate protein phosphorylation in the brain cytosol. In human brain microvascular endothelial cells and pericytes, methylmercury can lower protein tyrosine phosphatase 1B activity. Alteration in protein phosphorylation is concentration-, cell type- and developmental stage-dependent. Methylmercury exposure upregulates protein kinase A (PKA) activity. This in turn causes neural hyperactivity and neuronal degeneration. Symptoms of methylmercury toxicity share some symptoms with that of Parkinson's disease and genomic study reveals similarity in the signaling pathways in PD pathogenesis. Having strong affinity, mercury binds easily with thiol and selenol groups to impair protein function. It inhibits glutathione reductase, glutathione peroxidases, and superoxide dismutases to decrease the antioxidant enzymatic activities. Methylmercury modulates mitogen-activated protein kinases{(MAPKs) like extracellular signal-regulated kinase 1/2 (ERK1/2), c-Jun N-terminal kinase (JNK) and $p38^{MAPK}$)}, protein kinase A (PKA) and protein kinase C (PKC) activity. Methylmercury modulates the cytochrome P450 activity by inhibiting

TCDD-mediated induction of CYP1A1 activity by post-translational modification (Ke et al. 2019).

Methylmercury is reported to alter gene transcription via both upregulation and downregulation of genes encoding for proteins related to oxidative stress responses in *Caenorhabditis elegans* and RNA splicing via spliceosome in human neuroblastoma respectively. In marmoset, methylmercury is found to target apolipoprotein E. Glutathione enzyme plays an important role in antioxidant defenses, and being in large quantity they act as the prime target for ROS formed due to methylmercury toxicity. Methylmercury results in glutathione depletion a result of the pro-oxidative events caused by methylmercury. Glutathione peroxidase is a group of peroxide detoxifying enzymes, selenoprotein in nature. Methylmercury selectively targets selenol-containing molecules than thiol-containing molecules reducing the activity of glutathione peroxidase, also a pro-oxidative event in methylmercury neurotoxicity (Ke et al. 2019).

4.9 Control of Mercury Pollution

4.9.1 Removal of Mercury from the Air

Removing mercury from the air can be a challenging task due to its volatile nature and potential health risks. However, several methods and technologies are available for reducing mercury levels in the air.

Adsorption: Sulfur-impregnated activated carbon is very effective in adsorbing mercury vapor from the air. Air is passed through a bed of activated carbon when the mercury vapor molecules adhere to the carbon's surface. This method is commonly used in air purifiers, industrial air pollution control systems, and mercury emission control equipment (Otani et al. 1988). Other adsorbents such as manganese oxide-impregnated gamma alumina, $CuCl_2$-impregnated neutral alumina/artificial zeolite have also been reported (Du et al. 2014; Yao et al. 2018).

Wet Scrubbing: commonly used in industrial settings, it involves passing the contaminated air through a liquid solution, usually containing chemicals that react with and capture mercury. The most common flue gas desulfurization process is the wet limestone scrubbers, proven more effective in reducing Hg^0 emissions (Gonzalez et al. 1999).

Chemical Absorption: typically used in industrial settings, certain chemicals, such as potassium permanganate or sodium sulfide, can react with mercury vapor to form stable compounds that can then be removed from the air.

Electrostatic Precipitation: use an electrical charge to capture and collect airborne particles, including mercury. The charged particles are attracted to collector plates or electrodes and can be removed for proper disposal. While cleaning SO_2 produced in the smelting of metalliferous sulfide by electrostatic precipitation, the precipitated material is often found to contain mercury (Newman and Palmer 1978).

Filtration Systems: High-efficiency particulate air (HEPA) filters or specialized mercury filters can be used to effectively capture mercury particles in the air rather than mercury vapor.

4.9.2 Treatment Methods for Mercury Removal from Wastewater

There are several treatment methods available for removing mercury from wastewater, including:

Chemical precipitation: is most widely used for both groundwater and wastewater. Sulfides, ferric sulfates, and calcium hydroxide are more commonly used chemicals to precipitate mercury. The development and use of a Mg_2Al layered double hydroxide supported iron sulfide composite (FeS@Mg_2Al-LDH) showed high selectivity for mercury (Wang et al. 2020).

Coagulation and flocculation: Chemical coagulants such as ferric chloride, ferric sulfate, and polyaluminium chloride were reported to remove up to 97% of inorganic mercury and 80% of methylmercury in the presence of dissolved organic matters (DOM) in a DOM-coagulant mediated process in natural waters (Henneberry et al. 2011).

Adsorption: activated carbon is very promising in treating wastewater. The activity can be augmented by adding specific functional groups to form polyacrylate-modified carbon composite (PAMC) (Al-Yaari and Saleh 2022).

Membrane filtration: involves using membranes to physically separate the mercury from the wastewater. There are quite a few types of membrane filtration, including microfiltration, ultrafiltration, nanofiltration, and reverse osmosis were reported to be effective in removing mercury from the wastewater of oil refineries (Urgun-Demirtas et al. 2012). Endothermic mercury removal was reported using a membrane made up of a matrix of poly(vinylalcohol) crosslinked by gaseous dibromoethane that immobilizes poly(vinylimidazole) chains (Bessbousse et al. 2010).

Electrochemical treatment: involves using an electrical current to oxidize or reduce the mercury in the wastewater which can then be removed through sedimentation or filtration. Maximal mercury removal efficiency was reported using galvanized iron as cathode and magnesium alloy with magnesium, aluminium, and mild steel as anode (Vasudevan et al. 2012).

Nanotechnology and Mercury Remediation: Many of the techniques stated above are either high energy demanding, or have high implementation costs or are not universally applicable. Among the novel methods use of nanoparticles received considerable attention and has been attempted owing to their large surface area, excellent adsorption capacity, faster diffusion rate, marked chemical reactivity, and so on. The use of carboxymethyl cellulose and chitosan were already used as nano-adsorbent materials. Magnetite and maghemite nanoparticles are observed to be small scavengers of mercury. Currently, organic nanoparticles also show potentiality in mercury adsorption. Modification can enhance the performance of nano-adsorbents for mercury remediation to increase the adsorption capacity and selectivity. This can involve using functional groups such as thiol, amine, or carboxyl groups to enhance the affinity for mercury, or incorporating other materials such as graphene oxide or carbon nanotubes to increase the surface area and reactivity (Liu et al. 2022).

One of the key strategies for developing effective nano-adsorbents for mercury remediation is to synthesize materials with tailored surface properties and high adsorption capacity using a range of methods, such as sol–gel synthesis, hydrothermal synthesis, template-assisted synthesis, and characterizing the materials using techniques such as scanning electron microscopy (SEM), X-ray diffraction, and Fourier transform infrared spectroscopy.

Optimization of the process can improve the performance of nano-adsorbents for mercury remediation through optimization of the process parameters such as pH, contact time, temperature, and concentration. This can involve conducting batch or column studies to determine the optimal conditions for maximum adsorption capacity and efficiency.

Scale-up and deployment: Once an effective nano-adsorbent material has been developed, it is important to scale up the process to ensure that it can be applied in real-world situations. This can involve developing methods for large-scale synthesis, testing the material under realistic conditions, and ensuring that the process is cost-effective and practical for widespread deployment.

4.9.3 Environmental Mercury Bioremediation and Its Toxicity

Mercury is the 16th rarest element but its use and abuse have thrown us a great environmental challenge. The conventional techniques discussed above are expensive and hence bioremediation seems to be a sustainable, eco-friendly, and cost-effective process to remove mercury from contaminated sites. Environmental mercury bioremediation is the use of microorganisms such as bacteria, fungi, or algae or their enzymes naturally capable of breaking down or transforming mercury, while others can be genetically engineered or adapted to do so in a range of environments such as soil, sediment, and water. Bioremediation can be used as a stand-alone technique or in combination with other physical or chemical treatments. The process usually targets to precipitate and immobilize metal contaminants and also compartmentalizes metals to the parts of the environment to reduce their impending damage.

Mechanisms of mercury bioremediation: Mercury bioremediation can occur through four categories of enzymatic transformations—reduction, breakdown, or demethylation of organomercury compounds like methylmercury, methylation, and oxidation. Both reduction and demethylation are catalyzed by enzymes and proteins. The microbial resistance operon (mer) encodes for such enzymes and proteins and HgO is produced as a result. Such mer operon-based mercury detoxification caters to the intracellular reduction of Hg^{2+} to HgO (non-toxic) by mercuric reductase enzyme. (Mercuric reductase enzyme is related to glutathione reductase enzyme.). Later, HgO diffuses out from the cell. The HgO produced is observed to be retained in the bed bioreactors. The mer operon provides proteins merT, merP, and merC uptake which then transport Hg^{2+} in the cytosol. Mercuric reductase enzyme induces NADPH-mediated reduction of Hg^{2+} into HgO within minutes. Mercuric reductase enzyme is related to glutathione reductase enzyme. This reaction is energy intensive and hence the bacteria in a bioreactor need supplement nutrients. Another mer operon containing the merB gene encodes lyase enzyme in the broad spectrum mercury-resistant host bacteria that can demethylate organomercury compounds. Worth mentioning that Gram +ve and Gram –ve organisms have this highly conserved mer operon generally on the plasmids or transposable elements. This directs toward an ancient evolutionary origin. Methylation is slow and not a detoxification process. The process of oxidation of Hg is by the hydroxiperoxidases, as observed in *E. coli* (Wagner-Döbler 2003).

Pseudomonas veroni showed promising results of mercuric ions' intracellular uptake and volatilization when immobilized in a xanthan gum-based biopolymer and further coated into natural zeolite granules. The ions then diffuse out from the cell in the air. The *P. veroni* cells remain viable even after intercontinental transport (McCarthy et al. 2017). *Pseudomonas aeruginosa* and *Bacillus licheniformis* under optimized condition were found to exhibit bioremediation capacity; *Pseudomonas aeruginosa* being highly efficient (Kotwal et al. 2018). In this regard, *Lentinus edodes*, *U. lactuca*, and *Typha domingensis* also show great promise for mercury removal (Rani et al. 2021).

Optimization and monitoring: Bioremediation can be a promising approach for mercury remediation, but it is important to carefully consider the potential risks and toxicity concerns to the processes which need to be optimized and monitored to maximize effectiveness and minimize toxicity. This can involve testing different microorganisms or enzyme systems, optimizing environmental conditions such as pH and temperature, and monitoring the effectiveness of the remediation over time.

4.9.4 AI Technology to Detect and Control Environmental Mercury

Machine Learning (ML): ML algorithms can be trained to analyze large datasets of environmental samples and identify patterns or correlations related to mercury contamination. ML models can learn from existing data to make predictions or classifications, enabling the development of predictive models for mercury detection. Deep learning, a subset of ML, can be trained to analyze large datasets of mercury-related information, including spectroscopic data, chemical compositions, or historical monitoring data, to identify unique signatures or anomalies associated with mercury contamination (Vishal Rathee).

Remote sensing technologies, such as satellite imagery and aerial sensors, can capture data on environmental parameters that may be associated with mercury contamination, such as vegetation health, water quality, or land cover. AI techniques, including image recognition and pattern analysis, can be employed to process and analyze remote sensing data for identifying potential mercury hotspots or affected areas.

AI can be applied to analyze data from sensor networks designed specifically for mercury detection and include various types of sensors, such as optical sensors, electrochemical sensors, or gas chromatographs. AI algorithms can process the sensor data in real time, detect mercury levels, and provide alerts or warnings when mercury concentrations exceed predetermined thresholds. AI can help integrate data from multiple sources, such as monitoring stations, field measurements, and laboratory analyses, to create comprehensive mercury pollution maps or models. By combining data from different sensors and sources, AI algorithms can provide a more accurate and holistic understanding of mercury distribution and sources in the environment. AI can aid in visualizing and interpreting complex mercury-related data. By leveraging AI techniques, data visualization tools can provide interactive and intuitive representations of mercury pollution data, enabling decision-makers and stakeholders to better understand the information and make informed decisions.

These AI technologies, when integrated with appropriate data collection systems and domain expertise, can enhance our ability to detect and monitor mercury in the environment more efficiently and effectively. They can assist in identifying pollution sources, prioritizing remediation efforts, and informing policy decisions to mitigate mercury contamination.

4.9.5 Comprehensive Approach to the Control of Mercury Pollution (Rhee 2015)

Various strategies and actions at different levels can operate toward controlling mercury pollution.

Reduction of mercury emissions into the environment is crucial and accomplished by implementing stricter regulations and emission standards for industries, such as coal-fired power plants, waste incinerators, and industrial processes that use or produce mercury. Employing technologies like mercury scrubbers, filters, and advanced combustion techniques can help minimize mercury emissions.

Phasing out mercury-containing products by substituting mercury-containing products with the use of safer alternatives is an effective way to reduce mercury pollution by eliminating its use in batteries, switches, thermometers, fluorescent lamps, and certain medical devices.

Controlling mercury in artisanal and small-scale gold mining (ASGM) by promoting the use of mercury-free techniques like gravity concentration and cyanide leaching can help mitigate the environmental impact. Providing training and technical assistance to miners, as well as establishing appropriate regulations and formalization processes, are crucial in reducing mercury use in this sector.

Proper waste management of mercury-containing waste and ensuring proper disposal is vital. This includes establishing collection programs for mercury-containing products at the end of their life cycle and implementing safe recycling and disposal methods.

Regular monitoring and assessment of mercury pollution are essential to evaluate the effectiveness of control measures and identify areas of concern. This involves monitoring air, water, and soil quality for mercury levels and conducting studies on mercury exposure in humans and wildlife. The gathered data can guide decision-making and help prioritize actions for pollution control.

Given the global nature of mercury pollution, international cooperation and regulations on mercury use are crucial. The Minamata Convention on Mercury, adopted on 10 October 2013 and enforced on 16 August 2017, aims to control and reduce mercury emissions and releases globally. The convention has been ratified by over 120 countries. The convention addresses the whole life cycle of mercury, together with its mining, production, use, release, storage, and disposal. It focuses on reducing mercury emissions and releases to the environment, promoting mercury-free alternatives, and protecting vulnerable populations. The convention includes provisions to control the supply and trade of mercury, intending to reduce the global mercury supply and limit its availability for harmful uses. Parties to the convention are required to develop and implement national plans to reduce mercury emissions. The convention aims to phase out or restrict the use of mercury in various products and processes, including batteries, fluorescent lamps, switches, and dental amalgam. The convention emphasizes the proper storage, treatment, and disposal of mercury waste and contaminated sites to prevent further contamination and protect human health and the environment. The convention promotes international cooperation, capacity building, sharing best practices, exchanging information and technology transfer, and finance to support the implementation of mercury reduction measures, mostly in developing nations and countries with economies in transition. Parties are required to submit regular reports on their mercury emissions and reduction efforts. The convention establishes a global monitoring program to assess the efficacy of measures and monitor the levels of mercury in the environment and human populations (Chemicals).

Public awareness, education, and outreach programs are also essential to engage communities, industries, and policymakers in efforts to control mercury pollution. By adopting a multi-faceted approach and implementing these strategies, it is possible to minimize mercury pollution and protect human health and the environment.

4.10 Conclusion

It has been accepted that mercury pollution is not a regional crisis; moreover, it is very difficult to tackle single-handedly. Thus several national governments all over the world have found it advantageous to collectively participate irrespective of national borders to deal with the undesirable impacts of mercury on health and the environment in a particular sub-region or region and also to deal with the transboundary pollution. Living organisms including humans are exposed to mercury from a variety of sources. In air and rain water mercury ranges between 2–10 ng/m^3 and 5–100 ng/m^3 respectively. The WHO recommends a maximum Hg^0 concentration of 0.2 μg/m^3 in ambient air from long-term exposure (Jitaru and Adams 2004). Under natural conditions, Mercury levels in groundwater and surface water are less than 0.5 μg/l. Mineral deposits may increase their level in groundwater. On average, the daily dietary intake of mercury varies from 2 to 20 μg. WHO also recommends the total methylmercury intake within 1.6 μg/kg of body weight/week for a minimal risk (Jitaru and Adams 2004; Organization 2005).

Specifically, Minamata Convention, 2013 is negotiated to tackle mercury use and release. Provisions for controlling mercury pollution are also found in several other international agreements under the auspices of the UN, namely The 1979 Convention on Long-Range Transboundary Air Pollution (LRTAP) and 1998 Aarhus Protocol on Heavy Metals, the OSPAR Convention, 1992 (Convention for the Protection of the Marine Environment of the North-East Atlantic), The 1992 Helsinki Convention (HELCOM or Convention on the Protection of the Marine Environment of the Baltic Sea area), The 1989 Basel Convention on the Control of Transboundary Movements of Hazardous Wastes and their Disposal, The 1998 Rotterdam Convention on the Prior Informed Consent Procedure for Certain Hazardous Chemicals and Pesticides in International Trade, The 2001 Stockholm Convention on Persistent Organic Pollutants, and so forth.

A Global Mercury Partnership was established in 2005 by the UNEP Governing Council Decision 23/9 IV(USEPA). Besides UNEP and the World Bank, the International Agency for Research on Cancer (IARC), International Labor Organization (ILO), International Programme on Chemical Safety (IPCS), Organization of the Economic Cooperation and Development (OECD) organizations also take leading steps in tackling this global menace by adopting and implementing several action plans such as.

- The Arctic Council Action Plan (ACAP); the Arctic Council collaborates with the Norwegian Institute for Water Research (NIVA) evaluates the threat of mercury release in the Arctic environment.

- The Asia–Pacific Mercury Monitoring Network measures the mercury in the rainfall (USEPA).
- The Great Lakes Bi-national Toxics Strategy is an action strategy developed by both Canada and the USA to minimize or exterminate the POPs (Shimizu and Gulezian 1998).
- In 1998, the New England Governors/Eastern Canada Premiers (NEG-ECP) adopted a regional Mercury Action Plan (Smith and Trip 2005).
- The Nordic Environmental Action Plan for reducing mercury release (Environment).
- The North American Region Action Plan on Mercury, a regional strategy agreed upon by Canada, Mexico, and the USA for establishing a Commission on Environmental Cooperation pertaining to the regulation of toxic substances including mercury (Mercury).
- The North Sea Conference provides a political framework for the broad estimation of the measures required to safeguard the North Sea (Commission).
- Governor Blagojevich's Mercury Pollution Reduction Plan to reduce mercury releases and improve public health (Gov 2006).

According to a UNEP study, mercury pollution in India is increasing at an alarming level due to the discharge of industrial effluents, especially chlor-alkali, with mercury ranging between 0.058 and 0.268 mg/l, far more than the prescribed standards. Both groundwater and surface water mercury are detected in Delhi, Mumbai, Vapi, Ankleshwar, Bhopal, Singhrauli, Ganjam, Dhanbad, Howrah, and others. The Central Pollution Control Board (CPCB) and the Indian Institute of Technology (IIT) conduct monitoring programs to assess mercury levels in air, water, and biota. Several laws as given in the table below have been implemented to regulate the use, handling, and disposal of mercury and mercury-containing products (Board 2009). An Indian legislation related to protection against Mercury is described in Table 4.8. India has developed a National Implementation Plan (NIP) as an initial step for the implementation of the Stockholm Convention on POPs. This action plan commits to the establishment of inventories on the manufacture, use, trade, stockpiles, wastes, and places contaminated by chemicals enlisted in the annex of the convention including mercury. The NIP outlines strategies and actions to reduce mercury use and emissions, promote mercury-free alternatives, and enhance monitoring and reporting systems (GEF).

The substitution of mercury with mercury-free alternatives is possibly one of the steps against mercury release into the environment and protecting health. Increased awareness, intensive research, and innovation in the field of healthcare, devices, and industries are required to come up with sustainable solutions for our future. A table stating such alternatives is stated in Table 4.9.

Table 4.8 Indian legislation related to protection against Mercury

Acts	Comments
The Water Act, 1974	Prohibits the discharge of pollutants like mercury
The Water Cess Act, 1977	Specify the maximum amount of water to be used for caustic soda production by the mercury cell process
The Environment Protection Act, 1986	Standards for mercury emission by emitting industries are given
The Hazardous Waste Rules, 1989, its amendments; The Manufacture, Storage, and Import of Hazardous Chemicals Rules, and The E-Waste Management Rules	Mercury is mentioned as waste
The Municipal Solid Waste Rules, 2000, and its amendments	Standards for mercury given for groundwater, compost, and leachates
The Central Insecticides Act, 1968	Ban on the import of many mercury-based agrochemicals
The Workmen's Compensation Act, of 1923 and its amendments	Mercury-related diseases included
The Factories Act, 1948 and its amendment	The permissible limit for mercury exposure is mentioned. The level of mercury in the work zone environment in ambient air at 0.01 mg/Nm3 {8 h, time-weighted average (TWA) concentration} and 0.03 mg/Nm3 as short-term exposure limit of 15 min The National Institute for Occupational Safety and Health Administration (NIOSH) provides a recommended exposure limit for Hg vapor of 0.05 mg/m3 as a TWA for up to 10 h work/day and 40 h/week
The Public Liability Insurance Act, 1991	Provision for mercury-related reimbursement of medical expenses

Table 4.9 Substitutions of mercury in products

Product/application	Mercury use	Substitutes
Thermometers	Mercury-in-glass thermometers	Digital thermometers, infrared thermometers
Blood Pressure Devices	Mercury sphygmomanometers	Aneroid sphygmomanometers, digital blood pressure monitors
Switches and Relays	Mercury switches and relays	Solid-state switches, electronic relays
Dental Amalgams	Dental fillings	Composite resin fillings, glass ionomer fillings
Batteries	Mercury-containing batteries (e.g., button)	Mercury-free batteries, such as alkaline or lithium-based
Lighting	Compact fluorescent lamps (CFLs)	Light-emitting diode (LED) bulbs, energy-efficient CFLs
Laboratory Equipment	Mercury-based laboratory instruments	Digital equipment, non-mercury alternatives
Antiseptics	Merbromin (Mercurochrome)	Non-mercury antiseptics, such as iodine or alcohol-based
Fungicides/Pesticides	Mercury-based fungicides and pesticides	Alternative fungicides/pesticides recommended by experts
Electrical Components	Mercury-containing switches and relays	Solid-state switches, electronic relays

Acknowledgements The author acknowledges the support received from Fr. Dejus J.R. (for flowcharts and illustrations) and Dr. Arijit Basu (for scholarly advice).

References

Al-Tikriti K, Al-Mufti A (1976) An outbreak of organomercury poisoning among Iraqi farmers. Bull World Health Organ 53(Suppl):15

Al-Yaari M, Saleh TA (2022) Mercury removal from water using a novel composite of polyacrylate-modified carbon. ACS Omega 7(17):14820–14831

ATSDR U (1999) Toxicological profile for mercury (update). US Agency for Toxic Substances and Disease Registry, Atlanta, GA

Attar AM, Kharkhaneh A et al (2012) Serum mercury level and multiple sclerosis. Biol Trace Elem Res 146:150–153

Balali-Mood M, Naseri K et al. (2021) Toxic mechanisms of five heavy metals: mercury, lead, chromium, cadmium, and arsenic. Front Pharmacol 227

Berg T, Fjeld E et al (2006) Atmospheric mercury in Norway: contributions from different sources. Sci Total Environ 368(1):3–9

Bessbousse H, Rhlalou T et al (2010) Mercury removal from wastewater using a poly (vinylalcohol)/poly (vinylimidazole) complexing membrane. Chem Eng J 164(1):37–48

Bharti R, Wadhwani KK et al (2010) Dental amalgam: an update. J Conservative Dent JCD 13(4):204

Bhawan P, Nagar EA (2020) Central pollution control board

Board CPC (2009) Mercury-environmental implications and toxicity
Branch UC (2008) The global atmospheric mercury assessment: sources, emissions and transport. UNEP-Chemicals, Geneva
Bravo AG, Cosio C (2020) Biotic formation of methylmercury: a bio–physico–chemical conundrum. Limnol Oceanogr 65(5):1010–1027
Chang LW, Hartmann HA (1972) Ultrastructural studies of the nervous system after mercury intoxication: II. Pathological changes in the nerve fibers. Acta Neuropathol 20:316–334
Chemicals U. Minamata Convention on mercury
Commission O. North sea conferences
Cooke Andrews J (2006) Mercury speciation in the environment using X-ray absorption spectroscopy. Recent Dev Mercury Sci 1–35
Date SS, Parks JM et al (2019) Kinetics of enzymatic mercury methylation at nanomolar concentrations catalyzed by HgcAB. Appl Environ Microbiol 85(13):e00438-e419
Du W, Yin L et al (2014) Catalytic oxidation and adsorption of elemental mercury over $CuCl_2$-impregnated sorbents. Ind Eng Chem Res 53(2):582–591
Dufault R, LeBlanc B et al (2009) Mercury from chlor-alkali plants: measured concentrations in food product sugar. Environ Health 8(1):1–6
Environment R. M. o. t. Norway's Action Plan for Reducing Mercury Releases
Esdaile LJ, Chalker JM (2018).The mercury problem in artisanal and small-scale gold mining. Chem–A Eur J 24(27):6905–6916
Fisher JF, Organization WH (2003) Elemental mercury and inorganic mercury compounds: human health aspects. World Health Organization
Fitzgerald W, Lamborg C (2003) Geochemistry of mercury in the environment. Treatise Geochem 9:612
Fitzgerald WF, Engstrom DR et al (2005) Modern and historic atmospheric mercury fluxes in northern Alaska: global sources and Arctic depletion. Environ Sci Technol 39(2):557–568
GEF. Development of a National Implementation Plan in India as a First Step to Implement the Stockholm Convention on Persistent Organic Pollutants (POPs)
Gonzalez E, Livengood C et al (1999) Elemental mercury removal using a wet scrubber. Argonne National Lab., IL (US)
Gov I (2006) Governor Blagojevich signs new law to reduce toxic mercury emissions
Gray JE, Hines ME (2006) Mercury: distribution, transport, and geochemical and microbial transformations from natural and anthropogenic sources. Appl Geochem 21(11):1819–1820
Gustin MS, Lindberg SE et al (2008) An update on the natural sources and sinks of atmospheric mercury. Appl Geochem 23(3):482–493
Gworek B, Dmuchowski W et al (2020) Mercury in the terrestrial environment: a review. Environ Sci Eur 32(1):1–19
Halbach S, Clarkson T (1978) Enzymatic oxidation of mercury vapor by erythrocytes. Biochimica et Biophysica Acta (BBA)-Enzymol 523(2):522–531
Henneberry YK, Kraus TE et al (2011) Removal of inorganic mercury and methylmercury from surface waters following coagulation of dissolved organic matter with metal-based salts. Sci Total Environ 409(3):631–637
Horowitz HM, Jacob DJ et al (2014) Historical mercury releases from commercial products: global environmental implications. Environ Sci Technol 48(17):10242–10250
Jaishankar M, Tseten T et al (2014) Toxicity, mechanism and health effects of some heavy metals. Interdiscip Toxicol 7(2):60
Jasinski SM (1995) The materials flow of mercury in the United States. Resour Conserv Recycl 15(3–4):145–179
Jitaru P, Adams F (2004) Toxicity, sources and biogeochemical cycle of mercury. Journal de Physique IV (Proceedings), EDP sciences
Kahrizi F, Salimi A et al (2016) Repeated administration of mercury intensifies brain damage in multiple sclerosis through mitochondrial dysfunction. Iran J Pharm Res: IJPR 15(4):834

Ke T, Gonçalves FM et al (2019) Post-translational modifications in MeHg-induced neurotoxicity. Biochimica et Biophysica Acta (BBA)-Mol Basis Dis 1865(8):2068–2081

Keating MH, Mahaffey K et al (1997) Mercury study report to congress, volume 1. Executive summary, Environmental Protection Agency, Research Triangle Park, NC (United States)

Kotwal D, Shewale N et al (2018) Bioremediation of mercury using mercury resistant bacteria. Res J Life Sci Bioinform Pharm Chem Sci 4(2)

Lakshmanan S, Murugesan T (2014) The chlor-alkali process: work in progress. Clean Technol Environ Policy 16:225–234

Lamborg CH, Fitzgerald WF et al (2002) A non-steady-state compartmental model of global-scale mercury biogeochemistry with interhemispheric atmospheric gradients. Geochim Cosmochim Acta 66(7):1105–1118

Langford N, Ferner R (1999) Toxicity of mercury. J Hum Hypertens 13(10):651–656

Liu J, Shi J-Z et al (2008) Mercury in traditional medicines: is cinnabar toxicologically similar to common mercurials? Exp Biol Med 233(7):810–817

Liu Y, Chen H et al (2022) Detection and remediation of mercury contaminated environment by nanotechnology: progress and challenges. Environ Pollut 293:118557

Magos L, Halbach S et al (1978) Role of catalase in the oxidation of mercury vapor. Biochem Pharmacol 27(9):1373–1377

Masur LC (2011) A review of the use of mercury in historic and current ritualistic and spiritual practices. Altern Med Rev 16(4):314–320

Matta G, Gjyli L (2016) Mercury, lead and arsenic: impact on environment and human health. J Chem Pharm Sci 9(2):718–725

McCarthy D, Edwards GC et al (2017) An innovative approach to bioremediation of mercury contaminated soils from industrial mining operations. Chemosphere 184:694–699

Mercury N. A. T. F. o. North American Regional Action Plan on Mercury

Nandiyanto A, Ragadhita R et al (2023) Research trend on the use of mercury in gold mining: Literature review and bibliometric analysis. Moroccan J Chem 11(1):11–11 (2023) 2001–2019

Newman O, Palmer D (1978) Collection of atomic mercury by electrostatic precipitators. Nature 275(5680):526–527

Nriagu JO (1989) A global assessment of natural sources of atmospheric trace metals. Nature 338:47–49

Nriagu J, Becker C (2003) Volcanic emissions of mercury to the atmosphere: global and regional inventories. Sci Total Environ 304(1–3):3–12

Nriagu JO (1979) Biogeochemistry of mercury in the environment. Elsevier/North-Holland Biomedical Press

Ogata M, Aikoh H (1983) Oxidation of metallic mercury by catalase in relation to acatalasemia.[Mice; man]. Ind Health (Kawasaki, Jpn.);(Japan) 21(4)

Organization WH (2005) Mercury in drinking-water

Otani Y, Kanaoka C et al (1988) Removal of mercury vapor from air with sulfur-impregnated adsorbents. Environ Sci Technol 22(6):708–711

Pacyna E, Pacyna J et al (2001) European emissions of atmospheric mercury from anthropogenic sources in 1995. Atmos Environ 35(17):2987–2996

Pacyna EG, Pacyna JM et al (2006) Mercury emissions to the atmosphere from anthropogenic sources in Europe in 2000 and their scenarios until 2020. Sci Total Environ 370(1):147–156

Pamphlett R, Kum Jew S (2018) Inorganic mercury in human astrocytes, oligodendrocytes, corticomotoneurons and the locus ceruleus: implications for multiple sclerosis, neurodegenerative disorders and gliomas. Biometals 31(5):807–819

Rani L, Srivastav AL et al (2021) Bioremediation: an effective approach of mercury removal from the aqueous solutions. Chemosphere 280:130654

Ravichandran M (2004) Interactions between mercury and dissolved organic matter—a review. Chemosphere 55(3):319–331

Rhee S-W (2015) Control of mercury emissions: policies, technologies, and future trends. Energy Emission Control Technol 1–15

Rustagi N, Singh R (2010) Mercury and health care. Indian J Occup Environ Med 14(2):45
Sánchez-Alarcón J, Milić M et al (2021) Genotoxicity of mercury and its derivatives demonstrated in vitro and in vivo in human populations studies. Systematic review. Toxics 9(12):326
Seigneur C, Vijayaraghavan K et al (2004) Global source attribution for mercury deposition in the United States. Environ Sci Technol 38(2):555–569
Selin NE, Selin H (2006) Global politics of mercury pollution: the need for multi-scale governance. Rev Eur Commun Int Environ Law 15(3):258–269
Shetty SK, Lin C-J et al (2008) Model estimate of mercury emission from natural sources in East Asia. Atmos Environ 42(37):8674–8685
Shimizu R, Gulezian G (1998) The great lakes binational toxics strategy: Canada-United states strategy for the virtual elimination of persistent toxic substances. Can Water Resourc J 23(1):77–83
Siblerud R, Mutter J (2020) A hypothesis and additional evidence that mercury may be an etiological factor in multiple sclerosis. J Multiple Sclerosis 7(3):1–7
Siblerud R, Mutter J et al (2019) A hypothesis and evidence that mercury may be an etiological factor in Alzheimer's disease. Int J Environ Res Public Health 16(24):5152
Smith CM, Trip LJ (2005) Mercury policy and science in northeastern North America: the mercury action plan of the New England governors and eastern Canadian premiers. Ecotoxicology 14(1–2):19–35
Spencer A (2000) Dental amalgam and mercury in dentistry. Aust Dent J 45(4):224–234
Stein ED, Cohen Y et al (1996) Environmental distribution and transformation of mercury compounds. Crit Rev Environ Sci Technol 26(1):1–43
Tabatabaei S-A, Ariafar S et al (2022) Toxic environmental factors and multiple sclerosis: a mechanistic view
Ullrich SM, Tanton TW et al (2001) Mercury in the aquatic environment: a review of factors affecting methylation. Crit Rev Environ Sci Technol 31(3):241–293
Urgun-Demirtas M, Benda PL et al (2012) Achieving very low mercury levels in refinery wastewater by membrane filtration. J Hazard Mater 215:98–107
USEPA. International actions for reducing mercury emissions and use
Vasudevan S, Lakshmi J et al (2012) Optimization of electrocoagulation process for the simultaneous removal of mercury, lead, and nickel from contaminated water. Environ Sci Pollut Res 19:2734–2744
Vishal Rathee SSB Development of machine learning-based system for mercury (Hg) detection. Neuro Quantol 20(9):2119–2124
Wagner-Döbler I (2003) Pilot plant for bioremediation of mercury-containing industrial wastewater. Appl Microbiol Biotechnol 62:124–133
Wang L, Wang M et al (2020) Enhanced removal of trace mercury from surface water using a novel Mg_2Al layered double hydroxide supported iron sulfide composite. Chem Eng J 393:124635
Yao T, Duan Y et al (2018) Investigation of mercury adsorption and cyclic mercury retention over $MnOx/\gamma$-Al_2O_3 sorbent. Chemosphere 202:358–365

Chapter 5
Effects of Mercury: Neurological and Cellular Perspective

Khushbu Kumari and Gyanendra Bahadur Chand

5.1 Introduction

Earth's crust harbours a plethora of potentially toxic elements known to be harmful for cellular life. Natural and anthropogenic factors expose these dangerous elements to atmosphere and land–water system. In many cases, it enters the body through respiratory and food chain route causing severe threat to health and life of humans and other organisms. These metallic elements are Hg, As, Pb, Cd, Cr, Fe, Mn, Ni and Cu. With the advancement in modern technologies, the proportion of toxic metals and their compounds in drinking water and food items exceeded the maximum permissible limit. This paper reviews the prehistoric usage of mercury and related toxicity in human beings. The health effect of different important heavy metals is described below.

Chinese were the pioneer users of mercury who mined it in 1100 BC to get red substance (mercury sulphide) called Chinese red, vermilion, or cinnabar. Phoenicians (people around Mediterranean Sea) too used this material in 415 BC. It was named 'Quick Silver' by Aristotle in 400 BC in Greece. The naming of the planets by the Mesopotamians was based on the names of metals, and Mercury is still the name of the planet nearest to the sun. Mercury and its derivatives have been utilised for more than three thousand years as dental amalgam, antiparasitic, antipruritic, diuretic, antiseptic, antisyphilitic and alternative medicines. Mercury was used to treat syphilis from 1550 to 1940 until penicillin was discovered by Fleming (Ozuah 2000). One of the specialised branches of *Ayurveda 'rasashaastra'* is the science of mercury (Savrikar and Ravishankar 2011). Mercury sulphide commonly called as Cinnabar has been extensively used by Chinese and Indian medical practices for

K. Kumari (✉)
P.G. Department of Zoology, Patna University, Patna, Bihar, India
e-mail: drkhushbukumari22@gmail.com

G. B. Chand
Aquatic Toxicology Lab, Department of Zoology, Patna University, Patna, India

N. Kumar (ed.), *Mercury Toxicity*, Environmental Science and Engineering,
https://doi.org/10.1007/978-981-99-7719-2_5

the treatment of pneumonia, syphilis, insomnia, high fever, nervous disorder, tongue paralysis and deafness (Kamath et al. 2012). Since the toxic effect of mercury has been explored and its usage is minimised, still a lot of traditional medicine contains significant amount of mercury. Research carried out in the US about the availability of traditional medicine containing mercury showed that one-fifth of both US- and India-made Ayurvedic medications sold online had mercury, lead and arsenic levels (Saper et al. 2008).

Mercury is the natural element available is Earth's crust, air and water. Natural phenomenon of volcanic eruptions, weathering of rocks and forest fire exposes mercury to the environment and constitutes the geogenic sources of mercury. Burning of coal is the largest contemporary source of anthropogenic mercury released in air and terrestrial surfaces. Of these 71% are straightway released in air, remain there for a week to a year, transported to several kilometres away by wind to finally settle down on land surfaces and/or water (Sprovieri et al. 2010; Streets et al. 2018). ASGM (Artisanal and small-Scale Gold Mining) utilises mercury amalgamation during gold extraction from its ore. This process too releases a significant amount of mercury in environment. Mercury is mixed with the ore of gold and amalgamation is formed. Mercury is evaporated to get the residual gold. This is the simple and cost-effective process to get the precious gold and maintain the livelihood in the country of Uganda in African continent (Keane et al. 2023). Domestic release of mercury is through fluorescent lamp, thermometer breakage, and some types of batteries. Presently 9000 tonnes of mercury are released in air, land and water annually (UNEP 2013a). Mercury is extracted commercially from cinnabar, a mineral which has 86% concentration of mercury.

Mercury is bioaccumulated (higher concentration in aquatic plants and animals as compared to water) and biomagnified in higher food chain (Fig. 5.1). Gastrointestinal absorption of metallic mercury is low, but once it enters the human body it is oxidised to mercuric salts such as mercuric chloride. Mammalian kidney is the main site for mercuric compound decomposition. Mercury poisoning is related to neuro toxin, which results in distortion of gait and motor co-ordination (Clarkson and Magos 2006). Reproductive anomalies in mammals and other vertebrates are also related with mercuric poisoning.

Minamata Convention on Mercury was held in Kumamato, Japan in 2013 to discuss and regulate the use of toxic mercury. Minamata bay in Japan had a tragic fate (neurological disorder) in 1950s when its residents were unintentionally poisoned by eating mercury (methyl mercury) contaminated fish. A factory in Minamata discharged its industrial effluent to sea containing methyl mercury, a byproduct in the manufacture of acetaldehyde. In 1965 another case was detected in Niigata, Japan in Agano River basin. During 1971–72 in Iraq, bread contaminated with methyl mercury was reported and the reason was fungicide used in wheat seed (Millar 2022). The main aim of the Convention is to protect the public health as well as the environment from anthropogenic emissions and releases of mercury and its compounds (UNEP 2013a). Article 7 (A separate article) is dedicated to Artisanal and Small-Scale Gold Mining (ASGM) in Minamata Convention to work on it, as this process is among

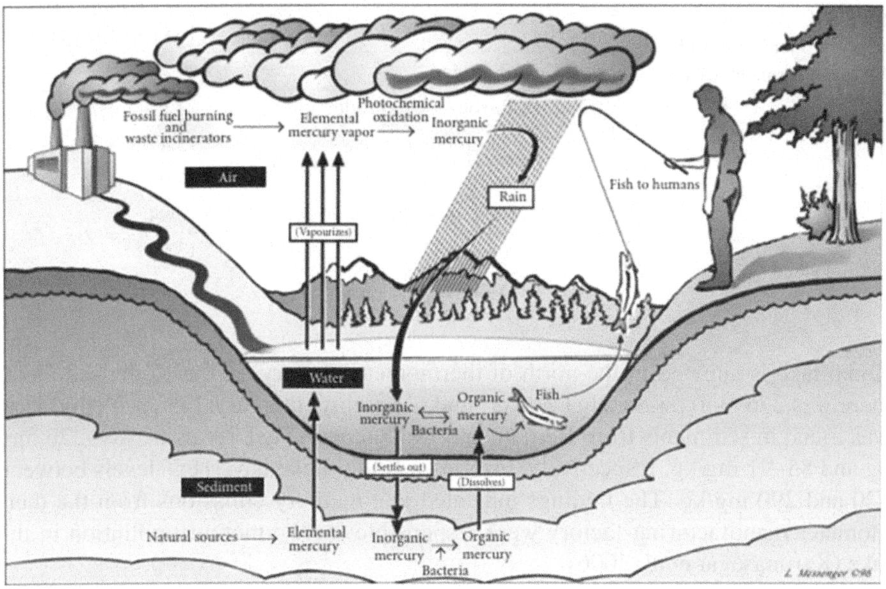

Fig. 5.1 Mercury biogeochemical cycle

the largest emitters of mercury in atmosphere. India ratified Minamata convention in 2018.

One of the main sources of mercury emissions into the environment is combustion of coal. India is the world's second-largest producer and user of coal. In India, coal-fired power plants (CFPPs) are the main source of mercury emissions, accounting for more than half of all atmospheric emissions (Burger Chakraborty et al. 2013). Coal is available in abundance in India and growing energy demands require more power plants to be set up. Minamata Convention Article 8 recognises Coal-fired power plants (CFPP) as one of the greatest sources of mercury emissions. India has currently 2,04,080 MW (2022) of installed coal-fired power plants (Government of India, Ministry of Power). Additional 62,200 MW of power plants (coal-fired) are under construction. The average level of mercury in Indian coal is 0.14 g/t, with a range of 0.003 to 0.34 g/t. With estimated emissions of 144.7 tonnes of Hg per year, India is the second-largest mercury emitter in the world. Other sectors which are mercury emitters can be summarized in Table 5.1 (UNEP 2013b).

South Indian Hill city of Kodaikanal (Tamil Nadu) witnessed a severe mercury poisoning for its worker in a thermometer manufacturing plant and an undisclosed amount of money was paid to them to calm down the dispute. The 591 workers campaigned for 15 long years for compensation from Hindustan Unilever Limited. It brought attention worldwide and finally it was settled in 2016. The compensation was paid to the workers but the damage it has already done to environment will have long-term impacts on the region. The amount of mercury found in soil was 25 mg/kg that is 250 times higher than International permitted limits (Agnihotri 2016).

Table 5.1 Amount of mercury released in India through various processes

Sector	Tonnes of Hg/year
1. Coal burning	89.4
2. Non-ferrous metal production	22.5
3. Cement production	13.4
4. Ferrous metal production	1.9
5. Chlor-alkali production	0.94
6. E-waste	13.6

Source UNEP (2022)

Kodai lake is situated in the north of thermometer factory. In the Kodai sediment, there was 276–350 mg/kg HgT and around 6% methyl mercury. Lesser methylation was found in sediments from Berijam and Kukkal, with HgT levels of 189–226 mg/kg and 85–91 mg/kg, respectively. Fish from Kodai Lake have HgT levels between 120 and 290 mg/kg. The findings indicated that mercury emissions from the thermometer manufacturing factory were responsible for the mercury pollution in the lake (Karunasagar et al. 2006).

In India, alarming rate of mercury is observed in water, industrial effluents, and fishes. India has an industrial effluent discharge range of 0.058 to 0.268 mg/l that contains mercury. This exceeds the mandated Indian and WHO norms of 0.001 mg/l (for drinking water) and 0.01 mg/l by a significant margin (for industrial effluents). This has led to severe contamination of fishes in the respective areas. Fish stock at Kolkata, Mumbai, Karwar (Karnataka) and North Koel River (Jharkhand) have shown high levels of mercury (Table 5.2).

High level of mercury is detected in soil and ground water in water basin of Eastern India, which have chlor-alkali plants. They are the largest consumer of mercury in India. This sector produces chlorine and sodium hydroxide. An estimated approx. 9 tonnes of toxins were released in air during 1997–2000 (Banerjee, n.d.). The rice plant analysed in the region did not show significant concentration of mercury, but the plants like cabbage and Amaranthus showed high mercury level viz. 8.91 mg/kg. Some of the aquatic plants with high bio concentration of mercury included *Marsilea* spp., *Jussiea repens, Spirodela polyrhiza, Paspalum scrobiculatam, Monochoria hastata, Pistia stratiotes, Hygrophila schulli, Eichhornia crassipes* and *Bacopa monniera. Chloris barbata, Cynodon dactylon, Cyperus rotundu,* and *Croton bonplandianum*, among wild terrestrial plants, were seen flourishing on

Table 5.2 Region of India showing water species mercury contamination

Region	Species	Observed total (mg) of Hg/kg dry weight
1. Maharashtra	Fish Crabs	0.03–0.82 1.42–4.94
2. Karwar (Karnataka)	Oyster	0.18–0.54
3. North Koel (Jharkhand)	Fish	300–350

highly polluted soil that contained mercury concentrations as high as 557 mg/kg (Lenka et al. 1992).

Most of the studies related to mercury toxicity in India showed that the mercury levels in blood, urine and breast milk of residents of integrated steel plants or those who lived nearby had rates of up to 30 times higher than those from distant areas (Pervez et al. 2010; Sahu et al. 2014a; Sharma and Pervez 2005). Vulnerable populations are often the marginal sections of the society (women and children) consisting primarily of waste segregators, rag pickers and factory workers' families or living close to the factories (Mondal et al. 2016). As high as 1 ppm of mercury in fairness cream and other cosmetic items have been detected in India (1 ppm = 103 μg/kg; regulatory limit) which is cause of concern for cosmetic industry (Pramanik et al. 2021). Chronic exposure due to mercury contamination has serious negative consequences on health and living organisms as its impact is extremely vulnerable to the entire ecosystem. The usage of fossil fuels must be reduced and one must rely more on renewable sources of energy to minimise the impact of serious impacts of mercury (Das et al. 2023).

High levels of mercury concentration in water and humans have been reported in Singrauli town (Sonbhadra district) of Uttar Pradesh in India. The district lodges thermal power plants and coal mines. Analysis of the drinking water samples showed high mercury concentration within a range of 3 to 26 μg/L (3–26 times the allowable limit) in 20% sample water. The mercury in soil samples had 0.5–10.1 mg/kg concentration. The average concentrations of mercury in human hair, blood and nails were found to be 7.4 mg/kg, 34 μg/L and 0.8 mg/kg respectively. The average mercury concentration in the blood of these persons were 45 and 28 μg/L in the case of men and women respectively which is far greater than the US Environmental Protection Agency's safe standard of 5.8 g/L (USEPA) (Sahu et al. 2014b). Obra and Anpara localities showed higher concentration of arsenic and mercury (Ahamad et al. 2021). In a study conducted at Kanpur the particulate mercury (HgP) observed in PM_{10} was approximately 100 to 4340 pg m^{-3}, with mean concentration 776 ± 846 pg m^{-3} during the sampling period. Concentration of mercury in air was examined by some of the researchers in Kanpur, UP, India. During the winter, significant Mercury Particulate (HgP) concentrations were found, with the maximum concentration observed in March (4340 pg m^{-3} on 4th march) and lowest during summer (100 pg m^{-3} on 14th June, in the year 2007) (Guo et al. 2020). $PM_{2.5}$ and PM_{10} are the major air pollutants in city and along with dust it carries toxic heavy metals too.

In 2020, maximum permitted limits of 1.0 mg/kg for total mercury and 0.25 mg/kg MeHg in milk and honey (and other foods) were published by the Food Safety and Standards Authority of India (FSSAI). For the separation and detection of Hg species in food samples, an Agilent 1260 HPLC coupled to the 7850 ICP-MS offers an efficient and affordable approach (Jain et al. 2021). India is the largest producer of milk (22% of global production) followed by the US, China, Pakistan and Brazil. India ranked 7th in the world for honey production in the year 2019 (FAO, UN). The use of untreated sewage and industrial effluents for irrigational purposes led to danger of toxic metal contamination. High toxicity of agricultural land contaminates the food crops and grasses which is later consumed by the cattle. In this regard, a robust mechanism of mercury detection in food items will pave a way for greater safety in food products.

Idol immersion is an important point source of water body pollution. Idols are constructed from a range of non-biodegradable and biodegradable materials including wood, paper, hay, clay, plaster of Paris, fabric, bamboo, adhesive material, thermocol, paints and others. They are also embellished with various pigments and colours (Joshi et al. 2017). The paints used for colouring idols contain toxic metals like lead and mercury. Few specific studies have been done on water quality after idol immersion with reference to mercury level in the water. River Budhabalanga, Balasore, Odisha, India showed increase in mercury level after 10 days of idol immersion (0.067 mg/L) while the prescribed limit is 0.001 mg/L of mercury as per BIS and ICMR standards (Das et al. 2012). Namami Ganga is the flagship programme of Government of India to clean Ganga and develop riverfronts along its cities. River front has been developed at Patna and several measures have been taken to clean Ganga. Sewage Treatment Plants (STPs) are made to treat the sewage of the city. Manual cleaning of riverbank is done frequently. Measures have been taken to create a separate idol immersion pool in Patna at Barharwa Ghat on river Ganga (Das et al. 2020).

Domestic sewage is an important source of toxic metal contamination of rivers. Sludge generated from STP (sewage treatment plants) contains the toxic material. Sewage sludge contains high amount of organic matter along with toxic heavy elements. Since sewage sludge is an important source of mercury and other toxic materials, its handling requires utmost care. Usage of contaminated sludge as manure is equally dangerous because it will again enter the food chain. Water discharged from sewage treatment plants need to be monitored for presence of heavy metals before using it for irrigation purpose. Flue gases escaping from incinerators also contain toxic metals harmful for human health.

We observe that the toxic element once released in ecosystem continues to cycle from land–water-air to living cells damaging one form or another. So, the ultimate solution is to minimise its occurrence at the point source and rely more on renewable sources of energy rather than fossil fuels.

5.2 Impact of Mercury on Floral Community

The Minamata Convention on Mercury (2013) aims to reduce anthropogenic mercury (Hg) risks to humans and the environment worldwide (Outridge et al. 2018). Researchers found that Hg has potential toxic impacts on plants even at very low concentration (Li et al. 2017). Numerous studies revealed that the vegetations cultivated near mercury sources may contain high concentration of Hg. Coal-fired power plants account for releasing maximum mercury to the environment. In a study, some samples of vegetables and grains collected within 10 km distance from power plants had exceeded the allowable upper limit of Hg content when compared with samples purchased from grocery stores away from power plants. In addition, it was also observed that if the fly ashes were rinsed off from leaf surfaces of plants, the mercury content was found to be decreased significantly (Li et al. 2017).

In the ecosystem, the atmospheric mercury uptake by vegetation accounts for 60–90%. Among vascular plants, the highest mercury concentration is found in the leaves in comparison to any other plant tissues. In woody biomass, it has been observed that the atmospheric Hg was first taken up by leaves, which is then translocated within plants, contributing unquantified source of Hg deposition. Hg taken up by roots accounts for minor transfer of mercury content in tissues above ground representing recycling of Hg within soil, whereas the Hg transportation from foliage to roots increases environmental Hg uptake and its deposition. The Hg deposition by non-vascular plants such as mosses and lichens exceed that of vascular plants and hence usually considered when quantifying environmental mercury deposition (Zhou et al. 2021). Due to presence of barrier to Hg transportation from roots of plants to foliage part, the Hg uptake from soil accounts for less absorption through roots. Thus, only modest increase of Hg from plant root is accountable even in areas of high Hg soil concentration (Patra and Sharma 2000).

Hg is considered as a toxic element as it has no beneficial impacts on animals and plants. So, it is regarded as ‘main threat’ and acts as a pollutant of the air, water and soil (Asati et al. 2016). It accounts for growth retardation in plants, decreases photosynthetic pigment and total biomass (Jameer Ahammad et al. 2018) and put many negative impacts on them (Natasha et al. 2020).

A study of mercury treatment among four species of plants producing non-edible oil demonstrated a considerable reduction in the leaf tissues leading to decline of the leaf thickness in majority of species. The anatomical structure of plant leaves had also shown the sign of Hg stress as revealed by decrease in leaf thickness, lower epidermis, upper epidermis, spongy tissues and palisade tissues. At the same time, decrease in chlorophyll content were observed in all four species with remarkable increase of MDA content of leaves (Hamim et al. 2019).

Mercury contamination in the soil has several negative impacts on plants viz. reduction in growth and metabolism of plants (Asati et al. 2016; Patra and Sharma 2000), transpiration rate, photosynthesis, chlorophyll synthesis (Marrugo-Negrete et al. 2016; Teixeira et al. 2018a, b) and water uptake. Besides, increased lipid peroxidation was also reported due to increased contamination of mercury in the soil (Cho and Park 2000). The impact of a number of stress-indicating antioxidant enzymes such as peroxidase (POD), superoxide dismutase (SOD) and ascorbate peroxides (APX) has been shown to be curtailed at higher mercury concentration (Zhou et al. 2021; Manikandan et al. 2015; Magarelli and Fostier 2005; Israr et al. 2006).

Hg-exposed plants exhibited significant reduction of dry and fresh biomass. The decrease in biomass, morphological symptoms and shoot growth rate were observed as the result of mercury’s toxic impact on uptake of minerals and plant metabolism (Jameer Ahammad et al. 2018). According to a number of studies, heavy metal serves as stressor in plants, causing physiological constraints that lowers plant vigour and restrict plant growth (Schützendübel et al. 2001).

Heavy metals generally affect the living organism at cellular level by damaging and blocking important biomolecules such as enzymes and polynucleotides, denaturing or inactivating proteins, by displacement of metal ions from molecules (e.g., Mg from chlorophyll) ultimately disrupting the cell membrane or organelles (Patra et al. 2004). Hg concentration induces phytotoxicity by changing the permeability of cell membrane, affinity to react with phosphate group, having high affinity to react with sulfhydryl (SH) group, and by disturbing functions of critical and non-protected proteins (Patra et al. 2004; Patra and Sharma 2000).

Mercury concentration induces injury in seeds as well as reduces its viability. Hg disturbs the sulfhydryl group, disrupting its stability and affecting germination of seeds and development of embryo (specially –SH ligands rich tissues). In Zea mays, mercuric chloride has also been seen to inhibit the seedlings' gravimetric response and diminish primary root elongation (Patra et al. 2004).

Microalgae are primary producers and may serve as a prominent entry site for the metal into the marine food web. They are among the phytoplankton that serve as foundation of the food web in marine environments (Faucheur et al. 2014). These microorganisms can be used as bioindicators of pollution due to their vulnerability, as they can bioaccumulate trace metals and phytoremediation strategies can be developed to reduce contamination in aquatic habitats (Renuka et al. 2015). Mercury shows great affinity for cysteine residues in proteins and binds via the thiol functional group. Upon mercury exposure, essential processes in plant cells and algae are disrupted due to inhibition or activation of various enzymes (Nesci et al. 2016).

The deleterious impacts of genetic materials have been observed due to mercury toxicity. Numerous DNA reactive sites are exposed due to nature of easily deformable outer electron shell of mercury. Patra et al. (2004) concluded effect of mercury depends on the time of exposure and concentration in plants leading to marked effects in S-phase that induces damages resulting in severe clastogenicity.

5.3 Impact of Mercury in Faunal Community

Mercury can bioaccumulate in animal tissues. The organic form of mercury viz. MeHg (Methylmercury) exists more in living organisms than any other form because it has tendency to get absorbed quickly while getting excreted at slower rate. Animals at the top of the food chain may have larger concentrations of mercury than those at the bottom, indicating that its concentrations increase as one moves up the food chain. This is also known as trophic bio-magnification (Chen et al. 2018; Rice et al. 2014; Ruus et al. 2017).

MeHg is regarded as bioindicator of mercury levels due to the direct relationship between aquatic biota and the environment (Condini et al. 2017). Methylmercury (MeHg) is thought to be the most hazardous form and is subjected to extensive research (Rice et al. 2014; Ruus et al. 2017). Researchers stated that MeHg can cross the blood–brain barrier and accumulate in the CNS (central nervous system) of humans and fish (Aschner et al. 2000), which damages the brain neurologically

(Berntssen et al. 2003). Additionally, MeHg is known to have harmful impact on the kidneys, CNS, cardiovascular, immunological and gastrointestinal systems (Holmes et al. 2009). MeHg is regarded as a highly neurotoxic substance that predominantly affects glial cells and causes oxidative stress and neuroinflammation. Additionally, this seems to have negative impact on the reproduction, kgenome and different physiological systems in both animals and humans (Rice et al. 2014).

The toxic impacts of mercury also affect the microorganisms present in the Hg-concentrated soil (Ha et al. 2017; Ruus et al. 2017). In a research, it was found that among soil microorganisms, bacterial communities are more affected by Hg toxicity, when compared to fungal communities. Even with the least bioavailable Hg in the soil, the abundance of the mercuric reductase gene revealed a decreased ability of bacteria to detoxify mercury. In soils having high concentration of mercury, researchers discovered a wide variety of Hg-responsive species, however they were not restricted to any one soil type or taxonomic group (Frossard et al. 2017).

Small organisms can be found at the base of the food chain, which extends to larger predatory fish like sharks, tilefish, king mackerel and swordfish as well as marine mammals (Ullrich et al. 2001). Methylmercury concentrations in large predatory fish and marine mammals can be up to 100,000 times greater than those in the surrounding aquatic environment. People who frequently consume seafood are therefore more likely to be exposed to concentrations of mercury, which has been correlated to damage to the neurological, immunological and cardiovascular systems as well as the brain, lungs and kidneys (Dabeka et al. 2004).

The rise in the price of gold on the global market as well as other socioeconomic factors are driving the expansion of artisanal small-scale gold mining (ASGM), which is responsible for significant mercury (Hg) emissions into the air and rivers in the Global South. Hg has the potential to endanger both animal and human populations by hastening the degradation of neotropical freshwater environments. Researchers investigated the possible reason of Hg contamination in fishes living in regions with high biodiversity values and growing ASGM-dependent human populations living in Peru's Madre de Dios. ASGM activity was hypothesised to be the major reason for Hg contamination upon investigation of local water quality, environmental Hg exposure and by studying trophic levels of fishes. It was observed that the level of mercury contamination is positively related with ASGM activity. The amount of mercury accumulation was found more in higher trophic level with less amount of dissolved oxygen (DO). Additionally, it was reported that top carnivorous animals and people of neotropical region depending on freshwater ecosystem are facing ASGM-dependent mercury toxicity (Barocas et al. 2023).

5.4 Impact of Mercury Toxicity in Humans

Humans have utilised animals and their products as a source of nutrition for ages. Fishes have been universally used as rich source of protein with low fat content along with minerals and vitamins. Cattles are also good source of nutrition providing dairy products and meat as well. Though some aquatic or terresterial animals may get exposed to mercury and serves as source of mercury for human since the metal has tendency to bioaccumulate in the living tissues. Accumulation of mercury doesn't have negative impact only on environment rather it harms animals and populations consuming these animals and their products. Mercury can't be removed by cooking process so it is a matter of great concern that meat and dairy products might bioaccumulate dangerous heavy metals like mercury as these products have significant role in human nutrition. Consumption of these sources of mercury viz. contaminated milk and meat enters humans leading to many health consequences in pregnant women, developing embryos and infants. Mercury induces malformation in embryo so it is also called embryocidal and teratogenic agent. Additionally, due to its mutagenic and carcinogenic nature people exposed to it may develop risk of cancer, deafness and blindness and permanent neurological anomalies (Noto 2021).

Mercury has ability to induce pathophysiological changes in the hypothalamus–pituitary–adrenal and gonadal axis that might have an impact on reproductive function by ways of changing hormonal levels of androgens, progesterone, inhibin, oestrogen, follicle-stimulating hormone (FSH) and luteinizing hormone (LH) in the blood (Davis et al. 2001; Schrag and Dixon 1985). Both men and women experience infertility as a result of elevated mercury levels (Dickman et al. 1998). Mercury adversely effects on testicular weight, epididymal sperm count and spermatogenesis in males (Boujbiha et al. 2009). Additionally, evidence suggests that mercury has potential to induce erectile dysfunction (Schrag and Dixon 1985).

Studies have proven that mercury prevents the anterior pituitary in females from releasing FSH and LH, which can influence the level of oestrogen and progesterone causing painful or irregular menstruation, ovarian dysfunction, tilted uterus and premature menopause (Chen et al. 2006). Mercury is strongly associated with menstruation disorders including short, lengthy or irregular cycles, abnormal bleeding and painful periods (Davis et al. 2001).

Mercury is linked with foetal toxicity, which can be observed in spontaneous abortions, miscarriages, low birth weights and stillbirths in addition to severe reproductive problems. Mercury is known to pass through the placenta, where it may interfere with embryonic brain growth, leading to psychomotor retardation and cerebral palsy in later stages of development (Castoldi et al. 2001; Myers and Davidson 1998). Mercury exposure during pregnancy has been related with numerous birth abnormalities in neonates including defects in neural tube, abnormal craniofacial feature, delayed growth and others (Yoshida 2002).

Humans have also been subjected to embryo-pathogenic effects due to methylmercury. MeHg is a substance that quickly crosses the placenta and harms the developing foetus's brain. Foetal autopsies revealed a broad hypoplasia of the cerebellum, a reduction in the amount of nerve cells in the cerebral cortex, a significant loss in total weight of brain, improper migration of nerve cells and an abnormal arrangement of the brain's centres and layers (Choi et al. 1978; Mottet et al. 1985).

Peripheral motor neurons, autonomic ganglia, ganglia, brain and spinal cord are thought to bind mercury strongly immediately after its entrance to the body tissues.

Although the neurological system is the main organ where mercury is stored, symptoms could still develop in a variety of other organ systems due to mercury's temporary and persistent systemic dispersion. Furthermore, researchers suggest that a person's genetic heritage may differ in mercury toxico-kinetics (Gundacker et al. 2010). It is universally acknowledged that occupational chronic exposure to high mercury vapour concentrations can lead to high mercury level accumulation in the brain (Falnoga et al. 2000) and other tissues as the inorganic mercury deposited in brain following exposure of mercury vapour has long half-life. In addition to the brain (Guzzi et al. 2006), metallic mercury gets accumulated in the muscles (Björkman et al. 2007), adrenals (Berlin and Ullberg 1963), liver (Danscher et al. 1990), kidneys (Guzzi et al. 2006; Kumari 2021), thyroid (Björkman et al. 2007), myocardium (Frustaci et al. 1999), skin (von Burg 1995), breast (Crespo-López et al. 2009), sweat gland, pancreas, enterocytes, lungs, salivary glands, testes and prostate (Berlin and Ullberg 1963) and may be linked with dysfunction of these organs. Additionally, mercury shows strong affinity towards sulfhydryl groups that affect T cell activity and binding sites present on the surface of T cells (Hultman and Pollard 2022). Breast milk contains mercury, which rapidly accumulates in the placenta and embryonic organs (Vimy et al. 1990).

Cardiomyopathy has also been associated with abnormal deposition of mercury in the heart. Researchers showed that the average level of mercury deposited in the cardiac tissue of people who died of idiopathic dilated cardiomyopathy was 22,000 times more in their cardiac tissues than those people who died of other types of heart diseases (Haffner et al. 1991). Angina or chest pain can also be brought on by mercury poisoning, especially in people under the age of 45 (Frustaci et al. 1999). The cardio-protective effect of paraoxonase 1 may be inhibited by MeHg, according to in vitro investigations (Drescher et al. 2014). There is strong evidence connecting mercury to haemolytic and aplastic anaemia, since mercury and iron may compete with each other for binding with haemoglobin, which interrupts synthesis of haemoglobin. Additional evidences have revealed that mercury may contribute for the development of leukaemia, Hodgkin's disease, mononucleosis and other diseases in addition to anaemia (Pyszel et al. 2005).

The pituitary, thyroid, adrenal glands and pancreas may be disrupted by low exposure levels of mercury, which can have an impact on the endocrine systems of both people and animals (Minoia et al. 2009). Mercury is assumed to affect endocrine function by inhibiting one or more essential steps or enzymes in biosynthesis of hormones, as is the case with synthesis of adrenal steroid and suppression of 21-hydroxylase. Mercury is also thought to impair hormone-receptor binding (Iavicoli et al. 2009). By inactivating *S*-adenosyl-methionine, mercury can also prevent the breakdown of catecholamines, leading to a build-up of adrenaline, ptyalism (hyper salivation), hyperhidrosis, hypertension and tachycardia (Clifton 2007).

The diagnosis of autism is characterised by difficulties in social interaction, language and communication, a need for consistency and regularity, sensory anomalies and aberrant movements (Solt and Bornstein 2010). Some researchers have suggested that cases of autism may be associated with mercury poisoning because mercury can produce immunological, neurological, sensory, motor and behavioural dysfunctions, that are comparable with features characterising or linked with autism (Bhardwaj et al. 2009). MeHg in the diet has a significant effect on the overall mercury level in the brain in a population that consumes fish. Therefore, distinction between mercury species is required to assess how exposure to dental amalgam and fish consumption impacts the level of mercury concentrations in the brain (Björkman et al. 2007).

The thyroid gland regulates the metabolism, protein synthesis and hormone sensitivity. The thyroid exhibits a tendency to accumulate mercury, similar to the pituitary. By occupying iodine-binding sites, mercury inhibits or modifies the function of hormones, impairing body temperature regulation and causing depression, hypothyroidism and thyroid inflammation (Wada et al. 2009).

The negative consequences of mercury are extremely harmful for the pancreas. Mercury may combine with the three sulphur-binding sites of insulin thereby interfering with normal biological processes and disrupting the regulation of blood glucose levels (Chen et al. 2006). Teenagers and other vulnerable groups seem to be greatly influenced by low levels of pituitary function induced by mercury toxicity which are ultimately linked to depression and suicidal thoughts in them. Mercury poses direct impact on neurohypophysis leading to frequent urination and elevated blood (McGregor and Mason 1991).

5.5 Mercury-Induced Cytotoxicity

Because of its affinity for sulfhydryl and thiol groups, mercury exposure causes changes in macromolecular structure at the cellular level in addition to causing DNA damage (Wang and Jia 2005). Additionally, it has been demonstrated that mercury causes mitochondrial dysfunction and oxidative stress (Lund et al. 1993), which might enhance lipid peroxidation and affect calcium homeostasis (Peraza et al. 1998). Mercury has the ability to act as a catalyst for Fenton-type reactions, which is known to raise levels of reactive oxygen species (ROS) (Peraza et al. 1998). Four basic processes that can be attributed to mercury-mediated genotoxicity are—oxidative stress and free radical generation, effect on DNA repair pathways, impact on microtubules and its direct association with DNA molecules (Crespo-López et al. 2009).

Although the exact process that leads to put harmful impact of mercury on the cardiovascular system is not known yet, it is thought to induce an increase in oxidative stress. Free radical generation is increased by mercury exposure, possibly as a result of mercury's involvement in the Fenton reaction along with declining activity of antioxidant enzymes like glutathione peroxidase (Genchi et al. 2017). Reactive oxygen species (ROS) are produced as a result of mercury toxicity and these ROS are capable of destroying crucial metabolic pathways, which ultimately results in cell death (Barón-Sola et al. 2021; Ajitha et al. 2021). MeHg has been shown to stimulate reactive oxygen species (ROS), including hydroxyl radical, superoxide radical, oxygen singlet, hydrogen peroxide, peroxyl radical, nitric oxide and alkyl radical, which can damage the cells. The doses necessary to cause these responses vary according to the concentration of MeHg available in the environment (Ercal et al. 2001).

The production of ROS is a natural phenomenon against any stress. The preservation of a healthy thiol state is primarily indicated by the GSH/GSSG ratio, which is essential for shielding cells from oxidative damage. The key enzymes involved in reducing glutathione disulphide (oxidised glutathione; GSSG) and detoxifying peroxides are glutathione peroxidase (GPx) and glutathione reductase (GR), respectively (Dringen 2000). Even in the absence of considerable changes in GSH levels or the GSH/GSSG ratio, MeHg can result in oxidative stress due to its direct interaction with nucleophilic protein groups (Farina et al. 2010). Another notable in vivo investigation revealed that prenatal MeHg exposure prevented the antioxidant GSH system from developing normally in the first few weeks following birth (Stringari et al. 2008).

The negative charge produced by the respiratory chain of the mitochondrial matrix (Go and Jones 2013) allows the inorganic Hg^{2+} to penetrate mitochondria and disorganise oxidative phosphorylation (OXPHOS) (Dröse et al. 2014). Mercury gets localised in protein content of mitochondria (Mieiro et al. 2015), because of its higher affinity to -SH group of cysteine residue of protein (Mailloux et al. 2015; van Iwaarden et al. 1992), leading to alteration in protein structure.

Mercury was found to be intracellularly linked to membrane organelles viz. mitochondria, Golgi complex, nuclear envelopes, endoplasmic reticulum and lysosomes, according to a biochemical and electron-microscopic histochemical study with trace levels of mercury within the nucleus (Chang 1977). Mitochondrial malfunction coupled with decreased ATP was reported. It was capable to induce oxidative stress with rise in lipid peroxidation and reduction in glutathione. Thrombosis, dyslipidemia, loss of endothelial function, mitochondrial anomalies, dysfunction of vascular smooth muscle were observed as negative indication on vascular system against toxic impacts of mercury. Mercury has potency to generate negative impact on omega-3 fatty acid and immune response in fishes. Mercury-induced therapeutic consequences include considerable atherosclerosis, proteinuria condition having renal anomalies, CVA, MI, CHD and hypertension. Studies provide strong and logical correlation between biochemical, pathological and functional medicines (Houston 2005).

The cytotoxic impact of mercury can be best studied using histo-pathological biomarkers, since these changes provide a quick way to identify the influence of irritants in the various cells, tissues and organs (Bernet et al. 1999). In vivo study of the toxic impacts of mercuric chloride on kidney and liver of a fresh water air-breathing fish *Clarias batrachus* have shown prominent anomalies in the histological architecture of liver and kidney in piscine model (Kumari 2021). Various histopathological anomalies encountered in renal corpuscles were shrinkage and constriction of glomerular tufts, appearance of hyalinised, necrotic and avascular glomeruli, and their fusion, inflammation of podocytes and widening of urinary space as well. Besides, increased incidence of necrotic tubule with hyalinized and degenerated renal epithelial cells, infiltration of oedematous fluid, lymphocytes, plasma cells and debris of cytoplasmic organelles in the lumen of proximal and distal convoluted tubule, collecting tubule, massive fibrosis of arcuate artery and arcuate vein, congestion of inter-tubular space with fibrous tissues, neoplastic plasma cells, other inflammatory cells, etc. (hydronephrotic kidney) were reported. The electron micrographs of renal tissues of mercury-treated fish showed discontinuity in nuclear membrane, thick deposition of heterochromatin at periphery of inner lamella, increased incidence of apoptotic vesicle, reduction in the number of cytoplasmic organelles, congregation and fusion of degenerating mitochondria with dissolved cristae, loosened and inflamed cisternae of RER, clumping of lumen of DCT, PCT and more specifically CT with bulk of inflammatory cells, pus, fibrous clots, fusiform vesicles, appearance of autophagic and apoptotic vesicles, etc. (Kumari 2021).

Light photomicrographs of treated liver showed constriction in central vein due to infiltration of eosinophilic inclusions, plasma cells, hemorrhagic clots and edematous fluid and deposition of fibrous tissues around it. Besides appearance of swollen and hydropic hepatocytes with bridging necrosis, congested sinusoids filled with bile, perivenular fibrosis and plugging of portal vein by hemorrhagic clots and eosinophilic inclusions, hepatocellular necrosis, neoplasia, hepatoblastoma and chronic venous congestion were prominently marked (Kumari 2021). The worldwide pollutant mercury has a significant negative impact on many facets of human health. It has hazardous impact on many physiological systems in humans such as urinary system, immunological system, cardiovascular system, skin, endocrine system and respiratory system (Chen anf Driscoll 2018).

5.6 Mercury-Induced Neurotoxicity

It was revealed that mercury rapidly penetrates and damages the blood–brain barrier, causing the system to become dysfunctional. It was discovered that the granule cells in the cerebellum and sensory neurons in the spinal ganglia were especially susceptible to mercury poisoning in experimental mice. According to ultrastructural investigations, coagulative type of degeneration was primarily seen in organic mercury poisoning while vacuole type of degeneration of the neurons was primarily linked to inorganic mercury intoxication (Chang 1977).

Methylmercury (MeHg) is deadly poisonous substance to CNS (central nervous system). Prenatal phase of life is assumed to be most vulnerable time (Davidson et al. 2000). MeHg prevents essential brain growth processes such as neuronal cell division and its migration (Sager et al. 1984). The earliest examples of prenatal poisoning were documented during the methylmercury outbreaks in Japan, describing abnormal features in the brain of foetus. The cytoarchitecture of the human prenatal brain was disrupted, and it appeared to be more vulnerable to methylmercury than the adult brain. Blindness, ataxia and mental retardation were all signs of prenatal exposure, while moderate exposure was linked to motor skill impairment, high exposure caused severe mental retardation and mortality (Davidson et al. 2000; Myers et al. 2003).

Biochemical and cytochemical tests revealed that mercury-intoxicated mice had much lower neuronal RNA and protein production. Reduction in protein production was thought to cause cell death in these neurons. In chronic mercuric dichloride intoxication, RNA levels of neurons were restored. This study may indicate that these animals are more resistant to the damaging effects of mercury. Mercury-poisoned animals also had enzymatic system disruptions in the glycolytic pathway. Mercury poisoning was discovered to be especially dangerous to granule cells in the cerebellum and sensory neurons of spinal ganglia. Ultrastructural studies revealed that inorganic mercury poisoning was related with vacuolar neurodegeneration, whereas organic mercury poisoning was associated with coagulative neurodegeneration. Nerve fibre degeneration was also found (Chang 1977).

In an experiment, prolonged low doses of inorganic mercury exposure in adult rats resulted in deposition of mercury in the brain parenchyma coupled with oxidative stress and damage of cell via cytotoxicity leading to cell death, which impairs motor function (Fig. 5.2). The motor cortex exhibited cytotoxicity and apoptosis as a result of chronic $HgCl_2$ exposure, demonstrating that inorganic mercury has lethal impact on central nervous system. Although other brain regions, such as basal ganglia and cerebellum, may also be involved in the detrimental impact of exposure, the motor cortex may be particularly important in the motor deficits brought on by inorganic mercury. On inorganic mercury exposure, astrocyte and neuronal cellular densities in the motor cortex were decreased. Additionally, research findings showed that abnormalities in behaviour were connected to both cellular and molecular damage (Teixeira et al. 2018a, b). Another toxicological investigation had already shown that the decline in both astrocyte and neuronal numbers is typically linked to the functional damage observed in behavioural testing (Flores-Montoya et al. 2015; Maodaa et al. 2016).

Rice et al. (2014a), have considered MeHg as a highly neurotoxic chemical that mainly damages glial cells, induces oxidative stress and triggers neuroinflammation. Gender is a crucial aspect to consider when investigating MeHg-exposed animals and analysing motor defects. An unexpected observation from mice exposed to methyl mercury-containing drinking water as revealed that female mice were found to be more resistant to MeHg-induced motor disability than male mice. In line with this finding, two weeks of ad libitum exposure to MeHg in drinking water resulted in a severe motor dysfunction in male adult Swiss mice but not in females (Malagutti et al. 2009). Malagutti et al. (2009), also added that sex steroids have neuroprotective properties because 17-estradiol co-administration prevented male mice from the neurotoxic effects induced by oral exposure of methyl mercury. Large myelinated fibres were found to me more vulnerable in comparison to small neuronal cells (Chang 1977).

A study was done on 80 community members with median age of 57 years native to Grassy Narrows First Nation, Ontario, Canada, in which 55% of members were women. The water of their territorial region had mercury contamination by means of industrial discharge. These people experienced long-term exposure to methyl mercury due to consumption of contaminated fishes. In this study, the mercury-exposed visual characteristics were documented between 1970 to 1997 in which all

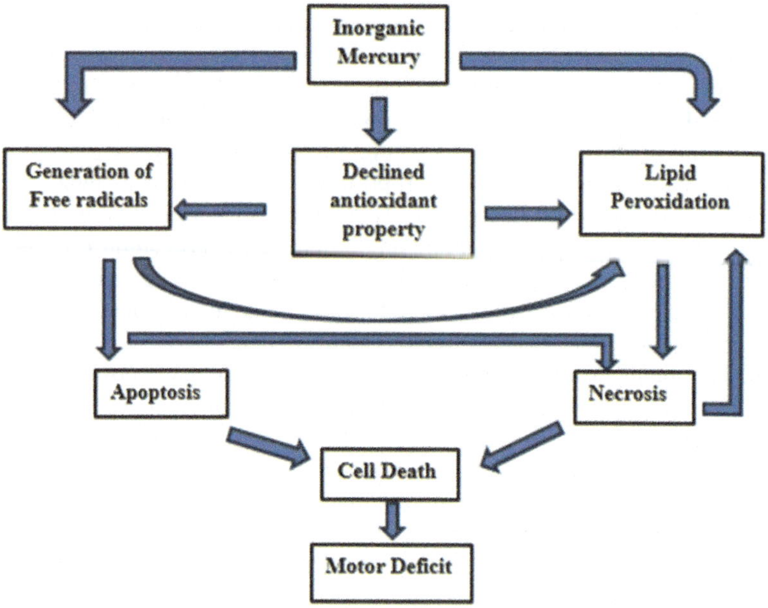

Fig. 5.2 Model of oxidative stress brought on by inorganic mercury deposits

the community members had their eyes and vision tested for optical coherence tomography (OCT), automated visual field, sensitivity to colour and contrast and visual acuity. As a result, it was observed that community members with chronic exposure to mercury had lost contrast sensitivity, impaired colour vision and significant loss of visual fields (Tousignant et al. 2023).

References

Agnihotri S (2016) Kodaikanal mercury contamination: why unilever is paying settlement to its 591 workers. India Today

Ahamad A, Janardhana Raju N, Madhav S, Gossel W, Ram P, Wycisk P (2021) Potentially toxic elements in soil and road dust around Sonbhadra industrial region, Uttar Pradesh, India: Source apportionment and health risk assessment. Environ Res 202:111685. https://doi.org/10.1016/j.envres.2021.111685

Ajitha V, Sreevidya CP, Sarasan M, Park JC, Mohandas A, Singh ISB, Puthumana J, Lee JS (2021) Effects of zinc and mercury on ROS-mediated oxidative stress-induced physiological impairments and antioxidant responses in the microalga *Chlorella vulgaris*. Environ Sci Pollut Res 28:32475–32492

Asati A, Pichhode M, Nikhil K (2016) Effect of heavy metals on plants. Int J Appl Innov Eng Manage 5(03):56–66. https://doi.org/10.13140/RG.2.2.27583.87204

Aschner M, Yao CP, Allen JW, Tan KH (2000) Methylmercury alters glutamate transport in astrocytes. Neurochem Int 37(2):199–206. https://doi.org/10.1016/S0197-0186(00)00023-1

Banerjee S (n.d.) Mercury pollution in India

Barocas A, Vega C, Alarcon Pardo A, Araujo Flores JM, Fernandez L, Groenendijk J, Pisconte J, Macdonald DW, Swaisgood RR (2023) Local intensity of artisanal gold mining drives mercury accumulation in neotropical oxbow lake fishes. Sci Total Environ 886:164024. https://doi.org/10.1016/j.scitotenv.2023.164024

Barón-Sola Á, Toledo-Basantes M, Arana-Gandía M, Martínez F, Ortega-Villasante C, Dučić T, Yousef I, Hernández LE (2021) Synchrotron radiation-fourier transformed infrared microspectroscopy (μSR-FTIR) reveals multiple metabolism alterations in microalgae induced by cadmium and mercury. J Hazard Mater 419:126502. https://doi.org/10.1016/j.jhazmat.2021.126502

Berlin M, Ullberg S (1963) Accumulation and retention of mercury in the mouse. Arch Environ Health Int J 6(5):610–616. https://doi.org/10.1080/00039896.1963.10663449

Bernet DB, Schmidt H, Meier W, Burkhardt-Holm P, Wahli T (1999) Histopathology in fish: proposal for a protocol to assess aquatic pollution. J Fish Dis 22:25–34

Berntssen MHG, Aatland A, Handy RD (2003) Chronic dietary mercury exposure causes oxidative stress, brain lesions, and altered behaviour in Atlantic salmon (Salmo salar) parr. Aquatic Toxicol (amsterdam, Netherlands) 65(1):55–72. https://doi.org/10.1016/s0166-445x(03)00104-8

Bhardwaj A, Kar JP, Thakur OP, Srivastava P, Sehgal HK (2009) Electrical characteristics of PbSe nanoparticle/Si heterojunctions. J Nanosci Nanotechnol 9(10):5953–5957. https://doi.org/10.1166/jnn.2009.1254

Björkman L, Lundekvam BF, Lægreid T, Bertelsen BI, Morild I, Lilleng P, Lind B, Palm B, Vahter M (2007) Mercury in human brain, blood, muscle and toenails in relation to exposure: an autopsy study. Environ Health 6(1):30. https://doi.org/10.1186/1476-069X-6-30

Boujbiha MA, Hamden K, Guermazi F, Bouslama A, Omezzine A, Kammoun A, El Feki A (2009) Testicular toxicity in mercuric chloride treated rats: association with oxidative stress. Reprod Toxicol (Elmsford, N.Y.) 28(1):81–89. https://doi.org/10.1016/j.reprotox.2009.03.011

Burger Chakraborty L, Qureshi A, Vadenbo C, Hellweg S (2013) Anthropogenic mercury flows in India and impacts of emission controls. Environ Sci Technol 47(15):8105–8113

Castoldi AF, Coccini T, Ceccatelli S, Manzo L (2001) Neurotoxicity and molecular effects of methylmercury. Brain Res Bull 55(2):197–203. https://doi.org/10.1016/s0361-9230(01)00458-0

Chang LW (1977) Neurotoxic effects of mercury—a review. Environ Res 14(3):329–373. https://doi.org/10.1016/0013-9351(77)90044-5

Chen YW, Huang CF, Tsai KS, Yang RS, Yen CC, Yang CY, Lin-Shiau SY, Liu SH (2006) Methylmercury induces pancreatic beta-cell apoptosis and dysfunction. Chem Res Toxicol 19(8):1080–1085. https://doi.org/10.1021/tx0600705

Chen CY, Driscoll CT (2018) Integrating mercury research and policy in a changing world. Ambio 47(2):111–115. https://doi.org/10.1007/s13280-017-1010-y

Chen MM, Lopez L, Bhavsar SP, Sharma S (2018) What's hot about mercury? Examining the influence of climate on mercury levels in Ontario top predator fishes. Environ Res 162:63–73. https://doi.org/10.1016/j.envres.2017.12.018

Cho U-H, Park J-O (2000) Mercury-induced oxidative stress in tomato seedlings. Plant Sci 156(1):1–9. https://doi.org/10.1016/S0168-9452(00)00227-2

Choi BH, Lapham LW, Amin-Zaki L, Saleem T (1978) Abnormal neuronal migration, deranged cerebral cortical organization, and diffuse white matter astrocytosis of human fetal brain: a major effect of methylmercury poisoning in utero. J Neuropathol Exp Neurol 37(6):719–733. https://doi.org/10.1097/00005072-197811000-00001

Clarkson T, Magos L (2006) The toxicology of mercury and its chemical compounds. Crit Rev Toxicol 36:609–662. https://doi.org/10.1080/10408440600845619

Clifton JC (2007) Mercury exposure and public health. Pediatric Clin North America 54(2):237–269, viii. https://doi.org/10.1016/j.pcl.2007.02.005

Condini MV, Hoeinghaus DJ, Roberts AP, Soulen BK, Garcia AM (2017) Mercury concentrations in dusky grouper Epinephelus marginatus in littoral and neritic habitats along the Southern

Brazilian coast. Mar Pollut Bull 115(1):266–272. https://doi.org/10.1016/j.marpolbul.2016.12.006

Crespo-López ME, Macêdo GL, Pereira SID, Arrifano GPF, Picanço-Diniz DLW, do Nascimento JLM, Herculano AM (2009) Mercury and human genotoxicity: critical considerations and possible molecular mechanisms. Pharmacol Res 60(4):212–220. https://doi.org/10.1016/j.phrs.2009.02.011

Dabeka R, McKenzie AD, Forsyth DS, Conacher HBS (2004) Survey of total mercury in some edible fish and shellfish species collected in Canada in 2002. Food Addit Contam 21(5):434–440. https://doi.org/10.1080/02652030410001670184

Danscher G, Hørsted-Bindslev P, Rungby J (1990) Traces of mercury in organs from primates with amalgam fillings. Exp Mol Pathol 52(3):291–299. https://doi.org/10.1016/0014-4800(90)90070-T

Das KK, Panigrahi T, Panda RB (2012) Idol immersion activities cause heavy metal contamination in river Budhabalanga, Balasore, Odisha, India. Int J Mod Eng Res 2(6):4540–4542

Das BK, Singh R, Kumar R, Pathak A (2020) Assessing the efficacy of idol's immersion water pool, a case of barharwa ghat, Patna, India. Pollut Res 39(4):1131–1136

Das BK, Kumari K, Kumar S, Kush A (2023) Impacts of mercury toxicity in aquatic ecosystem : a review. Eur Chem Bull 12(10):1476–1482. https://doi.org/10.48047/ecb/2023.12.si10.00174

Davidson PW, Palumbo D, Myers GJ, Cox C, Shamlaye CF, Sloane-Reeves J, Cernichiari E, Wilding GE, Clarkson TW (2000) Neurodevelopmental outcomes of seychellois children from the pilot cohort at 108 months following prenatal exposure to methylmercury from a maternal fish diet. Environ Res 84(1):1–11. https://doi.org/10.1006/enrs.2000.4084

Davis BJ, Price HC, O'Connor RW, Fernando R, Rowland AS, Morgan DL (2001) Mercury vapor and female reproductive toxicity. Toxicol Sci off J Soc Toxicol 59(2):291–296. https://doi.org/10.1093/toxsci/59.2.291

Dickman MD, Leung CK, Leong MK (1998) Hong Kong male subfertility links to mercury in human hair and fish. Sci Total Environ 214:165–174. https://doi.org/10.1016/s0048-9697(98)00062-x

Drescher O, Dewailly E, Diorio C, Ouellet N, Sidi EAL, Abdous B, Valera B, Ayotte P (2014) Methylmercury exposure, PON1 gene variants and serum paraoxonase activity in Eastern James Bay Cree adults. J Eposure Sci Environ Epidemiol 24(6):608–614. https://doi.org/10.1038/jes.2013.96

Dringen R (2000) Metabolism and functions of glutathione in brain. Prog Neurobiol 62(6):649–671. https://doi.org/10.1016/s0301-0082(99)00060-x

Dröse S, Brandt U, Wittig I (2014) Mitochondrial respiratory chain complexes as sources and targets of thiol-based redox-regulation. Biochimica et Biophysica Acta (BBA) - Proteins and Proteomics 1844(8):1344–1354. https://doi.org/10.1016/j.bbapap.2014.02.006

Ercal N, Gurer-Orhan H, Aykin-Burns N (2001) Toxic metals and oxidative stress part I: mechanisms involved in metal-induced oxidative damage. Curr Top Med Chem 1(6):529–539. https://doi.org/10.2174/1568026013394831

Falnoga I, Tušek-Žnidarič M, Horvat M, Stegnar P (2000) Mercury, selenium, and cadmium in human autopsy samples from idrija residents and mercury mine workers. Environ Res 84(3):211–218. https://doi.org/10.1006/enrs.2000.4116

Farina M, Rocha JBT, Aschner M (2010) Oxidative stress and methylmercury-induced neurotoxicity. John Wiley & Sons, Indianapolis, pp 357–385

Le Faucheur S, Campbell PG, Fortin C, Slaveykova VI (2014) Interactions between mercury and phytoplankton: speciation, bioavailability, and internal handling. Environ Toxicol Chem 33:1211–1224

Flores-Montoya MG, Alvarez JM, Sobin C (2015) Olfactory recognition memory is disrupted in young mice with chronic low-level lead exposure. Toxicol Lett 236(1):69–74. https://doi.org/10.1016/j.toxlet.2015.04.013

Frossard A, Hartmann M, Frey B (2017) Tolerance of the forest soil microbiome to increasing mercury concentrations. Soil Biol Biochem 105:162–176. https://doi.org/10.1016/j.soilbio.2016.11.016

Frustaci A, Magnavita N, Chimenti C, Caldarulo M, Sabbioni E, Pietra R, Cellini C, Possati GF, Maseri A (1999) Marked elevation of myocardial trace elements in idiopathic dilated cardiomyopathy compared with secondary cardiac dysfunction. J Am Coll Cardiol 33(6):1578–1583. https://doi.org/10.1016/s0735-1097(99)00062-5

Genchi G, Sinicropi MS, Carocci A, Lauria G, Catalano A (2017) Mercury exposure and heart diseases. Int J Environ Res Public Health 14(1). https://doi.org/10.3390/ijerph14010074

Go Y-M, Jones DP (2013) The redox proteome*. J Biol Chem 288(37):26512–26520. https://doi.org/10.1074/jbc.R113.464131

Gundacker C, Gencik M, Hengstschläger M (2010) The relevance of the individual genetic background for the toxico-kinetics of two significant neurodevelopmental toxicants: mercury and lead. Mutat Res 705(2):130–140. https://doi.org/10.1016/j.mrrev.2010.06.003

Guo J, Ram K, Tripathee L, Kang S, Huang J, Chen P, Ghimire PS (2020) Study on mercury in PM10 at an urban site in the central indo-gangetic plain: seasonal variability and influencing factors. Aerosol Air Qual Res 20(12):2729–2740. https://doi.org/10.4209/aaqr.2019.12.0630

Guzzi G, Grandi M, Cattaneo C, Calza S, Minoia C, Ronchi A, Gatti A, Severi G (2006) Dental amalgam and mercury levels in autopsy tissues: food for thought. Am J Forensic Med Pathol 27(1)

Ha E, Basu N, Bose-O'Reilly S, Dórea JG, McSorley E, Sakamoto M, Chan HM (2017) Current progress on understanding the impact of mercury on human health. Environ Res 152:419–433. https://doi.org/10.1016/j.envres.2016.06.042

Haffner HT, Erdelkamp J, Göller E, Schweinsberg F, Schmidt V (1991) Morphological and toxicological findings after intravenous injection of metallic mercury. Deutsche medizinische Wochenschrift (1946) 116(36):1342–1346. https://doi.org/10.1055/s-2008-1063756

Hamim H, Mutyandini A, Sulistyaningsih YC, Putra HF, Saprudin D, Setyaningsih L (2019) Effect of mercury on growth, anatomy and physiology of four non-edible oil-producing species. Asian J Plant Sci 18(4):164–174. https://doi.org/10.3923/ajps.2019.164.174

Holmes P, James KAF, Levy LS (2009) Is low-level environmental mercury exposure of concern to human health? Sci Total Environ 408(2):171–182. https://doi.org/10.1016/j.scitotenv.2009.09.043

Houston MC (2005) The role of mercury and cadmium heavy metals in vascular disease, hypertension, coronary heart disease, and myocardial infarction. In: 13th international symposium of the institute for functional medicine, pp 128–133

Hultman PA, Pollard KM (2022) Chapter 24—Immunotoxicology of metals (G. F. Nordberg & M. B. T.-H. on the T. of M. (Fifth E. Costa, (eds)). Academic Press, pp. 543–564. https://doi.org/10.1016/B978-0-12-823292-7.00022-X

Iavicoli I, Fontana L, Bergamaschi A (2009) The effects of metals as endocrine disruptors. J Toxicol Environ Health. Part B, Crit Rev 12(3):206–223. https://doi.org/10.1080/10937400902902062

Israr M, Sahi S, Datta R, Sarkar D (2006) Bioaccumulation and physiological effects of mercury in *Sesbania drummondii*. Chemosphere 65(4):591–598. https://doi.org/10.1016/j.chemosphere.2006.02.016

van Iwaarden PR, Driessen AJM, Konings WN (1992) What we can learn from the effects of thiol reagents on transport proteins. Biochimica et Biophysica Acta (BBA)—Rev Biomembranes 1113(2):161–170. https://doi.org/10.1016/0304-4157(92)90037-B

Jain V, Kapadnis G, Ghosh B (2021) Fast and accurate specification of mercury in milk and honey food products by HPLC-ICP-MS. Agilent Technologies, Inc.

Jameer Ahammad S, Sumithra S, Senthilkumar P (2018) Mercury uptake and translocation by indigenous plants. Rasayan J Chem 11(1):1–12. https://doi.org/10.7324/RJC.2018.1111726

Joshi A, Shivhare N, Patel N, Khan S (2017) Surface water quality assessment during idol immersion. Int J Eng Sci Res Technol 6:413–419

Kamath SU, Pemiah B, Sekar RK, Krishnaswamy S, Sethuraman S, Krishnan UM (2012) Mercury-based traditional herbo-metallic preparations: a toxicological perspective. Arch Toxicol 86(6):831–838. https://doi.org/10.1007/s00204-012-0826-2

Karunasagar D, Balarama Krishna MV, Anjaneyulu Y, Arunachalam J (2006) Studies of mercury pollution in a lake due to a thermometer factory situated in a tourist resort: Kodaikkanal, India. Environ Pollut 143(1):153–158. https://doi.org/10.1016/j.envpol.2005.10.032

Keane S, Bernaudat L, Davis K, Stylo M, Mutemeri N, Singo P, Twala P, Mutemeri I, Nakafeero A, Etui I (2023) Mercury and artisanal and small-scale golding mining: review of global use estimates and considerations for promoting mercury-free alternatives. Ambio 52. https://doi.org/10.1007/s13280-023-01843-2

Kumari K (2021) Assessment of haematological and certain biochemical changes in fresh water air breathing fish *Clarias batrachus* (Linn.) at different exposure levels of mercury

Lenka M, Panda KK, Panda BB (1992). Monitoring and assessment of mercury pollution in the vicinity of a chloralkali plant. IV. Bioconcentration of mercury in in situ aquatic and terrestrial plants at Ganjam, India. Arch Environ Contam Toxicol 22(2):195–202. https://doi.org/10.1007/BF00213285

Li R, Wu H, DIng J, Fu W, Gan L, Li Y (2017) Mercury pollution in vegetables, grains and soils from areas surrounding coal-fired power plants. Sci Rep 7:1–9. https://doi.org/10.1038/srep46545

Lund BO, Miller DM, Woods JS (1993) Studies on Hg(II)-induced H_2O_2 formation and oxidative stress in vivo and in vitro in rat kidney mitochondria. Biochem Pharmacol 45(10):2017–2024. https://doi.org/10.1016/0006-2952(93)90012-l

Magarelli G, Fostier AH (2005) Influence of deforestation on the mercury air/soil exchange in the Negro River Basin, Amazon. Atmos Environ 39(39):7518–7528. https://doi.org/10.1016/j.atmosenv.2005.07.067

Mailloux RJ, Yumvihoze E, Chan HM (2015) Superoxide produced in the matrix of mitochondria enhances methylmercury toxicity in human neuroblastoma cells. Toxicol Appl Pharmacol 289(3):371–380. https://doi.org/10.1016/j.taap.2015.11.001

Malagutti KS, da Silva AP, Braga HC, Mitozo PA, Soares dos Santos AR, Dafre AL, de Bem AF, Farina M (2009) 17β-estradiol decreases methylmercury-induced neurotoxicity in male mice. Environ Toxicol Pharmacol 27(2):293–297. https://doi.org/10.1016/j.etap.2008.11.005

Manikandan R, Sahi SV, Venkatachalam P (2015) Impact assessment of mercury accumulation and biochemical and molecular response of *Mentha arvensis*: a potential hyperaccumulator plant. Sci World J 2015:715217. https://doi.org/10.1155/2015/715217

Maodaa SN, Allam AA, Ajarem J, Abdel-Maksoud MA, Al-Basher GI, Wang ZY (2016) Effect of parsley (*Petroselinum crispum*, Apiaceae) juice against cadmium neurotoxicity in albino mice (*Mus musculus*). Behav Brain Funct BBF 12(1):6. https://doi.org/10.1186/s12993-016-0090-3

Marrugo-Negrete J, Durango-Hernández J, Pinedo-Hernández J, Enamorado-Montes G, Díez S (2016) Mercury uptake and effects on growth in *Jatropha curcas*. J Environ Sci 48:120–125. https://doi.org/10.1016/j.jes.2015.10.036

McGregor AJ, Mason HJ (1991) Occupational mercury vapour exposure and testicular, pituitary and thyroid endocrine function. Hum Exp Toxicol 10(3):199–203. https://doi.org/10.1177/096032719101000309

Mieiro CL, Pardal M, Duarte A, Pereira E, Palmeira CM (2015) Impairment of mitochondrial energy metabolism of two marine fish by in vitro mercuric chloride exposure. Mar Pollut Bull 97(1):488–493. https://doi.org/10.1016/j.marpolbul.2015.05.054

Millar H (2022) What to know about Minamata disease? Medical News Today

Minoia C, Ronchi A, Pigatto P, Guzzi G (2009) Effects of mercury on the endocrine system. Crit Rev Toxicol 39(6):538; author reply 539. https://doi.org/10.1080/10408440903057029

Mondal NK, Roychoudhury S, Mukherjee S, Siddique S, Banerjee M, Slaughter MS, Lahiri T, Ray MR (2016) Increased risk of cardiovascular disease in premenopausal female ragpickers of Eastern India: involvement of inflammation, oxidative stress, and platelet hyperactivity. Mol Cell Biochem 419:193–203

Mottet NK, Shaw CM, Burbacher TM (1985) Health risks from increases in methylmercury exposure. Environ Health Perspect 63:133–140. https://doi.org/10.1289/ehp.8563133

Myers GJ, Davidson PW (1998) Prenatal methylmercury exposure and children: neurologic, developmental, and behavioral research. Environ Health Perspect 106(Suppl 3):841–847. https://doi.org/10.1289/ehp.98106841

Myers GJ, Davidson PW, Cox C, Shamlaye CF, Palumbo D, Cernichiari E, Sloane-Reeves J, Wilding GE, Kost J, Huang L-S, Clarkson TW (2003) Prenatal methylmercury exposure from ocean fish consumption in the Seychelles child development study. The Lancet 361(9370):1686–1692. https://doi.org/10.1016/S0140-6736(03)13371-5

Natasha, Shahid M, Khalid S, Bibi I, Bundschuh J, Khan Niazi N, Dumat C (2020) A critical review of mercury speciation, bioavailability, toxicity and detoxification in soil-plant environment: ecotoxicology and health risk assessment. Sci Total Environ 711:134749. https://doi.org/10.1016/j.scitotenv.2019.134749

Nesci S, Trombetti F, Pirini M, Ventrella V, Pagliarani A (2016) Mercury and protein thiols: stimulation of mitochondrial F1FO-ATPase and inhibition of respiration. Chemico-Biol Interact 260:42–49. https://doi.org/10.1016/j.cbi.2016.10.018

Noto H (2021) Mercury contamination of fish, cattle and its public health impacts: a review. 12

Outridge PM, Mason RP, Wang F, Guerrero S, Heimbürger-Boavida LE (2018) Updated global and oceanic mercury budgets for the united nations global mercury assessment 2018. Environ Sci Technol 52(20):11466–11477. https://doi.org/10.1021/acs.est.8b01246

Ozuah PO (2000) Mercury poisoning. Curr Prob Pediat 30(3):91–99. https://doi.org/10.1067/mps.2000.104054

Patra M, Bhowmik N, Bandopadhyay B, Sharma A (2004) Comparison of mercury, lead and arsenic with respect to genotoxic effects on plant systems and the development of genetic tolerance. Environ Exp Bot 52(3):199–223. https://doi.org/10.1016/j.envexpbot.2004.02.009

Patra M, Sharma A (2000) Mercury toxicity in plants. Bot Rev 66(3):379–422. https://doi.org/10.1007/BF02868923

Peraza MA, Ayala-Fierro F, Barber DS, Casarez E, Rael LT (1998) Effects of micronutrients on metal toxicity. Environ Health Perspect 106(Suppl 1):203–216. https://doi.org/10.1289/ehp.98106s1203

Pervez S, Koshle A, Pervez Y (2010) Study of spatiotemporal variation of atmospheric mercury and its human exposure around an integrated steel plant, India. Atmos Chem Phys 10(12):5535–5549

Pramanik S, Kumar M, Qureshi A (2021) Mercury in skin-care products in India and consumer exposure risks. Regul Toxicol Pharmacol 121:104870. https://doi.org/10.1016/j.yrtph.2021.104870

Pyszel A, Wróbel T, Szuba A, Andrzejak R (2005) Effect of metals, benzene, pesticides and ethylene oxide on the haematopoietic system. Med Pr 56(3):249–255

Renuka N, Sood A, Prasanna R, Ahluwalia AS (2015) Phycoremediation of wastewaters: a synergistic approach using microalgae for bioremediation and biomass generation. Int J Environ Sci Tech 12:1443–1460

Rice KM, Walker EM, Wu M, Gillette C, Blough ER (2014a) Environmental mercury and its toxic effects. In: Journal of preventive medicine and public health, vol 47, issue 2. Korean Society for Preventive Medicine, pp 74–83. https://doi.org/10.3961/jpmph.2014.47.2.74

Ruus A, Hjermann DØ, Beylich B, Schøyen M, Øxnevad S, Green NW (2017) Mercury concentration trend as a possible result of changes in cod population demography. Mar Environ Res 130:85–92. https://doi.org/10.1016/j.marenvres.2017.07.018

Sager PR, Aschner M, Rodier PM (1984) Persistent, differential alterations in developing cerebellar cortex of male and female mice after methylmercury exposure. Dev Brain Res 12(1):1–11. https://doi.org/10.1016/0165-3806(84)90170-6

Sahu R, Saxena P, Johnson S, Mathur HB, Agarwal HC (2014a) Mercury pollution in the Sonbhadra district of Uttar Pradesh, India, and its health impacts. Toxicol Environ Chem 96(8):1272–1283

Saper RB, Phillips RS, Sehgal A, Khouri N, Davis RB, Paquin J, Thuppil V, Kales SN (2008) Lead, mercury, and arsenic in US- and Indian-manufactured ayurvedic medicines sold via the internet. JAMA 300(8):915–923. https://doi.org/10.1001/jama.300.8.915

Savrikar S, Ravishankar B (2011) Introduction to 'rasashaastra'-the iatrochemistry of ayurveda. Afr J Tradit Complement Altern Med 8(5S)
Schrag SD, Dixon RL (1985) Occupational exposures associated with male reproductive dysfunction. Annu Rev Pharmacol Toxicol 25:567–592. https://doi.org/10.1146/annurev.pa.25.040185.003031
Schützendübel A, Schwanz P, Teichmann T, Gross K, Langenfeld-Heyser R, Godbold DL, Polle A (2001) Cadmium-induced changes in antioxidative systems, hydrogen peroxide content, and differentiation in Scots pine roots. Plant Physiol 127(3):887–898. https://doi.org/10.1104/pp.010318
Sharma R, Pervez S (2005) Toxic metals status in human blood and breast milk samples in an integrated steel plant environment in Central India. Environ Geochem Health 27:39–45
Solt I, Bornstein J (2010) Childhood vaccines and autism--much ado about nothing? Harefuah 149(4):251–260
Sprovieri F, Pirrone N, Ebinghaus R, Kock H, Dommergue A (2010) A review of worldwide atmospheric mercury measurements. Atmos Chem Phys 10:8245–8265. https://doi.org/10.5194/acp-10-8245-2010
Streets DG, Lu Z, Levin L, ter Schure AFH, Sunderland EM (2018) Historical releases of mercury to air, land, and water from coal combustion. Sci Total Environ 615:131–140. https://doi.org/10.1016/j.scitotenv.2017.09.207
Stringari J, Nunes AKC, Franco JL, Bohrer D, Garcia SC, Dafre AL, Milatovic D, Souza DO, Rocha JBT, Aschner M, Farina M (2008) Prenatal methylmercury exposure hampers glutathione antioxidant system ontogenesis and causes long-lasting oxidative stress in the mouse brain. Toxicol Appl Pharmacol 227(1):147–154. https://doi.org/10.1016/j.taap.2007.10.010
Teixeira FB, de Oliveira ACA, Leão LKR, Fagundes NCF, Fernandes RM, Fernandes LMP, da Silva MCF, Amado LL, Sagica FES, de Oliveira EHC, Crespo-Lopez ME, Maia CSF, Lima RR (2018b) Exposure to inorganic mercury causes oxidative stress, cell death, and functional deficits in the motor cortex. Front Mol Neurosci 11:125. https://doi.org/10.3389/fnmol.2018.00125
Teixeira DC, Lacerda LD, Silva-Filho EV (2018) Foliar mercury content from tropical trees and its correlation with physiological parameters in situ. Environ Pollut 242:1050–1057. https://doi.org/10.1016/j.envpol.2018.07.120
Tousignant B, Chatillon A, Philibert A, Da Silva J, Fillion M, Mergler D (2023) Visual characteristics of adults with long-standing history of dietary exposure to mercury in grassy narrows first nation, Canada. Int J Environ Res Public Health 20(6). https://doi.org/10.3390/ijerph20064827
Ullrich SM, Tanton TW, Abdrashitova SA (2001) Mercury in the aquatic environment: a review of factors affecting methylation. Crit Rev Environ Sci Technol 31(3):241–293. https://doi.org/10.1080/20016491089226
UNEP (2013a) Minimata convention on mercury
UNEP (2013b) Technical background report for the global mercury assessment
Vimy MJ, Takahashi Y, Lorscheider FL (1990) Maternal-fetal distribution of mercury released from dental amalgam fillings. Am J Physiol 258(4):939–945
von Burg R (1995) Inorganic mercury. J Appl Toxicol 15(6):483–493. https://doi.org/10.1002/jat.2550150610
Wada H, Cristol DA, McNabb FMA, Hopkins WA (2009) Suppressed adrenocortical responses and thyroid hormone levels in birds near a mercury-contaminated river. Environ Sci Technol 43(15):6031–6038. https://doi.org/10.1021/es803707f
Wang L, Jia G (2005) Progress in developmental toxicity of methylmercury. Wei sheng yan jiu = J Hygiene Res 34(5):633–635
Yoshida M (2002) Placental to fetal transfer of mercury and fetotoxicity. Tohoku J Exp Med 196(2):79–88
Zhou J, Obrist D, Dastoor A, Jiskra M, Ryjkov A (2021) Vegetation uptake of mercury and impacts on global cycling. Nat Rev Earth Environ 2(4):269–284. https://doi.org/10.1038/s43017-021-00146-y

Part III
Sustainable Mitigation Strategies and Solutions

Chapter 6
Mercury Adsorption Using Biowaste Biochar: A Green Technology Approach

Abudu Ballu Duwiejuah, Ziblim Abukari Imoro, Ammal Abukari, Iddrisu Abdul-Mumeen, and Abubakari Zarouk Imoro

6.1 Introduction

Mercury (Hg) has been named as one of the dangerous "ten leading compounds of concern" (WHO 2017). The total amount of anthropogenic mercury emissions into the atmosphere in 2015 was around 2220 tonnes (UNEP 2018). Due to its toxicity, mercury has been a source of worry for public health around the world (Raju et al. 2019). The top Hg-emitting nations are China, Indonesia, India, Colombia, South Africa, Russia, Ghana, and the United States of America as 56% (1095 Mg/year) of the anthropogenic emissions come from these nations (UNEP 2013). The amount of mercury that is collected worldwide in soil ranged between 250 to 1000 Gg (Obrist et al. 2018). When mercury from ores, fossil fuels, and mineral deposits is released into the atmosphere, it can be very mobile and deposit on topsoil, water, and

A. B. Duwiejuah (✉)
Department of Biotechnology and Molecular Biology, Faculty of Biosciences, University for Development Studies, Nyankpala Campus, Tamale, Ghana
e-mail: aduwiejuah@uds.edu.gh

Z. A. Imoro
Department of Biodiversity Conservation and Management, University for Development Studies, Nyankpala Campus, Tamale, Ghana

A. Abukari
Department of Forestry and Forest Resources Management, Faculty of Natural Resources and Environment, University for Development Studies, Nyankpala Campus, Tamale, Ghana

I. Abdul-Mumeen
Department of Biochemistry, Faculty of Bioscience, University for Development Studies, Nyankpala Campus, Tamale, Ghana

A. Z. Imoro
Department of Environment, Water and Waste Engineering, University for Development Studies, Nyankpala Campus, Tamale, Ghana

N. Kumar (ed.), *Mercury Toxicity*, Environmental Science and Engineering,
https://doi.org/10.1007/978-981-99-7719-2_6

bottom sediments (Xu et al. 2015). The primary source of mercury emissions into the atmosphere is coal combustion (Sitko et al. 2021).

Mercury is a trace element that is bioaccumulative, volatile, and persistent. It significantly affects both the environment and human health (Feng et al. 2021). Acute exposure to mercury may also result in psychological issues such as anxiety, insomnia, and depression (Gall et al. 2015). Every day 14,000 people die as a result of water contamination, which is a problem that affects the entire world. Hence, the need to remediate mercury contaminated environments (Chaudhry and Malik 2017).

Adsorption was recognised as an economically effective, easily operated, and less-second pollution way amongst various approaches for the eradication of these water contaminants (Zhu et al. 2019). Agricultural residues can be utilised for water treatment. The mechanisms of adsorption, reduction, oxidation, and desorption are used in removal technologies. Containment and stabilisation which stop mercury migration by physical trapping or chemical complexation, respectively, are the most extensively used immobilisation strategies (He et al. 2015). These technologies are aimed at removing Hg from polluted media or changing harmful Hg species into less hazardous ones (Lewis et al. 2016). New materials and technologies for Hg cleanup are being developed as a result of more recent research (Xu et al. 2019a). Recent research works have looked closely at novel materials, particularly those with high surface areas, significant porosities, and active adsorption sites (Kumar et al. 2019). Other factors including generation process, stability, and reusability require significant consideration in addition to adsorption capacity, which is the main factor determining the effectiveness of an adsorbent (Abraham et al. 2018).

Innovative techniques like adsorption have proven to be more economical and eco-friendlier than traditional remediation technologies. The most efficient and eco-friendly method for removing Hg from water is adsorption using biowaste biochar. The purpose of this chapter was to review the application of biowaste biochar techniques for mercury removal via adsorption, factors influencing mercury adsorption, the role of biochar in Hg remediation technologies, water treatment technologies, innovative approaches for the treatment of mercury, regeneration and economic challenges of biowaste-derived adsorbents, management of post-adsorption materials, green economy framework, and challenges and future research directions.

6.2 Mercury, Chemistry, and Behaviour of Mercury

The consumption of mercury-contaminated water negatively impacts the neurological, gastrointestinal, and renal systems. Mercury reacts with thiols in wastes containing sulphated proteins in these organ systems. The consumption of mercury-contaminated water thus impairmental and neurological functioning (Poste et al. 2018). The WHO has set a limit of 1 and 5 g/L for Hg ions (Hg^{2+}) in drinkable water and wastewater discharge, respectively (Li et al. 2017).

Mercuric sulphide (HgS) ore is used to recover the metallic form of mercury (Donatello et al. 2012). The most basic form of mercury, known as mercury(II)oxide

(HgO), is dangerous to humans and the environment. Cl^-, OH^-, S^{2-} and the organic ligands and the functional groups that contain sulphur are the main complexes that Hg forms in nature (Xu et al. 2015). Mercury undergoes both reduction and oxidation during biotic activities in the natural environment, and these processes alter the speciation of Hg (Tan et al. 2004), which in turn influences the uptake of Hg (Xu et al. 2015).

The interconversion of Hg's many forms determines how it behaves biologically. For instance, inhaled HgO vapour is quickly converted to various forms in the lungs and mucosal membranes (Kabata-Pendias 2010). Mercury is found in a liquid state below room temperature and atmospheric pressure because its outermost 5d orbital is fully filled with electrons and it does not readily share its valence electrons. As a result, at room temperature, the link between the Hg atoms is exceedingly weak and easily broken making Hg quite reactive. In organic molecules, the carbon-Hg link is chemically stable (Alloway 2013) mainly due to colinear bond formation around the mercury atom. On this basis, the solubility of different mercury compounds in water varies. In water, Hg0 is insoluble, HgS and Hg(I) chloride are less soluble, whilst Hg(II) chloride is easily soluble. In water, the inorganic form of mercury is methylated, creating extremely poisonous MeHg in aquatic environments. The methylation of mercury in water is facilitated by the presence of a bacterium species (Pseudomonas) in fish. MeHg eventually penetrates the aquatic ecosystem's food chain when it is created (WHO 2004).

6.3 Occurrence and Sources of Mercury in the Environment

Mercury can be discovered in the soil, water, and air. Mercury sulphide is the most widely used type of mercury as an ore. Large amounts of naturally occurring cinnabar can be found in the United States in Texas, California, Nevada, Utah, Oregon, Arkansas, and Idaho. Many different types of rocks, including shale and granite, contain trace quantities of inorganic mercury (Massachusetts Department of Environmental Protection 1996). Mercury concentrations in the atmosphere are often higher than they are elsewhere in the vicinity of active volcanoes and substantial anthropogenic sources. In the United States, coal-burning power stations are the major source of airborne mercury emissions, contributing around 45% of total domestic mercury emissions (U.S. EPA 2006).

Both natural and man-made sources introduce mercury into environmental media. In the physical world, there are two main sources of mercury: (a) Sources from nature: Hg is a naturally occurring element that can be found in the earth's crust as Hg0 or HgS (Kostova et al. 2013). (b) Anthropogenic mercury discharges into the environment amount to roughly 2 gigatons a year, and human activity has resulted in an increase in surface soil mercury contamination of around 86 gigatons (UNEP 2018). The main sources of Hg pollution are burning coal, mining, metallurgy, processing chlor-alkali, and other industries (Liu et al. 2022). Mercury is prevalent in human hair and is absorbed by crops like rice due to soil pollution. Due to the harmful effects of either short-term or long-term mercury species on human health and aquatic ecology, Hg abatement from the aqueous medium is a critical environmental management endeavour (Abbas et al. 2018). Furthermore, mercury in water has become a serious problem as a result of the world's rapid industrial development. Hg(II) is ranked as the sixth most toxic chemical on the list of dangerous compounds and is considered one of the most pervasive and dangerous metals in aqueous environments (Abbas et al. 2018). Other sources of Hg in the environment include the oil refineries, plastics sector, pulp industry, cement industry, and a number of other industries (Kabiri et al. 2015). Mercury can also be released after use from fluorescent lamps and mercury cells. Decontaminating or collecting mercury from spent fluorescent lamps' end caps, phosphor powder, and glass can release mercury into the environment.

The amount of mercury recovered varies on whether dry or wet treatment is employed. The electroleaching procedure used by Sorbal et al. (2006) to remove 99% of the mercury from used fluorescent bulbs and spent fluorescent lamps recovered 81% of the mercury (Ozgur et al. 2016). Additionally, 95% of the mercury was recovered using a photocatalytic technique in conjunction with a sodium hypochlorite extraction solution (Bussi et al. 2010). In a recent study by Al-Ghouti et al. (2016), mercury leaching efficiency from fluorescent lamps was calculated using a microwave-assisted approach, which showed an efficiency of 76.4%. Mercury can be found in a variety of media at waste sites, including soil, groundwater, sediment, sludge, and leachate. Hospitals, clinical laboratories, and the pharmaceutical industry all produce waste streams that contain thimerosal, an organic form of mercury, which has been employed as a preservative in many medications.

Mercury can be released via volcanic activity or rock outgassing into the atmosphere (Martín and Nanos 2016). Geothermal activity is another natural way that Hg can enter the environment (Gustin 2003). Anthropogenic sources: types of anthropogenic sources of mercury: point sources and diffuse sources. The diffuse sources are landfills, mine wastes, and sewage sludge-amended fields, whereas the point sources are coal-fired and incinerators (Alloway 2013). The amount of Hg released into the atmosphere because of the combustion of fossil fuels like coal was predicted to be around 475 metric tonnes in 2010 (Streets et al. 2018). According to Zhang et al. (2014), the smelting and mining of ores are significant sources of anthropogenic air pollution because they release mercury into the environment. The use of fungicides and pesticides that contain mercury in agricultural fields is dangerous and poses a serious health risk to people when they consume food tainted with Hg. The main human-caused sources of Hg in the environment include the use of Hg in electrical switches, medicines, thermal sensing devices, and dental amalgam preparations (Chen et al. 2010).

Mercury pollution affects the environment, people's health, and all living organisms and the ecological processes they depend on (Wang et al. 2012; Spahić et al. 2019). Human exposure to MeHg through food is the main risk associated with mercury (Pavlish et al. 2003). The primary organs in which MeHg accumulates in humans are the kidneys, liver, and brain (central nervous system) (Genthe et al. 2018); other side effects include vision abnormalities, hearing loss, speech loss, and death in certain circumstances. As a result of the exposure to MeHg, it may be argued that fetal life is more subtle than adult life (Alloway 2013). The target organ for inhaled Hg vapour is predominantly the brain, and the toxic effects of mercury depend on the rate and dose of exposure to various forms of mercury (Beckers and Rinklebe 2017). The measurement of Hg levels in hair, blood, and urine in people allows for the estimation of Hg exposure. With an increase in soil Hg concentration, the soil fertility declines (Alloway 2013). The most typical signs of Hg toxicity in plants include poor root development, growth inhibition, and failure of metabolic activities such as the respiration rate and the generation of chlorophyll pigment during photosynthesis and low yield. Also, plants are unable to absorb K^+ ions because of the buildup of mercury in root tissue (Kabata-Pendias 2010).

6.4 Mechanism of Mercury Toxicity

According to the potential mechanism of mercury toxicity at the cellular level, mercury can harm cells by obstructing vital components like enzymes and polynucleotides. Mercury binding on the outer membrane has an impact on ion transport as well as cell membrane disruption (Azevedo and Rodriguez 2012). According to Patra et al. (2004), mercury reacts very favorably with both phosphate and sulphydryl (–SH) groups. Mercury also interferes with non-enzymatic antioxidants (non-protein thiols and glutathione), affecting their antioxidant properties. Mercury interaction with –SH groups (tissues rich in –SH ligands) results in the formation of S-Hg-S bridges, which disrupts group stability and hinders seed germination (Alloway 2013). Although studies have shown that Hg can create lethal defects in the genetic makeup of animals and plants, Hg genotoxic effects are extremely rare (Alloway 2013). Inside cells, Hg ions make covalent connections with the reactive spots on deoxyribonucleic acid (DNA) molecules.

Around the world, efforts have been made to solve the growing water shortage, lower water body pollution, and water-related difficulties, leading to developments in water treatment technology (Tang et al. 2021; Yu et al. 2022). A workable solution is the preparation of appropriate materials to achieve or maintain adequate or elimination effects on a variation of water-based refractory compounds.

6.4.1 Adsorption

When compared to precipitation/coprecipitation, adsorption is more likely to be impacted by the properties of the media and pollutants besides mercury. With these technologies, smaller systems typically have cheaper operating and maintenance expenses and less operator skill. Adsorption is frequently utilised as a polishing process for the effluent from bigger systems and for relatively smaller systems when mercury is the only pollutant to be treated (Yu et al. 2016). One of the most significant and practical ways to remove Hg from solid surfaces is through the adsorption process (Yu et al. 2016). For the effective removal of Hg, a number of sorbents have been utilised, including carbon nanotubes (Habuda-Stanić and Nujić 2015), mesoporous carbon (Anbia and Dehghan 2014), activated carbon (Lu et al. 2014), and iron oxides (Figueira et al. 2011). The physical characteristics of the adsorbent (pore volume, pore size, surface area, and pore number) may have an impact on the adsorption capacity, which may result in an increase or a decrease in the effectiveness of Hg removal. For instance, the effectiveness of removing Hg is slowed when activated carbon is sulphurised because fewer micropores are created (Yu et al. 2016).

Colloidal nanoparticles were employed as adsorbents to remove mercury from aqueous solutions by Ojea-Jiménez et al. (2012). For the removal of Hg0, low-cost adsorbents such as copper oxide, copper chloride, and synthetic zeolite have also been utilised (Yu et al. 2016). A maximum adsorption capacity of 140.85 mg/g was demonstrated by the bacterial species Paecilomyces catenlannulatus for the removal of Hg at the optimal pH of 7 (Li et al. 2013). The adsorption efficacy may be impacted by specific adsorbent characteristics such as surface area, pore size distribution, and polarity (Yu et al. 2016).

Adsorption is the most widely used technique amongst the several approaches intended to remove Hg from aqueous solution (Abbas et al. 2018). According to Rocha et al. (2016), the production of chelates is the main mechanism of adsorption, and the adsorbents typically have a large surface area and high porosity. To adsorb mercury ions Hg(0) and Hg^{2+} from water, Mondal et al. (2019) created covalent triazine nanospheres with thioether functionalisation. The excellent adsorption capacity was seen (813 mg/g for Hg(0) and 1253 mg/g for Hg^{2+}), and the findings of the kinetic investigation show a quick adsorption rate (Mondal et al. 2019). In a different study, Abbas et al. (2018) created a novel mesoporous conjugate sorbent on the basis of pentasil zeolite, and 172.6 mg/g was the maximum adsorption capacity. Bao et al. (2017) extracted Hg(II) using magnetic nanoparticles coated with silica from wastewater, and they also discovered that mercury ions bind to imine (C–NH–) groups on the nanoparticle's surface. Derivatives of chitosan were utilised as adsorbents in another instance. The primary adsorption mechanism is the chelation of Hg(II) with the chitosan nitrogen atoms (Kyzas and Kostoglou 2015).

The gas-phase elemental mercury Hg(0) removal also frequently employs the adsorption mechanism (Abraham et al. 2018). The "hard and soft acid base" (HSAB) theory states that Hg preferentially forms complexes with soft ligands like sulphur to produce stable, insoluble molecules. It was discovered that an adsorbent's high sulphur content helped it to be more effective in adsorbing Hg(0) (Abraham et al. 2018). It is important to highlight that stability and reusability should be carefully taken into account when using adsorbents for Hg removal. Thermal desorption has been proven to be a successful method for remediating Hg-contaminated soil, waste, or sludge (Zhao et al. 2018). There is no need to remove the polluted environmental media, especially the soil, hence in situ thermal desorption is recommended. In this method, heat is transferred into the soil using thermal conductive heating devices (He et al. 2015).

Mercury separates from the soil as a result of the heating process' rise in mercury volatility. Mercury comes in many different forms in soils, including HgS, HgO, $HgCl_2$, and mercury connected to organic materials (Wang et al. 2012). When the heating temperature is above 600 °C, these Hg species are volatilised and the treatment can achieve an acceptable decontamination level (Zhao et al. 2018). The cost of this technique is projected to be 480 USD/t soil (He et al. 2015), and it has a wide range of full-scale and pilot-scale applications (Wang et al. 2012). The high energy cost of conventional thermal desorption is one of its main drawbacks and thus, there is a need for further research to identify low-temperature techniques of desorption.

6.5 Techniques for Mercury Removal

Simple forms of mercury or its compounds are dangerous to people and bad for the environment. Mercury is very mobile and cannot be broken down. The most harmful form of mercury, methylmercury, has a high capacity for accumulation. Despite initiatives to reduce Hg(II) waste, human activity continues to produce wastewater that include Hg(II) in many developing countries. Examples include effluents from many gold mining areas (Marrugo-Negrete et al. 2017), coal power plants (Chen et al. 2022), biomedical (Clyde Vincent 2016), paper and pulp industries (Sharma et al. 2022), and even groundwater in historically polluted sites (Richard et al. 2016). Novel materials such as polymers (Huang et al. 2015), functionalised clays (Wang et al. 2020), and composites of these materials (Bai and Hong 2021) are being developed to treat such contaminated waters. Although these sorbents have received extensive research, their widespread use in low-income nations may present economic difficulties. A less expensive carbonaceous sorbent, such as biochar, might be an option for treating Hg(II)-contaminated waterways. Agricultural and wood waste can be converted into biochar, which supports environmentally friendly restoration strategies.

High operational and maintenance costs, toxic sludge production, effluent production, chemical consumption, the inability to reuse Hg, and the difficulty of processes involving multiple steps are all common drawbacks for most of the existing technologies (Lecler et al. 2018). As a result, it has been suggested that adsorption, specifically employing inexpensive, simple-to-prepare adsorbents, may be a cost-effective method for the remediation of mercury in water at low concentrations (Al-Ghouti et al. 2019). The adsorption approach provides a number of advantages over other procedures, including the convenience of use, high removal effectiveness, which may reach 90 to 99% (Al-Ghouti et al. 2019), and simplicity of design. Activated carbon is one of the most often used sorbents for the removal of contaminants in wastewater.

One of the most popular adsorbents for removing pollutants and treating wastewater is activated carbon. The usage of activated carbon is restricted by its exorbitant cost, which makes it more challenging to extract mercury from an aqueous medium

Table 6.1 Merits and demerits of mercury removal techniques

Removal technique	Advantages	Disadvantages
Adsorption	Low cost, wide pH range, easy operation conditions, and metal binding capacities are high	Selectivity is low and waste products are produced
Ion exchange	High removal efficiency, fast kinetics, and high capacity of treatment	Secondary pollution is caused costly and selectivity is low
Chemical precipitation	Capital cost is low and operation is simple	Sludge generation and sludge disposal need extra operational costs
Membrane filtration	Requires small space and low pressure and separation selectivity is high	The process is complex, membrane fouling is expensive, and permeate flux is low
Flotation	More concentrated sludge is produced, with high metal selectivity and removal efficiency	High initial capital cost, maintenance, and operation cost

(Igwe and Abia 2007). According to Arias et al. (2017), the generally used methods for extracting mercury either require preliminary activation or involve numerous procedures for the synthesis of the sorbent material, which results in ineffective or expensive green technologies. Therefore, it is necessary to develop a new class of adsorbing materials that can overcome the disadvantages and shortcomings of the ones now in use to extract mercury from liquid solutions. Merits and demerits of mercury removal techniques are described in Table 6.1

6.6 Water Treatment Technologies

Coprecipitation/precipitation is the method most frequently employed to treat mercury-polluted water amongst the initiatives for mercury treatment that have been discovered. Compared to other water treatment technologies, this one is less susceptible to being impacted by the properties of the media and pollutants (U.S. Environmental Protection Agency 2007). Adsorption and membrane filtration are two alternative Hg treatment technologies; however, they are less frequently used than precipitation/coprecipitation because both are more susceptible to being impacted by the properties of the media and non-mercury pollutants.

Adsorption is typically utilised more frequently for relatively smaller systems, when Hg is the sole pollutant to be treated, and as a polishing process for effluent from larger systems (U.S. Environmental Protection Agency 2007). Because membrane filtration tends to produce more residuals than other mercury removal procedures, it is utilised less frequently. Two pilot-scale studies that were found for this report used bioremediation to remediate Hg. In one experiment, bacteria transformed mercury ions into elemental mercury, which was subsequently adsorbed onto a carbon molecule. In the other study, wastewater was treated using a combination of anaerobic and aerobic biological processes to remove mercury.

Low-cost biowaste has been used to remove heavy metals from wastewater in distant and underdeveloped areas. To recover metal cations, dyes, and insecticides from water, tomato peels were researched by Mallampatia and Valiyaveettil in 2012. Their finding were that tomato peels were effective at adsorbing cationic dyes as well as metal cations like Pb^{2+} and Ni^{2+}. According to Mallampatia and Valiyaveettil (2012), the electrostatic interaction between the positive adsorbents and the HCOO and OH functional groups found on the surface of the peels is the mechanism for adsorption. Methyl orange was eliminated from aqueous solutions by Mittal et al. (2007) using bottom ash and de-oiled soya extracts. In comparison to bottom ash at 3.62 mg/g, the de-oiled soya extracts demonstrated greater adsorption capabilities of 16.66 mg/g. Prior to needing regeneration, the de-oiled soya and bottom ash sorbed up to 99.80% and 98.61%, respectively. In both instances, it was possible to renew up to 98% of the adsorbents by simply washing the columns with sodium hydroxide (Bhatnagar et al. 2010). Bhatnagar and associates investigated the efficacy of lemon peels in removing heavy metals from wastewater using column stacks. They discovered that before the adsorbent (lemon peels) became saturated, could eliminate cobalt ions from 210 L of synthetic effluent using 1 kg of lemon peels, 22 mg/g was the adsorption capacity. They concluded that it would be more affordable to treat wastewater for heavy metals using these inexpensive organic waste peels rather than the 10 times more expensive activated carbon.

Specific functional groups such as CeCl, CeO, CeI, and C]O proved to be the activated adsorption sites and increased the Hg(0) removal effectiveness (Liu et al. 2018; Wang et al. 2018). Following pyrolysis, biochars can undergo physical activation using non-thermal plasma (Wang et al. 2018) or a microwave (Shen et al. 2015) to boost the quantity of active sites, improve pore structure, and introduce particular functional groups. The removal of Hg(0), however, has been found to be greatly aided by chemical activation and chemisorption (Li et al. 2015b). Utilising NH_4Cl to modify biochar was shown to be the best modification method due to its low cost and high efficiency (Xu et al. 2019b). Due to the production of CeCl on the biochars surfaces which may convert Hg(0) to $HgCl_2$ or other HgeCl complexes, Hg(0) was removed from the environment by NH_4Cl-impregnated biochars. The assumption that the majority of the combined CeCl groups on the surface of biochars were converted to ionic Cl during the process of chemisorption was proven correct by XPS analysis (Li et al. 2015b). In addition to NH_4Cl, additional chemicals including NH_4Br, KI, KCl, and KBr might also be used to modify biochars to improve their ability to remove Hg(0) (Liu et al. 2018), because halogen is a key element in chemisorption.

The impacts of biochars modified with Mn-Ce mixed oxides (Yang et al. 2017) and CuOx and CeO_2 (Xu et al. 2018) which are thought to be potential catalysts for the oxidising of Hg(0), were also studied in a number of researches. Magnetic biochars (MBC) have received a lot of interest lately for treating flue gas Hg(0). According to Yang et al. (2016), the C]O group serves as an electron acceptor, facilitating the electron transfer during Hg(0) oxidation. Additionally, the production of Hg-Fe_3O_4 is caused by the Hg(0) reaction using the Fe_3O_4 as adsorption/oxidation site in MBC.

The removal of Hg(0) using a biochar-based flue gas is a promising strategy, however the effectiveness of the removal will be impacted by other flue gas constituents. According to research (Xu et al. 2019b), large quantities of H_2O and SO_2 prevented the Hg(0) removal whilst O_2 and NO in the flue gas enhanced the process. Therefore, a new modification technique that can reduce the impact of other flue gas constituents is required.

6.7 Application of Biowaste Biochar

The creation and development of research disciplines in waste management, agriculture, the environment, materials, and energy were facilitated by the invention of biochar. New uses for biowaste biochar enable the development of a fresh strategy for solving environmental issues. The use of biochar and how it relates to its characteristics (Lu et al. 2020). For successful and efficient application, biochar structure and surface reactivity should be taken into account. In light of this, the creation of effective and affordable alternatives to adsorbents is a work with significant practical applicability (Jia et al. 2018). As a solid byproduct of biowaste biomass pyrolysis, it has high surface chemistry and a complicated pore structure. There has been much international research on the application of biochar to remove combustion contaminants (Xu et al. 2019b).

6.8 Factors Influencing Mercury Adsorption

6.8.1 Effect of pH Value on Hg^{2+} Adsorption

When the pH of the solution is raised, the density of the surface charge decreases, which may lead to a reduction in the electrostatic repulsion between the surface of the adsorbent and the positively charged Hg^{2+} (Silva et al. 2010). Another factor to take into account when analysing the impact of pH on mercury adsorption is Hg^{2+} speciation in the solution. Zhang et al. (2005) demonstrated that, in the absence of chelating agents, $Hg(OH)_2$ and Hg^{2+} coexist in solutions with pH values between 3 and 5. Mercury at pH values below 3 is the dominant species, whereas $Hg(OH)_2$ at pH values above 5 is the dominant species (Zhang et al. 2005). Furthermore, $HgCl_2$ is the main Hg^{2+} species at pH values lower than 4 (Fatoni et al. 2015).

Additionally, Arias et al. (2017) discovered that $HgCl_2$ pH values between 3.5 and 5.5 are the dominating species for, $HgCl_2$ or Hg(OH)Cl are the major species between 5.5 and 6.5, and $Hg(OH)_2$ or $HgCl_2$ are the dominant species at pH values above 6.5. The close resemblance between the pHPZC and pH value of the adsorbent, which results in practically zero net charge and no discernible electrostatic effect in comparison to pH 4 (Powell et al. 2005), could be the cause of the modest decrease in removal percentage. Additionally, as was already indicated, the electrostatic attraction between the adsorbent surface at basic circumstances and the negatively charged species caused the adsorption to decrease. When pH rose from pH 2 to pH 10, the proportion of Hg^{2+} removed increased from 76.8 to 93.7% (Al-Ghouti et al. 2019). According to Al-Ghouti et al. (2019), this is caused by the fact that at low pH values, the H^+ ions concentration in the solution is high, which will compete with Hg^{2+} on the binding sites of the surface of the adsorbent and result in low Hg^{2+} ions binding on the adsorbent active sites. The Hg^{2+} ions' ability to bind to the adsorbent surface will rise as pH is raised whilst H^+ ions' concentration in the solution will decrease (Khraisheh et al. 2004).

6.8.2 Effect of Initial Concentration of Hg^{2+} on the Adsorption Process

The impacts of initial concentration on mercury adsorption onto biochar show that as the concentration rises, so does the quantity of mercury adsorbed on the material. The adsorbents' empty surface sites are responsible for this. The adsorption capacity displayed a consistent adsorption trend beyond 5 mg/g (Al-Ghouti et al. 2019). It was noted that continuous increases were seen between 3 and 5 mg/g. This might be because there are more active sites on the surface of the adsorbent than there are Hg^{2+} ions in the solution. According to Samra (2014), the RDP adsorption capacity rises with the initial metal concentration as a result of the metal's increased diffusion in the boundary layer, which raises the RDP's adsorption capacity. However, it is confirmed that when the reaction was carried out at 35 °C, the amount of Hg being adsorbed was linear until an initial concentration of 5 mg/L, after which it displayed a constant trend, indicating that there were insufficient active sites present for molecules to spontaneously bind to the adsorbent surface. Additionally, on dispersed active sites on the surface of the adsorbent, mercury and carbon interact (Karatza et al. 2000).

6.8.3 Biowastes as Adsorbents

Materials made from waste have drawn a lot of interest lately since they are so economical. Compared to disposal in a landfill or burning, using these materials in the process of remediation is seen to be more appealing and eco-friendlier (Li et al. 2015a). According to the source of the feedstock, these materials can be broadly classified into two groups: those made from industrial waste and those made from biomass. Biochar and activated carbon that is formed from biomass are examples of biomass-derived materials. The pyrolysis of biomass produces biochar, a type of black carbon. According to Beckers et al. (2019), it is a brand-new porous material with a large surface area that is beneficial for immobilising and adsorbing mercury.

A thorough review article displays a variety of biowaste materials, including stems, coconut, mango, and banana peels as well as fruit husks, shells, and husks (Bhatnagar and Sillanpää 2010). The various functional group's presence such as the hydroxyl groups inherent in the chemical makeup of the waste and capable to bind pollutants in water, is a typical characteristic.

The waste is frequently chemically or physically altered to increase the surface area and reactive surface groups, hence improving the adsorption capability. The waste is frequently physically or chemically altered to increase the surface area and reactive surface groups, hence improving the adsorption capability. According to Acharya et al. (2018), chemical modification of biowaste adsorbents aids in the removal of soluble organic compounds, improving the chelating effect which is particularly important for metal adsorption by removing these compounds from the mix. The quantity of agricultural waste in the area provides an easily accessible raw material that can be utilised to purify water. For this sector to grow further, performance must be enhanced but expenses must remain under control. Due to their widespread and plentiful availability around the world, agricultural wastes have drawn particular attention (El-Ramady et al. 2020). Adsorbents with a low cost can be used to remediate contaminated wastewater. Before they can be employed in numerous applications, research is needed to change the structure and performance of many cheap adsorbents since they have a poor adsorption capacity (Pourrahim et al. 2020).

Environmental issues associated with the disposal and management of green wastes can be lessened by using agricultural wastes as sorbents. The creation of new adsorbents for the heavy metals removal from water that are based on solid waste (such as plant biomass) may be an excellent substitute to lower treatment costs (Grenni et al. 2019). According to Malakahmad et al. (2016), a wide range of green wastes, including corn cobs, rice husks, wheat bran, tree leaves and bark, and aquatic weeds, have been investigated for their ability to bind metals. Though, there are important challenges in its regeneration, making the expensive treatment of tainted water a particular issue for developing nations. Therefore, it is necessary to choose inexpensive, locally accessible materials with excellent adsorption capacities. Recycling of biowastes as an alternative to disposal and incineration has been inspired by the conservation of natural resources and the efficient use of non-renewable fuels. The need for novel eco-friendly remediation techniques that make use of plant or vegetable wastes is expanding.

6.9 Role of Biochar in Hg Remediation Technologies

Although biochar is an efficient adsorbent for absorbing diverse organic and inorganic pollutants from different media (such as wastewater or flue gas), its expensive cost (about 135,000 USD/t) prevents its widespread application (Sajjadi et al. 2018). The biochar surface is affected by the chemistry of the water as well; higher alkaline pH upsurges the negative surface charge (Tan et al. 2020) and cations in the solution can lessen the biochar negative surface charge (Chen et al. 2017). NOM can plug tiny pores and adsorb them onto the surface. It is anticipated that coexisting ions and ternary interactions between biochar, NOM, and Hg(II) will affect how successfully biochar removes Hg(II). These interconnected mechanisms have received relatively little research. Before implementing innovative carbonaceous sorbents on a large scale, it is essential to comprehend how complicated water chemistry affects Hg(II) immobilisation. Therefore, it is promising to produce biochar utilising waste as the feedstock. Through the chemical activation of wood sawdust, Kazemi et al. (2016) developed activated carbon, with a 1789 m^2/g surface area. In a different study, Sajjadi et al. (2018) synthesised activated carbon from pistachio wood waste with pore volumes and surface areas of 1448 m^2 per gram and 0.901 cm^3 per gram, respectively. Due to the significant annual production, industrial waste is causing a lot of concern. Attari et al. (2017) synthesised zeolite from coal fly ash, and the average effectiveness of Hg^{2+} removal stretched 94% in a concentration of 10 mg/L. Red mud is another by-product of the Bayer process, which is used to produce alumina, and its annual global production has stretched to 160 million tonnes (Yang et al. 2018). The elemental Hg can be removed from flue gas using modified red mud after being impregnated with potassium halides. The oxidation of the trapped Hg(0) is aided by the halides (such as KI and KBr).

Although waste-derived materials show considerable promise for use in the remediation of Hg, they may also include toxic metals and other pollutants (Hua et al. 2017), and more research is needed to assess the possible dangers of their application. A few investigations looked into how biochars affected the immobilisation of Hg. According to O'Connor et al. (2018), non-toxic elemental S was applied to modify rice husk, increasing the material's Hg^{2+} adsorption capacity by 73%, to 67 mg/g. The dosage of 5% (dry wt.) also resulted in a 99.3% reduction in the amount of freely accessible Hg in leachates when compared to untreated soil. Other immobilisation amendments, such as activated carbon and activated clay, were also used for the remediation of methylmercury in addition to biochar (Gilmour et al. 2018). Selenium nanoparticles have proven to be successful at immobilising Hg(0) (Wang et al. 2019). According to Wang et al. (2017), selenium nanoparticles transformed 45.8%–57.1% and 39.1%–48.6%, respectively, of the soil's Hg(0) into the insoluble mercuric selenide (HgSe). Since the mechanisms of Hg remediation are still poorly understood, the inclusion of biochar and other immobilisation modifications is novel.

6.10 Green Approaches for Treatment of Mercury

Innovative methods for mercury treatment have been tested at the bench and in pilot programmes, and they have the potential to be used extensively. These include in situ thermal desorption, air stripping, phytoremediation, and nanotechnology. The amount of mercury removed from an aqueous waste stream in a pilot-scale test of the nanotechnology, thiol-Self Assembled Monolayers on Mesoporous Silica (SAMMS), was 97.4% after the first treatment and 99.4% after two additional (U.S.

Environmental Protection Agency 2007). The potential of eastern cottonwood trees to remove mercury has also been evaluated in the field; however, the study's findings were not yet available when the research for this article was being done. Another innovation being tested for its capacity to remove mercury from water is air stripping. More than 94% of the mercury was removed by stannous chloride at doses greater than 0.011 mg/L, and the remaining total mercury was reduced to concentrations below 10 ng/L (U.S. Environmental Protection Agency 2007). Low dosages of stannous chloride, however, demonstrated minimal mercury elimination

6.11 Regeneration and Economic Challenges of Biowaste-Derived Adsorbents

According to Zhou et al. (2019), the process of desorption can lead to the use of previously used adsorbent, reducing waste and lowering capital and operating expenses. Thermodynamic, acidic (HCl, H_2SO_4, phosphoric acid (H_3PO_4), nitric acid (HNO_3), NaOH, vacuum, organic solvent (methanol), and biological procedures are a few of the common desorption techniques (Hassan and Carr 2021).

In addition to reusing or regenerating the adsorbent, choosing a proper adsorbent, especially at large sizes, is essential for assuring an effective and affordable treatment technique. According to reports, powdered activated carbons are unsuitable for industrial applications because of their high costs, long production times, and difficult recovery procedures (Moosavi et al. 2020). This may result in high energy usage as a result of ineffective procedures. However, when mercury has been removed from wastewater, post-treatment of effluent and adsorbent that contains pollutants is required. Stability and immobilisation are the two potential methods for the safe disposal of Hg. For instance, the end effluent can be used as a binder material in the concrete technology or immobilisation (Saha et al. 2020). According to Dahiru et al. (2018), after 5 usages, the banana peel adsorbent's effectiveness dropped to 64%. Despite the fact that there have been many studies on the creation of activated carbon obtained from biowaste, there have been relatively little attempts at the life cycle analysis and techno-economic assessment of these applications. To maintain removal rates, reduce costs, and encourage the long-term regeneration of adsorbents, this indicates the need for the creation of more durable adsorbents.

6.12 Management of Post-Adsorption Materials

The adsorbent can be handled in a number of ways after use, including regeneration, reusing, and safe disposal. A chelating desorbing agent, a salt desorbing agent, an alkali desorbing agent, or thermal regeneration are just a few of the methods that can be used to regenerate (Yang et al. 2020). The following techniques can also be used to regenerate organic pollutants: ozonation, microwave-assisted regeneration, chemical regeneration, microbiological regeneration, photo-assisted oxidation, thermal regeneration, and electrochemical oxidation (Naghizadeh et al. 2017). The process makes the adsorbent unnecessary after numerous adsorption-regeneration cycles with the same contaminant. The used adsorbent can either be burned, recycled, or dumped in a landfill. Used adsorbents with hazardous elements may need to be stabilised or solidified before being disposed of in landfills (Paudyal et al. 2020), which will increase the cost of the life cycle analysis of the adsorbent. By appropriately disposing of the adsorbent and reusing it in other applications, you can increase its sustainability.

6.13 Challenges and Future Research Directions

Despite some serious drawbacks, traditional remediation techniques have been used extensively for mercury removal. First off, large-scale applications are hampered by the high cost of various techniques such as thermal desorption and activated carbon adsorption (Gilmour et al. 2018). Long-term monitoring of the stability of mercury should be carried out when adopting technologies that are appropriate for use on large sites to stabilise and contain mercury in soil (Wang et al. 2012).

Adsorption is the main method by which new remediation materials remove Hg. Numerous researches have looked at a variety of materials with high specific areas, plenty of pores, and active mercury sites. Although characteristics of sorbents, including their adsorption capacity, pore size, and specific area have been thoroughly investigated, it is recommended that future research focus on the partition coefficient of Hg species to make comparing the adsorption effectiveness of various materials easier.

Additionally, the effectiveness of adsorption may be impacted by competing metal ions that are present (Leus et al. 2017). Waste-derived materials are used for the adsorption of mercury to lower the cost. However, these materials might include additional dangerous components that pose a risk to your health. Reuse is a different technique to maximise the cost-efficiency of emerging biowaste. Many new materials have shown considerable potential for reuse in studies that have looked at the regeneration of adsorbents, whereas minerals cannot be regenerated whilst retaining their sorption activity.

6.14 Conclusion

Mercury can be found in a variety of media at waste sites, including soil, groundwater, sediment, sludge, and leachate. Pollution of Hg in water is related to possible toxicity to the environment and living organisms. An adsorption technique has been applied to remove Hg from polluted water and wastewater using biochar produced from different biowastes. The physical characteristics of the adsorbent (pore volume, pore size, surface area, and pore number) may have an impact on the adsorption capacity, which may result in an increase or a decrease in the effectiveness of Hg removal. Recycling of biowastes as an alternative to disposal and incineration has been inspired by the conservation of natural resources and the efficient use of non-renewable fuels. Innovative techniques have proven to be more economical and eco-friendlier than traditional remediation technologies like adsorption. The most efficient and eco-friendly method for removing Hg from water is adsorption using biowaste biochar. With a green technological approach using biowaste biochar can transform mercury and its compounds into less hazardous forms to guarantee mercury absence in the environment and water systems.

References

Abbas K, Znad H, Awual MR (2018) A ligand anchored conjugate adsorbent for effective mercury(II) detection and removal from aqueous media. Chem Eng J 334:432–443

Abraham AM, Kumar SV, Alhassan SM (2018) Porous sulphur copolymer for gasphase mercury removal and thermal insulation. Chem Eng J 332:1–7

Acharya J, Kuma U, Mahammed Rafi P (2018) Removal of heavy metal ions from wastewater by chemically modified agricultural waste material as potential adsorbent—a review. Int J Curr Eng Technol 8:526–530. https://doi.org/10.14741/ijcet/v.8.3.6

Al-Ghouti M, Abuqaoud R, Abu-Dieyeh M (2016) Detoxifcation of mercury pollutant leached from spent fluorescent lamps using bacterial strains. Waste Manage 49:238–244

Al-Ghouti MA, Da'ana D, Abu-Dieyeh M, Khraisheh M (2019) Adsorptive removal of mercury from water by adsorbents derived from date pits. Sci Rep 9:15327. https://doi.org/10.1038/s41598-019-51594-y

Alloway BJ (2013) Heavy metals in soils: trace metals and metalloids in soils and their bioavailability, vol 22. Springer Science & Business Media

Anbia M, Dehghan R (2014) Functionalized CMK-3 mesoporous carbon with 2-amino-5-mercapto-1, 3, 4-thiadiazole for Hg(II) removal from aqueous media. J Environ Sci 26:1541–1548

Arias F, Beneduci A, Chidichimo F, Furia E, Straface S (2017) Study of the adsorption of mercury(II) on lignocellulosic materials under static and dynamic conditions. Chemosphere 180:11–23

Attari M, Bukhari SS, Kazemian H, Rohani S (2017) A low-cost adsorbent from coal fly ash for mercury removal from industrial wastewater. J Environ Chem Eng 5:391–399

Azevedo R, Rodriguez E (2012) Phytotoxicity of mercury in plants: a review. J Bot 2012:1–6

Bai Y, Hong J (2021) Preparation of a novel millet straw biochar-bentonite composite and its adsorption property of Hg^{2+} in aqueous solution. Materials 14:1117. https://doi.org/10.3390/ma14051117

Bao S, Li K, Ning P, Peng J, Jin X, Tang L (2017) Highly effective removal of mercury and lead ions from wastewater by mercaptoamine-functionalisedsilicacoated magnetic nano-adsorbents: behaviours and mechanisms. Appl Surf Sci 393:457–466

Beckers F, Rinklebe J (2017) Cycling of mercury in the environment: sources, fate, and human health implications: a review. Crit Rev Environ Sci Technol 47:693–794

Beckers F, Awad YM, Beiyuan J, Abrigata J, Mothes S, Tsang DCW, Ok YS, Rinklebe J (2019) Impact of biochar on mobilization, methylation, and ethylation of mercury under dynamic redox conditions in a contaminated floodplain soil. Environ Int 127:276–290

Bhatnagar A, Sillanpää M (2010) Utilization of agro-industrial and municipal waste materials as potential adsorbents for water treatment—a review. Chem Eng J 157:277–296. https://doi.org/10.1016/j.cej.2010.01.007

Bhatnagar A, Minocha A, Sillanpää M (2010) Adsorptive removal of cobalt from aqueous solution by utilizing lemon peel as biosorbent. Biochem Eng J 48:181–186. https://doi.org/10.1016/j.bej.2009.10.005

Bussi J, Cabrera MN, Chiazzaro J, Canel C, Veiga S, Florencio C, Dalchiele EA, Belluzzi M (2010) The recovery and recycling of mercury from fluorescent lamps using photocatalytic techniques. J Chem Technol Biotechnol 85:478–484. https://doi.org/10.1002/jctb.2313

Chaudhry FN, Malik M (2017) Factors affecting water pollution: a review. J Ecosyst Ecogr 7:225–231

Chen X, Xia X, Wu S, Wang F, Guo X (2010) Mercury in urban soils with various types of land use in Beijing, China. Environ Pollut 158:48–54

Chen M, Wang D, Yang F, Xu X, Xu N, Cao X (2017) Transport and retention of biochar nanoparticles in a paddy soil under environmentally-relevant solution chemistry conditions. Environ Pollut 230:540–549. https://doi.org/10.1016/j.envpol.2017.06.101

Chen H, Zhan L, Gu L, Zhou H, Feng Q, Wu H, Zhao S, Yang L (2022) Fate and distribution of mercury during the desulfurization wastewater evaporation process. Fuel 318:123609. https://doi.org/10.1016/j.fuel.2022.123609

Clyde Vincent JA (2016) Comparative study of heavy metal contamination at common biomedical waste treatment and disposal sites (Incineration and Deep Burial) in Mumbai, Maharashtra, India. Int J Health Sci Res 6:415–419

Dahiru M, Zango ZU, Haruna MA (2018) Cationic dyes removal using low-cost banana peel biosorbent. Am J Mater Sci 8:32–38

Donatello S, Fernández-Jiménez A, Palomo A (2012) An assessment of Mercury immobilisation in alkali activated fly ash (AAFA) cements. J Hazard Mater 213:207–215

El-Ramady H, El-Henawy A, Amer M, Omara AED, Elsakhawy T, Elbasiouny H, Elbehiry F, Abou Elyazid D, El-Mahrouk M (2020) Agricultural waste and its nano-management: mini review. Egypt J Soil Sci 60(4):349–366. https://doi.org/10.21608/ejss.2020.46807.1397

Fatoni A, Koesnarpadi S, Hidayati N (2015) Synthesis, Characterization of cellulose modifed with 2-mercaptobenzothiazole and its adsorption to Cu(II) ion in aqueous solution. Indonesian J Chem 15:194

Feng XF, Long RX, Wang LL, Liu CC, Bai ZX (2021) A review on heavy metal ions adsorption from water by layered double hydroxide and its composites. Sep Purif Technol 120099. https://doi.org/10.1016/j.seppur.2021.120099

Figueira P, Lopes CB, Daniel-da-Silva AL, Pereira E, Duarte AC, Trindade T (2011) Removal of mercury(II) by dithiocarbamate surface functionalized magnetite particles: application to synthetic and natural spiked waters. Water Res 45:5773–5784

Gall JE, Boyd RS, Rajakaruna N (2015) Transfer of heavy metals through terrestrial food webs: a review. Environ Monit Assess 187:201

Genthe B, Kapwata T, Le Roux W, Chamier J, Wright CY (2018) The reach of human health risks associated with metals/metalloids in water and vegetables along a contaminated river catchment: South Africa and Mozambique. Chemosphere 199:1–9

Gilmour C, Bell T, Soren A, Riedel G, Riedel G, Kopec D, Bodaly D, Ghosh U (2018) Activated carbon thin-layer placement as an in situ mercury remediation tool in a Penobscot River salt marsh. Sci Total Environ 621:839–848

Grenni P, Caracciolo AB, Mariani L, Cardoni M, Riccucci C, Elhaes H, Ibrahim MA (2019) Effectiveness of a new green technology for metal removal from contaminated water. Microchem J 147:1010–1020. https://doi.org/10.1016/j.microc.2019.04.026

Gustin MS (2003) Are mercury emissions from geologic sources significant? A status report. Sci Total Environ 304:153–167

Habuda-Stanić M, Nujić M (2015) Arsenic removal by nanoparticles: a review. Environ Sci Pollut Res 22:8094–8123

Hassan MM, Carr CM (2021) Biomass-derived porous carbonaceous materials and their composites as adsorbents for cationic and anionic dyes: a review. Chemosphere 265:129087

He F, Gao J, Pierce E, Strong PJ, Wang H, Liang L (2015) *In situ* remediation technologies for mercury-contaminated soil. Environ Sci Pollut Res 22:8124–8147

Hua YM, Heal KV, Friesl-Hanl W (2017) The use of red mud as an immobiliser for metal/metalloid-contaminated soil: a review. J Hazard Mater 325:17–30

Huang Y, Du JR, Zhang Y, Lawless D, Feng X (2015) Removal of mercury(II) from wastewater by polyvinylamine-enhanced ultrafiltration. Sep Purif Technol 154:1–10. https://doi.org/10.1016/j.seppur.2015.09.003

Igwe J, Abia A (2007) Adsorption isotherm studies of Cd(II), Pb(II) and Zn(II) ions bioremediation from aqueous solution using unmodifed and EDTA-modifed maize cob. Eclética Química 32:33–42

Jia L, Fan BG, Li B, Yao YX, Huo RP, Zhao R, Qiao XL, Jin Y (2018) Effects of pyrolysis mode and particle size on the microscopic characteristics and mercury adsorption characteristics of biomass char. BioResources 13(3):5450–5471. https://doi.org/10.1537/biores.13.3.5450-5471

Kabata-Pendias A (2010) Trace elements in soils and plants. CRC Press, Washington DC

Kabiri S, Tran DNH, Azari S, Losic D (2015) Graphene-diatom silica aerogels for efficient removal of mercury ions from water. ACS Appl Mater Interfaces 7:11815–11823

Karatza D, Lancia A, Musmarra D, Zucchini C (2000) Study of mercury absorption and desorption on sulfur impregnated carbon. Exp Thermal Fluid Sci 21:150–155

Kazemi F, Younesi H, Ghoreyshi AA, Bahramifar N, Heidari A (2016) Thiol-incorporated activated carbon derived from fir wood sawdust as an efficient adsorbent for the removal of mercury ion: batch and fixed-bed column studies. Process Saf Environ Prot 100:22–35

Khraisheh M, Al-Ghouti M, Allen S, Ahmad M (2004) Effect of pH, temperature, and molecular size on the removal of dyes from textile effluent using manganese oxides-modified diatomite. Water Environ Fed 76:2655–2663

Kostova I, Vassileva C, Dai S, Hower JC, Apostolova D (2013) Influence of surface area properties on mercury capture behaviour of coal fly ashes from some Bulgarian power plants. Int J Coal Geol 116:227–235

Kumar N, Fosso-Kankeu E, Ray SS (2019) Achieving controllable MoS_2 nanostructures with increased interlayer spacing for efficient removal of Pb(II) from aquatic systems. ACS Appl Mater Interfaces 11:19141–19155

Kyzas GZ, Kostoglou M (2015) Swelling-adsorption interactions during mercury and nickel ions removal by chitosan derivatives. Sep Purif Technol 149:92–102

Lecler MT, Zimmermann F, Silvente E, Masson A, Morèle Y, Remy A, Chollot A (2018) Improving the work environment in the fluorescent lamp recycling sector by optimizing mercury elimination. Waste Manage 76:250–260

Leus K, Perez JPH, Folens K, Meledina M, Van Tendeloo G, Du Laing G, Van Der Voort P (2017) UiO-66-(SH)(2) as stable, selective and regenerable adsorbent for the removal of mercury from water under environmentally-relevant conditions. Faraday Discuss 201:145–161

Lewis AS, Huntington TG, Marvin-Dipasquale MC, Amirbahman A (2016) Mercury remediation in wetland sediment using zero-valent iron and granular activated carbon. Environ Pollut 212:366–373

Li Z, Wu L, Liu H, Lan H, Qu J (2013) Improvement of aqueous mercury adsorption on activated coke by thiolfunctionalization. Chem Eng J 228:925–934

Li G, Shen B, Lu F (2015a) The mechanism of sulfur component in pyrolyzed char from waste tire on the elemental mercury removal. Chem Eng J 273:446–454

Li G, Shen B, Wang Y, Yue S, Xi Y, An M, Ren K (2015b) Comparative study of element mercury removal by three bio-chars from various solid wastes. Fuel 145:189–195

Li R, Wu H, Ding J, Fu W, Gan L, Li Y (2017) Mercury pollution in vegetables, grains and soils from areas surrounding coal-fired power plants. Sci Rep 7(1):46545

Liu Z, Yang W, Xu W, Liu Y (2018) Removal of elemental mercury by bio-chars derived from seaweed impregnated with potassium iodine. Chem Eng J 339:468–478

Liu Z, Xu Z, Xu L, Buyong F, Chay TC, Li Z, Cai Y, Hu B, Zhu Y, Wang X (2022) Modifed biochar: synthesis and mechanism for removal of environmental heavy metals. Carbon Res 1:8. https://doi.org/10.1007/s44246-022-00007-3
Lu X, Jiang J, Sun K, Wang J, Zhang Y (2014) Influence of the pore structure and surface chemical properties of activated carbon on the adsorption of mercury from aqueous solutions. Mar Pollut Bull 78:69–76
Lu L, Yu W, Wang Y, Zhang K, Zhu X, Zhang Y, Wu Y, Ullah H, Xiao X, Chen B (2020) Application of biochar-based materials in environmental remediation: from multilevel structures to specific devices. Biochar 2(1):1–31. https://doi.org/10.1007/s42773-020-00041-7
Malakahmad A, Tan S, Yavari S (2016) Valorization of wasted black tea as a low-cost adsorbent for nickel and zinc removal from aqueous solution. J Chem. https://doi.org/10.1155/2016/5680983
Mallampatia R, Valiyaveettil S (2012) Application of tomato peels as an efficient adsorbent for water purification-alternative biotechnology? RSC Adv 2:9914–9920. https://doi.org/10.1039/c2ra21108d
Marrugo-Negrete J, Enamorado-Montes G, Durango-Hernández J, Pinedo-Hernández J, Díez S (2017) Removal of mercury from gold mine effluents using Limnocharis flava in constructed wetlands. Chemosphere 167:188–192. https://doi.org/10.1016/j.chemosphere.2016.09.130
Martín JAR, Nanos N (2016) Soil as an archive of coalfired power plant mercury deposition. J Hazard Mater 308:131–138
Massachusetts Department of Environmental Protect (MADEP) (1996) Mercury: forms, fate & effects (Chapter 2). http://www.mass.gov/dep/toxics/stypes/hgch2.htm
Mittal A, Malyiya A, Kaur D, Mittal J, Kurup L (2007) Studies on the adsorption kinetics and isotherms for the removal and recovery of methyl orange from wastewaters using waste materials. J Hazard Mater 148:229–240. https://doi.org/10.1016/j.jhazmat.2007.02.02
Mondal S, Chatterjee S, Mondal S, Bhaumik A (2019) Thioether-functionalized covalent triazine nanospheres: a robust adsorbent for mercury removal. ACS Sustain Chem Eng 7:7353–7361
Moosavi S, Lai CW, Gan S, Zamiri G, AkbarzadehPivehzhani O, Johan MR (2020) Application of efficient magnetic particles and activated carbon for dye removal from wastewater. ACS Omega 5:20684–20697
Naghizadeh A, Momeni F, Derakhshani E (2017) Efficiency of ultrasonic process in the regeneration of graphene nanoparticles saturated with humic acid. Desalin Water Treat 70:290–293
Obrist D, Kirk JL, Zhang L, Sunderland EM, Jiskra M, Selin NE (2018) A review of global environmental mercury processes in response to human and natural perturbations: changes of emissions, climate, and land use. Ambio 47:116–140
O'Connor D, Peng T, Li G, Wang S, Duan L, Mulder J, Cornelissen G, Cheng Z, Yang S, Hou D (2018) Sulfur-modified rice husk biochar: a green method for the remediation of mercury contaminated soil. Sci Total Environ 621:819–826
Ojea-Jiménez I, López X, Arbiol J, Puntes V (2012) Citrate-coated gold nanoparticles as smart scavengers for mercury(II) removal from polluted waters. ACS Nano 6:2253–2260
Ozgur C, Coskun S, Akcil A, Beyhan M, Üncü IS, Civelekoglu G (2016) Combined oxidative leaching and electrowinning process for mercury recovery from spent fuorescent lamps. Waste Manage 57:215–219
Patra M, Bhowmik N, Bandopadhyay B, Sharma A (2004) Comparison of mercury, lead and arsenic with respect to genotoxic effects on plant systems and the development of genetic tolerance. Environ Exp Bot 52:199–223
Paudyal H, Ohto K, Kawakita H, Inoue K (2020) Recovery of fluoride from water through adsorption using orange-waste gel, followed by desorption using saturated lime water. J Mater Cycles Waste Manage 22:1484–1491
Pavlish JH, Sondreal EA, Mann MD, Olson ES, Galbreath KC, Laudal DL, Benson SA (2003) Status review of mercury control options for coal-fired power plants. Fuel Process Technol 82:89–165
Poste AE, Pastukhov MV, Braaten HFV, Qzersky T, Moore M (2018) Past and present mercury accumulation in the Lake Baikal seal: temporal trends, effects of life history, and toxicological implications. Environ Toxicol Chem 37(5):1476–1486

Pourrahim S, Salem A, Salem S, Tavangar R (2020) Application of solid waste of ductile cast iron industry for treatment of wastewater contaminated by reactive blue dye via appropriate nano-porous magnesium oxide. Environ Pollut 256:113454. https://doi.org/10.1016/j.envpol.2019.113454

Powell KJ, Brown PL, Byrne RH, Gajda T, Hefter G, Sjöberg S, Wanner H (2005) Chemical speciation of environmentally significant heavy metals with inorganic ligands. Part 1: The Hg^{2+}–Cl^-, OH^-, $CO_3{}^{2-}$, $SO_4{}^{2-}$, and $PO_4{}^{3-}$ aqueous systems (IUPAC Technical Report). Pure Appl Chem 77(4):739–800

Raju A, Singh A, Srivastava N, Singh S, Jigyasu DK, Singh M (2019) Mapping human health risk by geostatistical method: a case study of mercury in drinking groundwater resource of the central Ganga alluvial plain, Northern India. Environ Monit Assess 191:298

Richard JH, Bischoff C, Biester H (2016) Comparing modeled and measured mercury speciation in contaminated groundwater: importance of dissolved organic matter composition. Environ Sci Technol 50(14):7508–7516

Rocha LS, Almeida Â, Nunes C, Henriques B, Coimbra MA, Lopes CB, Silva CM, Duarte AC, Pereira E (2016) Simple and effective chitosan based films for the removal of Hg from waters: equilibrium, kinetic and ionic competition. Chem Eng J 300:217–229

Saha A, Basak BB, Ponnuchamy M (2020) Performance of activated carbon derived from *Cymbopogon winterianus* distillation waste for scavenging of aqueous toxic anionic dye Congo red: comparison with commercial activated carbon. Sep Sci Technol 55:1970–1983

Sajjadi SA, Mohammadzadeh A, Tran HN, Anastopoulos I, Dotto GL, Lopičić ZR, Sivamani S, Rahmani-Sani A, Ivanets A, Hosseini-Bandegharaei A (2018) Efficient mercury removal from wastewater by pistachio wood wastes-derived activated carbon prepared by chemical activation using a novel activating agent. J Environ Manage 223:1001–1009

Samra S (2014) Biosorption of Pb^{2+} from natural water using date pits: a green chemistry approach. Mod Chem Appl 2

Sharma P, Iqbal HMN, Chandra R (2022) Evaluation of pollution parameters and toxic elements in wastewater of pulp and paper industries in India: a case study. Case Stud Chem Environ Eng 5:100163. https://doi.org/10.1016/j.cscee.2021.100163

Shen B, Li G, Wang F, Wang Y, He C, Zhang M, Singh S (2015) Elemental mercury removal by the modified bio-char from medicinal residues. Chem Eng J 272:28–37

Silva H, Ruiz S, Granados D, Santángelo J (2010) Adsorption of mercury(II)from liquid solutions using modifed activated carbons. Mat Res 13:129–134

Sitko R, Musielak M, Serda M, Talik E, Zawisza B (2021) Thiosemicarbazide-grafted graphene oxide as superior adsorbent for highly efficient and selective removal of mercury ions from water. Sep Purif Technol 254:117606. https://doi.org/10.1016/j.seppur.2020.117606

Sorbal L, Yallouz AV, Fernandes AL (2006) Treatment of mercury bearing fuorescent lamps by using electrochemical process. CETEM—Centre for Mineral Technology

Spahić MP, Manojlović D, Tančić P, Cvetković Ž, Nikić Z, Kovačević R, Sakan S (2019) Environmental impact of industrial and agricultural activities to the trace element content in soil of Srem (Serbia). Environ Monit Assess 191:133

Streets DG, Lu Z, Levin L, Schure AF, Sunderland EM (2018) Historical releases of mercury to air, land, and water from coal combustion. Sci Total Environ 615:131–140

Tan Y, Mortazavi R, Dureau B, Douglas MA (2004) An investigation of mercury distribution and speciation during coal combustion. Fuel 83:2229–2236

Tan Z, Yuan S, Hong M, Zhang L, Huang Q (2020) Mechanism of negative surface charge formation on biochar and its effect on the fixation of soil Cd. J Hazard Mater 384:121370

Tang H, Wang J, Zhang S, Pang H, Wang X, Chen Z, Li M, Song G, Qiu M, Yu S (2021) Recent advances in nanoscale zero-valent iron-based materials: characteristics, environmental remediation and challenges. J Clean Prod 319:128641

U.S. Environmental Protection Agency (2007) Treatment technologies for mercury in soil, waste, and water. Office of Superfund Remediation and Technology Innovation Washington, DC, p 20460

UNEP (2013) Global mercury assessment 2013: sources, emissions, releases and environmental transport. United Nations Environment Programme, Geneva
UNEP (2018) Global mercury assessment. United Nations Environment Programme, Geneva
U.S. EPA (2006) Binational toxics strategy mercury progress report. From Alexis Cain, EPA Region 5, to the Binational Toxics Strategy Mercury Workgroup. http://www.epa.gov/region5/mercury/progress06.pdf
Wang J, Feng X, Anderson CW, Xing Y, Shang L (2012) Remediation of mercury contaminated sites–a review. J Hazard Mater 221:1–18
Wang T, Liu J, Zhang Y, Zhang H, Chen WY, Norris P, Pan WP (2018) Use of a non-thermal plasma technique to increase the number of chlorine active sites on biochar for improved mercury removal. Chem Eng J 331:536–544
Wang X, Wang S, Pan X, Gadd GM (2019) Heteroaggregation of soil particulate organic matter and biogenic selenium nanoparticles for remediation of elemental mercury contamination. Chemosphere 221:486–492
Wang L, Li X, Tsang DCW, Jin F, Hou D (2020) Green remediation of Cd and Hg contaminated soil using humic acid modified montmorillonite: Immobilization performance under accelerated ageing conditions. J Hazard Mater 387:122005. https://doi.org/10.1016/j.jhazmat.2019.122005
WHO (2004) Guidelines for drinking-water quality, vol 1. World Health Organization
WHO (2017) Ten chemicals of major health concern. World Health Organization
Xu J, Bravo AG, Lagerkvist A, Bertilsson S, Sjöblom R, Kumpiene J (2015) Sources and remediation techniques for mercury contaminated soil. Environ Int 74:42–53
Xu W, Adewuyi YG, Liu Y, Wang Y (2018) Removal of elemental mercury from flue gas using CuOx and CeO_2 modified rice straw chars enhanced by ultrasound. Fuel Process Technol 170:21–31
Xu W, Pan J, Fan B, Liu Y (2019a) Removal of gaseous elemental mercury using seaweed chars impregnated by NH_4Cl and NH_4Br. J Clean Prod 216:277–287
Xu Y, Luo G, He S, Deng F, Pang Q, Xu Y, Yao H (2019b) Efficient removal of elemental mercury by magnetic chlorinated biochars derived from co-pyrolysis of $Fe(NO_3)$3-laden wood and polyvinyl chloride waste. Fuel 239:982–990
Yang J, Zhao Y, Ma S, Zhu B, Zhang J, Zheng C (2016) Mercury removal by magnetic biochar derived from simultaneous activation and magnetization of sawdust. Environ Sci Technol 50:12040–12047
Yang W, Liu Y, Wang Q, Pan J (2017) Removal of elemental mercury from flue gas using wheat straw chars modified by Mn-Ce mixed oxides with ultrasonic-assisted impregnation. Chem Eng J 326:169–181
Yang W, Hussain A, Zhang J, Liu Y (2018) Removal of elemental mercury from flue gas using red mud impregnated by KBr and KI reagent. Chem Eng J 341:483–494
Yang X, Debeli DK, Shan G, Pan P (2020) Selective adsorption and high recovery of La^{3+} using graphene oxide/poly (*N*-isopropyl acrylamide-maleic acid) cryogel. Chem Eng J 379:122335
Yu JG, Yue BY, Wu XW, Liu Q, Jiao FP, Jiang XY, Chen XQ (2016) Removal of mercury by adsorption: a review. Environ Sci Pollut Res 23:5056–5076
Yu S, Tang H, Zhang D, Wang S, Qiu M, Song G, Fu D, Hu B, Wang X (2022) MXenes as emerging nanomaterials in water purifcation and environmental remediation. Sci Total Environ 811:152280
Zhang F, Nriagu J, Itoh H (2005) Mercury removal from water using activated carbons derived from organic sewage sludge. Water Res 39:389–395
Zhang H, Chen J, Zhu L, Yang G, Li D (2014) Anthropogenic mercury enrichment factors and contributions in soils of Guangdong Province, South China. J Geochem Explor 144:312–319
Zhao T, Yu Z, Zhang J, Qu L, Li P (2018) Low-thermal remediation of mercury contaminated soil and cultivation of treated soil. Environ Sci Pollut Res 25:24135–24142
Zhou Y, Lu J, Zhou Y, Liu Y (2019) Recent advances for dyes removal using novel adsorbents: a review. Environ Pollut 252:352–365

Zhu Y, Fan W, Zhou T, Li X (2019) Removal of chelated heavy metals from aqueous solution: a review of current methods and mechanisms. Sci Total Environ 678:253–266. https://doi.org/10.1016/j.scitotenv.2019.04.416

Chapter 7
Removal of Mercury from Wastewater by Different Natural Biomasses

Rakesh Pant, Umangjit Kaur, Sunidhi Shreya, Nirmal Patrick, and Amit Gupta

7.1 Introduction

The increasing human population causes a drastic change in the environment. Due to its bioaccumulation and prolonged persistence properties, mercury (Hg(II)) is regarded as among the most significant poisonous heavy metals. It has a negative impact on individuals' health when exposed to it over an extended period of time. High amounts of mercury are released into the air, soil, and water due to anthropogenic activities like fuel use and gold mining (Fig. 7.1). The World Health Organization (WHO) and the United States Environmental Protection Agency (USEPA) have thus determined that the maximum tolerable amounts of Hg(II) in drinking water are, respectively, 0.002 mg L1 and 0.001 mg L1. Water must be purified of Hg(II) due to its toxicity (Giraldo et al. 2020). There are several effective methods for removing Hg(II) from wastewater in order to comply with legally acceptable limits, although ion exchange technology, reverse osmosis, and chemical precipitation are the most popular ones. However, such methods work efficiently, but they are costly, produce a lot of sludge, and produce a lot of byproducts. Adsorption is one of the most intriguing methods because of its high effectiveness, affordability, and simplicity of usage in wastewater treatment facilities. It is necessary to produce highly adsorbent, particular, and affordable materials (Awad et al. 2018).

Due to their diversity, microbes and microalgae can work together to effectively cleanse wastewater and recover nutrients. Wastewater-derived microalgal biomass has a wide range of applications as a renewable feedstock for the production of essential biochemical, biofuel, and animal feed. Microalgae are particularly good at

R. Pant · U. Kaur · S. Shreya · A. Gupta (✉)
Department of Microbiology, Graphic Era (Deemed to Be) University, Dehradun, UK 248001, India
e-mail: dr.amitgupta.bt@geu.ac.in; amitvsbt@gmail.com

R. Pant · N. Patrick
Department of Quality, Eureka Forbes Ltd Lal Tappar, Dehradun, UK 248140, India

N. Kumar (ed.), *Mercury Toxicity*, Environmental Science and Engineering,
https://doi.org/10.1007/978-981-99-7719-2_7

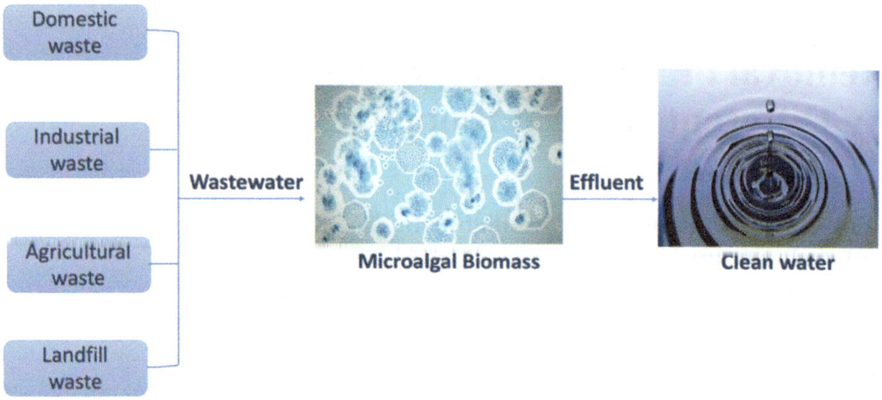

Fig. 7.1 Source of waste water and its treatment

using minerals from the water to feed themselves as they develop (Aditya et al. 2022). Continuous wastewater dumping without proper treatment might result in major environmental issues. The referred to as eutrophication phenomenon represents one of the main issues connected to the continual release of wastewater into lakes and rivers. The term "eutrophication" is broad and refers to a significant issue in the nation's estuaries. The occurrence of eutrophication materializes when the surroundings get enriched with vitamins and minerals, increased plant and algae growth is what leads to toxic dead zones, algal blooms, and the death of fish in estuaries and coastal waterways. An excess of algae and plants is the first sign of eutrophication, which then triggers a series of events in the ecosystem. When the surplus algae and plant stuff finally break down, it releases a lot of carbon dioxide. This results in ocean acidification, a process where seawater's pH decreases. The production of shells in bivalve molluscs can be prevented by acidification, which also retards fish and shellfish growth. This results in lower harvests and more costly seafood for both commercial and recreational fisheries due to the decreased catch (Gonçalves et al. 2017). During the tertiary treatment stage, phosphorus and nitrogen are mostly removed from wastewater. The biological processes that involve anaerobic digestion, nitrification, and denitrification are among the most often utilized techniques. To reach the nutrient levels recognized by EU regulation, numerous anaerobic, nitrification, and denitrification cycles are necessary. These techniques also need many tanks and inner recycles of activated sludge, which raises the price, complexity, and energy requirements of the process as a whole. The elimination of nitrogen and phosphorus can also be accomplished chemically, for example, by precipitation using aluminum and iron salts. However, these techniques are expensive and generate a lot of sludge that needs to be further treated since it contains chemical contaminants (Boelee et al. 2011).

Treatment with the help of photosynthetic microorganisms helps to reduce the side effects of tertiary wastewater treatment plants. Since these eubacteria need a high amount of nitrogen and phosphorous to grow, they absorb these nutrients from the wastewater. These microbes help in the efficient removal of phosphorus and nitrogen from the wastewater of various sources. Another advantage of using microalgae for nutrient removal is recycling nitrogen and phosphorus via the generation of fertilizers, biofuels, animal feed, and even pharmaceuticals from biomass.

Metals with a relatively high density compared to water are referred to as heavy metals. Most of the heavy metals are naturally found on the earth's crust. However, many heavy metals are the result of anthropogenic processes like metal processing in petroleum, refineries, mining, etc. Despite having a few benefits, heavy metals cause a drastic effect on the biological system by damaging various organelles like mitochondria (the powerhouse), cell membranes, nuclei, and lysosomes, and also damage many enzymes that aid in cell repair. The mutation is one of the hazardous effects of trace elements on human health, which can even lead to apoptosis of cells. All heavy metals like mercury, arsenic, cadmium, lead, and chromium are known to have high toxicity thereby causing multiple organ failure. Many molecular components of heavy metal-induced toxicity and carcinogenicity are involved, some of which are not well characterized or understood. However, it is recognized that each metal has distinct characteristics and physicochemical attributes that give rise to its own toxicological action mechanism (Fig. 7.2). This review examines the molecular processes of the toxicity, genotoxicity, and carcinogenicity of mercury as well as their environmental presence, usage, and production, possible exposure of humans (Tchounwou et al. 2012).

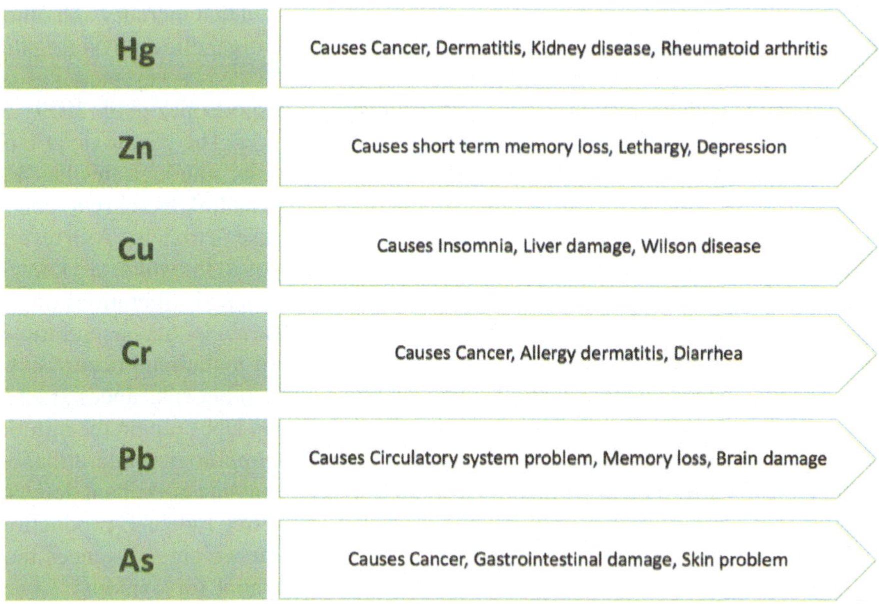

Fig. 7.2 Heavy metals and health problems associated with them

7.2 Occurrence of Mercury

Fertilizers, domestic waste inputs, and groundwater infiltration are some examples of where mercury enters the water stream (Fig. 7.3). Mercury is a metal that is found in three forms i.e. element, organic, and inorganic form. Environmental and occupational contexts can both expose people to inorganic mercury salts. Mining, the production of electrical equipment, and the use of mercury in chemical and metal processing are among the professions that provide a greater chance of risk exposure to mercury and its salts (Hargreaves et al. 2016). Mercury is a common environmental toxin and contaminant that causes serious changes in bodily tissues and a variety of negative health consequences. Mercury is present in the environment in a variety of chemical forms that can harm both people and animals. These consist of organic mercury compounds, inorganic mercuric, mercuric (Hg^{+1}), and elemental mercury vapor (Hg^0). Since mercury is so common in the environment, it is impossible for anybody to escape being exposed to it. This includes people, plants, and animals. The study of metal-induced carcinogenicity has attracted a lot of attention from the public health community. Carcinogenesis is often divided into three stages: commencement, promotion, advancement, and metastasis. Latest research has shown that additional molecular processes, including oncogene amplification, recombination, signal transduction, and transcription activation, also make up major contributors to the onset of carcinogenesis, despite the reality that DNA mutations, that may trigger cancer development or retard tumor suppression, were previously believed to be essential variables. Research has demonstrated that mercury and other hazardous metals have an unfavorable influence on cellular organelles' physiological functioning. Additionally, mounting evidence points to ROS as a key mediator of metal-induced cellular responses and carcinogenesis (Crespo-López et al. 2009). It is very debatable whether exposure to mercury causes cancer. The genotoxic potential of mercury has been established by certain investigations, although this has not always been the case according to other research. Studies that link mercury to genotoxicity have identified oxidative stress as the molecular cause of toxicity. As a result, it has been demonstrated that mercury causes the creation of ROS, which are known to damage DNA in cells and start carcinogenic processes. Genetic alterations might result from these free radicals' direct interaction with nucleic acids. At concentrations as low as 0.5 M, inorganic mercury was exhibited to elicit mutational occurrences in eukaryotic cell lines, despite the fact that mercury-containing substances are not mutagenic in bacterial experiments. The proteins involved in DNA repair, the mitotic spindle, and chromosomal segregation may also undergo morphological alterations as a result of these free radicals (Pinheiro et al. 2008). Cells contain antioxidant defense systems that work to rectify and prevent excessive ROS (free radicals) production in order to counteract these consequences. Low molecular weight substances that defend cells by chelating mercury and lowering its potential for oxidative stress include vitamins C and E, glutathione, melatonin, catalase, superoxide dismutase, glutathione reductase, and glutathione peroxidase.

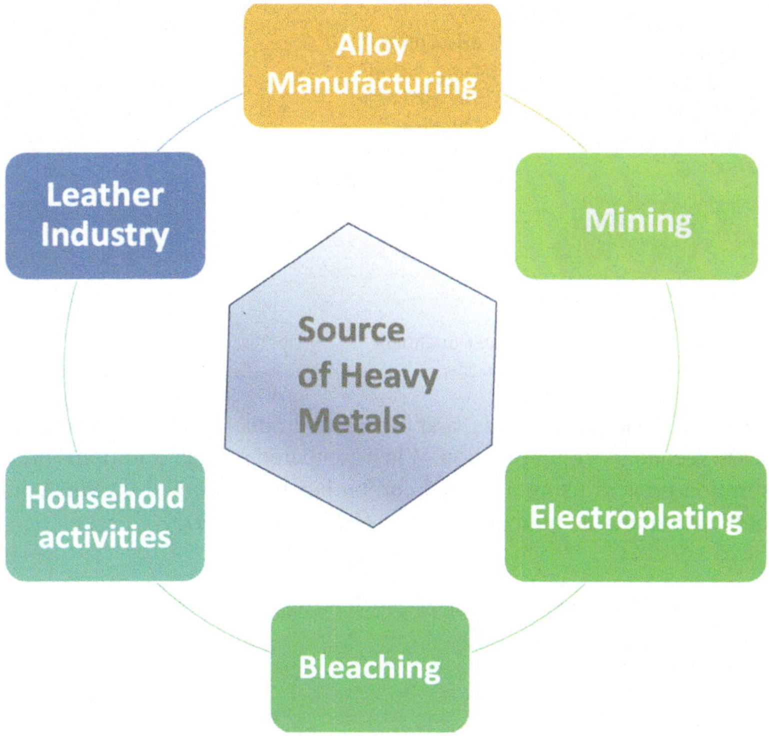

Fig. 7.3 Source of heavy metals

7.3 Removal of Mercury Using Biochar

Mercury is a hazardous element for the environment. Through food chains, it progressively bio-accumulates in the aquatic environment before reaching humans where it causes a variety of neurological diseases and problems. Carbon molecule is known as the most effective material to remove mercury from wastewater. Which biochar is one of the best components? Biochar is a rich source of carbon atoms which is derived using pyrolysis of biomass waste. Bio-oil (60–75 wt%), bio-char (15–20 wt%), and non-condensable gases (10–20 wt%) are the three products generated. The composition of the feedstock affects yields. This biochar, which has a high carbon content and little ash, can serve as an affordable and acceptable source of activated carbons, which might be used as an adsorbent to remove mercury from coal-fired power plants. Wastewater mercury removal methods have been established and put to use. Ion exchange, sulfide precipitation, adsorption, alum and iron coagulation, electrodeposition, and biological methods are the most often employed techniques for eliminating mercury from wastewater treatment procedures (Fig. 7.4 and Table 7.1). Yet, the majority of these traditional methods require a lot of time and money, their

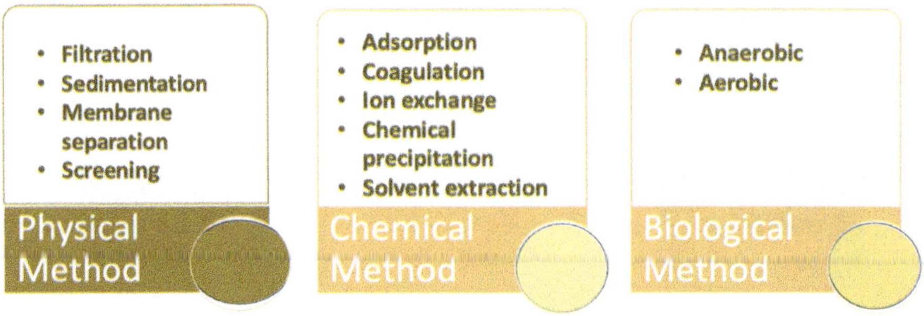

Fig. 7.4 Various methods of wastewater treatment

treatment effectiveness is unstable, and the waste product produced during the reaction leads to secondary contamination. When compared to other traditional processes, the adsorption process using activated carbon has been deemed the most practical and profitable among the many procedures (Park et al. 2019).

Table 7.1 Traditional method to draw out mercury from wastewater

Conventional method for the eradication of mercury from wastewater	Advantages	Disadvantages
Conventional coagulation	Can be used in the removal of high concentrations of mercury	Does not provide the desired result
Ion exchange	This method is metal-specific, also resins can be reused	High maintenance cost, brine once used must be changed
Membrane filtration	Helps in the separation of heavy metals and organic compounds. There are different sizes of filters available to separate different sizes of impurities	The process is highly expensive and the flow rate is slow
Chemical precipitation	It is non-metal selective, cheap, and reasonably easy	Sludge formation is high, metal precipitation is sluggish, and settling is poor
The reverse osmosis	Process proceeds by simultaneous separation and concentration at normal temperature	High cost, high pressure can damage equipment, high risk of clogging
Electrochemical treatment	Reduces the production of sludge, specific in nature	High capital investment, sensitive to changes in pH
Adsorption by biomass	Is environmentally friendly, easily available, and cost-effective	Slow process
Adsorption by biochar	Effective, easy handling	High capital cost, slow reaction

In recent studies, scientists have modeled biochar to absorb more amount of heavy metals like mercury. Activating agents are used to produce these modifications which is costly. This activation seeks to produce porosity, surface area, and a variety of additional functional groups. These properties are greatly impacted by the adsorbent material's preparation conditions, such as the kind of biomass, activation duration, activation temperature, and the impregnation ratio (chemical agent: biomass). As a result, the difficulty in making adsorbents is in producing carbonaceous materials that are extremely specialized and appropriate for certain purposes (Njoku et al. 2014). Due to its efficiency, affordability, and environmental friendliness, adsorption is regarded as one of the greatest methods for decontaminating water. It is a method that is reliable enough to fulfill industry requirements for stringent runoff standards and water reuse. Adsorption essentially involves the mass transfer of a metal ion from the solution to the surface of the sorbent, where it forms attached by physical and chemical bonds. Only in a low-temperature environment with the proper pH balance can physical adsorption take place (Abdel-Raouf and Abdul-Raheim 2017).

Apart from having many advantages, there are disadvantages too. To overcome those disadvantages, biochar has received a lot of attention lately as a cost-effective adsorbent. There is an alternative process of forming biochar, i.e., by the use of agricultural and industrial waste which is comparatively cheaper. Modified and pristine biochar has high adsorption properties of attracting heavy metals, dyes, and many inorganic salts. Environmental changes also affect mercury adsorption with the help of biochar but its characteristics are still unrecognized (Tan et al. 2016). Recent research on the features of mercury adsorption on post-treated sulfurized biochar utilizing calcium polysulfide and element S have been published, however, they are also economically inefficient, like the previously described sulfurized activated carbon. Additionally, no research has been done on the properties and processes of mercury adsorption by sulfurized biochar that has been pre-treated with element S. Prior studies on the triggering of activated carbon for mercury adsorption relied heavily on the reactions of activated carbon with H_2S, SO_2, sulfuric acid, along with salts of thionyl and mercaptoethanol (Park et al. 2019). Additionally, while some studies found that sulfurized activated carbon and carbon nanotubes were more effective in adsorbing mercury, they did not give much information on the interactions among the sulfur functional groups and mercury throughout the adsorption process. On the other hand, because of their poor wetting properties, interference with water permeability, and lack of adequate economic benefits, elemental S and S-containing minerals like pyrite are not employed in wastewater treatment. Additionally, the environment's Sulfate-reducing bacteria may produce more extremely hazardous methylmercury as a result of their breakdown, increasing Sulfate levels (Zhao et al. 2020).

7.4 Application of Graphene in the Elimination of Mercury from Wastewater

Graphene is a nanoparticle made up of carbon atoms that assist in eliminating heavy metals from wastewater. Several layers of graphite sum up to make graphene. It has a unique 2-D structure and various different physic-chemical properties. The more functional groups attached to graphene, the more its performance. Recent studies focus on the use of graphene for the removal of impurities from water and air. Therefore, there are various modifications made in graphene to increase its adsorption capacity of heavy metals that are dangerous pollutants in the ecosystem. They discovered that functional groups with oxygen have an important function in the sorption of metal ions on the oxidized graphene (GO) surface. Adsorption highly depends on the pH of water. For maximum adsorption, it is found to have a pH of 6 and a temperature of 30 °C. Among the different types of carbon allotropes of carbon, graphene is the one that shows the maximum elimination of trace elements from wastewater. The findings demonstrated that the pH and ionic strength had little role in the adsorption of the tested metals on the graphene oxide membranes. Adsorption equilibrium was reached quickly, and the isotherms of adsorption followed Langmuir's model effectively (Oke et al. 2023).

Graphene oxide membranes may be renewed over six times with negligible capacity for adsorption damages, according to an analysis of the adsorption/desorption cycles. The functional groups on the surface of graphene oxide may be altered by adding organic compounds, which will increase the adsorption of various materials. In comparison to graphene oxide and activated carbon, the quantity of Hg^{2+} ions adsorbed on the surface of the modified graphene oxide by Gao et al. was six times more. A combination of FE-SEM, UV–vis, and FT-IR methods was used to characterize the produced non-composites. In contrast to graphene oxide, the graphene oxide-porphyrin combination exhibits superior Hg(II) ion adsorption. It was found that the composite with an EDTA solution has an extended lifespan and can renew itself fast. SEM, TEM, and AFM methods were used to examine the structure of the resultant graphene. Raman spectroscopy was used to evaluate structural alterations, while FT-IR and UV–vis analysis were used to examine functional groups. The maximal ability to absorb mercury, which is higher than any other, was made possible by the large surface area of graphene.

7.5 Process of Adsorption

The action of an atom or an ion from a liquid or gaseous bulk phase adhering to a solid surface is known as adsorption (Fig. 7.5). Adsorbents are the solids utilized in this instance to adsorb the molecule or ion. Rarely are liquids utilized as adsorbents. Adsorbate ought not to penetrate the structure of the adsorbent since adsorption primarily is a surface activity that only affects the adsorbent's surface (Russo et al.

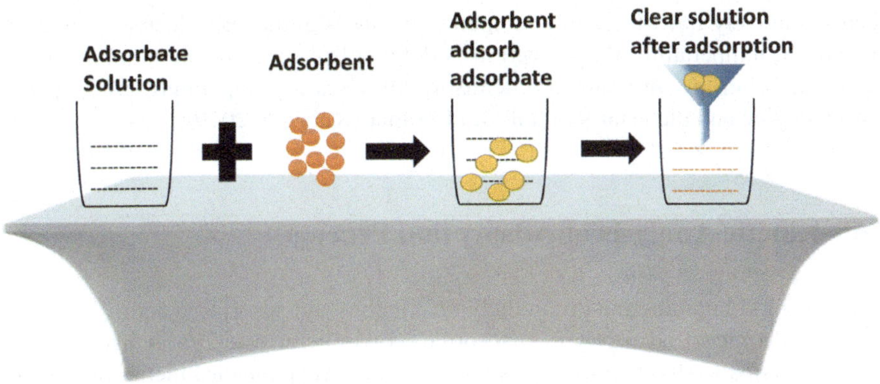

Fig. 7.5 Process of adsorption

2020). The exchange of mass and adsorption of a molecule of carbon from an aqueous phase onto a solid surface (adsorbent) is the basic idea behind carbon adsorption. This method of making activated carbon is done solely to manufacture very porous molecules of carbon with a significant internal surface. The aforementioned porous structure captures and holds metals as well as organic and inorganic compounds. Adsorption takes place when a pollutant has a restricted ability to dissolve in waste, a higher carbon affinity than trash, or both. Adsorption is a result of the adsorbent's surface particles having unequal or residual attraction factors which are not in a comparable form as the bulk adsorbent's atoms, which are in a balance with all the forces. Atoms on the adsorbent's surface have an energy that is far greater than atoms within do. The binding of an adsorbate to its surface on an adsorbent requires surface energy, defined as the excess power per unit area of a surface (Rathi and Kumar 2021). The rate of adsorption increases with the amount of the adsorbent's surface area per unit mass under a certain setting. Adsorption kinetics and entropy are additional main factors to take into account.

Adsorption may be categorized into two types: adsorption by matter and adsorption by chemicals, according to how firmly the adsorbate adheres to the adsorbent surface. During physical adsorption, also known as physisorption, adsorbate bonds to the adsorbent surface due to weak forces including electrostatic pull and van der Waals forces. Chemisorption, whereby the adsorbate is attached to the adsorbent surface through powerful covalent bonds, is another term for chemical adsorption. Chemisorption is gradual and usually entails the creation of monolayers on the outermost layer of the adsorbent, in comparison with physisorption, which frequently leads to the production of multilayers on the adsorbent's surface (Kecili and Hussain 2018). At temperatures below or close to an adsorbate's threshold temperature, physisorption emerges as a reversible technique and is extremely effective. Chemisorption, in contrast to physisorption, typically occurs at temperatures much higher than the critical temperature. When conditions are favorable, both adsorptions may occur simultaneously or one after the other. Adsorption is exothermic, which means it occurs relatively quickly at low temperatures and significantly decreases with an increase in temperature. Physisorption is defined by a drop in the amount of free energy and entropy. In contrast, chemisorption rises as temperature rises up to a certain degree and then starts to fall off after that (Xia et al. 2019).

7.6 Kinetic Analysis of Adsorption Process

According to Langmuir and Freundlich, the kinetics and adsorption equilibria determine the degree of adsorption. Adsorption isotherm for research in batches. The Lagergren equation has been used in some instances to explain the kinetics of ongoing compacted bed adsorption. The kinetic variables are crucial for understanding the adsorption process, predicting the rate at which Hg(II) is eliminated from aqueous solutions, and pinpointing the adsorption rate-determining step. This may be due to the fact that there are several potentially active points (porosity and functional groups) accessible during the adsorption process. The considerable decrease in Hg(II) in the context of SBH, where the surface area is considerably less than that of SBZ and CSB, is caused by chemical interactions between the metal action and various functional groups on the surface of a molecule. This outcome is in line with that of, who demonstrated the impact of various functional groups on the adsorption of Hg(II), showing that biochar with a lower surface area is more effective at adsorbing Hg(II) than activated carbon with a higher surface area (Tovar et al. 2019).

7.7 Physiochemical Characteristics

There are various parameters that play an important role in the absorption of heavy metals. This includes temperature, the concentration of adsorbent, pH, time of contact, etc.

- **Effect of temperature**—Optimum temperature helps enhance absorption. It has been noted that the rate of removal varies with an increase or decrease in temperature depending on the adsorption process. Adsorbate is more soluble in the solvent than adsorbent as a result of the solubility of the solute increasing proportionately with temperature, which inhibits adsorption (Saravanan et al.).
- **Effect of Concentration of Adsorbent**—Adsorbent dosage is a crucial factor in controlling the accessibility and availability of adsorption sites. Adsorption has been seen to increase as the adsorbent dosage is increased. However, the mobility of the ion decreases with the increased dose of adsorbent in the solution, which causes a decrease in the rate of adsorption.
- **Effect of pressure**—Adsorption intensity and pressure are closely correlated; as adsorbate pressure rises, so does adsorption. There are two examples here. Adsorption increases quickly with pressure when the pressure is lessened, however

as the pressure is greater and gets near to the vapor pressure, the adsorption starts approaching a limiting value.

- **Effect of initial concentration**—The movement and polarity of the ions found in the solutions affect how much adsorption occurs. The oppositely charged sites on the surfaces and pores of tiny molecules of the adsorbents are where ions (cations or anions) tend to be adsorbed. Adsorption consistently exhibits a decreasing trend with increasing starting adsorbate concentrations in the solution. At initial metal ion levels that were more diluted, the adsorption was shown to be more favorable.
- **Effect of contact time**—The adsorption efficiency specifically improves with more contact hours. At first, the amount of adsorbate that is adsorbed onto the adsorbent surface grows quickly, but after a while, the process stops and reaches a constant value.
- **Effect of pH**—The extent of adsorption is influenced by the pH of a solution. The highly strong adsorption of H and OH ions influences the method of adsorption by separating groups of function on the active sites of the adsorbent surface, which changes the rate of the reaction and the state of equilibrium characteristics. It is typically reported that an adsorbent preferentially adsorbs anions at lower pH owing to the presence of H ions, but at greater pH levels, it is liable for action adsorption because of the influence of OH e ions (Saravanan et al. 2020).

7.8 Removal of Mercury Using Algal Biomass

Chlorella vulgaris is recently been used to remove heavy metals such as mercury from wastewater. Mercury is removed by the process of biosorption at inappropriate temperature, pressure, and pH. The maximum absorption of mercury from algal biomass is at pH 6. Mercury enters aquatic environments through various processes like mining, electroplating, etc. This heavy metal thus, then enters the plants or the human body in various ways leading to a negative impact on our body. Because of the hazardous effects of mercury, it is very important to eliminate these metals. Algae due to its vast availability is now being used to remove mercury from wastewater. Various different microalgal strains were tested again for mercury removal. Out of which *Chlorella vulgaris* showed the best result. These algae are able to grow everywhere and serve a key part in the food chain, and preserving the oxygen supply on our planet. According to numerous studies, the existence of polysaccharides, proteins, or lipids on the surface of cell walls that contain the functional groups' amines, hydroxyls, carboxyls, and sulfhydryl, which can serve as binding sites for metals, may explain why algal cells have such a high capacity for binding with metals (Kumar et al. 2019). The purpose of this research is to examine *C. vulgaris* capacity to eliminate mercury from aqueous solutions. The groups of molecules responsible for the chemical reaction of the metal (Hg(II)) with algae are identified. A study was made where hazardous heavy metal (Hg) from industrial effluents was treated using a low-cost adsorbent. Algal biomass (*C. vulgaris*) was examined as a potential bio adsorbent using FTIR (Fourier transform infrared spectroscopy) to examine different functional groups, SEM (Scanning electron microscope) with an EDX (Energy dispersive spectroscopy) to examine morphological characteristics, elemental examination, and physical, chemical, and thermal variables. In a batch system, the impacts of various operational parameters including pH, starting Hg(II) ion concentration, contact duration, and adsorbent dose were examined. To correlate the experimental findings, several kinetic models and adsorption isotherms were looked at. The findings of that study will help to understand how *C. vulgaris* absorbs metals and may help us design possible biosorbents that have a significant capacity for absorbing heavy metals like mercury from aquatic environments (Kumar et al. 2020).

7.9 Removal of Mercury from Wastewater Using Agricultural Waste

The escalating level of production and the associated stress on the environment has created a demand for powerful adsorbents. Numerous efforts have been made to solve this issue by chemically treating or altering these wastes in order to extract useful elements. Agricultural waste products made of lingo-cellulosic materials are a plentiful supply of biomass that may be able to sorb metals. For the removal of heavy metals in wastewater, these substances could be a viable alternative. There are a number of agricultural wastes like rice, coffee, wool, tea waste, straw, peanut hull, coconut husk, and many more, which can be used as an absorbent to remove mercury from the wastewater. Lignin and cellulose make up the majority of agricultural waste products in most cases. Hemicellulose, extractives, lipids, proteins, simple sugars, starches, water, hydrocarbons, ash, and several other substances that include a wide range of functional groups that are reactive are additional components. The most likely possibilities for chemical alteration by processes like ethoxylation, carboxy methylation, admiration, etc., are cellulose and starch.

7.10 Conclusion

This review entails the removal of mercury from wastewater using absorbents like biochar and graphene. In order to address economic and environmental problems, the green chemistry method focuses on substituting synthetic polymers with substances that occur naturally by optimizing the use of biopolymers in various industrial, biological, and medicinal applications. In this regard, a detailed study of the use of biopolymers in water purification was conducted. The complex surface of the adsorbents under study produces a variety of physical and chemical interactions that are reflected in the isothermal and kinetic parameters, demonstrating that the process of adsorption requires numerous stages.

References

Abdel-Raouf MS, Abdul-Raheim ARM (2017) Removal of heavy metals from industrial waste water by biomass-based materials: a review. J Pollut Eff Cont 5(1):1–13

Aditya L, Mahlia TI, Nguyen LN, Vu HP, Nghiem LD (2022) Microalgae-bacteria consortium for wastewater treatment and biomass production. Sci Total Environ 838(155871):250–266

Awad FS, AbouZied KM, Abou El-Maaty WM, El-Wakil AM, Samy ES (2018) Effective removal of mercury (II) from aqueous solutions by chemically modified graphene oxide nanosheets. Arab J Chem 13(1):2659–2670

Boelee NC, Temmink H, Janssen M, Buisman CJN, Wijffels RH (2011) Nitrogen and phosphorus removal from municipal wastewater effluent using microalgal biofilms. Water Res 45(18):5925–5933

Crespo-López ME, Macêdo GL, Pereira SI, Arrifano GP, Picanco-Diniz DL, do Nascimento JLM, Herculano AM (2009) Mercury and human genotoxicity: critical considerations and possible molecular mechanisms. Pharmacol Res 60(4):212–220

Giraldo S, Robles I, Ramirez A, Flórez E, Acelas N (2020) Mercury removal from wastewater using agroindustrial waste adsorbents. SN Appl Sci 2:1–17

Gonçalves AL, Pires JCM, Simões M (2017) A review on the use of microalgal consortia for wastewater treatment. Algal Res 24:403–415

Hargreaves AJ, Vale P, Whelan J, Constantino C, Dotro G, Cartmell E (2016) Mercury and antimony in wastewater: fate and treatment. Water Air Soil Pollut 227:1–17

Kecili R, Hussain CM (2018) Mechanism of adsorption on nanomaterials. In: Nanomaterials in chromatography. Elsevier, pp 89–115

Kumar M, Goswami L, Singh AK, Sikandar M (2019) Valorization of coal fired-fly ash for potential heavy metal removal from the single and multi-contaminated system. Heliyon 5:1–9

Kumar M, Singh AK, Sikandar M (2020) Biosorption of Hg(II) from aqueous solution using algal biomass: kinetics and isotherm studies. Heliyon 6(1):1–10

Njoku VO, Islam MA, Asif M, Hameed BH (2014) Utilization of sky fruit husk agricultural waste to produce high quality activated carbon for the herbicide bentazon adsorption. Chem Eng J 251:183–191

Oke EA, Oluyinka OA, Afolabi SD, Ibe KK, Raheem SA (2023) Latest insights on technologies for halides and halogenated compounds extraction/abatement from water and wastewater: challenges and future perspectives. J Water Process Eng 53(103724):1–13

Park JH, Wang JJ, Zhou B, Mikhael JE, DeLaune RD (2019) Removing mercury from aqueous solution using sulfurized biochar and associated mechanisms. Environ Pollut 244:627–635

Pinheiro MCN, Macchi BM, Vieira JLF, Oikawa T, Amoras WW, Guimarães GA, Do Nascimento JLM (2008) Mercury exposure and antioxidant defenses in women: a comparative study in the Amazon. Environ Res 107(1):53–59

Rathi BS, Kumar PS (2021) Application of adsorption process for effective removal of emerging contaminants from water and wastewater. Environ Pollut 280(116995):1–13

Russo V, Hmoudah M, Broccoli F, Iesce MR, Jung OS, Martino Di Serio MD (2020) Applications of metal organic frameworks in wastewater treatment: a review on adsorption and photo degradation. Front Chem Eng 2(581487):1–13

Saravanan A, Kumar PS, Renita AA (2018) Hybrid synthesis of novel material through acid modification followed ultrasonication to improve adsorption capacity for zinc removal. J Clean Prod 172:92–105

Saravanan A, Jeevanantham S, Kumar PS, Varjani S, Yaashikaa PR, Karishma S (2020) Enhanced Zn(II) ion adsorption on surface modified mixed biomass–*Borassus flabellifer* and *Aspergillus tamarii*: equilibrium, kinetics and thermodynamics study. Ind Crops Prod 153(112613):1–15

Tan G, Sun W, Xu Y, Wang H, Xu N (2016) Sorption of mercury(II) and atrazine by biochar, modified biochars and biochar based activated carbon in aqueous solution. Biores Technol 211:727–735

Tchounwou PB, Yedjou CG, Patlolla AK, Sutton DJ, Luch A (2012) Molecular, clinical and environmental toxicology. Mol Clin Environ Toxicol 3:133–164

Tovar AK, Godínez LA, Espejel F, Ramírez-Zamora RM, Robles I (2019) Optimization of the integral valorization process for orange peel waste using a design of experiments approach: production of high-quality pectin and activated carbon. Waste Manage 85:202–213

Xia M, Chen Z, Li Y, Li C, Ahmad NM, Cheema WA, Zhu S (2019) Removal of Hg(II) in aqueous solutions through physical and chemical adsorption principles. RSC Adv 9(36):20941–20953

Zhao B, O'Connor D, Shen Z, Tsang DC, Rinklebe J, Hou D (2020) Sulfur-modified biochar as a soil amendment to stabilize mercury pollution: an accelerated simulation of long-term aging effects. Environ Pollut 264:114687

Chapter 8
Microbial Remediation of Mercury: An Overview

Marwa Eltarahony, Eman Ibrahim, Ghada Hegazy, and Amira Sabry

Abbreviations

ATSDR	Agency for Toxic Substances and Disease Registry
EPA	US Environmental Protection Agency
GMOs	Genetically modified organisms
EPS	Extracellular polymeric substances
FTIR	Fourier transform infrared analysis
TEM	Transmission electron microscope
SEM	Scanning electron microscope
EDX	Energy Dispersive X-ray
SRB	Sulfate-reducing bacteria
MICP	Microbial-induced carbonate precipitation
AMF	Mycorrhizal fungi
PGP	Plant growth promoting
OM	Organic matter

M. Eltarahony (✉) · E. Ibrahim
Environmental Biotechnology Department, Genetic Engineering and Biotechnology Research Institute (GEBRI), City of Scientific Research and Technological Applications (SRTA-City), New Borg El-Arab City 21934, Alexandria, Egypt
e-mail: m_eltarahony@yaoo.com; meltarahony@srtacity.sci.eg

G. Hegazy
National Institute of Oceanography & Fisheries, NIOF, Cairo, Egypt

Bioprocess Development Department, Genetic Engineering and Biotechnology Research Institute (GEBRI), City of Scientific Research and Technological Applications (SRTA-City), New Borg El-Arab City 21934, Alexandria, Egypt

A. Sabry
Protein Research Department, Genetic Engineering and Biotechnology Research Institute (GEBRI), City of Scientific Research and Technological Applications (SRTA-City), New Borg EL-Arab 21934, Alexandria, Egypt

N. Kumar (ed.), *Mercury Toxicity*, Environmental Science and Engineering,
https://doi.org/10.1007/978-981-99-7719-2_8

DNA	Deoxyribonucleic acid
NMR	Nuclear magnetic resonance spectroscopy
CFA	Coal fly ash
MOFs	Metal–organic frameworks
COFs	Covalent organic frameworks

8.1 Introduction

Mercury (Hg) is metallic element that was ranked among "big three" heavy metals in the list of hazardous materials according to the Agency for Toxic Substances and Disease Registry (ATSDR 2015). Owing to its inert nature, nonbiodegradability, toxicity, and long-lasting in the atmosphere, its presence represents a risky stalemate that jeopardizes all living creatures and their ambient ecosystem (Kim et al. 2015). Despite its release in the environment could be from natural roots (e.g., oceanic emission, volcanic eruption, photoreduction, and degassing from minerals (combustion of organic substances), the majority of its aquatic and terrestrial genesis was assigned to anthropogenic activities. As well known, Hg is the main ingredient in different medical devices (e.g., oesophageal dilators, sphygmomanometers, dental amalgams, etc.), electrical apparatus (e.g., batteries, switches, etc.), measuring tools (e.g., psychrometers, thermometers, flow meters, manometers, barometers, hydrometers, etc.), as insecticides, herbicides, fungicides. Besides, its role in paper pulps, painting industry, steel industry, and chloro-alkali process are documented. Hereby, all previous activities undoubtedly contribute to global Hg tainting besides mining and fossil fuel incineration (Balan et al. 2018; Amin et al. 2022).

Mercury characterizes by its odorless nature, tolerance to a broad range of temperatures, and coexists in both liquid and vapor phases and also in different organic and inorganic states, namely, elemental (Hg^0), mercurous (Hg_2^{2+}), mercuric (Hg^{2+}), monomethyl mercury, ethyl mercury or dimethyl mercury, etc. Each form of Hg possesses its own discerned physiochemical properties, environmental attitude, and biotoxicity (He et al. 2015). Remarkably, (Mahbub et al. 2017a) reported that alkylated compounds of mercury are severe neurotoxins; however, Amin et al. (2022) reported that the inorganic form of mercury (Hg^{2+}), which is commonly present form in the environment, is the most toxic due to its superior affinity to cysteine moiety of protein, more soluble in lipids and highly accessible through biological membranes. On the other hand, Saranya et al. (2017) documented that the association of mercury with chloride, hydroxide, sulfide, and oxide groups plays a crucial role in elevating mercury poisoning symptoms. Generally, Minamata's disease is considered the most popular disease caused by mercury besides other gastrointestinal, hematological, renal, cardiovascular, and neurological disorder, which had been detected (Amin et al. 2022) (Fig. 8.1).

Whatever the mercury state in the environment, US Environmental Protection Agency (EPA) determined the limits of mercury by 2.0 μg/L in water and the

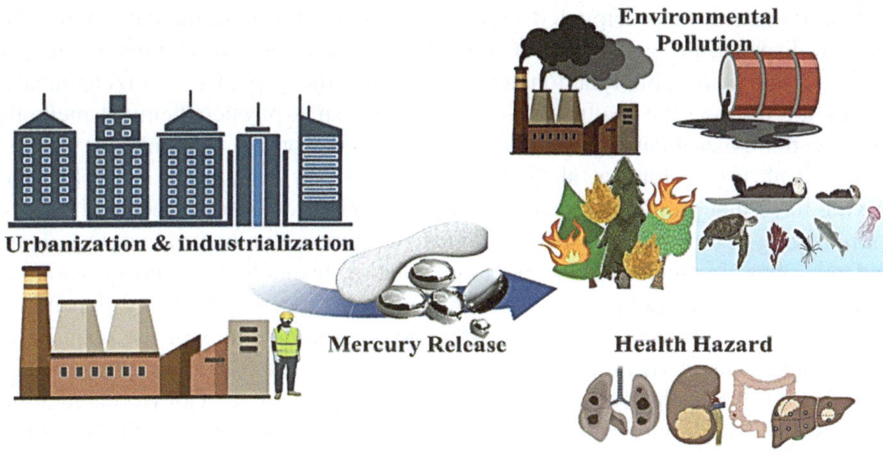

Fig. 8.1 Anthropogenic activities and mercury release in environment causing severe human injuries besides air, aquaculture, and agricultural pollution

range of soil 6.6–3600 mg/kg according to the land use (Mahbub et al. 2017a). The bioaccumulative traits of mercury compounds through the food chain lead to biomagnifications and inherent toxicity; hereby, imperative necessity for scavenging such elements and reducing its pathogenicity via several remediation approaches. Conventionally, the hazardous wastes were manipulated through their digging up in isolated landfills or capped sites. Nonetheless, numerous shortcomings emerged beginning from handling/transportation and safety requirements, passing through precise finding of proper landfill sites, which are potentially cost processes, ending with the possibility for pollutant migration elsewhere, which is risky and entails monitoring and strict maintenance of the landfill barriers long into the future (Gong et al. 2018). Therefore, physicochemical techniques were developed, to compensate the drawbacks of isolation method, through degrading the pollutant completely or transforming it into innocuous matter. However, the type of contaminated site enforces technologists to employ specific approaches. Namely, in ground water and wastewater polluted with mercury, the precipitation method is the most commonly applied via utilizing coagulants such as sodium sulfide and lignin derivatives followed by filtration or clarification. Also, adsorption process proved its effectiveness in cleaning up Hg^{2+} contaminated water (O'rear et al. 2014). Whereas, soil washing, stabilization/solidification, vitrification, thermal treatment, and electrokinetic recovery are considered being the most efficient means of soil treatment via employing acid/alkali chelating agents/surfactants, stabilizing agents (e.g., lime, ceramics, zeolites, Portland cement, fly ashes, sulfur polymer, aluminosilicates, metal oxides, bentonites, biosolids and animal manure activated carbon, biochar, clay minerals, phosphates, etc.), high temperature with low pressure (350 °C/1 atm pressure), and a low intensity direct current (Gong et al. 2018).

Unfortunately, the precipitated sludge that requires further multistage-treatment prior to disposal, the generation of fouling or plugging, the possibility for leachability, diminishing soil functionality/quality, and the capital cost of chemicals/energy are deemed the major obstacles that limit such physicochemical methods to be at the experimental phase for field application or even greenhouse studies (Gong et al. 2018; Taha et al. 2023). Arguably, such limitations symbolize the driving force that promoted researchers and biologists to harness green technology for removing or transforming mercury in natural bioremediation processes by plants and microbes. Interestingly, the phytoremediation technology found an ecologically sound in cleaning up of several contaminants, especially mercury, depending on biochemical, physical, biological, and microbial interactions of the plants. Via several mechanisms including, phytostimulation, phytostabilization, phytoextraction, rhizofiltration, phytovolatilization, and phytodegradation, the phytoremediation process could be implemented by various plant species (Verma 2021). Nevertheless, the type/physicochemical properties of contaminant, choice/bioavailability of hyperaccumulators phytoremediator plant species, the ingathering of contaminate in the edible parts of fruit and vegetable crop, the slow rate of growth process, plant seasonal variation and handling/disposing of contaminated plants are the substantial constraints that handicap the extended application of phytoremediation (Farraji et al. 2016). Building on this previous knowledge, the microbial manipulation of heavy metals has piqued the interest of technologists and researchers to find a cost-effective, sustainable, easy solution for mercury removal. Hence, in the current chapter, the microbial strategies are detoxifying mercury via different microbial groups and under various microbial growth conditions would be discussed. Besides, the remedy methods, genetic system, enzymatic pathways, and the hybridization of microbes with advanced approaches would be also addressed (Fig. 8.2).

8.2 Microbial Pathways in Mercury Remediation

A plethora of microbial species possess the capability to detoxify a vast array of metal contaminants, by the virtue of their versatile metabolic activities. Remarkably, various species, either indigenous, genetically modified (GMOs), or exogenously introduced, could exert more than pathway in containment metals and metalloids simultaneously and restrict their availability in contamination site, even the dead cells could participate more or less in detoxification process. Generally, mercury-remediating microbes symbolize by their tolerance and low sensitivity to the toxicity of mercury ions. Remarkably, the binding of metals on the cell wall or internally by intracellular proteins (e.g., phytochelatins, metallothioneins, siderophores, etc.), enzymatic conversion of metals, reduced metal uptake, modifying uptake system and utilizing effective efflux systems are the common strategies by which microbes could resist heavy metals (Tarekegn et al. 2020; Tarfeen et al. 2022). Thereby, the biosorption/adsorption, bioprecipitation, biotransformation (bio-reduction, bio-oxidation, methylation, demethylation), bioaccumulation/

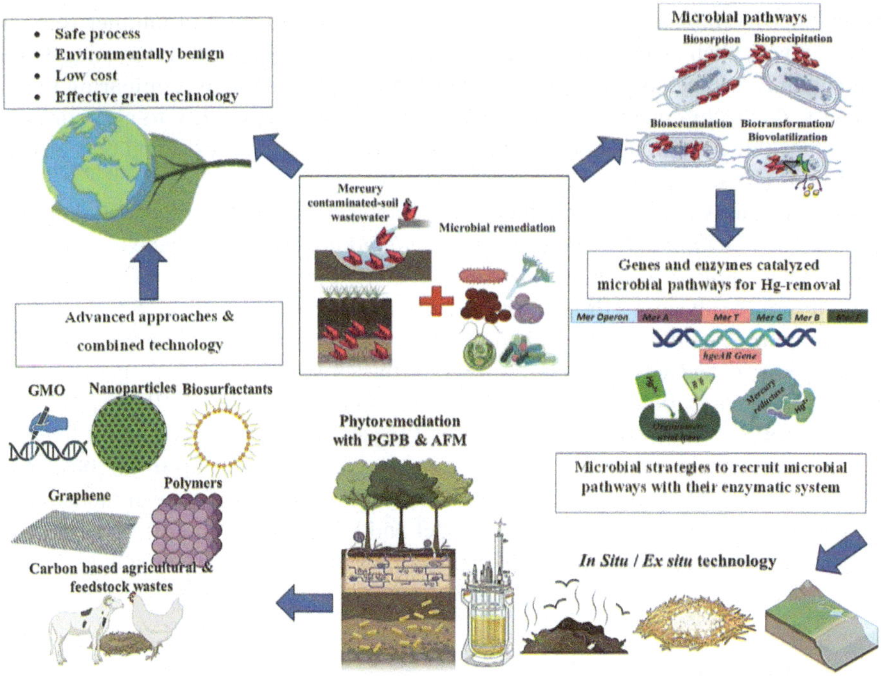

Fig. 8.2 Schematic diagram summarizes Hg-bioremediation by various microbial groups through different remediating pathways that were catalyzed by different enzymatic systems expressed from varied genes and their recruitment in environmental sited via in situ and ex situ technologies, besides the advanced materials in combined approaches that would ameliorate the performance of microbes to guarantee the success of detoxification process

sequestration, bioleaching, and biovolatilization are the most widely mechanisms utilized by microbes in bioremediation process.

8.2.1 Biosorption/Adsorption

As a surface phenomenon based mainly on the cellular surface traits, the adsorption of mercury takes place by both live and dead biomass. In fact, heavy metal ions like mercury were trapped passively on the microbial surface (i.e., avoiding energy requirement) and bound by different physical and chemical interactions (e.g., Van der Waals, electrostatic, covalent bonding, and ion exchange) to negatively charged surface groups of phosphates, carboxylates, sulfates, hydroxyl, and amides. Such functional groups constitute the main cell wall ingredients of proteins, polysaccharides, extracellular polymeric substances (EPS), and lipids, which allow external adsorption of mercury; however, in some resistant microbes, metal ions could pass through porins and reside in the periplasmic space creating a potent binding with the cellular membranes; facilitating by such way the activation of another remediating strategy such as internal sequestration or biotransformation. Interestingly, the dead biomass seemed being more effective in biosorption strategy, comparing to live biomass, due to insensitivity to higher concentrations, adaptability to alterations in environmental conditions, and the unnecessity for adjusting nutritional and growth conditions (Jin et al., 2018; Tarfeen et al. 2022).

In this context, a study conducted by Balan et al. (2018) reported that Hg-tolerant *Pseudarthrobacter oxydans* and *Pseudomonas frederiksbergensis* succeeded in removing 2 ppm of Hg under experimental conditions. By the aid of Fourier transform infrared analysis (FTIR), they explained that nitro compound, alkynes, alkenes, alcoholic alkyl halide, primary amines, aliphatic and aromatic amines, alkanes, carboxylic acid, and amide groups represented the active ligand to Hg that enable both bacterial species to tolerate and immobilize Hg. On the other hand, *Rhizopus oryzae* and *Aspergillus niger* removed about 90% of Hg at 10 and 100 ppm by their live and dead biomass; revealing that the higher sorption performance (up to 90.38%) was implemented by *R. oryzae* (dead cells) at 100 ppm. Besides, FTIR analysis and kinetic studies (Pseudo-second-order kinetic model and Langmuir isotherm) reflected that the chemisorption process happened on the homogenous surface (Anuar et al. 2020).

8.2.2 Bioaccumulation

The metabolic-dependent active absorption and infiltration of contaminants internally by living biomass to the middle of the cell is known as intracellular accumulation. It commences by adsorption of contaminant externally followed by its uptake through the phospholipid bilayers of living biomass. This process takes place with the aid of active transporters and protein channels in a process mimic that occurs for internalization of essential ions such as K^+, Na^{2+}, Mg^{+2}, and Ca^{2+} using ion pumps and passive diffusion mechanisms (Tarfeen et al. 2022). Despite bioaccumulation process being time-consuming, relative to biosorption, the removal rate could be enhanced easily by adjusting the reaction conditions as revealed by (Jin et al. 2018; Tarekegn et al. 2020). The intracellular accumulation could be described as a toxicokinetic process that influenced by the sensitivity of living organisms to the contaminants and based on their concentrations and microbial physiology. However, the accumulative microbes characterized by their distinguished capability to transform and modify the toxicity of the sequestered contaminant to be less toxic, by other additional pathways while remaining inside the cellular compartments (Tarekegn et al. 2020). Notably, two mercury-tolerant bacterial strains isolated from gold mining tailings in Indonesia were identified as *Fictibacillus nanhainensis* and *Bacillus toyonensis* exhibited their potential accumulative performance for mercury by more than 81% removal capacity (Nurfitriani et al. 2020). In the same sense, Tazaki and Asada (2007) found that bacteria resident in Geita (small gold mine pond near Lake Victoria, Tanzania), accumulate mercury through EPS as visualized by transmission electron microscope (TEM). Whereas, a white rot fungus *Phlebia floridensis* trapped about 70–84% of mercury, which induced morphological and textural alterations in the bioremediated hyphae as depicted by scanning electron microscope (SEM) and energy dispersive X-ray (EDX) (Sharma et al. 2022).

8.2.3 Bioprecipitation

It could also be called biomineralization or biocrystallization (Tarekegn et al. 2020). It involves the conversion of heavy metals or metalloids from their soluble states to insoluble states such as sulfides, hydroxides, phosphates, and carbonates. The microbial growth, metabolic activity, and various enzymatic systems mediate such a process by liberating microbial metabolites such as organic acids, EPS, and electron donors, which thereafter change the surrounding environment chemistry to that favor the precipitation. Interestingly, bioprecipitation process relies fundamentally on the environmental changes generated by microbial activity like alterations in pH and redox potential changes (Jeyakumar et al. 2023). Undoubtedly, no one can deny the pivotal role of sulfate-reducing bacteria (SRB) in immobilizing heavy metals by producing their sulfides (Vitor et al. 2015; Zhang and Wang 2016). Wherein, Groudev et al. (2014) reported that indigenous SRB-dwelling cinnamonic forest soil stimulated the mobility of Zn, Cu, and Cd and precipitated them as insoluble metals sulfides; however, pertinent studies on mercury are scarce. Notwithstanding that, a study mediated by Pan-Hou and Imura (1981) found that *Clostridium cochlearium* was able to form HgS anaerobically. Besides, HgS was formed by *Klebsiella aerogenes* NCTC418 after its cultivation in continuous aerobic culture in the presence of $HgCl_2$ (2 μg/mL) and the author confirmed the elevation of cellular sulfide upon the existence of mercuric ions (Aiking et al. 1985) On the other hand, Håkansson et al. (2008) hybridized electrokinetic remediation with the metabolic activity of SRB, in which contaminated soil from a chlor-alkali industry encompassed mercury (100 mg/kg) treated with iodide/iodine complexing agent and exposed to electric field. The complexes of mercury iodide reacted with H_2S in water solution, which generated by the action of SRB and resulted in mercury precipitation in mercury sulfide crystals.

Moreover, mercury precipitation in hydroxide form is also rare. Nonetheless, microbial-induced carbonate precipitation (MICP) seemed to be a promising tool in mercury remediation where, the activity of nitrate reductase enzyme of *Proteus mirabilis 10B*, either under aerobic or anaerobic conditions, entrapped about 322 and 309 of mercury, in their oxide forms (i.e., HgO and Hg_2O), in calcite matrix during 168 and 186 hrs., respectively, in an investigation conducted by Eltarahony et al. (2020). On the other hand, the ureolytic strains of *Metschnikowia pulcherrima* and *Raoultella planticola* transformed the soluble form of mercury (350 ppm) completely into insoluble forms of $CaHgO_2$, HgO and Hg_2O within 102 hrs., which also encapsulated inside $CaCO_3$ trap (Eltarahony et al. 2021); yet, the denitrification and ureolysis processes mediate the precipitation stage through elevating pH and alkalinity of solution. Strikingly, MICP is a proficient technique that remediates several heavy metals and nuclides in their carbonate form (Kim et al. 2021; Wang et al. 2023). Also, the precipitation of heavy metals in hydroxide form was also detected by several research groups (Chan et al. 2009; Li et al. 2019), but in case of mercury, it was not reported.

8.2.4 Bioleaching/Biomining

It contradicts the bioprecipitation, in which it involves the extraction of metals from their ores by dissolution of their insoluble minerals to soluble form by the catalysis of organic acids released by the acidophilic microbes (Gong et al. 2018; Jeyakumar et al. 2023). Broadly, as reported by Tarekegn et al. (2020), metals are present in the environment in sulfide and oxide forms, such processes could be catalyzed by various microbial genera such as *Acidithiobacillus ferrooxidans, Acidithiobacillus thiooxidans, Leptospirillum ferrooxidans, Aspergillus* sp., *Alternaria* sp., *Mucor* sp., *Penicillium* sp., *Rhizopus* sp. and *Cladosporium* sp. (Jin et al. 2018). Bioleaching process or biological leaching was employed mainly in biohydrometallurgy and is implemented through one of three pathways, including acidolysis, complexolysis, and redoxolysis. Importantly, acidolysis entails the bioleaching microbe characterized by its ability to produce low molecular weight organic/inorganic acids (e.g., oxalic acid, gluconic acid, citric acid, sulfuric acid, etc.) during their metabolic activity to assure the recovery and mobilization of required metal. While redoxolysis could happen either directly, by direct microbial oxidation of metal sulfide with electrons gained directly from the reduced minerals, or indirectly through the use of an oxidant such as Ferric iron that is generated via microbial oxidation of ferrous iron, which coexists naturally in the minerals. Whereas, complexolysis encompasses the complexation of metal by organic acids or metabolites like siderophores in a slower rate process than acidolysis (Okoh et al. 2018). However, some bacterial strains like *Citrobacter* sp. could exert bioleaching process through excreting free inorganic phosphate. Although bioleaching is a cost-effective process, relative to other chemical approaches, its application is restricted to Cu, Ur, and Au (Tarekegn et al. 2020); the sluggish leaching kinetics are the main impediments in mercury remediation from contaminated solid waste as reported by Xie et al. (2020).

8.2.5 Biovolatilization

Via such a strategy, the enzymatic system of the microbes mediates the conversion of metals from the soluble phase to the volatile phase. By the virtue of the cytoplasmic flavoenzyme mercuric reductase (MerA) and mercurial lyase (MerB), Hg-resistant microorganisms reduce Hg^{2+} to its volatile state Hg^0 (Gong et al. 2018; Tarfeen et al. 2022). As stated by Anthony (2014), Hg-resistance property is carried on transposons (Tn2) or conjugative plasmids (HgR), which enable the microbe to resist nor only mercury but also other heavy metals and even antibiotics. A broad spectrum of microorganism could eliminate the toxicity of organomercurials through volatilization after mercury uptake into cytoplasm through a specific transport system (MerT) or through auxiliary transporters MerC/MerF, followed by its reduction to elemental form, which volatilized immediately from the cells through passive diffusion system before re-oxidation to its divalent form occurs (Essa et al. 2002). In this regard, a mercury-resistant *Bacillus* sp. strain, which was isolated from molybdenum-lead mining soils in China, exhibited an appreciable resistance and removal rates to several heavy metals, especially mercury. The authors determined the Hg-adsorption rate and Hg-volatilization rate at 36 h, which reached 8.24% and 89.08%, respectively; reflecting the promising bioremediation potential of the strain in remediating 100 mg/L of mercury under nutrient availability (Yao et al. 2023). Whereas, a novel Hg(II)-volatilizing filamentous fungus *Penicillium* spp., DC-F11 was able to diminish mercury-phytotoxicity and total Hg in contaminated soil through multisystem collaborative process started with extracellular adsorption and precipitation followed by volatilization (Chang et al. 2020). Upon combining chemical extraction and microbial volatilization, Chen et al. (2018) got rid of 77% of Hg from field-tainted soil by using ammonium thiosulfate (0.5 M) as a first-step remediating process, thereafter, ≥81% of Hg^{2+} was reduced and volatilized by *Enterobacter cloacae*.

8.2.6 *Biotransformation*

8.2.6.1 Bioremediation via Redox State Change (Bioreduction/ Biooxidation)

It describes the bioremediation through switching the oxidative state of heavy metal to a different state with entirely varied physicochemical traits. Such alteration in oxidative state may influence metal mobility, toxicity, and bioavailability (Gong et al. 2018; Tarfeen et al. 2022). This approach seems to be effective in the case of metals whose toxicity differs with varied redox states; therefore, intensive attention should be paid before employing this method, in particular, in field applications. That is for evading transformation to a more hazardous or more mobile phase, which would impact adversely on the ecosystem and public health (Gong et al. 2018). As highlighted by Colombo et al. (2014), gaseous mercury (Hg^0) is highly mobile in groundwater and can accumulate easily in aquatic creatures. Hence, via such a study, the authors utilized anaerobic bacteria like *Geothrix fermentans* and facultative anaerobic bacteria such as *Shewanella oneidensis* and *Cupriavidus metallidurans* to oxidize dissolved mercury to its divalent state, which could be eliminated subsequently via complexation with sulfide and precipitation as insoluble HgS phase. Through X-ray absorption near edge structure (XANES) spectroscopy, the authors emphasized the covalently binding of Hg^{2+} with thiol moieties in both dead and live cells. In addition, Hg^0 oxidation by anoxic is commonly observed in waterlogged soils and anoxygenic sediments (Bouffard and Amyot 2009; Poulin et al. 2016). Regarding mercury reduction, it could be executed under different aeration conditions and either solitary on in combination with other bioremediating pathways. Wherein, an investigation performed by Wu et al. (2022) declared that the plant symbiotic *Metarhizium robertsii* reduced Hg^{2+} to its gaseous form by the catalysis of mercury reductase as a second step after demethylating methylmercury by the activity of methylmercury demethylase, by such combinatory pathways, the fungus curtailed the accumulation of mercury in the plants and promoted their cultivation in contaminated soils. On the other hand, phototrophic non-sulfur purple bacteria (e.g., *Rhodopseudomonas palustris, Rhodobactor capsulatus*, and *Rhodobacter sphaeroides*) reduced HgII under anaerobic photoheterotrophic incubation (Grégoire and Poulain 2016), hence, participating in Hg redox cycling. Interestingly, *Heliobacterium modesticaldum*, which is a member of spore-forming fermentative photoheterotrophs, was reported as an effective Hg reducer anaerobically through the pathway cometabolized by ferredoxin (i.e., reduced redox cofactor) rather than MerA reductase, which was not detected in this strain (Grégoire et al. 2018).

8.2.6.2 Mercury Methylation

It is a process dedicated to transfer a methyl group to mercury and the formation of methymercury ($MeHg^+$). Although $MeHg^+$ is more toxic than other Hg forms, but in this case, it may be considered as less toxic to the manipulating microorganism. Such process was executed predominantly by SRB such as *Desulfovibrio desulfuricans* under low availability of sulfate ions and anoxic incubation (Barkay et al. 2003; Wagner-Döbler, 2003). Yet, other microbial groups of methanogens (Gilmour et al. 2018), iron-reducing bacteria (Fleming et al. 2006), and some members of Chloroflexi and Firmicutes phyla also recently recognized as Hg methylators. Notably, *hgcAB* genes encode the corrinoid protein, which commences the initial stage in methylation process and was utilized to identify Hg methylators among a broad microbial spectrum in any contaminated habitats. As noticed by Lin et al. (2014), *Geobacter sulfurreducens*, which is metal dissimilating anaerobic bacteria, had the ability to reduce mercury when *hgcAB* gene was deleted; reflecting the presence of a physiological link between two pathways of Hg transformations. Interestingly, the investigations concerning mercury methylation by phototrophs are scarce and more/deep mechanistic studies are required (Grégoire and Poulain 2014). Nevertheless, (Franco et al. 2018) studied the potential role of *Nostoc paludosum* in detoxifying mercury by methylation and they found that the cyanobacterium removes mercury through reduction and volatilization in lieu of methylation.

8.2.6.3 Demethylation of Methylmercury

It is called also MeHg degradation, which describes the removal of methyl group from organomercurial compounds; utterly forming insoluble mercuric sulfide in the presence of hydrogen sulfide. Two main scenarios addressed demethylation process according to their final byproducts. Namely, in reductive demethylation, methane

(CH_4) and elemental Hg(0) are generated; yet, oxidative demethylation produces CO_2 and Hg(II). Under aerobic conditions and availability of mercury (μM), the reductive demethylation is preferential, while under anoxic circumstances and low existence of mercury (nM), oxidative is favored. Broadly, demethylation process is catalyzed by two successive enzymatic machines. It begins with the catalysis of organomercurial lyase (MerB) that cleaves the C–Hg bond generating CH_4 and Hg^{2+}, which is less toxic than methyl mercury by 100-times, followed by mercuric reductase (MerA), which yielding volatile Hg^0 after reduction (Lu et al. 2016).

Intriguingly, diverse microbial species among both prokaryotes and eukaryotes exerted their best performance in detoxification of MeHg via demethylation accompanying by other pathways till reaching the safest state. Lu et al. (2016) recruited *Geobacter bemidjiensis* in Hg detoxification. The data revealed that such iron-reducing bacterium mediated Hg transformations anaerobically through simultaneous accompanying strategies of MeHg yielding, degradation, Hg(II) reduction, and Hg(0) oxidation. The authors proposed that *G. bemidjiensis* employed a reductive demethylation strategy to degrade MeHg and transform it to a volatile phase by the virtue of MerB and MerA. Meanwhile, under phototrophic conditions, Kritee et al. (2017) studied thoroughly and deeply the demethylation process by *Isochrysis galbana* using Hg stable isotope. The data highlighted the production of a pool of isotopically HgII confirming the demethylation capacity of algal cells. In this context, Li et al. (2022a, b) examined the capacity of 15 marine microalgae (Diatoms (8 species), Dinoflagellates (4 species), Chlorophyta (2 species), and Chrysophyte) in mercury methylation and demethylation potential in natural environments. The authors found that all examined microalgae lack the ability to methylate inorganic Hg, while six species induced MeHg demethylation at an equivalent level with photodemethylation. Besides, they suggested that demethylation ability could be attributed to the extracellular phyco-secretions (i.e., photo-induce demethylation and thiol biomolecules) in association with bacterial activity.

8.3 Molecular Aspects-Mediating Microbial Remediation

The versatile resistance mechanisms mediated by microbes, as described in detail in the previous section, are attributed to their enzymatic systems expressed from specific genes (Christakis et al. 2021; Li et al. 2022a, b; Yu and Barkay 2022; Yadav et al. 2023).

8.3.1 mer *Operon-Mediated Inorganic Hg Reduction and Volatilization*

Hg(II) and organomercury compounds are detoxified to a volatile less-toxic form (Hg^0) by Hg resistance-mediated system (*mer*) (Priyadarshanee et al. 2022; Yu and Barkay 2022).

The *mer* operon is distributed widely among bacteria, archaea, and integrated into chromosomal DNA or on mobile genetic elements such as plasmids, transposons, and integrons (Krout et al. 2022; Yu and Barkay 2022). It consists of regulatory proteins such as MerR and MerD, inner membrane-spanning transporter proteins (e.g., MerC, MerE, MerF, MerG, and MerT) that transport Hg^{2+} to the cytoplasm for reduction by MerA, a protein with reductase activity and MerP (periplasmic Hg(II) scavenging protein) (Agarwal et al. 2019; Li et al. 2022a, b). Priyadarshanee et al. (2022) depicted a schematic representation of bacterial *mer* operon-mediated Hg detoxification system. MerR and MerD are dual-function transcriptional regulators tightly regulating the *mer* operon expression through binding to the *mer* operator/ promoter (O/P) region. They function as either activators or repressors in the absence or presence of Hg^{2+}.

Mercury resistance has been classified into two categories according to the *mer* determinants: narrow-spectrum and broad-spectrum. The broad-spectrum *mer* determinants (*merA* and *merB* genes) resist both organic and inorganic mercury compounds, in contrast to the narrow-spectrum *mer* determinants (*merA*), which only resist inorganic mercury (Agarwal et al. 2019; Priyadarshanee et al. 2022).

Upon exposure to ionic Hg^{2+}, the toxic heavy metal (Hg^{2+}) binds to MerP cysteine residues at positions 14 and 17, thus transferring the Hg^{2+} to the mercury-specific transporter MerT. Consecutively the Hg^{2+} bounds to MerT, and it is transferred directly to the MerA amino-terminal domain cysteine residues. Thenceforth, the Hg^{2+} is conveyed to the MerA (mercuric reductase, an NAD(P)H dependent flavin disulfide oxidoreductase) cysteine residues located in the active site then the Hg^{2+} is reduced into volatile less-toxic form (Hg^0). Eventually, the Hg^0 passively diffuses from the cellular environment (Zheng et al. 2018; Agarwal et al. 2019; Zhang et al. 2020; Priyadarshanee et al. 2022).

8.3.2 hgcA *Gene-Mediated Methylation*

Net production and the tremendous bioaccumulative nature of neurotoxic methylmercury (MeHg) in terrestrial and marine food webs are regulated by microbial processes of methylation and demethylation (Lin et al. 2021; Gionfriddo et al. 2023; Luo et al. 2023). Iron-reducing bacteria, sulfate-reducing bacteria, methanogenic archaea, and fermentative bacteria play a role in the conversion of inorganic mercury into MeHg, which primarily takes place under anaerobic conditions (Christakis et al. 2021; Cardona et al. 2022; Frey et al. 2022). *hgcAB* is a gene cluster that encodes the proteins, HgcA and HgcB, which are crucial for the methylation process. First, the HgcA, a corrinoid methyltransferase (encoded by the *hgcA* gene), is a member of the carbon monoxide dehydrogenase/acetyl-CoA synthase delta subunit family, which is involved notably in the methyl transfer reactions. It has a cytosolic corrinoid binding domain (CBD) which transfers the methyl group to Hg(II), and a transmembrane domain (TMD) for the Hg uptake and cellular MeHg efflux. As well, the HgcB, a dicluster ferredoxin (iron-sulfur cluster protein encoded by the *hgcB* gene) that contains three conserved cysteine residues at the C-terminus. It plays a pivotal role in methylation as an electron donor, thereby reducing the cobalt ion of HgcA besides binding and delivering the Hg(II) to HgcA (Yu and Barkay 2022; Gionfriddo et al. 2023; Lin et al. 2023; Luo et al. 2023).

Apart from Hg methylation, reductive demethylation, and oxidative demethylation are two mechanisms involved in biotic MeHg demethylation. The reductive demethylation occurs under oxic conditions, whereby the *mer* operon (*merB* gene) encodes the organomercurial lyase (MerB), which cleaves the C-Hg bond of organomercurials by protonolysis resulting in Hg(II), which is further reduced to generate methane and volatile elemental Hg. As for oxidative demethylation, it takes place mainly in anaerobes lacking *mer* operon whereas, the Hg^{2+}, CO_2, and CH_4 are the end products. However, further research is required to identify the genes mediating oxidative demethylation (Tiodar et al. 2021; Yu and Barkay 2022; Luo et al. 2023; Tada et al. 2023).

In addition to the above mechanisms, other pathways assigned for other alternative enzymatic systems were detected to detoxify mercury, especially in transgenic bacteria. As documented by (Shahpiri and Mohammadzadeh 2018), *mt-1* and *ppk* genes encoding metal-scavenging agents (i.e., metallothionein) and polyphosphate kinase adopt an intrinsic role in Hg resistance and accumulation. Interestingly, Ruiz et al. (2011) described that Hg sequestration is governed by metallothionein (*mt-1*) and polyphosphate kinase (*ppk*) genes, which are expressed in transgenic bacteria (*Escherichia coli*/pBSK-P16S-mt1-rpsT and pBSK-P16S-g10-ppk-rpsT). Similarly, Deng and Jia (2011) and Alcántara et al. (2018) reported that the expression of the metallothionein gene and polyphosphate synthesis aided in the Hg removal efficiency of the recombinant strain *Rhodopseudomonas palustris* and *Lactobacillus* sp. respectively.

8.4 Microbial Paradigms of Mercury Bioremediation

Irrespective of whether aerobic or anaerobic conditions, planktonic or aggregated biofilm, the remediating microbes are able to decontaminate Hg pollution using one or more of the previously mentioned pathways either sequentially or simultaneously run. The bacterial remediation, mycoremediation, and even phycoremediation were effectively achieved by a wide range of bacteria, fungi (unicellular or filamentous), and algae (microalgae or macroalgae) as summarized in Table 8.1.

Table 8.1 Examples of Hg-remediating microbes affiliated with different taxonomic groups and exhibiting remediating potentiality against Hg via different pathways

Microbial group		Species	Bioremediation mechanism	References
Aerobic	Bacteria	*Microbacterium oxydans* HG3, *Serratia marcescens* HG19	Extracellular precipitation	François et al. (2012)
		Ochrobactrum sp. strain HG16, *Lysinibacillus sp.* strain HG17 *Bacillus cereus* MM8	Biosorption by secreted exopolysaccharides (EPS) and accumulation of mercury as spherical deposits or amorphous aggregates	François et al. (2012)
Anaerobic		*Cupriavidus metallidurans* MSR33	Removed about 71% of Hg (II) by reduction and volatilization to Hg (0)	Bravo et al. (2020b)
		Enterobacter sp.	Precipitation of mercury (7.3 mg/ l) as nano-size particles in the cytoplasm as well as on the cell wall within 72 h incubation	Sinha, and Khare, (2012); Mahbub et al. (2017b)
Bacterial biofilm		*B. thuringiensis* PW-05	By sequestration and volatilization of >90% of inorganic mercury	Dash et al. (2014)
		P. putida SP-1	By complete volatilization at pH range of 8–9	Zhang et al. (2012); Mahbub et al. (2017c)
		Biofilm consisting of seven different species of Hg-resistant *Pseudomonas* spp.	The bacteria present in biofilms reduced Hg^{2+} (98% ~ 28.8 mg/Kg soil) to volatile Hg^0 in 8 months	Wagner-Döbler (2003); Mahbub et al. (2017c)
Consortium		Mixed cultures of *Pseudomonas aeruginosa* and *Bacillus subtilis* (1:1)	Biosorption by dead cells reached 90% pH 5, biomass concentration 2 mg/ml, and temperature 32 °C	Tarangini (2009)
Fungi	Unicellular	*Candida xylopsoci, Pichia kudriavzvii*	Co-precipitation of 95% mercury in the non-toxic form (HgS) within 36 h	Amin and Latif (2011)

(continued)

Table 8.1 (continued)

Microbial group		Species	Bioremediation mechanism	References
	Filamentous Ascomycetes	*Didymella glomerata* P2.16, *Fusarium. Oxysporum, Cladosporium* sp., *Phoma costaricensis, Sarocladium kiliense*	Biosorption by living biomass achieved Hg removal in the range of 47–62% after 12 h	Văcar et al. (2021)
		Metarhizium robertsii	Degradation of methylmercury (MeHg) by the demethylase MMD into divalent mercury (Hg^{2+}), that is subsequently reduced to elemental Hg through the Hg^{2+} reductase MIR	Wu et al. (2022)
Algae	Microscopic green algae	*Scenedesmus* sp. *Chlorella* sp. *Pleeurococcus* sp.	Adsorption by 64%, 83% and 86% of Hg in 20 days	Vela-García et al. (2019); Yan et al. (2022)
	Macroscopic green algae	*Cladophora* sp.	Biosroption of mercury by immobilized *Cladophora* sp. alga in alginate beads and silica gel	Mokone et al. (2018)

8.4.1 Extremophiles as Mercury Bioremediators

Extremophiles are microorganisms that possess attractive skills to tolerate high Metals and radionuclides levels, extreme physical (e.g., radiation, temperature), chemical (e.g., acidic/alkaline pH, salinity), and other climate-changing conditions. They can act as bioremediation minor factories (micro factories), by which their performance can even be enhanced and customized for metals and radionuclides elimination (Marques 2018). Extremophilic bacteria and Achaea can simultaneously evolve defense mechanisms against multiple and concurrent extrema (Rothschild and Mancinelli 2001). The secret of their advantageous traits lies behind their speedy-adapting transcriptional and translational scenarios that modulate, either by inhibition or activation, many responses such as anti-oxidative stress, metal-binding/transport, and membrane-permeability (Mukherjee et al. 2012; Dekker et al. 2016). Remarkably, extremophiles cell membranes possess a distinct structure/composition with an inner layer carrying a positive charge that regulates the function of metal transporters and minimizes the entrance of metals and protons, which ultimately manage acidity and metal toxicity (Zhang et al. 2016; Singh and Singh 2017).

Thus, thermophiles, halophiles, radiophiles, and polyextremophiles have been described as important microbial resources for metal bioremediation. Genome sequencing of extremophilic microorganisms such as *Sulfolobus solfataricus* (Schelert et al. 2013), *Laptospirillum ferriphilum* (Mi et al. 2011), and the extremely thermoacidophilic *Metallosphaera sedula* (Auernik et al. 2008) has harbored clusters coding the mercury resistance gene *mer*A. Meanwhile, *Acidithiubacillus ferroxidans* SUG 2–2 strain was examined for its ability to volatilize mercury in acidic soils contaminated with this metal (pH = 2.5) (Takeuchi et al. 2001; Giovanella et al. 2020). Strikingly, the metallophilic *Cupriavidus metallidurans* is able to evacuate heavy metals like Cd, Hg, and Cu; concomitantly degrading the poisonous organic contaminant toluene aerobically and anaerobically as being facultative anaerobe (Rojas et al. 2011; Bravo et al. 2020a; Millacura et al. 2018; Alviz-Gazitua et al. 2019).

On the other hand, *C. metallidurans* strain MSR33, which is a transconjugant derivative of metallophilic *C. metallidurans* CH_{34}, encoded the environmental plasmid pTP6 that authorized it not only to expand mercury resistance from 2 to 4-folds, relative to the pristine strain, but also reducing inorganic and organic mercurial compounds into Hg(0) under aerobic and anaerobic conditions (Rojas et al. 2011; Bravoa et al. 2020a). Consequently, it was utilized by (Bravo et al. 2020b) in remediating mercury-contaminated agricultural soil via a rotary drum bioreactor as ex situ technology. The data of this study recorded 82% elimination potential and revealed the fostering of nitrogen-fixing and nitrification processes by endogenous communities in the remediated soil. In the same sense, *Haloferax* sp. HA1, *Haloferax* sp. HA2, *Halobacterium* sp. HA3, and *Halococcus* sp. HA4 are hydrocarbon-utilizing halophilic archaea strains, isolated from a hypersaline coastal area of the Arabian Gulf, they showed resistance to mercury and were able to volatize 42.6, 46.2, 50.8, and 51.6% of Hg (II) (100 ppm), respectively after 8 days, under 45 °C as incubation temperature and 4 M solution (Al-Mailem et al. 2011).

8.4.2 Plant-Microbes Interaction in Mercury Remediation

The soil is inhabited by various microbial groups; remarkably, the mycorrhizal fungi that occupy rhizosphere are the most distinguished dwellers. Their role in absorption or adsorption of heavy metals triggered through their extending mycelia in the soil, which fosters the increment of plant roots surface area profoundly (Jinet al. 2018; Singh et al. 2021, 2022). Such endophytic mycorrhiza increases the plants' capability to withstand heavy metal ions through the production of siderophores, organic acids, and chelating agents. However, their ability to acidify the ambient medium and activate metal phosphates is also accounted as another means for fungi to synergize plants. Furthermore, the exopolysaccharides secreted by fungi, upon increasing heavy metal levels, could capture heavy metals on the surface of fungal cell walls, blocking by such way the mobility and bioavailability of heavy metals in the plant.

It is worth mentioning the pivotal role of arbuscular mycorrhizal fungi (AMF), which are mutualistic symbionts colonized in plant roots, in enhancing the uptake of water, micro, and macronutrients from the soil to the plants. Besides, AMF ooze a glycoprotein, called glomalin that consolidate the structure and physicochemical properties of the soil, which increases soil fertility (Herath et al. 2021). Remarkably, in a study conducted by Li et al. (2023), the phytoremediating role of AMF in the mercury (Hg) uptake by rice plants was investigated by using ^{199}Hg isotope. The results of pot trials highlighted the Hg content in rice associated with AMF group ranged from 52.82 to 96.42% lower than that in control group (without AMF inoculation). Let alone the tendency of Hg to accumulate in non-edible parts of the plant like the stems and leaves. Additionally, the accumulated Hg in the grains inoculated by AMF recorded only 20.19%, while AMF non-inoculated grains contained 48.07% of Hg. That resulted in ameliorating in the growth of rice indicated by increasing in some growth indices like biomass and antioxidant enzyme activities.

Interestingly, plant growth-promoting (PGP) bacteria also assist in phytoremediation process through two steps. Firstly, metals chelation and transformation in the soil and subsequently quenching their availability and facilitating the uptake of necessary soil-bound metals. Secondly, enhancing plant growth and further plant biomass by improving vital processes such as nitrogen fixation, phosphorus solubilization, iron sequestration, and production of 1-aminocyclopropane 1-carboxylic acid deaminase and other phytohormones (Gong et al. 2018). In this concern, Rafique et al. (2015) utilized a nitrogen-fixing bacterium, *Cronobacter* sp., which is root nodules symbiont, as a simultaneous biofertilizer and mercury bioremediator. Moreover, *Enterobacter. aerogenes* was used to bioremediate mercury and zinc (Ravikumar et al. 2007), chromium (Panda and Sarkar 2012), cadmium, and copper (Huang et al. 2005). This bacterium was found to have a symbiotic relationship with legume plants of *Vicia faba*, *Phaseolus vulgaris*, *Pisum sativum*, and non-legume plants of *Cucumis sativus* and *Lycopersicon esculentum* and help the plants to remediate the mercury in the soil (Sorkhoh et al. 2010).

8.5 Microbial Bioremediation Strategies of Hg

Through utilizing the previous microbial remediation mechanisms, which was achieved by various microbial groups as listed formerly, two distinct strategies can be applied in real environmental locations depending on the site characteristics, at which the process of bioremediation would be executed, including in situ and ex situ.

8.5.1 In-Situ *Bioremediation of Hg*

In this strategy, the decontamination process occurred at the polluted place/site itself by using the biological agent in the contaminated site. It encompasses the detoxification of sorbed and dissolved Hg in different places such as saturated soil, unsaturated soil, and groundwater using indigenous microbes either solitary or in a consortium. For example, in aquatic environments, Hg methylation can occur by microbial communities at anoxic–oxic conditions in the soils and sediments, which generally contain organic matter (OM) that is considered a main vector of MeHg and Hg transport from the catchments to the surface of the water. Different microorganisms have been identified as Hg methylators in these environments as mentioned by Schaefer et al. (2014), who detected Hg methylating microbial communities in the tropical swamp at southern Sweden and Florida. In addition, in situ technology involves the acceleration of detoxification process through adjustment and modifying the ambient conditions in place to be more appropriate for wild microorganisms to maximize their performance in the least period. As referred by He et al. (2015), this strategy is predominantly preferred due to its practical easiness, low possibility of pollutant transfer, less expensive and less destructive/intrusive to on-site ecological operations. Nonetheless, the depth of the contaminated site and pollutant concentration could configure one of this strategy's limitations (Kulshreshtha et al. 2014). Remarkably, in situ strategy could be fulfilled through several types as follows.

8.5.1.1 Bio-sparing Process

The bio-sparing process occurs by air injection through a pipe found below the table of water to enhance the indigenous microorganisms' growth by elevating oxygen concentration. Also, it is completely different from the bio-venting process in mixing both the groundwater and soil by air injection in the saturated area, which leads to the movement of volatile organic compounds from the saturated to the unsaturated area, this process is affected by the pollutant biodegradability and characteristics of the soil. It characterizes by easiness and flexibility in designing and constructing the system of air injection points (Jain et al. 2012).

8.5.1.2 Bio-venting Process

Bio-venting process in which the indigenous soil microorganisms can be stimulated to degrade the targeted contaminant by injecting a small amount of oxygen to increase microbial activity. The air injection occurred in the unsaturated area and also supplemented it with moisture and nutrients. This process could be more efficient in the case of anaerobic bioremediation, also mixing both oxygen and nitrogen will increase the remediation potency (Jain et al. 2012).

8.5.1.3 Bio-augmentation Process (Bio-vagnification)

Bio-augmentation process aims to increase the native microbiota either by introducing naturally occurring or genetically engineered microorganisms (GMO) to decontaminate the polluted site. This treatment usually uses a microbial consortium that has the ability to produce all the required degradative enzymes and pathways. This process is used to treat soil, ground, and wastewater (Jain et al. 2012). In this sense, the microbial biomass could be used as immobilized material in matrices like silica and alginate to be a suitable biosorbent with suitable porosity and strength. Intriguingly, utilizing Hg-resistant strains in immobilized form to decontaminate polluted sites is evident intensively in numerous laboratory-scale pilot studies (Pepi et al. 2013; Jafari et al. 2015).

However, Vidali (2001) declared the necessity of well-competing ability of exogenous microbes with indigenous populations to ensure the sustainability and successfulness of the bioaugmentation process. Notably, Mahbub et al. (2017a) removed about 60% of soil-bound Hg via bioaugmentation with improved growth of cucumber and lettuce in the bio-augmented soils. He and coworkers demonstrated that insufficient application of bioaugmentation in the soil could be attributed to several reasons including, poor abundancy of Hg in soil, coexistence of mixed contaminants, and improper nutrients supplements, which collectively interfere with biochemical potential and metabolic activity of Hg-remediating microorganisms. Meanwhile, the combination remediating therapy seems to be influential. Nakamura et al. (1999) merged both chemical leaching processes with seeding by *Pseudoalteromonas haloplaktis* for removing 85% of Hg content found in Minamata Bay sediments.

8.5.1.4 Bio-stimulation Process (or Accelerated Natural Attenuation)

This process modifies the polluted environment to stimulate indigenous microorganisms for enhancing the bioremediation. This can occur via circulating an inflow of extra nutrients and electron acceptors (e.g., nitrogen, carbon, oxygen, and phosphorus) through contaminated areas (Riseh et al. 2022). In this regard, Feng et al. (2014) demonstrated that uplifting the concentration of sulfate to 59.9 mg/L in inflow water promoted sulfate-reducing bacteria to enhance Hg methylation in the wetland ecosystem. Meanwhile, Winardi et al. (2020) performed a comparative study to remediate Hg from the soil in Ka-limantan Barat-Indonesia. The different groups of sampling plots were exposed to different in situ bioremediating technology. The design included aeration (bioventing), while biostimulation was implemented by nutrient addition and pH flocculation. The experiment was conducted during rainy and dry seasons to detect the seasonal variation effect. The finding of this comparative study unveiled the effective Hg-remediation accounted by 89% within 90 days under rainy conditions, neutral pH with nutrients addition.

8.5.1.5 Bio-attenuation Process (Natural Process)

Bio-attenuation process involves naturally occurring physical, chemical, and biological processes that decrease the toxicity, volume, mass, and contaminant concentration (Riseh et al. 2022). It could be implemented aerobically, anaerobically, and under simultaneous or sequential or both conditions. Despite its simplicity and lower cost, it suffers from some drawbacks like slow rate and effectiveness only in the case of simple or less complex contaminates. Nevertheless, its combination with other techniques (e.g., biostimulation and bioaugmentation) would boost its efficacy (Goswami et al. 2018).

8.5.2 Ex-Situ *Bioremediation of Hg*

The main concept of this strategy depends on the treatment of the contaminated site by soil excavation followed by transferring it away to another place to be remediated. As many biogenic processes that were mediated by microbial activity, the efficiency of this strategy count on different variables like pH, temperature, salinity, pollutants overload, and microbial biomass. This strategy includes five pathways as demonstrating:

8.5.2.1 Slurry-Phase Bioremediation

This technique depends on contaminated soil excavation and mixing it with water and then transporting the mixed soil to a bioreactor, followed by rubble and stone removal. The used water amount depends on the contamination concentration, type, biodegradation rate, and soil nature. Then, the soil can be separated by centrifugation and filtration, followed by soil drying and transferred to the original site (EPA 2003). In this context, Azoddein (2013) employed *Pseudomonas putida* (ATCC 49,128) in a field study using petroleum industrial plants (two different locations) in Peninsular, Malaysia contained 1000 ppm of Hg. The results revealed efficient removal recorded 90.5%, 97.27%, and after 96 h for point-1 and point-2, respectively; reflecting by such way the potentiality of such strain in remediating Hg from actual petroleum wastes. In a similar study conducted by Deckwer et al. (2004), mercury-contaminated wastewater was treated by Hg-resistant bacterial biomass in an aerated bioreactor and the data indicated the reduction of Hg^{2+} to volatile Hg^0 gas that was constrained in an activated carbon filter.

8.5.2.2 Solid Phase Bioremediation

This process includes three steps: soil excavation, followed by transferring the soil to piles, sometimes the soil contains agricultural, organic, and municipal wastes, followed by the biodegradation process stimulation by oxygen supplying through a pipes network to enhance the respiration of microorganisms; subsequently increase microbial activity. This technique requires a large area and takes a long time to be done and complete (Hyman and Dupont 2001).

8.5.2.3 Land-Farming

This technique focuses on indigenous microbe stimulation and their aerobic manipulation towards contaminants. It mediates by spreading on the soil surface by excavated soil supplementation with minerals and nutrients to stimulate the biodegradation process. It could be described as a superficial process that is restricted to the treatment of the top 10–35 cm of soil (Vidali 2001).

8.5.2.4 Biopiles

This technique is similar to the land-farming technique but differs in using aboveground piles and used pipes for air injection through the soil; thereby, it could be considered as a merge between composting and landfilling. This process is characterized by its low cost and complete control of aeration, temperature, and nutritional feed, this technique is applied in treating surface-contaminated environments and limiting volatilization of low molecular weight compounds (Verma 2022).

8.5.2.5 Composting Bioremediation

Composting bioremediation process is similar to land farming in employing contaminated soil excavation and indigenous microorganisms that were stimulated through nutrients feeding and air injection. The main difference lies behind the soil supplementation, which is nonhazardous additives such as animal manure and agricultural residues (e.g., hay, straw, corncobs, etc.). Such organic supplements aid in the eventual distribution of the oxygen through the soil, maintain the moisture content, enrich microbial populations and raise up the compost temperature (Vidali 2001). However, this technique is not suitable for volatile pollutants due to the periodic turning through the process (Hobson et al. 2005). Recently, the combination between composting and carbon-based materials like biochar could ameliorate the bioremediation process by expanding the surface area that is supported by various functional groups and also extending more nutrients and organic matter that facilitate and expedite the metabolic activities of microbes (Gong et al. 2018).

Albeit the suitability of ex-situ technology to scavenge the toxicity of various pollutants within a suitable time frame, excavation and pollutant transfer process remains the major obstacle, which thereafter increases both transfer cost and probability of cross-contamination. Anyway, Mahbub et al. (2017b, c) stated that both methods, namely, in situ and ex situ, are still in the experimental phase of field studies.

As the overall process entails accurate knowledge about the nature/concentration of contaminant and perceiving the physicochemical/local biogeochemical features of the contaminated sites and also appraise the multitasking functions of microbes that could be easily harmonized with any modification to achieve their goals. Hence, more investigations and researches are going on in this aspect (abo-Alkasem et al. 2023).

8.6 Advanced Approaches in Mercury Remediation

As a natural process, bioremediation process gains a lot of attention owing to several merits, including, safety, economic, easiness, and appreciable efficiency. That's besides the possibility to recover heavy metals, its low requirement of energy/temperature, comparing to other physicochemical means, and so less expensive operation cost. Additionally, the feasibility to be executed on site avoids by such a way disrupting of normal activities and transportation step that consequently leads to additional risk (Gupta et al. 2016; Volarić et al. 2021). But the geochemical conditions, nutrients availability, physicochemical properties of contaminant, and contaminant concentration are collectively controlling the microbial performance and facilitate/retard the clean-up process (Jeyakumar et al. 2023). Meanwhile, the slow growth rate of microbes, longer growth time, the potential of more persistent/toxic byproduct, and regulatory uncertainty are habitually the main drawbacks of microbial remediation, which triggers researchers and technologists to adopt modern and advanced tools to speed up bioremediation and augment its efficacy as referred by Vidali (2001) and Tripathi and Ram (2018).

8.6.1 Synthetic Biology and Genetically Engineered Microorganisms (GMOs)

As a naturally inspired process rather than artificially designed, the genetic exchange among microorganisms promoted the researchers to invest recombinant DNA technology in bioremediation. Microbes are engineered with intendedly inserted desired traits such as metal homeostasis, higher metabolic rate, tolerance of biotic/abiotic stressors and overexpression of meta-chelators, uptake regulator, transport, and degradative genes. Thereby, GMOs act as smart cell factories that utilize risky unwanted wastes in an enhanced manner in contaminated groundwater, soil, and active sludge (Volarić et al. 2021). Hence, Tay et al. (2017) cloned MerR promotor of *Shigella flexneri* plasmid to bacterial biofilm. Interestingly, MerR is responsible for curli nanofibers synthesis that facilitates sequestration of mercury. Nonetheless, the safe release of foreign modified organisms in the ecosystem still symbolizes a cryptic matter and may cause unmeasurable, unaccounted, and unreacted adverse impacts on the natural structure including functional microbial community composition and diversity alterations as highlighted by Sarao and Kaur (2021) Volarić et al. (2021). In this regard, Xue et al. (2022a, b) designed and developed a self-controlled genetic circuit of *Pseudomonas putida* KT2440 and *Escherichia coli* cells, respectively, that exhibit superior performance in mercury sensing and adsorption, followed by programmable killing stage by utilizing a cell suicide module.

8.6.2 *Metagenomics*

It is a technical term used to characterize the genetic profile of microorganisms in any ecosystem. It gives detailed information about the response of ecosystems members against environmental changes induced by any pollution by bestowing the sequence and functions of genomes concerning adaptive microorganisms in the site community (Malla et al. 2018; Jaiswal and Shukla 2020) via such advanced technology, the total DNA of any examined site (soil, wastewater, sludge, etc.) was extracted, which serve as DNA of all indigenous microbes collectively present in the examined site that terms site metagenome. Once, the extraction step was fulfilled, the construction of the DNA library followed to facilitate the screening of the target genes, which finalized by the intense expression of the target gene product (Volarić et al. 2021). In fact, Jaiswal et al. (2019) constructed *mer* operon metagenomic library of Panipat, which is one of the well-known sites contaminated with mercury, by utilizing *E. coli* as a host. The promising results indicated that the clones displayed the potentiality for mercury tolerance and volatilization by 90 ppm and 91.89%, respectively. Additionally, the efficiency of mercury remediation could be elevated by encapsulating the clones in polyacrylamide gel and alginate microspheres, which also enable their reusability.

8.6.3 *Biosurfactants*

Biosurfactants are surface-active compounds produced microbially and characterized by their amphiphilic nature (i.e., encompass both hydrophobic and hydrophilic moieties). They have been utilized recently as alternatives to synthetic surfactants by the dint of their biodegradability, biocompatibility, biosafety, bioavailability, specificity, withstanding extreme conditions, and higher surface and interfacial activity. Biosurfactants possess variable chemical structures; exhibiting a broad range of chelating capabilities with different metals (Jeyakumar et al. 2023). The biosurfactant produced by *Bacillus sp.* MSI 54 was characterized and its chemical structure of anionic nature lipopeptide was identified by FTIR and nuclear magnetic resonance spectroscopy (NMR). Its chelating capacity to mercury from fresh vegetables and wastewater was detected by atomic absorption spectroscopy, which was assessed by 75.5% (Ravindran et al. 2020).

8.6.4 *Combined Remediation*

To compensate the limitations of each sole method, it is recommended to hybridize two or more approaches together. Namely, utilize a consortium of different microorganisms, wherein each can remediate by a pathway differs from the pathway used by the others. Additionally, the amalgament between physical/chemical, physical/biological, or chemical/biological remediation approaches could be also grouped into the same concept (Gong et al. 2018). Interestingly, phyoextraction combined with electrokinetic remediation (Mao et al. 2016), chemical stabilization-assisted soil washing (Zhai et al. 2018), thermal treatment combined with a chelating agent, Ma et al. (2015) are promising paradigms on such combined biotechnology.

In this context, innumerable materials, fabricated either chemically or physically, were employed as immobilizing matrices for entrapping mercury-remediating microbes. In recent studies, innovative matrices varied in its chemical structure and physical properties, oscillated in its origin from natural to synthetic, and are exploited as ecofriendly, cost-effective, high surface area, porous-structure adsorbents that impregnated with microbial cells with degradative/remediating traits (Gong et al. 2018).

8.6.4.1 Carbon-Based Materials

Waste-derived materials, which represent an environmental burden, were directed to adopt bioremediation purposes (Beckers et al. 2019; Gong et al. 2018; O'Connor et al. 2018; Liu et al. 2022). However, such materials could be classified according to their feedstock into industrial waste-derived substance and biomass-derived materials (e.g., biochar, activated carbon, graphene, graphene oxide, etc.), while coal fly ash (CFA) is categorized among the most common industrial waste-derived substance (Wang et al. 2020; Liu et al. 2022). Strikingly, mercury-volatilizing bacteria like *Pseudomonas* sp. DC-B1 and *Bacillus* sp. DC-B2 immobilized on 4% sawdust biochar diminished mercury phytoavailability in lettuce shoots, roots, and in soil by 2.0–48.6%, 12.3–27.4%, and 24.8–57.8%, respectively, within 56 days without changing community compositions of the soil microbial ecosystem; reflecting the successful hybridization of bacteria-biochar as green' additives (Chang et al. 2019). Moreover, Yan et al. (2018) used graphene oxide as a carrying matrix for *Enterococcus avium* and the data revealed the improvement of remediating potential of this sulfate-reducing bacterium by accelerating the growth rate and maximizing the removal rate of sulfate and metal.

8.6.4.2 Polymers

Polymers, especially those that exhibit adsorptive capacity, also gained colossal popularity in remediation technology owing to their chemical stability, pore size, and considerable surface area. Acrylamide is one of the most common synthetic polymers that distinguished by its higher potential in adsorption of Hg(II); however, its microbial toxicity and environment-unfriendly restrict its utilization in immobilization of microbes (Wang et al. 2020). As a consequence, the attention was directed to employ natural polymers to guarantee the effectiveness of the remediation process in a sustainable manner. Wherein, chitosan polymer and its functionalized/modified co-polymer show multidimensional properties with tunable adsorptive capability, in particular to Hg as revealed by Goci et al. (2023). Upon synthesizing microbeads of chitosan/algal (*Cladophora* sp.) composite, the metals sorption capacity elevated more than each ingredient individually (Sargın et al. 2016). More so, McCarthy et al. (2017) entrapped the cells of *Pseudomonas veronii* in a xanthan gum-based biopolymer and coated them with zeolite granules. Such an innovative method employed combined remediation in a tripartite way (i.e., carbon-based material, polymer, and degradative bacteria), which exerted superior performance in response to mercury volatilization with increased viability for 16 weeks at least.

8.6.4.3 Nanomaterials

The miniaturization in dimensions and increasing the surface area of materials elevate their functionality, mechanical, electrical, chemical, and adsorptive features, which trigger the utilization of materials in nanoscale dimension sign for innovative products with promising applications. A vast array of nanomaterials (e.g., nanoparticles, nanocomposites, nanosheets, carbon nanotubes, etc.) emphasized their efficiency in Hg scavenging as stated by Wang et al. (2020). The recruitment of nanoadsorbents (e.g., porous silica, titania, etc.) as carriers to immobilize microorganisms proved its efficiency in recent years (Velkova et al. 2018). In this regard, the chitosan-coated Fe_3O_4 nanoparticles and TiO_2 nanoparticles were immobilized with *Saccharomyces cerevisiae* biomass to mitigate the toxicity of heavy metals (Choudhury et al. 2017; Peng et al. 2010). Similarly, Akhtar et al. (2021) hybridized the cells of *Bacillus cereus* and *Lysinibacillus macrolides* with ZnO nanoparticles on a rice crop irrigated with heavy metals contaminated water in a synergistic manner. Such hybridization maximized significantly the removal efficacy via the synergistic mechanism of both remediating bacterial consortium and nanoparticles. Furthermore, their hybridization enhanced the plant growth and its tolerance index, while lessened the bioaccumulation index and metallothionine content. On the other hand, Ozdemir et al. (2017) used nanodiamond as biosorbent carriers for thermophilic *Bacillus altitudinis* to eliminate Hg^{2+} along with other metals from food sample. The data unveiled the simultaneous preconcentration-separation of examined metals with 0.3 ml/min as an optimum flow rate under pH 6 and biosorption capacity assessed by 19.5 mg/g.

8.7 Concluding Remarks and Future Outlook

The continuous and regular disposing of various effluents containing mercury into water bodies increases the likelihood of their access and persistence in the food chain through agricultural crops and aquatic animals, leading then to the bioaccumulating and bioaugmentation in human bodies. However, traditional remediation technologies displayed some significant disadvantages, bioremediation techniques could compensate them in an environmentally friendly, least destructive, biosafe, and cost-efficient way. Bioremediation strategies seemed to be convenient to diverse environmental circumstances, via both in situ and ex situ approaches as explained herein. This chapter addressed the main principles, strategies, effectiveness of different microbial forms, and advanced tools of mercury microbial remediation, which had been studied. However, various microbial genera possess varied metabolic prerequisites and showed disparate efficiency in the bioremoval process, which can also differ contingent upon the nature of contaminated sites, concentration of mercury dumping off in the field and also seasonal changes. Albeit efficacy, more and deep investigations entail in the following aspects: (1) A key challenge is appropriate screening and selection of novel species that exhibit characteristic metabolic traits and advantageous physiological properties in accelerating mercury removal at both lab and commercial scales. Interestingly, the extremophilic dwellers (e.g., Archaea) are the promising and potent category recommended, owing to their metabolic versatility, adaptability, and tolerance, for xenobiotics remediation. (2) It is highly desired to employ novel OMICs tools (proteomics, metabolomics, genomics, transcriptomics, and fluxomics) combined with bioinformatics (e.g., in silico) and computational platforms. Such integrative ways of these new techniques could predict and optimize mechanism-based models to uplift the removal performance. (3) It is noteworthy to develop monitoring approaches to trace the stability of the remediated phase and residues of Hg in the contaminated field. (4) In this context, it is crucial also to monitor the performance of remediating microbes either GMO or native, especially Hg don't coexist in the environment solitary but among multiple pollutants either organic or inorganic, which actually influence adaptive behavior, removal rate, survival time of remediating microbes. (5) More comprehensive studies in combined remediation technology necessitate new porous crystalline nanobiosorbent materials such as nanobiosurfactants, metal–organic frameworks (MOFs), and covalent organic frameworks (COFs), which act as carriers or immobilization matrices. Functionalization and chemical modifications of such innovative biosorbents with various functional groups will instigate their chemical stability and adsorptive behavior. (6) The recovery, reusability, and stability of microbe-MOFs/COFs composites should be conducted more through and the impact of harsh environmental conditions like temperature, pH, and salinity should also be operated. (7) Translating the obtained results accurately to full-scale operation and perceiving the whole image, namely, industrial applications and field scale with a precise assessment of expenditure through collaborative groups of researchers, technologists, health specialists, governmental institutions encourage remediation companies to apply long-term sustainable approaches efficiently.

Acknowledgements The graphical illustrations were performed via biorender.com and without permission from biorender.com.

References

Abo-Alkasem M, Nemat H, Hassan M, Mostafa A (2023) Microbial bioremediation as a tool for the removal of heavy metals. Bull Natl Res Cent 47:31

Agarwal M, Rathore RS, Jagoe C, Chauhan A (2019) Multiple lines of evidences reveal mechanisms underpinning mercury resistance and volatilization by *Stenotrophomonas* sp. MA5 isolated from the Savannah River Site (SRS), USA. Cells 8:1–14

Aiking H, Govers H, Van't Riet J (1985) Detoxification of mercury, cadmium, and lead in *Klebsiella aerogenes* NCTC 418 growing in continuous culture. Appl Environ Microbiol 50(5):1262–1267

Akhtar N, Khan S, Rehman S, Rehman Z, Khatoon A, Rha E, Jamil M (2021) Synergistic effects of zinc oxide nanoparticles and bacteria reduce heavy metals toxicity in rice (*Oryza sativa* L.) plant. Toxics 9(5):113

Alcántara C, Coll-Marqués JM, Jadán-Piedra C, Vélez D, Devesa V, Zúñiga M, Monedero V (2018) Polyphosphate in *Lactobacillus* and its link to stress tolerance and probiotic properties. Front Microbiol 9:1–10

Al-Mailem DM, Al-Awadhi H, Sorkhoh NA, Eliyas M, Radwan SS (2011) Mercury resistance and volatilization by oil utilizing haloarchaea under hypersaline conditions. Extremophiles 15:39–44

Alviz-Gazitua P, Fuentes-Alburquenque S, Rojas LA, Turner RJ, Guiliani N, Seeger M (2019) The response of *Cupriavidus metallidurans* CH34 to cadmium involves inhibition of the initiation of biofilm formation, decrease in intracellular c-di-GMP levels, and a novel metal regulated phosphodiesterase. Front Microbiol 10:1499

Amin A, Latif Z (2011) Isolation and characterization of H2S producing yeast to detoxify mercury containing compounds. Int J Microbiol 2:517–525

Amin A, Naveed M, Sarwar A, Rasheed S, Saleem HG, Latif Z, Bechthold A (2022) *In vitro* and *in silico* studies reveal *Bacillus cereus* AA-18 as a potential candidate for bioremediation of mercury-contaminated wastewater. Front Microbiol 13:847806

Anthony E (2014) Bioremediation of mercury by biofilm forming mercury resistant marine bacteria. Doctoral dissertation. National Institute of Technology, Rourkela, India

Anuar SN, Othman F, Chay TC, Nasir NA (2020) Biosorption of mercury ion (Hg^{2+}) using live and dead cells of *Rhizopus oryzae* and *Aspergillus niger*: characterization, kinetic and isotherm studies. J Pure Appl Microbiol 14(3):1749–1760

Auernik KS, Maezato Y, Blum PH, Kelly RM (2008) The genome sequence of the metal-mobilizing, extremely thermoacidophilic archaeon *Metallosphaera sedula* provides insights into bioleaching-associated metabolism. Appl Environ Microbiol 74(3):682–692

Azoddein AAM (2013) Mercury removal from petroleum based industries wastewater by *Pseudomonas Putida* ATCC 49128 in membrane bioreactor. Doctoral dissertation, UMP

Balan BM, Shini S, Krishnan KP, Mohan M (2018) Mercury tolerance and biosorption in bacteria isolated from Ny-Ålesund, Svalbard. Arctic J Basic Microbiol 58(4):286–295

Barkay T, Miller SM, Summers AO (2003) Bacterial mercury resistance from atoms to ecosystems. FEMS Microbiol Rev 27:355–384

Beckers F, Awad YM, Beiyuan J, Abrigata J, Mothes S, Tsang DC, Ok YS, Rinklebe J (2019) Impact of biochar on mobilization, methylation, and ethylation of mercury under dynamic redox conditions in a contaminated floodplain soil. Environ Int 127:276–290

Bouffard A, Amyot M (2009) Importance of elemental mercury in lake sediments. Chemosphere 74(8):1098–1103

Bravo G, Vega-Celedón P, Gentina JC, Seeger M (2020a) Effects of mercury II on *Cupriavidus metallidurans* strain MSR33 during mercury bioremediation under aerobic and anaerobic conditions. Process 8(8):893

Bravo G, Vega-Celedón P, Gentina JC, Seeger M (2020b) Bioremediation by *Cupriavidus metallidurans* strain MSR33 of mercury-polluted agricultural soil in a rotary drum bioreactor and its effects on nitrogen cycle microorganisms. Microorganisms 8(12):1952

Cardona GI, Escobar MC, Acosta-González A, Marín P, Marqués S (2022) Highly mercury-resistant strains from different Colombian Amazon ecosystems affected by artisanal gold mining activities. Appl Microbiol Biotechnol 106:2775–2793

Chan CS, Fakra SC, Edwards DC, Emerson D, Banfield JF (2009) Iron oxyhydroxide mineralization on microbial extracellular polysaccharides. Geochim Cosmochim Acta 73(13):3807–3818

Chang J, Yang Q, Dong J, Ji B, Si G, He F, Li B, Chen J (2019) Reduction in Hg phytoavailability in soil using Hg-volatilizing bacteria and biochar and the response of the native bacterial community. Microbiol Biotechnol 12(5):1014–1023

Chang J, Shi Y, Si G, Yang Q, Dong J, Chen J (2020) The bioremediation potentials and mercury(II)-resistant mechanisms of a novel fungus *Penicillium* spp. DC-F11 isolated from contaminated soil. J Hazard Mater 396:122638

Chen SC, Lin WH, Chien CC, Tsang DC, Kao CM (2018) Development of a two-stage biotransformation system for mercury-contaminated soil remediation. Chemosphere 200:266–273

Choudhury PR, Bhattacharya P, Ghosh S, Majumdar S, Saha S, Sahoo GC (2017) Removal of Cr (VI) by synthesized titania embedded dead yeast nanocomposite: optimization and modeling by response surface methodology. J Environ Chem Eng 5(1):214–221

Christakis CA, Barkay T, Boyd ES (2021) Expanded diversity and phylogeny of *mer* genes broadens mercury resistance paradigms and reveals an origin for MerA among thermophilic Archaea. Front Microbiol 12:1–20

Colombo MJ, Ha J, Reinfelder JR, Barkay T, Yee N (2014) Oxidation of Hg(0) to Hg(II) by diverse anaerobic bacteria. Chem Geol 363:334–340

Deckwer WD, Becker FU, Ledakowicz S, Wagner-Döbler I (2004) Microbial removal of ionic mercury in a three-phase fluidized bed reactor. Environ Sci Technol 38(6):1858–1865

Dekker L, Arsène-Ploetze F, Santini JM (2016) Comparative proteomics of *Acidithiobacillus ferrooxidans* grown in the presence and absence of uranium. Res Microbiol 167(3):234–239

Deng X, Jia P (2011) Construction and characterization of a photosynthetic bacterium genetically engineered for Hg^{2+} uptake. Bioresour Technol 102:3083–3088

Eltarahony M, Zaki S, Abd-El-Haleem D (2020) Aerobic and anaerobic removal of lead and mercury via calcium carbonate precipitation mediated by statistically optimized nitrate reductases. Sci Rep 10(1):4029

Eltarahony M, Kamal A, Zaki S, Abd-El-Haleem D (2021) Heavy metals bioremediation and water softening using ureolytic strains *Metschnikowia pulcherrima* and *Raoultella planticola*. J Chem Technol Biotechnol 96(11):3152–3165

EPA (2003) EPA protocol for the review of existing national primary drinking water regulations, united states environmental protection agency office of water office of groundwater and drinking water standards and risk management division targeting and Analysis branch1200 pennsylvania avenue, NW (4607M) Washington, DC 20460

Essa AM, Macaskie LE, Brown NL (2002) Mechanisms of mercury bioremediation. Biochem Soc Trans 30:4

Farraji H, Zaman NQ, Tajuddin R, Faraji H (2016) Advantages and disadvantages of phytoremediation: a concise review. Int J Env Tech Sci. 2:69–75

Feng S, Ai Z, Zheng S, Gu B, Li Y (2014) Effects of dryout and inflow water quality on mercury methylation in a constructed wetland. Wat Air Soil Poll 225:1–1

Fleming EJ, Mack EE, Green PG, Nelson DC (2006) Mercury methylation from unexpected sources: molybdate-inhibited freshwater sediments and an iron-reducing bacterium. Appl Environ Microbiol 72(1):457–464

Franco MW, Mendes LA, Windmöller CC, Moura KA, Oliveira LA, Barbosa FA (2018) Mercury methylation capacity and removal of Hg species from aqueous medium by *cyanobacteria*. Wat Air Soil Poll 229:1–2

François F, Lombard C, Guigner JM, Soreau P, Brian-Jaisson F, Martino G, Vandervennet M, Garcia D, Molinier AL, Pignol D, Peduzzi J (2012) Isolation and characterization of environmental bacteria capable of extracellular biosorption of mercury. Appl Environ Microbiol 78(4):1097–1106

Frey B, Rast BM, Qi W, Stierli B, Brunner I (2022) Long-term mercury contamination does not affect the microbial gene potential for C and N cycling in soils but enhances detoxification gene abundance. Front Microbiol 13:1–18

Gilmour CC, Bullock AL, McBurney A, Podar M, Elias DA (2018) Robust mercury methylation across diverse methanogenic archaea. Mbio 9(2):e02403-e2417

Gionfriddo CM, Soren AB, Wymore AM, Hartnett DS, Podar M, Parks JM, Elias DA, Gilmour CC (2023) Transcriptional control of *hgcAB* by an ArsR-like regulator in *Pseudodesulfovibrio mercurii* ND132. Appl Environ Microbiol 89:1–23

Giovanella P, Vieira GA, Otero IVR, Pellizzer EP, de Jesus Fontes B, Sette LD (2020) Metal and organic pollutants bioremediation by extremophile microorganisms. Hazmats 382:121024

Goci MC, Leudjo, Taka A, Martin L, Klink MJ (2023) Chitosan-based polymer nanocomposites for environmental remediation of mercury pollution. Polymer J15(3):482

Gong Y, Zhao D, Wang Q (2018) An overview of field-scale studies on remediation of soil contaminated with heavy metals and metalloids: technical progress over the last decade. Water Res 147:440–460

Goswami M, Chakraborty P, Mukherjee K, Mitra G, Bhattacharyya P, Dey S, Tribedi P (2018) Bioaugmentation and biostimulation: a potential strategy for environmental remediation. J Microbiol Exp 6(5):223–231

Grégoire DS, Poulain AJ (2014) A little bit of light goes a long way: the role of phototrophs on mercury cycling. Metallomics 6(3):396–407

Grégoire DS, Poulain AJ (2016) A physiological role for HgII during phototrophic growth. Nat Geosci 9(2):121–125

Grégoire DS, Lavoie NC, Poulain AJ (2018) Heliobacteria reveal fermentation as a key pathway for mercury reduction in anoxic environments. Environ Sci Technol 52(7):4145–4153

Groudev S, Georgiev P, Spasova I, Nicolova M (2014) Decreasing the contamination and toxicity of a heavily contaminated soil by in situ bioremediation. J Geochem Explor 144:374–379

Gupta A, Joia J, Sood A, Sood R, Sidhu C, Kaur G (2016) Microbes as potential tool for remediation of heavy metals: a review. J Microbol Biochem Technol 8(4):364–372

Håkansson T, Suer P, Mattiasson B, Allard B (2008) Sulphate reducing bacteria to precipitate mercury after electrokinetic soil remediation. Int J Environ Sci Technol 5:267–274

He F, Gao J, Pierce E, Strong PJ, Wang H, Liang L (2015) In situ remediation technologies for mercury-contaminated soil. Environ Sci Poll Res 22:8124–8147

Herath BM, Madushan KW, Lakmali JP, Yapa PN (2021) Arbuscular mycorrhizal fungi as a potential tool for bioremediation of heavy metals in contaminated soil. World J Adv Res Rev 10(3):217–228

Hobson AM, Frederickson J, Dise NB (2005) CH_4 and N_2O from mechanically turned windrow and vermicomposting systems following in-vessel pre-treatment. Waste Manag Res 25(4):345–352

Huang Q, Chen W, Xu L (2005) Adsorption of copper and cadmium by Cu-and Cd-resistant bacteria and their composites with soil colloids and kaolinite. Geomicrobiol J 22(5):227–236

Hyma M, Dupont RR (2001) Groundwater and soil remediation: process design and cost estimating of proven technologies. American Society of Civil Engineers, Chicago

Jafari SA, Cheraghi S, Mirbakhsh M, Mirza R, Maryamabadi A (2015) Employing response surface methodology for optimization of mercury bioremediation by *Vibrio parahaemolyticus* PG02 in coastal sediments of Bushehr, Iran. CLEAN–Soil Wat Air Soil Poll 43(1):118–126

Jain AN, Udayashankara TH, Lokesh KS (2012) Review on bioremediation of heavy metals with microbial isolates and amendments on soil residue. Int J Sci Res 6:2319–7064

Jaiswal S, Shukla P (2020) Alternative strategies for microbial remediation of pollutants via synthetic biology. Front Microbiol 19:11

Jaiswal G, Singh R, Porwal S (2019) Bioremediation of mercury through encapsulation of the clone carrying meroperon. J Pure Appl Microbiol 13(1):553–561

Jeyakumar P, Debnath C, Vijayaraghavan R, Muthuraj M (2023) Trends in bioremediation of heavy metal contaminations. Environ Eng Res 28(4)

Jin Y, Luan Y, Ning Y, Wang L (2018) Effects and mechanisms of microbial remediation of heavy metals in soil: a critical review. Appl Sci 8(8):1336

Kim YM, Chung JY, An HS, Park SY, Kim BG, Bae JW, Han M, Cho YJ, Hong YS (2015) Biomonitoring of lead, cadmium, total mercury, and methylmercury levels in maternal blood and in umbilical cord blood at birth in South Korea. Int J Environ Res Public Health 12(10):13482–13493

Kim Y, Abdilla B, Yuan K, De Andrade V, Sturchio NC, Lee SS, Fenter P (2021) Replacement of calcium carbonate polymorphs by cerussite. ACS Earth Space Chem 5(9):2433–2441

Kritee K, Motta LC, Blum JD, Tsui MT, Reinfelder JR (2017) Photomicrobial visible light-induced magnetic mass independent fractionation of mercury in a marine microalga. ACS Earth Space Chem 2(5):432–440

Krout IN, Scrimale T, Vorojeikina D, Boyd ES, Rand MD (2022) Organomercurial lyase (*merB*)-mediated demethylation decreases bacterial methylmercury resistance in the absence of mercuric reductase (*merA*). Appl Environ Microbiol 88:1–14

Kulshreshtha A, Agrawal R, Barar M, Saxena S (2014) A review on bioremediation of heavy metals in contaminated water. IOSR J Environ Sci Toxicol Food Technol 8(7):44–50

Li Y, Li D, Song B, Li Y (2022a) The potential of mercury methylation and demethylation by 15 species of marine microalgae. Water Res 215:118266

Li X, Yang Z, Zhang G, Si S, Wu X, Cai L (2022b) Plasmid genomes reveal the distribution, abundance, and organization of mercury-related genes and their co-distribution with antibiotic resistant genes in gamma proteobacteria. Genes 13:1–13

Li XC, Yang ZZ, Zhang C, Wei JJ, Zhang HQ, Li ZH, Ma C, Wang MS, Chen JQ, Hu JW (2019) Effects of different crystalline iron oxides on immobilization and bioavailability of Cd in contaminated sediment. J Chem Eng 373:307–317

Li X, Zhou M, Shi F, Meng B, Liu J, Mi Y, Dong C, Su H, Liu X, Wang F, Wei Y (2023) Influence of arbuscular mycorrhizal fungi on mercury accumulation in rice (*Oryza sativa* L.): from enriched isotope tracing perspective. Ecotoxicol Environ Saf 255:114776

Lin H, Hurt RA Jr, Johs A, Parks JM, Morrell-Falvey JL, Liang L, Elias DA, Gu B (2014) Unexpected effects of gene deletion on interactions of mercury with the methylation-deficient mutant Δ hgcAB. Environ Sci Technol Lett 1(5):271–276

Lin H, Ascher DB, Myung Y, Lamborg CH, Hallam SJ, Gionfriddo CM, Holt KE, Moreau JW (2021) Mercury methylation by metabolically versatile and cosmopolitan marine bacteria. ISME J 15:1810–1825

Lin H, Moody ER, Williams TA, Moreau JW (2023) On the origin and evolution of microbial mercury methylation. Genome Biol Evol 15:1–12

Liu D, Yang X, Zhang L, Tang Y, He H, Liang M, Tu Z, Zhu H (2022) Immobilization of biomass materials for removal of refractory organic pollutants from wastewater. Int J Environ Res Public Health 19(21):13830

Lu X, Liu Y, Johs A, Zhao L, Wang T, Yang Z, Lin H, Elias DA, Pierce EM, Liang L, Barkay T (2016) Anaerobic mercury methylation and demethylation by *Geobacter bemidjiensis Bem*. Environ Sci Technol 50(8):4366–4373

Luo H, Cheng Q, He D, Sun J, Li J, Pan X (2023) Recent advances in microbial mercury methylation: a review on methylation habitat, methylator, mechanism, and influencing factor. Process Safe Environ Prot 170:286–296

Ma F, Peng C, Hou D, Wu B, Zhang Q, Li F, Gu Q (2015) Citric acid facilitated thermal treatment: an innovative method for the remediation of mercury contaminated soil. Hazmats 300:546–552

Mahbub KR, Krishnan K, Naidu R, Andrews S, Megharaj M (2017a) Mercury toxicity to terrestrial biota. Ecol Indic 74:451–462

Mahbub KR, Subashchandrabose SR, Krishnan K, Naidu R, Megharaj M (2017b) Mercury alters the bacterial community structure and diversity in soil even at concentrations lower than the guideline values. Appl Microbiol Biotechnol 101:2163–2175

Mahbub KR, Bahar MM, Labbate M, Krishnan K, Andrews S, Naidu R, Megharaj M (2017c) Bioremediation of mercury: not properly exploited in contaminated soils. Appl Microbiol Biotechnol 101:963–976

Malla MA, Dubey A, Yadav S, Kumar A, Hashem A, Abd_Allah EF (2018) Understanding and designing the strategies for the microbe-mediated remediation of environmental contaminants using omics approaches. Front Microbiol 4:9

Mao X, Han FX, Shao X, Guo K, McComb J, Arslan Z, Zhang Z (2016) Electro-kinetic remediation coupled with phytoremediation to remove lead, arsenic and cesium from contaminated paddy soil. Ecotoxicol Environ Saf 125:16–24

Marques CR (2018) Extremophilic microfactories: applications in metal and radionuclide bioremediation. Front Microbiol 9:1191

McCarthy D, Edwards GC, Gustin MS, Care A, Miller MB, Sunna A (2017) An innovative approach to bioremediation of mercury contaminated soils from industrial mining operations. Chemosphere 184:694–699

Millacura FA, Janssen PJ, Monsieurs P, Janssen A, Provoost A, Van Houdt R, Rojas LA (2018) Unintentional genomic changes endow *Cupriavidus metallidurans* with an augmented heavy-metal resistance. Genes 9(11):551

Mokone JG, Tutu H, Chimuka L, Cukrowska EM (2018) Optimization and characterization of *Cladophora* sp. alga immobilized in alginate beads and silica gel for the biosorption of mercury from aqueous solutions. Wat Air Soil Poll 229:1–14

Mukherjee A, Wheaton GH, Blum PH, Kelly RM (2012) Uranium extremophily is an adaptive, rather than intrinsic, feature for extremely *thermoacidophilic Metallosphaera* species. Proc Natl Acad Sci 109(41):16702–16707

Nakamura K, Hagimine M, Sakai M, Furukawa K (1999) Removal of mercury from mercury-contaminated sediments using a combined method of chemical leaching and volatilization of mercury by bacteria. Biodegradation 10:443–447

Nurfitriani S, Arisoesilaningsih E, Nuraini Y, Handayanto E (2020) Bioaccumulation of mercury by bacteria isolated from small scale gold mining tailings in Lombok, Indonesia. J Ecol Eng 21(6)

O'Connor D, Peng T, Li G, Wang S, Duan L, Mulder J, Cornelissen G, Cheng Z, Yang S, Hou D (2018) Sulfur-modified rice husk biochar: a green method for the remediation of mercury contaminated soil. Sci Total Environ 621:819–826

Okoh MP, Olobayetan IW, Machunga-Mambula SS (2018) Bioleaching, a technology for metal extraction and remediation: mitigating health consequences for metal exposure. Int J Develop Sustain 7:2103–2118

O'rear DJ, Cooper RE, Sheu FR, Belue JT (2014) Inventors; Chevron USA Inc, assignee. Process, method, and system for removing mercury from fluids. United States patent US 8,790,427

Ozdemir S, Kilinc E, Celik KS, Okumus V, Soylak M (2017) Simultaneous preconcentrations of CO^{2+}, Cr^{6+}, Hg^{2+} and Pb^{2+} ions by *Bacillus altitudinis* immobilized nanodiamond prior to their determinations in food samples by ICP-OES. Food Chem 215:447–453

Panda J, Sarkar P (2012) Bioremediation of chromium by novel strains *Enterobacter aerogenes* T2 and *Acinetobacter* sp. PD 12 S2. Environ Sci Poll Res 19:1809–1817

Pan-Hou HS, Imura N (1981) Role of hydrogen sulfide in mercury resistance determined by plasmid of *Clostridium cochlearium* T-2. Arch Microbiol 129:49–52

Peng Q, Liu Y, Zeng G, Xu W, Yang C, Zhang J (2010) Biosorption of copper(II) by immobilizing *Saccharomyces cerevisiae* on the surface of chitosan-coated magnetic nanoparticles from aqueous solution. JHM Lett 177(1–3):676–682

Pepi M, Focardi S, Tarabelli A, Volterrani M, Focardi SE (2013) Bacterial strains resistant to inorganic and organic forms of mercury isolated from polluted sediments of the Orbetello Lagoon, Italy, and their possible use in bioremediation processes. In: E3S web of conferences.1, 31002. EDP Sciences

Poulin BA, Aiken GR, Nagy KL, Manceau A, Krabbenhoft DP, Ryan JN (2016) Mercury transformation and release differs with depth and time in a contaminated riparian soil during simulated flooding. Geochim Cosmochim Acta 176:118–138

Priyadarshanee M, Chatterjee S, Rath S, Dash HR, Das S (2022) Cellular and genetic mechanism of bacterial mercury resistance and their role in biogeochemistry and bioremediation. JHM Lett 423:1–20

Rafique A, Amin A, Latif Z (2015) Screening and characterization of mercury-resistant nitrogen fixing bacteria and their use as biofertilizers and for mercury bioremediation. Pak J Zool 47(5):1271–1277

Ravikumar S, Williams GP, Shanthy S, Gracelin NAA, Babu S, Parimala PS (2007) Effect of heavy metals (Hg and Zn) on the growth and phosphate solubilising activity in halophilic *phosphobacteria* isolated from Manakudi mangrove. J Environ Biol 28(1):109–114

Ravindran A, Sajayan A, Priyadharshini GB, Selvin J, Kiran GS (2020) Revealing the efficacy of thermostable biosurfactant in heavy metal bioremediation and surface treatment in vegetables. Front Microbiol 10(11):222

Riseh RS, Vazvani MG, Hajabdollahi N, Thakur VK (2022) Bioremediation of heavy metals by *Rhizobacteria*. Appl Biochem Biotechnol 1–23

Rojas LA, Yáñez C, González M, Lobos S, Smalla K, Seeger M (2011) Characterization of the metabolically modified heavy metal-resistant *Cupriavidus metallidurans* strain MSR33 generated for mercury bioremediation. PLoS ONE 6(3):17555

Rothschild LJ, Mancinelli RL (2001) Life in extreme environments. Nat

Ruiz ON, Alvarez D, Gonzalez-Ruiz G, Torres C (2011) Characterization of mercury bioremediation by transgenic bacteria expressing metallothionein and polyphosphate kinase. BMC Biotechnol 11:1–8

Saranya K, Sundaramanickam A, Shekhar S, Swaminathan S, Balasubramanian T (2017) Bioremediation of mercury by *Vibrio fluvialis* screened from industrial effluents. Bio Med Res Int

Sarao LK, Kaur S (2021) Chapter-3 microbial bioremediation. New Vistas Microbial Sci 47

Sargın İ, Arslan G, Kaya M (2016) Efficiency of chitosan–algal biomass composite microbeads at heavy metal removal. React Funct Polym 98:38–47

Schaefer JK, Kronberg RM, Morel FMM, Skyllberg U (2014) Detection of a key Hg methylation gene hgcA, in wetland soils. Environ Microbiol Rep 6:441–447

Schelert J, Rudrappa D, Johnson T, Blum P (2013) Role of MerH in mercury resistance in the archaeon *Sulfolobus solfataricus*. Microbiol 159:1198

Shahpiri A, Mohammadzadeh A (2018) Mercury removal by engineered *Escherichia coli* cells expressing different rice *metallothionein isoforms*. Ann Microbiol 68:145–152

Sharma KR, Naruka A, Raja M, Sharma RK (2022) White rot fungus mediated removal of mercury from wastewater. Water Environ Res 94(7):10769

Singh A, Singh AK (2017) Haloarchaea: worth exploring for their biotechnological potential. Biotechnol Lett 39:1793–1800

Singh A, Kumari R, Yadav AN (2021) Fungal secondary metabolites for bioremediation of hazardous heavy metals. Recent Trends Mycol Res Environ Ind Perspect 2:65–98

Singh M, Jayant K, Bhutani S, Mehra A, Kaur T, Kour D, Suyal DC, Singh S, Rai AK, Yadav AN (2022) Bioremediation—sustainable tool for diverse contaminants management: current scenario and future aspects. J Appl Biol Biotechnol 10(2):48–63

Sinha A, Khare SK (2012) Mercury bioremediation by mercury accumulating *Enterobacter* sp. cells and its alginate immobilized application. Biodegradation 23:25–34

Sorkhoh NA, Ali N, Al-Awadhi H, Dashti N, Al-Mailem DM, Eliyas M, Radwan SS (2010) Phytoremediation of mercury in pristine and crude oil contaminated soils: contributions of rhizobacteria and their host plants to mercury removal. Ecotoxicol Environ Saf 73(8):1998–2003

Tada Y, Marumoto K, Iwamoto Y, Takeda K, Sakugawa H (2023) Distribution and phylogeny of mercury methylation, demethylation, and reduction genes in the Seto Inland Sea of Japan. Mar Pollut Bull 186:1–13

Taha A et al (2023) Bioremediation of heavy metals in wastewaters: a concise review. Egypt J Aquat Biol Fish 27(1):143–166

Takeuchi F, Iwahori K, Kamimura K, Negishi A, Maeda T, Sugio T (2001) Volatilization of mercury under acidic conditions from mercury-polluted soil by a mercury-resistant *Acidithiobacillus ferrooxidans* SUG 2–2. Biosci Biotechnol Biochem 65(9):1981–1986

Tarangini K (2009) Biosorption of heavy metals using individual and mixed cultures of *Pseudomonas aeruginosa and Bacillus subtilis*. Doctoral dissertation

Tarekegn M, Zewdu Salilih F, Ishetu AI (2020) Microbes used as a tool for bioremediation of heavy metal from the environment. Cogent Food Agric 6(1):1783174

Tarfeen N, Nisa KU, Hamid B, Bashir Z, Yatoo AM, Dar MA, Mohiddin FA, Amin Z, Ahmad RA, Sayyed RZ (2022) Microbial remediation: A promising tool for reclamation of contaminated sites with special emphasis on heavy metal and pesticide pollution: a review. Process 10(7):1358

Tay PK, Nguyen PQ, Joshi NS (2017) A synthetic circuit for mercury bioremediation using self-assembling functional amyloids. ACS Synth Biol 6(10):1841–1850

Tazaki K, Asada R (2007) Transmission electron microscopic observation of mercury-bearing bacterial clay minerals in a small-scale gold mine in Tanzania. Geomicrobiol J 24(6):477–489

Tiodar ED, Văcar CL, Podar D (2021) Phytoremediation and microorganisms-assisted phytoremediation of mercury-contaminated soils: challenges and perspectives. Int J Environ Res Public Health 18:1–37

Tripathi SM, Ram S (2018) Bioremediation of groundwater: an overview. Int J Appl Eng Res 13(24):16825–16832

Văcar CL, Covaci E, Chakraborty S, Li B, Weindorf DC, Frenţiu T, Pârvu M, Podar D (2021) Heavy metal-resistant filamentous fungi as potential mercury bioremediators. J Fungus 7(5):386

Vela-García N, Guamán-Burneo MC, González-Romero NP (2019) Efficient bioremediation of metallurgical effluents through the use of microalgae isolated from the amazonic and highlands of ecuador. Rev Int Contam Ambient 4:917–929

Velkova Z, Kirova G, Stoytcheva M, Kostadinova S, Todorova K, Gochev V (2018) Immobilized microbial biosorbents for heavy metals removal. Eng Life Sci 18(12):871–881

Verma A (2021) Bioremediation techniques for soil pollution: an introduction. IntechOpen, London, UK

Verma A (2022) Bioremediation techniques for soil pollution: an introduction. Biodegrad Technol Organ Inorgan Poll

Vidali M (2001) Bioremediation. An overview. Pure Appl Chem 73(7):1163–1172

Vitor G, Palma TC, Vieira B, Lourenço JP, Barros RJ, Costa MC (2015) Start-up, adjustment and long-term performance of a two-stage bioremediation process, treating real acid mine drainage, coupled with biosynthesis of ZnS nanoparticles and ZnS/TiO_2 nanocomposites. Miner Eng 89:18–85

Volarić A, Svirčev Z, Tamindžija D, Radnović D (2021) Microbial bioremediation of heavy metals. Hem Ind 75(2):103–115

Wagner-Döbler I (2003) Pilot plant for bioremediation of mercury-containing industrial wastewater. Appl Microbiol Biotechnol 62:124–133

Wang L, Hou D, Cao Y, Ok YS, Tack FM, Rinklebe J, O'Connor D (2020) Remediation of mercury contaminated soil, water, and air: a review of emerging materials and innovative technologies. Environ Int 134:105281

Wang L, Cheng WC, Xue ZF, Rahman MM, Xie YX, Hu W (2023) Immobilizing lead and copper in aqueous solution using microbial-and enzyme-induced carbonate precipitation. Front Bioeng Biotechnol 11:1146858

Winardi W, Haryono E, Sudrajat S, Soetarto ES (2020) *In situ* bioremediation strategies for the recovery of mercury-contaminated land in abandoned traditional gold mines in Indonesia. Biosaintifika: J Biol Educ 12(3):469–477

Wu C, Tang D, Dai J, Tang X, Bao Y, Ning J, Zhen Q, Song H, St. Leger RJ, Fang W (2022) Bioremediation of mercury-polluted soil and water by the plant symbiotic fungus *Metarhizium robertsii*. Proc Natl Acad Sci 119(47):e2214513119

Xie F, Dong K, Wang W, Asselin E (2020) Leaching of mercury from contaminated solid waste: a mini-review. Miner Process Extr Metall Rev 41(3):187–197

Xue Y, Du P, Shendi AA, Yu B (2022a) Mercury bioremediation in aquatic environment by genetically modified bacteria with self-controlled biosecurity circuit. J Clean Prod 337:130524

Xue Y, Qiu T, Sun Z, Liu F, Yu B (2022b) Mercury bioremediation by engineered *Pseudomonas putida* KT2440 with adaptationally optimized biosecurity circuit. Environ Microbiol 7:3022–3036

Yadav V, Manjhi A, Vadakedath N (2023) Mercury remediation potential of mercury-resistant strain *Rheinheimera metallidurans* sp. nov. isolated from a municipal waste dumping site. Ecotoxicol Environ Saf 257:1–12

Yan J, Ye W, Jian Z, Xie J, Zhong K, Wang S, Hu H, Chen Z, Wen H, Zhang H (2018) Enhanced sulfate and metal removal by reduced graphene oxide self-assembled *Enterococcus avium* sulfate-reducing bacteria particles. Bioresour Technol 266:447–453

Yan C, Qu Z, Wang J, Cao L, Han Q (2022) Microalgal bioremediation of heavy metal pollution in water: recent advances, challenges, and prospects. Chemosphere 286:131870

Yao H, Wang H, Ji J, Tan A, Song Y, Chen Z (2023) Isolation and identification of mercury-tolerant bacteria LBA119 from molybdenum-lead mining soils and their removal of Hg^{2+}. Toxics 11(3):261

Yu RQ, Barkay T (2022) Microbial mercury transformations: molecules, functions and organisms. In: Gadd GM, Sariaslani S (eds) Advances in applied microbiology. Elsevier, Amsterdam, Netherlands, pp 31–90

Zhai X, Li Z, Huang B, Luo N, Huang M, Zhang Q, Zeng G (2018) Remediation of multiple heavy metal-contaminated soil through the combination of soil washing and in situ immobilization. Sci Total Environ 635:92–99

Zhang M, Wang H (2016) Preparation of immobilized sulfate reducing bacteria (SRB) granules for effective bioremediation of acid mine drainage and bacterial community analysis. Miner Eng 92:63–71

Zhang W, Chen L, Liu D (2012) Characterization of a marine-isolated mercury-resistant *Pseudomonas putida* strain SP1 and its potential application in marine mercury reduction. Appl Microbial Biotechnol 93:1305–1314

Zhang X, Liu X, Liang Y, Fan F, Zhang X, Yin H (2016) Metabolic diversity and adaptive mechanisms of iron-and/or sulfur-oxidizing autotrophic acidophiles in extremely acidic environments. Environ Microbiol Rep 8(5):738–751

Zhang J, Zeng Y, Liu B, Deng X (2020) MerP/MerT-mediated mechanism: a different approach to mercury resistance and bioaccumulation by marine bacteria. JHM Lett 388:1–10

Zheng R, Wu S, Ma N, Sun C (2018) Genetic and physiological adaptations of marine bacterium *Pseudomonas stutzeri* 273 to mercury stress. Front Microbiol 9:1–14

Chapter 9
Remediation Strategies of Environmental Mercury: An Overview of Its Environmental Persistence, Associated Threats, and Health Impacts

Varun Dhiman

9.1 Introduction

Mercury is an exceptionally rare earth element since it only exists in very small amounts (0.08 ppm) in the Earth's crust (Scoullos et al. 2001). In addition to its uncommon abundance, it is classified as a global environmental pollutant that has harmed both human health and the environment by creating contamination hotspots through local and global diffusion mechanisms of both natural and artificial origin (Kumari et al. 2020). One such naturally occurring source that accounts for nearly half of the mercury emissions in the environment is volcanic eruption, while anthropogenic sources like stationary coal-based power plants, gold mining, non-ferrous metal smelters, cement production, different waste disposal methods, caustic soda production, steel and pig iron production, batteries, and numerous other sources account for 65%, 11%, 6.8%, 6.4%, 3.0%, 1.4%, 1.1%, and 2.0% of the mercury emissions in the environment, respectively (Pavithra et al. 2023). Mercury pollution in the environment is also a result of improper disposal of electronic waste (Rajesh et al. 2022). When compared to natural processes, human activity triples mercury exposure. Humans are exposed to methylmercury mostly through eating various foods, which alters body metabolism as contaminated water and soil resources redistribute the mercury in the food chain (Chemelo et al. 2023). Because of its enhanced bioaccumulation capacity, the ecosystem's living flora and fauna have suffered negative effects. Its position in the food chain is still up for debate, though. The environmental persistence and likely causes of its exposure to humans are depicted in Fig. 9.1.

V. Dhiman (✉)
Department of Environmental Sciences, Govt. Post Graduate College, Una, Himachal Pradesh 174303, India
e-mail: varundhiman79@gmail.com

N. Kumar (ed.), *Mercury Toxicity*, Environmental Science and Engineering,
https://doi.org/10.1007/978-981-99-7719-2_9

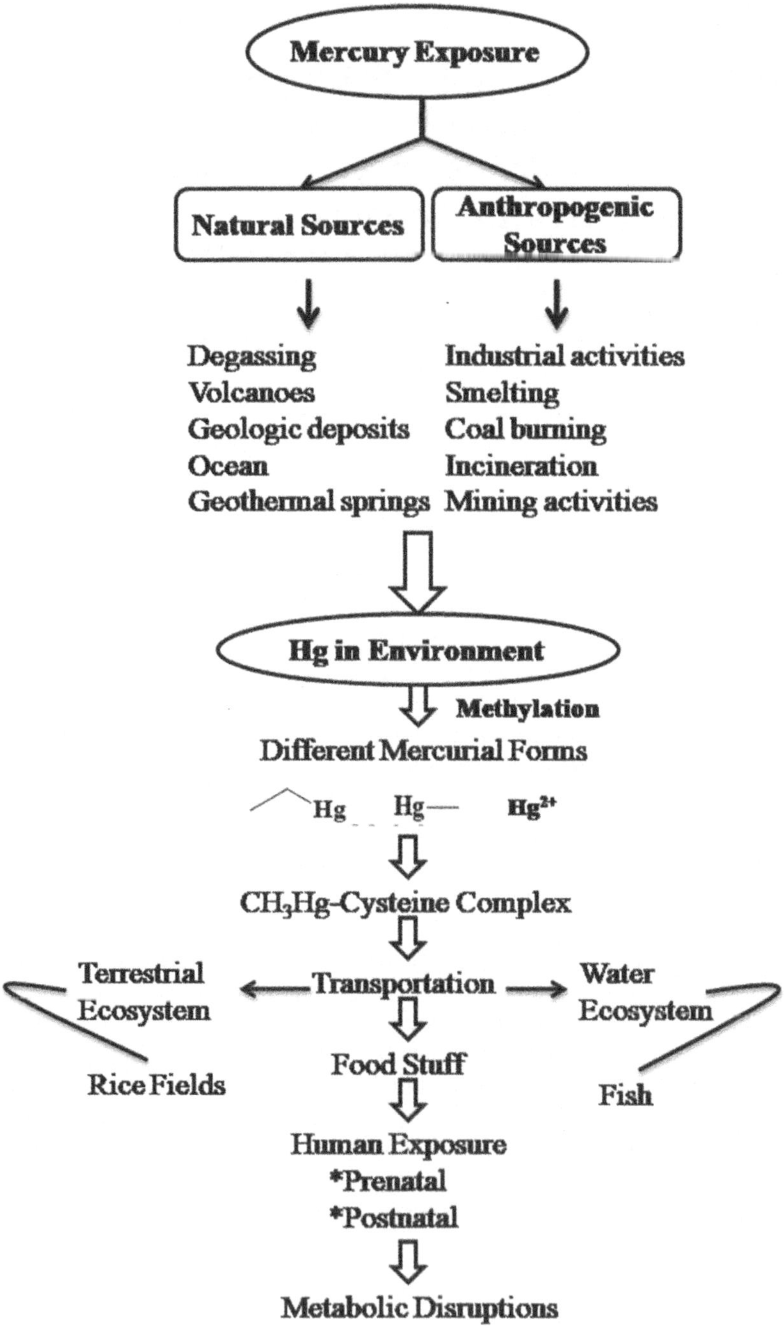

Fig. 9.1 Environmental persistence and probable roots of mercury exposure to human beings

In addition to its strong neurotoxicity, it altered the human body's immune, pulmonary, endocrine, hematological, embryonic, and renal systems. By inactivating the sulfur, its absorption into the human body further perturbs the operation of co-factors, many hormones, and enzymatic reactions.

9.2 Mercury Compounds

Mercury exists as a dimer of Hg^{2+} and has three oxidation states: metallic (0), mercuric (+1), and mercurous (+1) (Chethanakumar et al. 2023). The ability of the mercury, mercuric and mercurous forms to produce a wide variety of organic and inorganic mercury compounds is well recognized. Mercury reacts with substances like oxygen, sulfur, or other salts to form inorganic Hg compounds (Subeshan et al. 2023). While these compounds were present in nature, inorganic mercury compounds form covalent bonds with carbon atoms. 7–15% of these are absorbed by humans after gastrointestinal exposure. According to certain research, the skin can absorb it through the epidermis, hair follicles, sweat, and sebaceous glands (Park and Zheng 2012). These substances primarily affect the nephritic system because the kidney appears to contain the highest concentration of them (Gao et al. 2022). The mucosal barrier of the GI tract becomes damaged when elemental mercury is inhaled because it is quickly absorbed by the body (Rodríguez-Viso et al. 2022). The lungs absorbed 80% of the elemental mercury vapor, which swiftly diffused into blood and was then transported to all of the body's organs. The body's enzymatic and cellular activity is inhibited by mercury in the form of mercuric ions. Additionally, it causes oxidative stress and an immunological reaction in the body (Park and Zheng 2012). Different mercury compounds and their numerous characteristics have been discussed in Table 9.1.

9.3 Mercury in the Terrestrial Environment

Because of its greater level of volatility, mercury is known to be a very hazardous and persistent pollutant that is quickly mobilized in the environment (Singh et al. 2023). Hg is more easily carried within vast air masses due to its volatile nature. The scientific community is currently expanding its understanding of the sources, emissions, environmental cycling, human exposure, numerous routes, environmental persistence, and its effects on human health and biota. The Minimata Convention was adopted in 2013 primarily for environmental mercury pollution since the Hg pollution issue is so significant to the scientific and environmental communities (Joy and Qureshi 2023). Different mercury forms have significant biological and chemical activity. The health of humans and other living organisms is seriously endangered by the Hg contamination of the terrestrial environment. Its extremely high degree of toxicity earns it a spot among the top 10 most dangerous pollutants. A novel

Table 9.1 Mercury compounds and their numerous characteristics

Chemical name	Synonyms/ trade name	Synthesis	Formula	Atomic/ molecular mass	Melting point	Physical characteristics	Density/ solubility	Structure
Mercury(II) chloride	Abavit, TL-898 Hydrargyri dichloridum (homeopathy), Sulema (Russia), Agrosan	$C_2H_2 + HCl \rightarrow CH_2 = CHCl$	$HgCl_2$	271.52 g/ mol	276 °C	– White crystalline solid – Odorless – Metallic taste	5.4	$Cl^- \; Hg^{2+} \; Cl^-$
Mercury acetate	Mercuriacetate, Mercury diacetate,	$C_6H_5OH + Hg(OAc)_2 \rightarrow$ $C_6H_4(OH)$-2-HgOAc + HOAc $C_6H_4(OH)$-2-HgOAc + NaCl $\rightarrow$ $C_6H_4(OH)$-2-HgCl + NaOAc	$Hg(O_2CCH_3)_2$	318.7	178–180 (decomposes)	White crystalline in nature	3.27	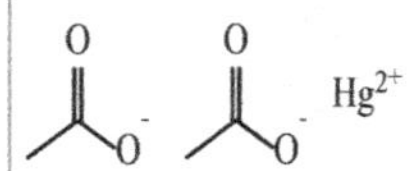
Mercuric oxide	Red mercury, Montroydite	$Hg(NO_3)_2$ + Hg (Pyrolysis) $\rightarrow$ HgO	HgO	232.66 g/ mol	580 °C	Black amorphous and insoluble powder	8.10 g/cm^3	Hg^{2+} O^2
Phenyl mercury acetate	Verdasan	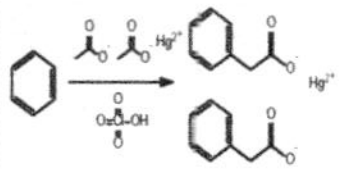	$C_8H_8HgO_2$	336.74 g/ mol	148–151 °C	lustrous crystals	Slightly soluble	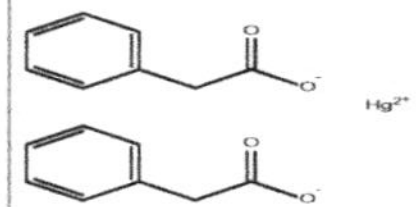
Methyl mercury chloride	Caspan, NSC 19,998, Caspan (VAN)	–	CH_3HgCl	251.08 g/ mol	338 °F	Odorless, white, crystalline powder	5.6 × 10^{-5} g/L at 25 °C (solubility), 13.534 g/cm at 25 °C (density)	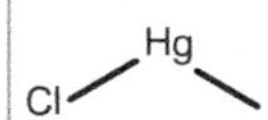
Mercury sulfide	Cinnabar Vermilion	HgS $\rightarrow$ Hg + S	HgS	232.66 g/ mol	580 C	Red/Black odorless solid	8.10 g/cm^3	Hg=S

(continued)

Table 9.1 (continued)

Chemical name	Synonyms/ trade name	Synthesis	Formula	Atomic/ molecular mass	Melting point	Physical characteristics	Density/ solubility	Structure
Mercury(II) stearate	*EINECS 211–458-8,*	$HgCl_2 + 2C_{17}H_{35}COONa \rightarrow Hg(C_{17}H_{35}COONa) + 2NaCl$	$C_{36}H_{70}HgO_4$	767.529	112.2 C	A yellow, granular, waxy substance	–	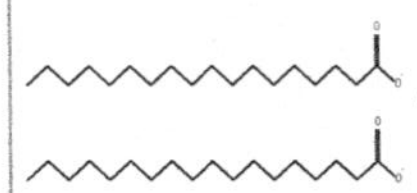
Mercury(I) nitrate	–	$Hg_2(NO_3)_2 + H_2O \rightleftharpoons Hg_2(NO_3)(OH) + HNO_3$ $Hg_2(NO_3)_2 \rightleftharpoons Hg + Hg(NO_3)_2$	$Hg_2(NO_3)_2$	525.19 g/mol (anhydrous)	Decomposes at 70 °C (dihydrate)	Hygroscopic powder with white or colorless crystals	4.8 g/cm^3 (dihydrate)	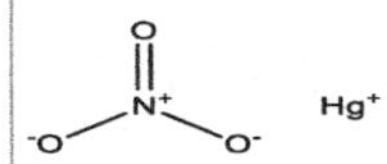
Mercury(II) cyanide		$HgO + 2\,HCN \rightarrow Hg(CN)_2 + H_2O$ $K_4Fe(CN)_6 + 3HgSO_4 \rightarrow 3Hg(CN)_2 + 2K_2O_4 + FeSO_4$ $Hg_2(NO_3)_2 + 2\,KCN \rightarrow Hg + Hg(CN)_2 + 2\,KNO_3$	$Hg(CN)_2$	252.63 g/mol	320 °C	colorless crystals/white powder	3.996 g/cm^3	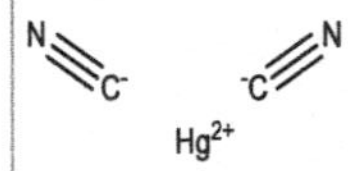
Mercuric amidochloride		$HgCl_2 + 2\,NH_3 \rightarrow HgCl(NH_2) + [NH_4]Cl$	$Hg(NH_2)Cl$	252.065 g/mol	–	White amorphous powder, white crystalline solid	5.56 g/cm^3	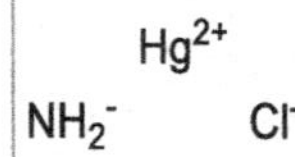
Mercuric dichromate	–	–	Cr_2HgO_7	416.58 g/mol	–	Red crystalline nature	–	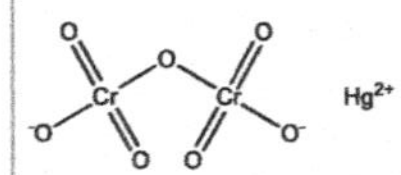
Methoxyethylmercuric acetate	Radosan, Landisan, Mema	–	$C_5H_{10}HgO_3$	318.72 g/mol	40–42 °C	Crystalline nature	–	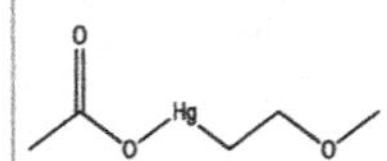

(continued)

Table 9.1 (continued)

Chemical name	Synonyms/ trade name	Synthesis	Formula	Atomic/ molecular mass	Melting point	Physical characteristics	Density/ solubility	Structure
Methoxyethyl mercury chloride	Agallol	–	C_3H_7ClHgO	295.13 g/mol	65 °C	White powdery with black granules	Less than 1 mg/mL at 68 °F (water solubility)	
Phenylmercuric propionate	Metasol 57, Methasol P-6	–	$C_9H_{10}HgO_2$	350.77 g/mol	80–81 °C	White solid but waxy	–	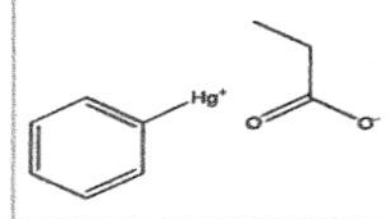

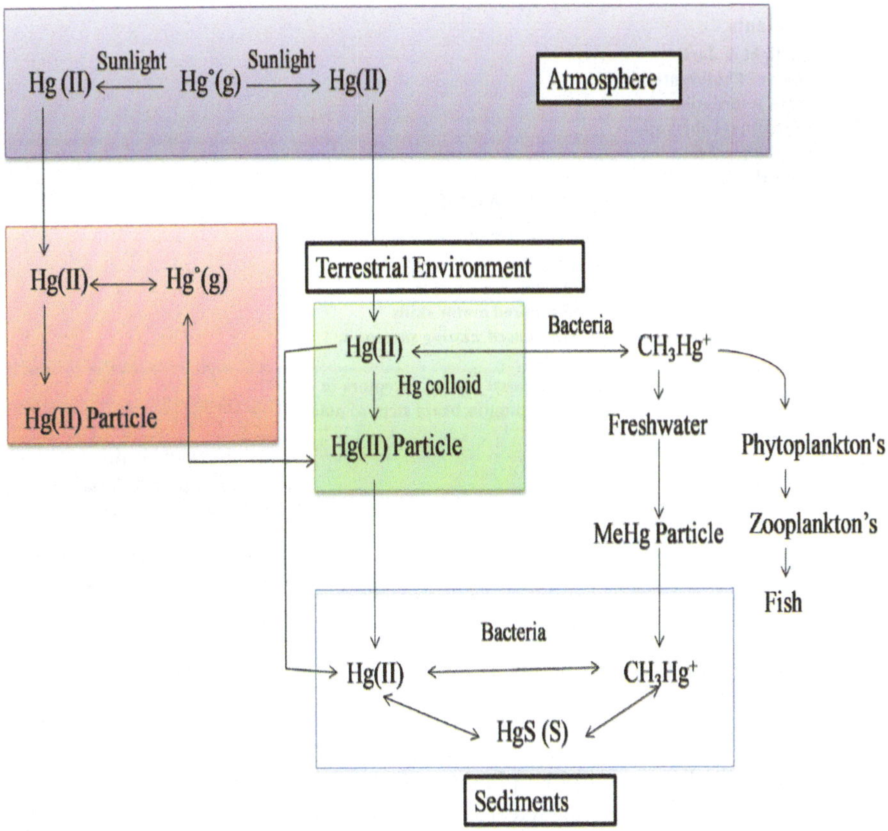

Fig. 9.2. Biogeochemical cycling of mercury in the environment (Adopted from *Source* Lasorsa et al. 2012)

understanding of Hg cycling has been investigated utilizing tree rings, ice cores, isotope measurements, and lake sediment analysis to identify the temporal trends of Hg in the environment (Gačnik and Gustin 2023). Mercury's biogeochemical cycling in the environment is seen in Fig. 9.2. (Lasorsa et al. 2012).

9.4 Associated Threats and Health Impacts

The toxicity of mercury in the environment and its many exposure pathways influence human health and have harmful effects on various environmental domains. It has been discovered that when the environment becomes more industrialized and subject to other anthropogenic activities, the concentration of mercury and CH_3Hg increases (Mason et al. 2000). The increase in concentration is seen in marine life,

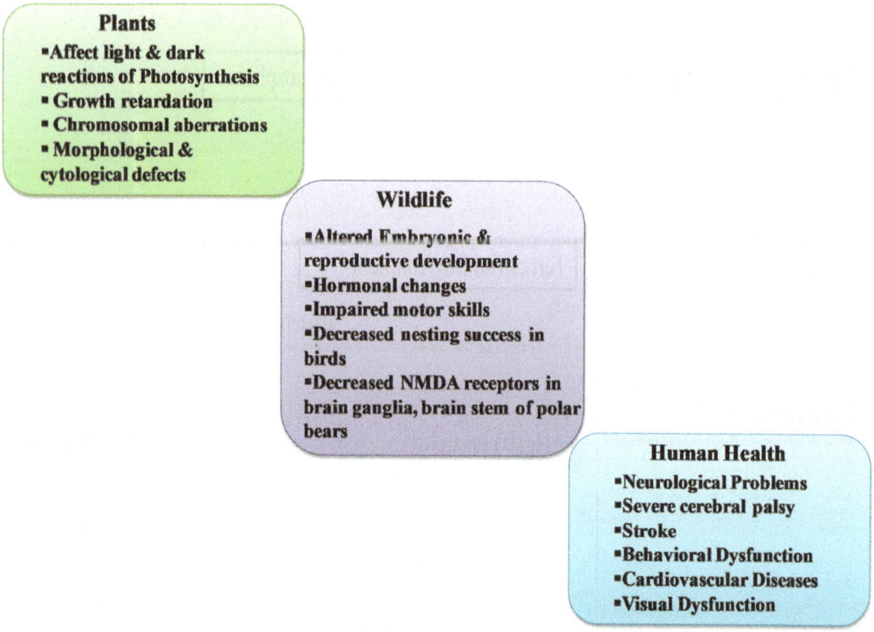

Fig. 9.3. Associated different health effects on humans and hazardous impacts on plants and wildlife (*Source* Driscoll et al. 2013)

including fish, shellfish, and phytoplankton. One of the main ways that people are exposed to CH_3Hg is by eating seafood and fish. Furthermore, rice grown close to artisanal Hg mining operations becomes contaminated with CH_3Hg as it builds up in the grain, which is a big health problem (Riaz et al. 2018). Infants can acquire neurological and other developmental abnormalities due to the current level of global Hg exposure (Al-Saleh et al. 2020). The frontal lobes become dysfunctional as a result of their exposure. Mechanistic effects of CH_3Hg exposure on heart functions have been discovered in recent investigations. The risk of cardiovascular illnesses, in particular myocardial infarction, is increased by its exposure (Downer et al. 2017). Figure 9.3 illustrates the relationship between various human health consequences and potentially harmful effects on flora and wildlife (Driscoll et al. 2013).

9.5 Redox Chemistry of Environmental Mercury

It was thought that O_3 and OH caused the mercury to oxidize, while Hg(II) and HO_2 (aq.) in clouds caused the mercury to reduce (Driscoll et al. 2013). However, new research indicates that these reactions are negligible in the atmosphere. Our ability to anticipate its availability and enhance its remediation mechanism is aided by redox chemistry. Mercury's circulation through several environmental spheres is strongly influenced by its redox reactions. Mercury's redox reactions have a significant impact on how it moves across many environmental spheres (Si et al. 2022). For instance, Hg(0) volatilization and Hg(II) methylation in natural waters may increase their level in the environment and further increase their environmental persistence. Potential oxidants in the atmosphere are represented by halogen atoms. In the total two-step exothermic reaction mechanism described below, they play an important role:

$$Hg(0) + Cl + M \rightarrow HgCl + M \quad (1)$$

$$HgCl + Y + M \rightarrow HgClY + M \quad (2)$$

Y = another radical; Cl = Efficient oxidant

In addition to Cl, bromine (Br) may play a significant role as an oxidant in the environment. The sources of Br in the environment are dimethyl-bromide (CH_2Br_2), methyl bromide (CH_3Br), and bromoform ($CHBr_3$), which undergo photolysis and oxidation (Lam et al. 2019). Figure 9.4 shows the reaction initiation of Hg(0) oxidation in a bromine-rich environment (Lam et al. 2019). Numerous variables, including pH, Hg(II)/NOM ratio and concentration, NOM structure, the presence of Cl^- and Ca^{2+} ions, and temperature conditions, all have an impact on the redox chemistry of environmental mercury (Si et al. 2022).

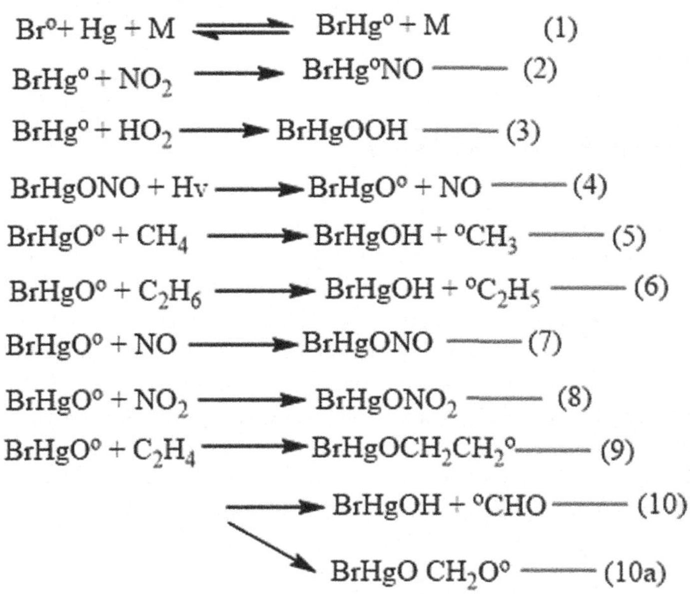

Fig. 9.4. Reactions showing initiation of Hg(0) oxidation in bromine-rich environment (*Source* Lam et al. 2019)

9.6 Remediation Strategies of Environmental Mercury

Inorganic mercury found in the environment is relatively immobile. However, it can be changed into a mobile state and made bioavailable for elimination through a variety of biological and chemical processes (O'Connor et al. 2019). There is always a chance of onsite leakage, and conventional methods of removal were rather expensive (Tiodar et al. 2021). The following are the remediation techniques.

9.6.1 Phytoremediation

Traditional remediation methods for mercury-contaminated areas have included moving, burying, and excavating the soil as well as the terrestrial environment (soil biota and sediments). But, throughout the remediation process, these activities typically cause the site to deteriorate and mercury leaks. Therefore, using phytoremediation is one of the most successful, economical, and environmentally friendly methods for reducing mercury leakage, enhancing soil structure, fostering biological activity, and lowering soil toxicant levels (González-Reguero et al. 2023). Some GM plants, including *Spartina alterniflora* (cordgrass), *Juncus* spp., *Oryza sativa,* and *Typha latifolia,* can mobilize mercury, extract Hg(II) and methyl mercury from the contaminated site, and convert into Hg(o), which is then sequestered into the shoot tissues of the used plants (Heaton et al. 1998).

9.6.2 Bacterial Bioremediation

Because many bacterial species exhibit a potential and reactive interface for the adsorption of the metal species and because their membranes serve as sites for enzymatic activities, this is one of the most effective remediation strategies for mercury removal (Priyadarshanee and Das 2021). To reduce or oxidize the metal ions, bacterial species take them up, exude them, or chelate with them (Dhiman and Pant 2022). When exposed to 100 L/ml of mercury for 42 h, the bacterial species *Vibrio fluvialis* is one of the crucial species that aid in the remediation of mercury by up to 60% (Saranya et al. 2017). Similar species remove 89.47% of Hg from mobile solutions at a concentration of 10 L/ml Hg. Some other species such as *Pseudomonas aeruginosa* 80%, *Brevibacterium casei* 70%, *Bacillus* spp. 68%, *Tetrahymena rostrate* 40% removal of Hg, respectively. The bacterial strain *B. thuringeinesis* PW-05 is one of the mercury-resistant species, isolated from marine environments that have the capability of volatilizing mercury which removes >90% of residual Hg (Dash et al. 2014).

9.6.3 Membrane Filtration

This is one of the important nanostructure membrane technology for mercury removal. This technique provides a selective layer for reaction components as well as catalytic support for heavy metals removal. Recently, scientists created multilayer membranes based on graphenes to segregate different ionic species. These nanomembranes use graphene oxide to improve contact between the membrane surface and the solute (Zunita 2021).

9.6.4 Soil Washing

Since the mercury complexes that are produced are insoluble, soil washing emerges as a key strategy for the efficient removal of mercury from the ecosystem (Richter and Flachberger 2010). Mercury forms strong bonds with the sulfides and organic compounds prevalent in the terrestrial environment, making it difficult to remove from the ecosystem (Horvat 2005). Some of the new-generation washing agents include biosurfactants, oxalic, citric, and potassium iodide (KI), EDTA, EDDS, DTPA, and EDTA are used in soil washing (Ifon et al. 2019). Physical and chemical separation of the pollutants occurs during soil washing. While contaminated soil is treated chemically using the necessary washing solutions made of water and related chemicals, physical separation uses a scrubber to separate coarse and fine-grained soil particles.

9.7 Summary and Concluding Remarks

Our understanding of mercury's probable sources, forms, and speciation, including its concentration in nature, has improved in light of recent breakthroughs in mercury remediation. The redox reactions involving ambient mercury offer a fresh viewpoint on how to comprehend its behavior. The methyl mercury concentration in the environment is measured by specifying the redox conditions. This will undoubtedly be beneficial in creating a collection of novel methods for remediating mercury in the environment. Nevertheless, in addition to the fact that mercury is recognized as a worldwide pollutant, it is necessary to establish mitigation plans based on regional and local considerations and to guarantee coordination between regional and global organizations to address the ongoing issue of mercury pollution. Further, It has been determined that a post-monitoring plan is necessary to assess the degree of efficacy of the mercury remediation strategy.

Acknowledgements The author is grateful to the Govt. Post Graduate College, Una, Himachal Pradesh for providing the necessary facilities for this work.

References

Al-Saleh I, Moncari L, Jomaa A, Elkhatib R, Al-Rouqi R, Eltabache C, Al-Rajudi T, Alnuwaysir H, Nester M, Aldhalaan H (2020) Effects of early and recent mercury and lead exposure on the neurodevelopment of children with elevated mercury and/or developmental delays during lactation: a follow-up study. Int J Hyg Environ Health 230:113629

Chemelo VS, Bittencourt LO, Nascimento PC, Paiva MF, Delbem ACB, Pessan JP, do Espírito Santo AR, Albuquerque ARL, Angélica RS, Crespo-Lopez ME (2023) Maternal methylmercury exposure during early-life periods adversely affects mature enamel structure of offspring rats at human exposure levels: a concern for oral health. Front Public Health 11

Chethanakumar, Budri M, Gudasi KB, Vadavi RS, Bhat SS (2023) Luminescent pyrene-based schiff base receptor for hazardous mercury(II) detection demonstrated by cell imaging and test strip. J Fluoresc 33:539–551

Dash HR, Mangwani N, Das S (2014) Characterization and potential application in mercury bioremediation of highly mercury-resistant marine bacterium *Bacillus thuringiensis* PW-05. Environ Sci Pollut Res 21:2642–2653

Dhiman V, Pant D (2022) π-π interaction: defining the role and relevance in environmental detoxification of heavy metals from soil. In: Shit PK, Adhikary PP, Bhunia GS, Sengupta D (eds) Soil health and environmental sustainability: application of geospatial technology. Springer International Publishing, Cham, pp 659–672

Downer MK, Martínez-González MA, Gea A, Stampfer M, Warnberg J, Ruiz-Canela M, Salas-Salvadó J, Corella D, Ros E, Fitó M, Estruch R, Arós F, Fiol M, Lapetra J, Serra-Majem L, Bullo M, Sorli JV, Muñoz MA, García-Rodriguez A, Gutierrez-Bedmar M, Gómez-Gracia E, PREDIMED Study Investigators (2017) Mercury exposure and risk of cardiovascular disease: a nested case-control study in the PREDIMED (PREvention with MEDiterranean Diet) study. BMC Cardiovasc Disord 17:9

Driscoll CT, Mason RP, Chan HM, Jacob DJ, Pirrone N (2013) Mercury as a global pollutant: sources, pathways, and effects. Environ Sci Technol 47:4967–4983

Gačnik J, Gustin MS (2023) Tree rings as historical archives of atmospheric mercury: a critical review. Sci Total Environ 898:165562

Gao Z, Wu N, Du X, Li H, Mei X, Song Y (2022) Toxic nephropathy secondary to chronic mercury poisoning: clinical characteristics and outcomes. Kidney Int Rep 7:1189–1197

González-Reguero D, Robas-Mora M, Probanza Lobo A, Jiménez Gómez PA (2023) Bioremediation of environments contaminated with mercury. Present and perspectives. World J Microbiol Biotechnol 39:249

Heaton ACP, Rugh CL, Wang N, Meagher RB (1998) Phytoremediation of mercury- and methylmercury-polluted soils using genetically engineered plants. J Soil Contam 7:497–509

Horvat M (2005) MERCURY. In: Worsfold P, Townshend A, Poole C (eds) Encyclopedia of analytical science, 2nd edn. Elsevier, Oxford, pp 545–557

Ifon BE, Togbé ACF, Tometin LAS, Suanon F, Yessoufou A (2019) Metal-contaminated soil remediation: phytoremediation, chemical leaching and electrochemical remediation. In: Metals in soil-contamination and remediation. IntechOpen London, pp 534–554

Joy A, Qureshi A (2023) Reducing mercury emissions from coal-fired power plants in India: possibilities and challenges. Ambio 52:242–252

Kumari S, Amit, Jamwal R, Mishra N, Singh DK (2020) Recent developments in environmental mercury bioremediation and its toxicity: a review. Environ Nanotechnol Monit Manage 13:100283

Lam KT, Wilhelmsen CJ, Dibble TS (2019) BrHgO· + C_2H_4 and BrHgO· + HCHO in atmospheric oxidation of mercury: determining rate constants of reactions with prereactive complexes and bifurcation. J Phys Chem A 123:6045–6055

Lasorsa BK, Gill GA, Horvat M (2012) Analytical methods for measuring mercury in water, sediment, and biota. Mercury in the environment: pattern and process 27–57

Mason RP, Laporte J-M, Andres S (2000) Factors controlling the bioaccumulation of mercury, methylmercury, arsenic, selenium, and cadmium by freshwater invertebrates and fish. Arch Environ Contam Toxicol 38:283–297

O'Connor D, Hou D, Ok YS, Mulder J, Duan L, Wu Q, Wang S, Tack FMG, Rinklebe J (2019) Mercury speciation, transformation, and transportation in soils, atmospheric flux, and implications for risk management: a critical review. Environ Int 126:747–761

Park J-D, Zheng W (2012) Human exposure and health effects of inorganic and elemental mercury. J Prev Med Public Health 45:344

Pavithra KG, SundarRajan P, Kumar PS, Rangasamy G (2023) Mercury sources, contaminations, mercury cycle, detection and treatment techniques: a review. Chemosphere 312:137314

Priyadarshanee M, Das S (2021) Biosorption and removal of toxic heavy metals by metal tolerating bacteria for bioremediation of metal contamination: a comprehensive review. J Environ Chem Eng 9:104686

Rajesh R, Kanakadhurga D, Prabaharan N (2022) Electronic waste: a critical assessment on the unimaginable growing pollutant, legislations and environmental impacts. Environ Challenges 7:100507

Riaz A, Khan S, Muhammad S, Liu C, Shah MT, Tariq M (2018) Mercury contamination in selected foodstuffs and potential health risk assessment along the artisanal gold mining, Gilgit-Baltistan, Pakistan. Environ Geochem Health 40:625–635

Richter RB, Flachberger H (2010) Soil washing and thermal desorption: reliable techniques for remediating materials contaminated with mercury. BHM Berg- Huettenmaenn Monatsh 155:571–577

Rodríguez-Viso P, Domene A, Vélez D, Devesa V, Monedero V, Zúñiga M (2022) Mercury toxic effects on the intestinal mucosa assayed on a bicameral in vitro model: possible role of inflammatory response and oxidative stress. Food Chem Toxicol 166:113224

Saranya K, Sundaramanickam A, Shekhar S, Swaminathan S, Balasubramanian T (2017) Bioremediation of mercury by vibrio fluvialis screened from industrial effluents. Biomed Res Int 2017:6509648

Scoullos MJ, Vonkeman GH, Thornton I, Makuch Z (2001) Mercury. In: Scoullos MJ, Vonkeman GH, Thornton I, Makuch Z, Scoullos MJ (eds) Mercury—Cadmium—lead handbook for sustainable heavy metals policy and regulation. Springer, Netherlands, Dordrecht, pp 11–69

Si L, Branfireun BA, Fierro J (2022) Chemical oxidation and reduction pathways of mercury relevant to natural waters: a review. Water 14:1891

Singh AD, Khanna K, Kour J, Dhiman S, Bhardwaj T, Devi K, Sharma N, Kumar P, Kapoor N, Sharma P, Arora P, Sharma A, Bhardwaj R (2023) Critical review on biogeochemical dynamics of mercury (Hg) and its abatement strategies. Chemosphere 319:137917

Subeshan B, Pham A, Murad MS, Asmatulu E (2023) Mercury recycling technologies in its' end-of-life management: a review. J Mater Cycles Waste Manage

Tiodar ED, Văcar CL, Podar D (2021) Phytoremediation and microorganisms-assisted phytoremediation of mercury-contaminated soils: challenges and perspectives. Int J Environ Res Public Health 18

Zunita M (2021) Graphene oxide-based nanofiltration for Hg removal from wastewater: a mini review. Membranes

Chapter 10
Mercury Remediation: Easing Biochar Approach

M. Kiranmai Reddy, K. V. N. Suresh Reddy, and Anima Sunil Dadhich

10.1 Introduction

Most of the elements that are found in the environment are either in elemental form, organic, or inorganic form; one such element is mercury (Hg). Due to its toxicity, mercury (Hg) is famous worldwide as a pollutant (Chen et al. 2018). The availability of mercury in the environment is either due to natural processes or through anthropogenic activities. The natural sources are forest fires, volcanic eruptions, geothermal emissions, and anthropogenic activities, including industrial emissions, mining and refining of mercury, and combustion of fossil fuels (O'Connor et al. 2019; Collado et al. 2016). It was evident that 80% of the mercury released into the environment is in a metallic state, most of which is released from human activities compared to natural release (Gupta et al. 2018). The metallic mercury will be in the form of white liquid, mostly used in barometers, thermometers, electric devices, and blood pressure devices; inorganic mercury will be formed with the combination of salts such as sulfur, oxygen, and chlorine; some might also act as fungicides. The organic form of mercury is available when it combines carbon forms into organic mercury, such as methyl, ethyl, and phenyl. The inorganic form of mercury is available, due to the action of anaerobic bacteria in the soil or sediment (Li et al. 2009), that can convert into a methylated group such as methyl mercury which tends to be biomagnified (Ullrich et al. 2001) in food chain finally causing a poisonous impact on the living organisms. Since the first report of the Minamata Bay incident in 1950, the world

M. K. Reddy (✉)
Department of Environmental Science, GITAM School of Science, GITAM (Deemed to Be University), Visakhapatnam, Andhra Pradesh 530045, India
e-mail: kmajji@gitam.edu

K. V. N. S. Reddy · A. S. Dadhich
Department of Chemistry, GITAM School of Science, GITAM (Deemed to Be University), Visakhapatnam, Andhra Pradesh 530045, India

N. Kumar (ed.), *Mercury Toxicity*, Environmental Science and Engineering,
https://doi.org/10.1007/978-981-99-7719-2_10

has been more focused on the impacts of mercury on human health as it accumulates in food. After this incident, much research was carried out to identify methyl mercury, and mostly, it was spotted in shellfish, fishes, and aquatic resources (García-Hernández et al. 2018). Apart from aquatic resources, it was also found in rice, as a primary dietary can threaten life (Cui et al. 2017) if contaminated.

One of the "hotspots" of methyl mercury recognized by researchers is soil, sediment, and water, which pose a direct source to crops and water sources (Wang et al. 2018a, b, c; Carrasco et al. 2011). Many researchers have reported mercury's impact on human health (Bank 2020; Jiang et al. 2021). Direct exposure to mercury is from plant sources, whereas indirect through the consumption of higher order organisms that leads to exposure of methyl mercury. Hence, many studies were conducted to find the techniques to remediate mercury from the soil, sediment, and water. The use of thermal treatment technologies has less advantageous to phytoremediation techniques; the latter is found to be eco-friendly, has easy operation, and has less damage to the structure of the soil. One such application is the amendment of the biochar in mercury-contaminated sites, have much beneficial due to its easy operation and high adsorption capacity (He et al. 2019).

Biochar is a carbonaceous material made from the thermal decomposition of plant material in the presence or absence of oxygen. Most studies indicated that biochar has a high surface area, porosity, more functional groups, and a high adsorption site (Tang et al. 2019). These mentioned advantages of biochar are believed to adsorb mercury (Lu et al. 2020). Many works in the past demonstrate that the use of biochar can reduce mercury concentration, as methyl mercury is removed from the solution (O'Connor et al. 2019); use of biochar in mercury-contaminated soil has reduced the bioavailability of the methyl mercury of Indian mustard (Yin et al. 2022) and rice (Man et al. 2021). These initial studies have shown that the mercury risk can be reduced by applying biochar in soils, sediments, and water. Thus, the present chapter summarizes the latest technology in applying biochar to remove mercury from environmental sources such as soil, sediment, and water. It will also discuss the potential limitations of using biochar and its feasibility. The chapter also highlights the future prospects of using biochar as an eco-friendly tool for the remediation of mercury-contaminated sites.

10.2 Biochar

Biochar is a carbonaceous product generated due to the thermal decomposition of plant and animal biomass in the minimum oxygen concentration.

10.2.1 Preparation and Sources

Many studies in the past have shown that wood and wood waste is the primary source for preparing biochar, as trees can grow in extreme conditions, thereby representing the availability of ample materials. Trees such as eucalyptus (Butphu et al. 2020), mulberry (Ali et al. 2019), and neem (Manoharan et al. 2022) have been used as raw materials for biochar production, which acts as an effective adsorbent for removing toxic substances from the polluted soils and sediments (Lu et al. 2015; Yang et al. 2021a, b). Even the waste generated from agriculture and forestry was used to prepare biochar, where biomass is used efficiently (Guo et al. 2020a, b, c). Being economical and easy to achieve, this type of biochar has grabbed the attention to control pollution (Zhang et al. 2013). The waste from corn cobs, shells of seeds, stalks of wheat, and rice husk was found to be effective in removing heavy metals such as mercury (O'Connor et al. 2018) with an adsorptive capacity of 73%. Most industrial waste consists of organic waste as sludge (Gopinath et al. 2021a, b) used to produce biochar, promoting wastewater treatment solutions as the circular economy. In one of the studies, biochar prepared from cork at 750 °C was found to adsorb heavy metals by 18.5 mg/g (Wang et al. 2020). Recent research focuses on algal biomass as biochar production for removing mercury from contaminated soils. Researchers found that the biochar made from the algal biomass has strong ion exchange properties found to be an effective agent in the remediation of contaminated sediments and soils (Yu et al. 2017).

10.2.1.1 Pyrolysis

Pyrolysis is the thermal degradation of biomass at higher temperatures of about 500–800 °C with very low or without oxygen. As pyrolysis proceeds, the carbon present in the biomass becomes less reactive and forms a stable chemical structure (Demirbas 2004). The temperature of the heating process and heating rate will determine the pore structure, surface properties, and characteristics of biochar (Sohi et al. 2010a). Pyrolysis has been categorized as fast and slow techniques, in which fast pyrolysis requires a shorter time and a higher yield of bio-oil, syngas, and high energy density (Liu et al. 2015) with a small amount of biochar whereas slow pyrolysis takes several minutes to hours to have the production of biochar rather than forming bio-oil and gas. During this process, with the increase in the temperature at about 400–500 °C the formation of fixed carbon in the biochar would be more (Shaheen et al. 2019). In this case, the fast pyrolysis technique is to be used efficiently to improve the biochar production with zeolite material catalyst (Akhil et al. 2021).

10.2.1.2 Gasification

It is a process where biomass is ignited in the presence of gases such as nitrogen, oxygen, carbon dioxide, air, and steam (Kamboo and Dutta 2015). In gasification, drying, dehydrogenation, combustion, and gasification occur to release gas and produce biochar (Loha et al. 2014). Due to the use of oxidation, the biochar's morphological structure, functional groups, and surface area differ from the pyrolysis process (Long et al. 2015).

10.2.1.3 Microwave Pyrolysis

Microwave radiations were used for heating the material where electromagnetic energy is converted to heat energy, known as microwave pyrolysis. Here, the heat is distributed uniformly through the material for biochar production. The use of microwave pyrolysis is more beneficial than convention in terms of a cost-effective, controlled process, uniform heating, and safe to use. Frequency and power output are the two crucial factors of microwave pyrolysis, where the frequency is maintained at 2.45 GHz and the temperature is between 400 to 600 °C (Omoriyekomwan et al. 2017). According to the study of Kostas et al. (2020) high temperature upgrades the thermal breakdown of raw materials and increases carbon content by 77.8%. For feedstock material, microwave pyrolysis has produced faster biochar but pore size increases with the increase in temperature and retention time (Antunes et al. 2017).

10.2.1.4 Flash Pyrolysis

It is a process where a thermochemical process has been used to produce good-quality biochar with low-quality feedstock. Here heating process occurs at a very high temperature in a few seconds. Reactors like entrainment can only manage such sort of high temperatures. When subjected to high heating rates and an intense radiant environment, the raw material particles in the reactor develop high velocities to impact the reactor heating surfaces, producing high-quality biochar. As per the author (Eibner et al. 2017), biochar yield has been identified as 14.89 wt% for palm kernel and 18.31 wt% for *Chlorella vulgaris* algae. The flash pyrolysis process occurs at high temperatures and heat, smoothens the char's surface, and increases the particle's spherical structure. Thus, having a good number of aromatic rings in the biochar.

10.2.2 Properties and Characterization

Before proceeding to the application of biochar in the bioremediation process, it is very much necessary to characterize the biochar for its further use. Two primary sources were documented standard protocol for biochar one is IBI (2015), and the second one, EBC (European Biochar Certificate), which provides guidelines for producing biochar from agricultural waste (Hans–Peter 2013). Furthermore, there is no specific guidelines have been published by international organizations on the characterization of biochar and its use in other fields. Hence, any technology can be used to characterize the biochar. However, sometimes, it might affect the strength and the consistency of the data due to poor availability of technology which often leads to improper characterization.

10.2.2.1 Elemental Analysis

This process comes under ultimate analysis when the biomass is subjected to a high temperature by pyrolysis (Gul et al. 2015). Biochar consists of minerals such as nitrogen, phosphorus, potassium, calcium, magnesium, and other elements, which this process can quantify. Authors (Singh et al. 2010) have studied that the carbon content of woody biomass was higher than poultry manure, but other minerals like N, P, S, K, and Ca are low. Poultry manure is rich in nutrients like potassium and phosphorus, but the total carbon content is low. The increase in the temperature in pyrolysis with biomass releases ash and carbon biochar (Xie et al. 2015), and sometimes nitrogen is also available due to the pyrolysis of lignocellulosic biochar. However, in the case of animal manure and sewage sludge, biochar shows less nitrogen content (Keiluweit et al. 2010; Hossain et al. 2011). In other studies, pyrolysis conditions such as temperature and residence time determine the increase of total potassium and phosphorus (Peng et al. 2011), iron, sulfur, manganese, calcium, silicon, and magnesium (Qambrani et al. 2017). It was also found that high temperature and more retention time also remove hydrogen and oxygen elements from the biomass due to deoxygenation and decarboxylation, which leads to the loss of volatile organic compounds and the H/C-O/C ratio also gets reduced (Chen et al. 2012; Zhao et al. 2018). This indicates that high temperatures applied on biomass for a longer period will form crystal-like graphite structures with high carbon content (Zhao et al. 2018; Cheng et al. 2020). As reported by Jindo et al. (2014), where the O/C ratio significantly decreased from rice straw to wood chips of an oak tree, indicating slow mineralization has been observed in the biomass of wood compared to the biomass of herbaceous plants and the structure of biochar would be more stable (Leng and Huang 2018).

10.2.2.2 Ash, Moisture, and Volatile Matter

The proximate analysis includes the determination of ash, moisture, and volatile matter. During biomass pyrolysis at high temperatures, ash, moisture, and volatile matter also get released apart from minerals. A standard protocol published by the American Society for Testing Materials (ASTM) is followed to determine proximate analysis (IBI 2015). Through this method, excessive grinding is not recommended as moisture content cannot be readily determined due to the loss during heating.

According to the ASTM format, the particle size distribution from 850 to 150 μm can have an important variation as biochar is passed after its grinding; to maintain a uniform distribution of particle size, 850 to 150 μm is recommended for easy comparison of diverse biochar. From the smaller sieve, volatile matter can be easily determined compared to large biochar particles, as these particles are covered with porous constituents. Regarding moisture content determination, ASTM recommends heating the biomass at 750 °C in crucibles for 10 min in a heated muffle furnace and cooling through a desiccator, and the weight measurement determines the moisture content (Ahmad et al. 2012).

10.2.2.3 Analysis of Physicochemical Parameters

Another important characteristic after ultimate and proximate analysis is physicochemical determination. The analysis of the physicochemical parameters of biochar depends on the material and conditions of the temperature (Ahmad et al. 2014) to assess the properties of biochar through analysis of different methods used for charcoal, compost, soil, and coal (Mukome and Parikh 2015) was in general adopted. Mostly, biochar exhibits alkaline on the pH scale due to the loss of oxygen at higher temperatures due to pyrolysis. An increase in the pyrolysis temperature improves the pH of the biochar, as shown in Fig. 10.1, and, when added to the soil, helps reduce the soil's acidity (Yuan et al. 2011). As pyrolysis begins, the specific surface area (SSA) increases, but at higher pyrolysis lowers the cation exchange capacity (CEC) (Weber and Quicker 2018). The increase in the pH value and specific surface area of biochar at higher temperature pyrolysis could balance for the less CEC value by addition of the higher CEC additions to the soil improves the characteristics of the soil (Gul et al. 2015). Hence, biochar has the property of both a charged surface and a specific surface area, which combines with heavy metals by adsorption reactions. As stated by Yuan et al. (2015) that increase in the temperature of sewage sludge has found that the ratio of O/C has lowered from 0.33 to 0.05, the concentration of volatile matter content decreased from 27.4 to 5.5%, and the specific surface area has increased from 14 to 26.70 m^2/g. It was identified from the study that the CEC of biochar has increased due to its activity to attract positive charge from the surface of the functional groups, which acts as a significant factor in remediating the heavy metal-contaminated soils (Kavitha et al. 2018). Moreover, the pore size and porosity depend on pyrolysis, which releases volatile matter at higher temperatures and enhances pore size.

10.2.3 Modifications of Biochar

Biochar has been found to be efficient in the remediation of pollutants (Gopinath et al. 2020; Madima et al. 2020). Apart from the remediation of pollutants, biochar

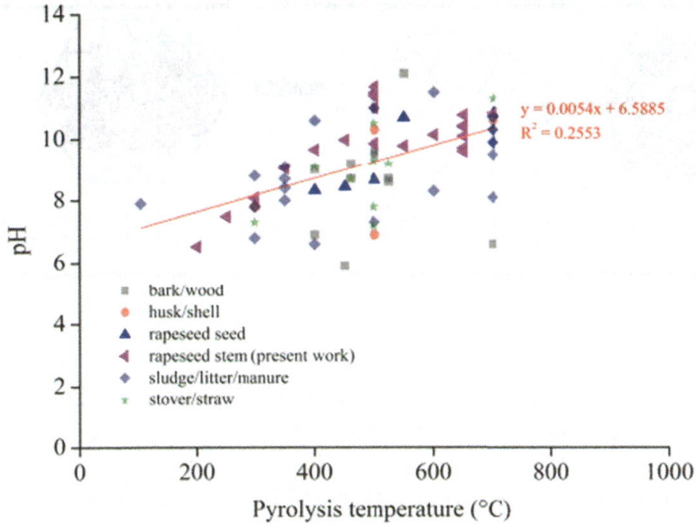

Fig. 10.1 Relationship between pyrolysis temperature, pH, and biochar of different sources (*Source* Zhao et al. 2018)

maintains the nutritional balance, physicochemical parameters, and biological characteristics of the soil and enhances the productivity of the crop (Gunarathne et al. 2019; Xiang et al. 2020). Again, it also depends on the feedstock material, biomass, sludge, grass, and temperature used during pyrolysis, activating the functional groups for mercury remediation. Though biochar has played a significant role in mercury adsorption in any state (elemental, inorganic, and organic), slight modifications will enhance the remediation capacity. Modified biochar has significantly improved its surface properties and characteristics of it. Biochar combined with any other material enhances the physical characteristics such as biochar density, particle size distribution, mechanical strength, organic and functional characteristics, and surface chemistry characteristics, thereby enhancing the remediation of any pollutant (Klasson et al. 2013). In a study, biochar prepared from biowaste was coated with nanoscale zero-valent iron, which enhanced the adsorption capacity by 2.2 times compared with the biochar prepared from pristine wood (Chen et al. 2019). When corn straw biochar is modified with Na_2S to remove Hg(II) from the aqueous solution, the sorption capacity of it has increased to 76.95% compared to the corn straw biochar (Tan et al. 2016). This gives the impression that the adsorption capacity of biochar and other components is greater than normal biochar content. During the study, biochar produced from the wheat straw was exposed to pyrolysis at 600 °C to non-thermal plasma with chlorine to activate chlorine sites on the biochar for mercury adsorption. This experiment was found efficient in removing Hg^0 to 80% due to the increased number of C–Cl groups on the surface of the biochar, which was found to be active sites for the adsorption of Hg^0 (Wang et al. 2018a, b, c), as plasma caused biochar to increase their pore size and improves the thermal stability of the modified biochar

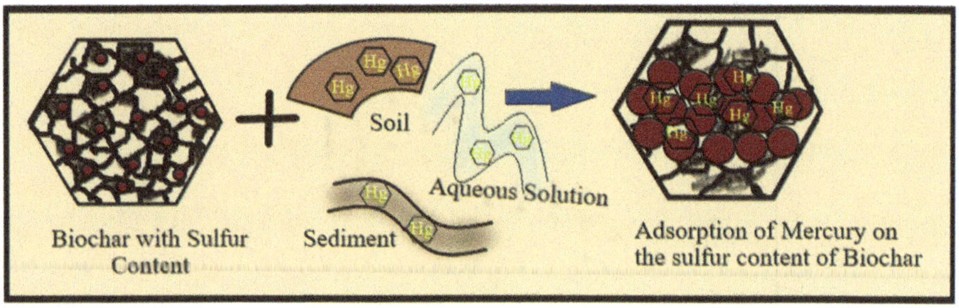

Fig. 10.2 Adsorption of Hg(II) on the sulfur modified biochar

(Fig. 10.2). To remove elemental mercury, H_2S was impregnated with biochar made of corn straw and exposed to plasma. Due to the exposure to plasma, there was an increase in carboxyl and sulfur-containing functional groups on the surface of the modified biochar; these carboxyl groups and C-S formed on the biochar helped to remove mercury in the form of HgS and HgO (Zhang et al. 2019a, b, c). Though biochar was found to be efficient in mercury removal, the removal rate has enhanced with the additions of the other materials to the biochar, as stated in Table 10.1.

Table 10.1 The efficiency of the biochar for the removal of the mercury

Biochar	Removal efficiency of mercury (%)	References
Ferric chloride-modified corn straw	90	Zhang et al. (2023)
Wood chips	70	Fotopoulou et al. (2015)
Sulfurized wood	75	Park et al. (2019)
Tobacco	100	Wang et al. (2018a, b, c)
Hardwood	80	Wang et al. (2019)
Sulfur-impregnated	55	Bailon et al. (2022)
Gooseberry leaves	96.4	Albishri and Yakout (2023)
Pine	57	Goñez-Rodríguez et al. (2021)
Tamarind flower	76.5	Pagala (2023)
Peanut and shea nut shells	100	Duwiejuah et al. (2022)
Rice shells	43	Xing et al. (2019)
Bamboo	73	Wang et al. (2019)
Sewage sludge	81.9	Zhang et al. (2019a, b, c)
Switch grass	80	Lui et al. (2018a)

10.2.4 Remediation Perspective

Biochar has been found to be an eco-friendly tool for the remediation of different pollutants in soil and water. As biochar is made of biological components, few are made of treated materials such as sludge and industrial waste, so its degradability in the environment is quite easy. Hence, in recent studies, biochar application has a promising role in the remediation of pollutants, particularly heavy metals like mercury, because of its long persistence in the environment. It is quite understood that the presence of mercury in elemental, inorganic, and organic forms becomes difficult or takes a very long time to degrade or sometimes enters into the food chain, causing a severe health hazard, so it is important to remediate. Thus, biochar helps immobilize the mercury in the soil or the aqueous solution. Malt spent rootlets were used to prepare biochar to adsorb Hg(II) from an aqueous solution. The biochar concentration of about 0.3 and 1 g/L at about 24-h contact time has removed Hg(II) with 71 and 100% accordingly. Through isotherm data, the maximum biochar sorption capacity of Hg(II) has reached 103 mg/g (Boutsika et al. 2014). In a study by Liu et al. (2016), about 36 different kinds of biochar were evaluated for their potential to remove mercury from aqueous solution; these 36 biochars were subjected to lower and higher pyrolysis temperatures and found that the biochar made after higher pyrolysis temperatures have adsorption capacity of mercury greater than 90% compared to lower pyrolysis. Similarly, when biochar from hickory chips and bagasse was used on the adsorption of Hg(II) along with activated carbon, it showed a higher adsorption capacity of biochar over activated carbon (Xu et al. 2016). Biochar prepared from grape stalk was modified with magnetic $MnFe_2O_4$ with double-layered hydroxide, activating the material's functionality to adsorb Hg(II). In 30 min of the contact time, the modified biochar showed 84% removal, which also satisfied pseudo-second-order and adsorption isotherms (Mahmoud et al. 2021). The elemental mercury was removed efficiently when biochar was modified with chlorine and polyvinyl chloride. Here magnetic biochar has provided C–Cl and C=O active sites, which trapped elemental mercury (Luo et al. 2019). The use of biochar as well as the modified biochar with additives, was found to remove mercury efficiently.

10.3 Mechanism of Interaction Between Mercury with Biochar from Soil/aqueous Solutions

There has been a major contribution in the application of biochar to the removal of mercury, either in inorganic mercury or methylmercury, through sorption from soil, water, and sediment (Boutsika et al. 2014; Shu et al. 2018; Wang et al. 2019; Lyu et al. 2020; Feng et al. 2020). The bioavailability of mercury or mobility in soils/sediments will be quite different from the bioavailability of water. The physical and chemical characteristics of biochar could influence mercury's biogeochemical properties when present in soil, sediment, and water (Yuan et al. 2017). Through research, several authors demonstrated that the presence of several substances in the soil or sediment and the interaction with the biogeochemical cycle is quite complex, so the biochar application to the soil and its action on the alteration of mercury is quite multifaceted (Xu et al. 2016; Yang et al. 2021a, b). As per the research, two mechanisms were identified. One is the sorption of mercury to biochar, which reduces the mobility and bioavailability of mercury. Secondly, biochar indirectly impacts the mobility and bioavailability of mercury with soil-biochar/aqueous solution–biochar interactions.

10.3.1 Straight Interface

The interaction between biochar and mercury lies in the surface composition and structure of the biochar. One of the important components of biochar is its surface area and porosity, which depends on the functional groups (Liu et al. 2015; Tan et al. 2016; Chaudhuri et al. 2021). The authors state many contact mechanisms between biochar and mercury, like precipitation, electrostatic interaction, complexation, and physisorption (Deng et al. 2020; Chaudhuri et al. 2022). As shown in Fig. 10.3 that the functional groups on the biochar are readily available to trap mercury from the soil or the sediment through complexation. In this mechanism, functional groups present on the biochar surface are rough to the complex formation of mercury. For instance, mercury sorption by biochar is prepared from bagasse to form a complex with carboxylic groups (COO^-) and phenolic hydroxyl (COH) that forms into ($-COO)_2Hg^{+2}$ and $(-O)_2Hg^{+2}$ (Xu et al. 2016). The wood-derived biochar has shown an interaction between Hg(II) with C=C and C=O to convert into Hg-π by sorption method (Park et al. 2019). In addition, mercury complexation with thiol groups was identified based on the analysis of X-ray absorption with mercury (Liu et al. 2016). The pyrolysis of Brazilian pepper biochar at 500 °C forms carboxylic and phenolic hydroxyl groups, which cleaves Hg(II) through complexation. The high surface area of the biochar with large porous complexation of mercury might be easy due to high sulfur content, which binds Hg(II). The descriptions by Park et al. (2019) and Liu et al. (2016) correlate as X-ray absorption analysis found that the mercury has been clung to the sulfur of biochar. In some instances when the sulfur content is low in biochar, mercury binds with other functional groups (Huang et al. 2019; Feng et al. 2020). Similarly, the studies on the removal of methylmercury through complexation as well as precipitation methods, carried by Shu et al. (2016), Wang et al. (2019) in which microcosm in anoxic incubation showed that the methylmercury concentration decreased significantly when bamboo-derived and rice straw biochar has applied. In another study, sodium sulfate was impregnated with corn straw biochar to remove Hg^{2+} where HgS precipitation formed from the aqueous solution (Tan et al. 2016). Similarly, in electrostatic interaction, the negatively charged ions will attract positively charged mercury ions to get remediated from the soil, sediment, or water. Due to the porous nature of the biochar, either inorganic mercury or methylmercury adsorbs on the surface of the functional groups of the biochar through complexation, precipitation, and ion exchange processes.

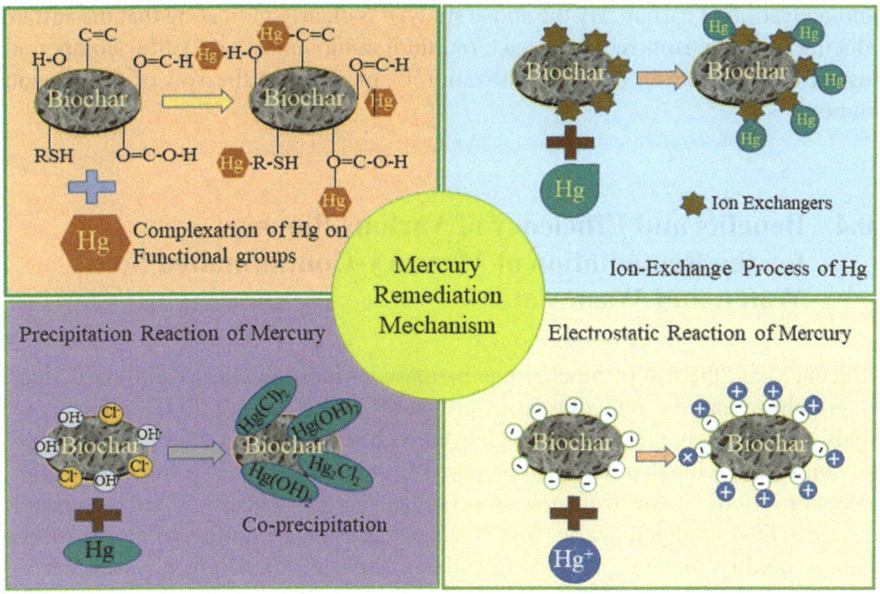

Fig. 10.3 Different mechanism process for the remediation of mercury

10.3.2 Ancillary Interactions

Due to the addition of biochar, the soil's physicochemical characteristics and biological properties, in turn, impact the biogeochemical cycle of mercury and other elements related to the mercury cycle (Joseph et al. 2015; Bandara et al. 2019). The alteration of redox biogeochemical processes influences the mobility and bioavailability of mercury in water, soils/sediment was studied very little. Few studies reported that the sulfate-reducing bacteria were actively involved in the indirect bioavailability of mercury (Sande 2016; Shu et al. 2016), here the biochar made from rice straw contains high sulfur content, which activates the sulfur concentration in the soil and elevates the growth of microflora of the soil which starts reducing the methylmercury in the absence of oxygen. Further, this study was confirmed by Liu et al. (2018a) that the changes in methylmercury state could be due to either activity of sulfate-reducing bacteria or iron-reducing bacteria in the early stage of incubation and later stages through the methanogenesis process. In another study, thiol compounds detected in the biochar applied to paddy soil could form complex with methylmercury to form into methylmercury thiol complexes, thereby crippling methylmercury in the soil (Xing et al 2020). Similarly, the investigations by Wang et al. (2020) confirmed that Hg indirectly gets immobilized when nano-activated carbon along with biochar is applied to rice crops; here mercury gets bound as nano-mercury sulfur in the soil; this is probably due to the reduction of sulfur oxide to reduced sulfur species at nano-activation of carbon. By the above study, it is clearly indicating that the sulfate-reducing bacteria, iron-reducing bacteria, thiol compounds, and sulfur compounds play an indirect role in the immobilization of mercury in the soil or the aqueous solutions.

10.4 Benefits and Efficiency of Various Biochars for the Remediation of Mercury-Contaminated Soils, Water, and Wastewater

In recent years the use of biochar has increased widely in every field, particularly in agriculture and the remediation of pollutants (Huang et al. 2021). Biochar is a highly porous, carbon-augmented achieved with less oxygen under biomass pyrolysis. Most of the biochar has been prepared from wood (bark of a tree, wood chips, pellets of wood), waste from the crop (stalks of rice, wheat and corn, nutshells, straw, rice husk), organic waste (sugar cane bagasse, paper sludge, olive waste), cow manure, poultry manure, and sewage sludge (Sohi et al. 2010). Physical adsorption (Zhang et al. 2019a) can remove mercury in small amounts, whereas modified biochars are highly efficient in adsorbing mercury; for example, metal oxides and halides embedded biochar oxidize Hg^0 to Hg^{2+}. In one study, walnut shell biochar was added with $FeCl_3$ to remove Hg^0 through chemical oxidation in which C=O, –COOH, metal ions, and metal oxides were chemically adsorbed oxygen and Cl^- has shown a synergistic effect on Fe_2O_3. This mechanism was also correlated with MnO_2 and CeO_2 (Yang et al. 2017). During the thermodynamic study, the adsorption of Hg^{2+} was very spontaneous and endothermic. The use of amino-grafted modified biochar, which has multifunctional groups on its surface with activation of carboxylic groups, was found to adsorb Hg(II) very efficiently from the aqueous solution (Bao et al. 2018). In another investigation, Pistachio wood waste was modified to test the mercury adsorption capacity from water in comparison with commercial activated carbon. The results showed that the pistachio wood waste adsorbed 202 mg/g, whereas commercial activated carbon with 66.5 mg/g indicates that wood chip waste contributes major adsorption capacity (Sajjadi et al. 2018). The biochar prepared from the sugarcane bagasse was efficient in removing mercury from the soil, but when it was modified with activating agents such as $ZnCl_2$ and H_3PO_4 under pyrolysis and chemical activation, it was found efficient in removing Hg(II) from aqueous solution (Giraldo et al. 2020). The complexation adsorption phenomenon has taken place for this Hg(II), and the functional groups such as –O, $-COO^-$, C=C, and C=O on the material surface have worked efficiently to bind Hg(II) at the adsorption of 90%. In another study, biochar from corn straw was fractionalized with hydroxyl-blocked carbon, carboxyl-blocked carbon, inorganic and organic carbon. Through this, the adsorption of Hg(II) was found to be from 22.4 to 77.6%, in which inorganic carbon showed the adsorption capacity of 92.63 mg/g and complexation of both hydroxyl and carboxyl groups showed about 71.6% of mercury adsorption (Guo et al. 2020a, b, c). After pyrolysis, the biochar made from the sewage sludge was utilized to remove mercury from the aqueous solution. The results showed that 90% of mercury removal was identified (Liu et al. 2018a, b, c) when the pyrolysis was carried out at 200 °C. Immobilization of mercury has been studied with the rice husk-prepared biochar at different temperatures of 300, 500, and 700 °C. In this study, the author tried to amend the biochar in sandy soil to reduce erosion and to immobilize mercury; the study showed that at different temperatures, biochar reduced runoff from 5.1 to 15.4% and soil loss from 43.5 to 77.2%. The biochar prepared at 300 °C was found to be efficient in immobilizing mercury with 48.8% (Shen et al. 2021). Similarly, rice husk-prepared biochar modified as activated carbon with KOH (Liu et al. 2020) had fine pore structure and good oxygen functional groups showed maximum adsorption of Hg(II) at 55.87 mg/g, and it maintained the removal capacity even after five regeneration cycles efficiently removing mercury ion from wastewater.

10.5 Strategies for Recovery and Recycling of Biochar Post-Mercury Removal

Recent studies suggest that the amendment of biochar is a potential approach to mitigate soil, water, and wastewater contamination through immobilization and adsorption of organic materials and heavy metals, particularly mercury in elemental, inorganic, and methyl forms. Most biochar applications involved in its formation, processes, mechanism, and effectiveness on the material need to be remediated (Guo et al. 2020a, b, c). The quality and characteristics of biochar mainly depend on the feedstock materials, sewage sludge from any industry, straw grass, and the conditions of pyrolysis temperature. The formed biochar from different raw materials shows diversified capabilities of sorption potentiality. For instance, soil-embedded biochar can immobilize Hg^0, IHg, and MeHg to reduce their mobility and bioavailability in the soil (Ippolito et al. 2017). Mercury adsorption occurs through complexation, precipitation, ion exchange, and electron attraction processes, as this adsorption efficiency depends on the surface area, pore size, and micro and hydro porosity of the biochar. Sometimes, adsorption has also been enhanced by adding certain compounds to the soil, which activates the soil's microflora to remediate indirectly. In many works about biochar, researchers have identified that the used biochar can be recycled in a number of cycles depending on the type of biochar prepared. Gupta et al. (2020), used biochar prepared from municipal solid waste for the adsorption of mercury, pyrolysis of biochar at 300 °C showed promising results of an adsorption rate of 48 mg/g of mercury at optimum conditions. It also satisfied pseudo-kinetic models and Langmuir and Freundlich isotherms, and the spent biochar was readily used for cement and brick manufacturing. In one more study, biochar derived from wheat straw prepared from ball-milled magnetic for removing Hg(II) at 700 °C had an adsorption capacity of 127.4 mg/g reused three more times for mercury removal by Li et al. (2020). Biochar derived from waste tea with modification of NH_4Cl

has been used to remove elemental mercury. The pyrolysis temperature recorded is about 120 °C for the absorption of H^0. The biochar recovered after mercury removal was used three times more for immobilization, according to Shen et al. (2017). In another investigation, rice straw biomass after anaerobic degradation was used to prepare biochar to remove mono and divalent mercury from the aqueous solution. Here it was found that due to the lignocellulosic nature of the material, it has a high surface area and rich pore structure, which adsorbed mercury easily, and the recovered biochar was further utilized for five more cycles (Liu et al. 2022). All these studies show that biochar can be reused and recycled a few times after being used for mercury adsorption. Hence, biochar is eco-friendly for immobilizing mercury to remediate mercury-contaminated soils and aqueous solutions.

10.6 Conclusions

The application of biochar could be an effective and practical approach to mitigate the threat of Hg transfer to the environment. This chapter has referred to the insights of the author's work as biochar has efficiently reduced the bioavailability and mobility of mercury. The increase in the use of biochar in many studies is due to its surface area, porosity, high surface activity, and low ash content, making it ideal for adsorption. Due to the availability of these properties in biochar, it has become an efficient and technical tool to use in the remediation of mercury in elemental, inorganic, and organic forms. In this chapter, we have highlighted various techniques and methods used to modify biochar and its efficiencies. Different mechanisms of mercury removal, such as adsorption, precipitation, ion exchange, and complexation, were also illustrated. There was a discussion on the direct and indirect interactions of mercury with biochar, including surface adsorption and functional groups' complexation. Furthermore, the benefits and efficiency of biochar, their properties, and processes were studied in detail with different improvements for mercury remediation. Moreover, it was also pointed out about the reuse and recycling of biochar after the remediation of mercury. Hence, biochar, a low-cost, eco-friendly product, can be amended for mercury remediation.

10.7 Future Perspective

As biochar was found to be an efficient tool in mercury remediation, very few studies were carried out for Hg-contaminated sediments. There have been some challenges and goals related to the biochar application that require proper investigation, including low dose application, smarter design model, combinations of biochar with other materials and their efficacy, effective measurement on prolonged use, and risk assessment. Further, the concept of waste valorization can be applied to the production of biochar and to improve the biochar mechanical extrusion process. More experiments were needed to investigate the ability of biochar and modified biochar through prolonged investigations on different soils and sediments. Focus has to be made on the preparation and modifications of the biochar with different materials as those should be environment friendly and will not release any toxicity to the soil; similarly, simple and fast monitoring methods should be adopted for better remediation. Effective application of these approaches will enhance the ability to abate the ecological risk from the mercury to the soil, sediment, and aqueous solution.

Acknowledgements The authors are thankful for the grant received from the Gandhi Institute of Technology and Management (Deemed to be University), Visakhapatnam in the form of (GITAM—Research Seed Grants—RSG) with ref. No-2022/0149 for the resources.

References

Ahmad M, Lee SS, Dou X, Mohan D, Sung JK, Yang JE, Ok YS (2012) Effects of pyrolysis temperature on soybean stover-and peanut shell-derived biochar properties and TCE adsorption in water. Bioresour Technol 118:536–544. https://doi.org/10.1016/j.biortech.2012.05.042

Ahmad M, Rajapaksha AU, Lim JE, Zhang M, Bolan N, Mohan D, Vithanage M, Lee SS, Ok YS (2014) Biochar as a sorbent for contaminant management in soil and water: a review. Chemosphere 99:19–33. https://doi.org/10.1016/j.chemosphere.2013.10.071

Akhil D, Lakshmi D, Kartik A, Vo DVN, Arun J, Gopinath KP (2021) Production, characterization, activation and environmental applications of engineered biochar: a review. Environ Chem Lett 19:2261–2297. https://doi.org/10.1007/s10311-020-01167-7

Albishri HM, Yakout AA (2023) Efficient removal of Hg(II) from dental effluents by thio-functionalized biochar derived from cape gooseberry (*Physalis peruviana* L.) leaves. Mater Chem Phys 295:127125. https://doi.org/10.1016/j.matchemphys.2022.127125

Ali A, Guo D, Jeyasundar PGSA, Li Y, Xiao R, Du J, Zhang Z (2019) Application of wood biochar in polluted soils stabilized the toxic metals and enhanced wheat (*Triticum aestivum*) growth and soil enzymatic activity. Ecotoxicol 184:109635. https://doi.org/10.1016/j.ecoenv.2019.109635

Antunes E, Jacob MV, Brodie G, Schneider PA (2017) Silver removal from aqueous solution by biochar produced from biosolids via microwave pyrolysis. J Environ Manage 203:264–272. https://doi.org/10.1016/j.jenvman.2017.07.071

Bailon MX, Chaudhary DK, Jeon C, Ok YS, Hong Y (2022) Impact of sulfur-impregnated biochar amendment on microbial communities and mercury methylation in contaminated sediment. JHM Lett 438:129464. https://doi.org/10.1016/j.jhazmat.2022.129464

Bandara T, Franks A, Xu JM, Bolan N, Wang HL, Tang CX (2019) Chemical and biological immobilization mechanisms of potentially toxic elements in biochar-amended soils. Crit Rev Environ Sci Technol 50(9):903–978. https://doi.org/10.1080/10643389.2019.1642832

Bank MS (2020) The mercury science-policy interface: history, evolution and progress of the Minamata convention. Sci Total Environ 722:137832. https://doi.org/10.1016/j.scitotenv.2020.137832

Bao J, Zheng H, Tufail H, Irshad S, Du J (2018) Adsorption-assisted decontamination of Hg(II) from aqueous solution by multi-functionalized corncob-derived biochar. RSC Adv 8:38425–38435. https://doi.org/10.1039/C8RA06622A

Boutsika LG, Karapanagioti HK, Manariotis ID (2014) Aqueous mercury sorption by biochar from malt spent rootlets. Water Air Soil Pollut 225:1–10. https://doi.org/10.1007/s11270-013-1805-9

Butphu S, Rasche F, Cadisch G, Kaewpradit W (2020) Eucalyptus biochar application enhances Ca uptake of upland rice, soil available P, exchangeable K, yield, and N use efficiency of sugarcane in a crop rotation system. JPNSS 183:58–68. https://doi.org/10.1002/jpln.201900171

Carrasco L, Benejam L, Benito J, Bayona JM, Díez S (2011) Methylmercury levels and bioaccumulation in the aquatic food web of a highly mercury-contaminated reservoir. Environ Int 37:1213–1218. https://doi.org/10.1016/j.envint.2011.05.004

Chaudhuri S, Sigmund G, Bone SE, Kumar N, Hofmann T (2022) Mercury removal from contaminated water by wood-based biochar depends on natural organic matter and ionic composition. Environ Sci Technol 56:11354–11362. https://doi.org/10.1021/acs.est.2c01554

Chen Z, Chen B, Chiou CT (2012) Fast and slow rates of naphthalene sorption to biochars produced at different temperatures. Environ Sci Technol 46:11104–11111. https://doi.org/10.1021/es302345e

Chen J, Dong J, Chang J, Guo T, Yang Q, Jia W, Shen S (2018) Characterization of an Hg(II)-volatilizing *Pseudomonas* sp. strain, DC-B1, and its potential for soil remediation when combined with biochar amendment. Ecotoxicol Environ Saf 163:172–179. https://doi.org/10.1016/j.ecoenv.2018.07.071

Chen L, Li F, Wei Y, Li G, Shen K, He HJ (2019) High cadmium adsorption on nanoscale zero-valent iron coated *Eichhornia crassipes* biochar. Environ Chem Lett 17:589–594. https://doi.org/10.1007/s10311-018-0811-y

Cheng S, Chen T, Xu W, Huang J, Jiang S, Yan B (2020) Application research of biochar for the remediation of soil heavy metals contamination: a review. Molecules 25:3167. https://doi.org/10.3390/molecules25143167

Collado S, Rosas I, González C, Diaz M (2016) Biodegradation of p-hydroxybenzoic acid by *Pseudomonas putida*. Desalin Water Treat 857:15230–15240. https://doi.org/10.1080/19443994.2015.1072584

Cui W, Liu G, Bezerra M, Lagos DA, Li Y, Cai Y (2017) Occurrence of methylmercury in rice-based infant cereals and estimation of daily dietary intake of methylmercury for infants. J Agric Food Chem 65:9569–9578. https://doi.org/10.1021/acs.jafc.7b03236

Demirbas A (2004) Determination of calorific values of bio-chars and pyro-oils from pyrolysis of beech trunk barks. JAAP 72:215–219. https://doi.org/10.1016/j.jaap.2004.06.005

Deng R, Huang D, Wan J, Xue W, Wen X, Liu X, Zhang Q (2020) Recent advances of biochar materials for typical potentially toxic elements management in aquatic environments: a review. J Clean Prod 255:119523. https://doi.org/10.1016/j.jclepro.2019.119523

Duwiejuah AB, Quainoo AK, Abubakari AH (2022) Simultaneous adsorption of toxic metals in binary systems using peanut and sheanut shells biochars. Heliyon 8(9):e10558. https://doi.org/10.1016/j.heliyon.2022.e10558

Eibner S, Margeriat A, Broust F, Laurenti D, Geantet C, Julbe A, Blin J (2017) Catalytic deoxygenation of model compounds from flash pyrolysis of lignocellulosic biomass over activated charcoal-based catalysts. Appl Catal B 219:517–525. https://doi.org/10.1016/j.apcatb.2017.07.071

Feng Y, Liu P, Wang Y, Liu W, Liu Y, Finfrock YZ (2020) Mechanistic investigation of mercury removal by unmodified and Fe-modified biochars based on synchrotron-based methods. Sci Total Environ 719:137435. https://doi.org/10.1016/j.scitotenv.2020.137435

Fotopoulou KN, Karapanagioti HK, Manariotis ID (2015) Biochar characteristics produced from food-processing products and their sorptive capacity for mercury and phenanthrene. In: EGU general assembly conference abstracts, p 9515

Fungai Mukome FND, Parikh SJ, Uchimiya SM, Chang SX, Bolan N (eds) (2015) Chemical, physical, and surface characterization of biochar. In: Book, production, characterization, and applications. CRC Press, Boca Raton, pp 67–96. https://doi.org/10.1201/b18920

García-Hernández J, Ortega-Vélez MI, Contreras-Paniagua AD, Aguilera-Márquez D, Leyva-García G, Torre J (2018) Mercury concentrations in seafood and the associated risk in women with high fish consumption from coastal villages of Sonora, Mexico. Food Chem Toxicol 120:367–377. https://doi.org/10.1016/j.fct.2018.07.029

Giraldo S, Robles I, Ramirez A, Flórez E, Acelas N (2020) Mercury removal from wastewater using agroindustrial waste adsorbents. SN Appl Sci 2:1–17. https://doi.org/10.1007/s42452-020-2736-x

Gopinath KP, Vo DV, Gnana Prakash D, Adithya Joseph A, Viswanathan S, Arun J (2020) Environmental applications of carbon-based materials: a review. Environ Chem Lett 19:557–82. https://doi.org/10.1007/s10311-020-01084-9

Goñez-Rodríguez L, Johs A, Lowe KA, Carter KE, Löffler FE, Mayes MA (2021) Evaluation of engineered sorbents for the sorption of mercury from contaminated bank soils: a column study. ESPR 28:22651–22663. https://doi.org/10.1007/s11356-020-12073-4

Gopinath A, Divyapriya G, Srivastava V, Laiju AR, Nidheesh PV, Kumar MS (2021a) Conversion of sewage sludge into biochar: a potential resource in water and wastewater treatment. Environ Res 194:110656. https://doi.org/10.1016/j.envres.2020.110656

Gopinath KP, Vo DVN, Gnana Prakash D, Adithya Joseph A, Viswanathan S, Arun J (2021b) Environmental applications of carbon-based materials: a review. Environ Chem Lett 19:557–582. https://doi.org/10.1007/s10311-020-01084-9

Gul S, Whalen JK, Thomas BW, Sachdeva V, Deng H (2015) Physico-chemical properties and microbial responses in biochar-amended soils: mechanisms and future directions. Agric Ecosyst Environ 206:46–59. https://doi.org/10.1016/j.agee.2015.03.015

Gunarathne V, Ashiq A, Ramanayaka S, Wijekoon P, Vithanage M (2019) Biochar from municipal solid waste for resource recovery and pollution remediation. Environ Chem Lett 17:1225–1235. https://doi.org/10.1007/s10311-019-00866-0

Guo X, Li M, Liu A, Jiang M, Niu X, Liu X (2020c) Adsorption mechanisms and characteristics of Hg^{2+} removal by different fractions of biochar. Water 12:2105. https://doi.org/10.3390/w12082105

Guo J, Gao Q, Jiang S (2020) Insight into dewatering behavior and heavy metals transformation during waste-activated sludge treatment by thermally-activated sodium persulfate oxidation combined with a skeleton builder—wheat straw biochar. Chemosphere 252:126542

Guo M, Song W, Tian J (2020b) Biochar-facilitated soil remediation: mechanisms and efficacy variations. Front Environ Sci 183. https://doi.org/10.3389/fenvs.2020.521512

Gupta S, Sireesha S, Sreedhar I, Patel CM, Anitha KL (2020) Latest trends in heavy metal removal from wastewater by biochar based sorbents. JWPE 38:101561. https://doi.org/10.1016/j.jwpe.2020.101561

Gupta RC, Milatovic D, Lall R, Srivastava A (2018) Mercury, veterinary toxicology (third edn) basic and clinical principles, pp 455–462. https://doi.org/10.1016/B978-0-12-811410-0.00031-3

Hans-Peter SA (2013) European biochar certificate guidelines for a sustainable production of biochar. European Biochar Found, p D3. http://www.bio-inspecta.ch/htm/dl_detail.htm?sprácheDe&idD105. Accessed 25 Dec 16

He L, Zhong H, Liu G, Dai Z, Brookes PC, Xu J (2019) Remediation of heavy metal contaminated soils by biochar: mechanisms, potential risks and applications in China. Environ Pollut 252:846–855. https://doi.org/10.1016/j.envpol.2019.05.151

Hossain MK, Strezov V, Chan KY, Ziolkowski A, Nelson PF (2011) Influence of pyrolysis temperature on production and nutrient properties of wastewater sludge biochar. J Environ Manage 92:223–228. https://doi.org/10.1016/j.jenvman.2010.09.008

Huang Y, Xia S, Lyu J, Tang J (2019) Highly efficient removal of aqueous Hg^{2+} and CH_3Hg^+ by selective modification of biochar with 3-mercaptopropyltrimethoxysilane. J Chem Eng 360:1646–1655. https://doi.org/10.1016/j.cej.2018.10.231

Huang WH, Lee DJ, Huang C (2021) Modification on biochars for applications: a research update. Bioresour Technol 319:124100. https://doi.org/10.1016/j.biortech.2020.124100

IBI (2015) Standardized product definition and product testing guidelines for biochar: that is used in soil. http://www.biochar-international.org/sites/default/files/IBI_Biochar_Standards_V2.1_Final.pdf. Accessed 25 Dec 16

Ippolito JA, Berry CM, Strawn DG, Novak JM, Levine J, Harley A (2017) Biochars reduce mine land soil bioavailable metals. J Environ Qual 46:411–419. https://doi.org/10.2134/jeq2016.10.0388

Jiang L, Zhang R, Zhang L, Zheng R, Zhong M (2021) Improving the regulatory health risk assessment of mercury-contaminated sites. J Hazard Mater 402:123493. https://doi.org/10.1016/j.jhazmat.2020.123493

Jindo K, Mizumoto H, Sawada Y, Sanchez-Monedero MA, Sonoki T (2014) Physical and chemical characterization of biochars derived from different agricultural residues. Biogeosciences 11(23):6613–6621. https://doi.org/10.5194/bg-11-6613-2014

Joseph S, Husson O, Graber ER, Lv Z, Taherymoosavi S, Thomas T, Nielsen S, Ye J, Pan GX, Chia C, Munroe P, Allen J, Lin Y, Fan XR, Donne S (2015) The electrochemical properties of biochars and how they affect soil redox properties and processes. Agronomy 5:322–340. https://doi.org/10.3390/agronomy5030322

Kambo HS, Dutta A (2015) A comparative review of biochar and hydrochar in terms of production, physico-chemical properties and applications. Renew Sustain Energy Rev 45:359–378. https://doi.org/10.1016/j.rser.2015.01.050

Kavitha B, Reddy PVL, Kim B, Lee SS, Pandey SK, Kim KH (2018) Benefits and limitations of biochar amendment in agricultural soils: a review. J Environ Manage 227:146–154. https://doi.org/10.1016/j.jenvman.2018.08.082

Keiluweit M, Nico PS, Johnson MG, Kleber M (2010) Dynamic molecular structure of plant biomass-derived black carbon (biochar). Environ Sci Techno 44:1247–1253. https://doi.org/10.1021/es9031419

Klasson KT, Ledbetter CA, Uchimiya M, Lima IM (2013) Activated biochar removes 100% dibromochloropropane from field well water. Environ Chem Lett 11:271–275. https://doi.org/10.1007/s10311-012-0398-7

Kostas ET, Durán-Jiménez G, Shepherd BJ, Meredith W, Stevens LA, Williams OS, Lye GJ, Robinson JP (2020) Microwave pyrolysis of olive pomace for bio-oil and bio-char production. Chem Engg J 387:123404. https://doi.org/10.1016/j.cej.2019.123404

Leng L, Huang H (2018) An overview of the effect of pyrolysis process parameters on biochar stability. Bioresour Technol 270:627–642. https://doi.org/10.1016/j.biortech.2018.09.030

Leng LJ, Yuan XZ, Huang HJ, Wang H, Wu ZB, Fu LH, Peng X, Chen XH, Zeng GM (2015) Characterization and application of bio-chars from liquefaction of microalgae, lignocellulosic biomass and sewage sludge. Fuel Process Technol 129:8–14. https://doi.org/10.1016/j.fuproc.2014.08.016

Li P, Feng XB, Qiu GL, Shang LH, Li ZG (2009) Mercury pollution in Asia: a review of the contaminated sites. J Hazard Mater 168:591–601. https://doi.org/10.1016/j.jhazmat.2009.03.031

Li R, Zhang Y, Deng H, Zhang Z, Wang JJ, Shaheen SM, Du J (2020) Removing tetracycline and Hg(II) with ball-milled magnetic nanobiochar and its potential on polluted irrigation water reclamation. J Hazard Mater 384:121095. https://doi.org/10.1016/j.jhazmat.2019.121095

Liu WJ, Jiang H, Yu HQ (2015) Development of biochar-based functional materials: toward a sustainable platform carbon material. Chem Rev 115:12251–12285. https://doi.org/10.1021/acs.chemrev.5b00195

Liu P, Ptacek CJ, Blowes DW, Landis RC (2016) Mechanisms of mercury removal by biochars produced from different feedstocks determined using X-ray absorption spectroscopy. J Hazard Mater 308:233–242. https://doi.org/10.1016/j.jhazmat.2016.01.007

Liu H, Xu F, Xie Y, Wang C, Zhang A, Li L, Xu H (2018a) Effect of modified coconut shell biochar on availability of heavy metals and biochemical characteristics of soil in multiple heavy metals contaminated soil. Sci Total Environ 645:702–709. https://doi.org/10.1016/j.scitotenv.2018.07.115

Liu P, Ptacek CJ, Blowes DW, Gould WD (2018b) Control of mercury and methylmercury in contaminated sediments using biochars: a long-term microcosm study. J Appl Geochem 92:30–44. https://doi.org/10.1016/j.apgeochem.2018.02.004

Liu T, Liu Z, Zheng Q, Lang Q, Xia Y, Peng N, Gai C (2018c) Effect of hydrothermal carbonization on migration and environmental risk of heavy metals in sewage sludge during pyrolysis. Bioresour Technol 247:282–290. https://doi.org/10.1016/j.biortech.2017.09.090

Liu P, Ptacek CJ, Blowes DW, Gould WD (2018d) Control of mercury and MeHg in contaminated sediments using biochars: a long term microcosm study. Appl Geochem 92:30–44. https://doi.org/10.1016/j.apgeochem.2018.02.004

Liu Z, Sun Y, Xu X, Qu J, Qu B (2020) Adsorption of Hg(II) in an aqueous solution by activated carbon prepared from rice husk using KOH activation. ACS Omega 5:29231–29242. https://doi.org/10.1021/acsomega.0c03992

Liu Z, Zhen F, Zhang Q, Qian X, Li W, Sun Y, Zhang L, Qu B (2022) Nanoporous biochar with high specific surface area based on rice straw digestion residue for efficient adsorption of mercury ion from water. Bioresour Technol 359:127471. https://doi.org/10.1016/j.biortech.2022.127471

Loha C, Gu S, De Wilde J, Mahanta P, Chatterjee PK (2014) Advances in mathematical modeling of fluidized bed gasification. Renew Sustain Energy Rev 40:688–715. https://doi.org/10.1016/j.rser.2014.07.199

Lu H, Li Z, Fu S, Méndez A, Gascó G, Paz-Ferreiro J (2015) Effect of biochar in cadmium availability and soil biological activity in an anthrosol following acid rain deposition and aging. Water Air Soil Pollut 226:1–11. https://doi.org/10.1007/s11270-015-2401-y

Lu N, Han JC, Wei Y, Li G, Sun YY (2020) The heavy metal adsorption capacity of stalk biochar in an aqueous phase. Appl Ecol Environ 18:2569–2579. https://doi.org/10.15666/aeer/1802_25692579

Luo J, Jin M, Ye L, Cao Y, Yan Y, Du R, Yoshiie R, Ueki Y, Naruse I, Lin C, Lee Y (2019) Removal of gaseous elemental mercury by hydrogen chloride non-thermal plasma modified biochar. J Hazard Mater 377:132–41. https://doi.org/10.1016/j.jhazmat.2019.05.045

Lyu H, Xia S, Tang J, Zhang Y, Gao B, Shen B (2020) Thiol-modified biochar synthesized by a facile ball-milling method for enhanced sorption of inorganic Hg^{2+} and organic CH_3Hg^+. J Hazard Mater 384:121357. https://doi.org/10.1016/j.jhazmat.2019.121357

Madima N, Mishra SB, Inamuddin I, Mishra AK (2020) Carbon-based nanomaterials for remediation of organic and inorganic pollutants from wastewater. A review. Environ Chem Lett 18:1169–1191. https://doi.org/10.1007/s10311-020-01001-0

Mahmoud ME, El-Bahy SM, Elweshahy SM (2021) Decorated Mn-ferrite nanoparticle@ Zn–Al layered double hydroxide@ Cellulose@ activated biochar nanocomposite for efficient remediation of methylene blue and mercury(II). Bioresour Technol 342:126029. https://doi.org/10.1016/j.biortech.2021.126029

Man Y, Wang B, Wang J, Slaný M, Yan H, Li P, El-Naggar A, Shaheen SM, Rinklebe J, Feng X (2021) Use of biochar to reduce mercury accumulation in *Oryza sativa* L: a trial for sustainable management of historically polluted farmlands. Environ Int 153:106527. https://doi.org/10.1016/j.envint.2021.106527

Manoharan T, Ganeshalingam S, Nadarajah K (2022) Mechanisms of emerging contaminants removal by novel neem chip biochar. Environ Adv 7:100158. https://doi.org/10.1016/j.envadv.2021.100158

O'Connor D, Peng T, Li G, Wang S, Duan L, Mulder J, Cornelissen G, Cheng Z, Yang S, Hou D (2018) Sulfur-modified rice husk biochar: a green method for the remediation of mercury contaminated soil. Sci Total Environ 621:819–826. https://doi.org/10.1016/j.scitotenv.2017.11.213

O'Connor D, Hou D, Ok YS, Mulder J, Duan L, Wu Q, Wang S, Tack FM, Rinklebe J (2019) Mercury speciation, transformation, and transportation in soils, atmospheric flux, and implications for risk management: a critical review. Environ Int 126:747–761. https://doi.org/10.1016/j.envint.2019.03.019

Omoriyekomwan JE, Tahmasebi A, Zhang J, Yu J (2017) Formation of hollow carbon nanofibers on bio-char during microwave pyrolysis of palm kernel shell. Energy Conver Manag 148:583–592. https://doi.org/10.1016/j.enconman.2017.06.022

Pagala B (2023) Characteristic evaluation of tamarind flower biomass for mercury biosorption: statistical approach. Biomass Convers Biorefin 1–16. https://doi.org/10.1007/s13399-023-03769-x

Park JH, Wang JJ, Zhou B, Mikhael JE, DeLaune RD (2019) Removing mercury from aqueous solution using sulfurized biochar and associated mechanisms. Environ Pollut 244:627–635. https://doi.org/10.1016/j.envpol.2018.10.069

Peng XYLL, Ye LL, Wang CH, Zhou H, Sun B (2011) Temperature-and duration-dependent rice straw-derived biochar: characteristics and its effects on soil properties of an Ultisol in southern China. Soil Tillage Res 112:159–166. https://doi.org/10.1016/j.still.2011.01.002

Qambrani NA, Rahman MM, Won S, Shim S, Ra C (2017) Biochar properties and eco-friendly applications for climate change mitigation, waste management, and wastewater treatment: a review. Renew Sustain Energy Rev 79:255–273. https://doi.org/10.1016/j.rser.2017.05.057

Sajjadi SA, Mohammadzadeh A, Tran HN, Anastopoulos I, Dotto GL, Lopičić ZR, Hosseini-Bandegharaei A (2018) Efficient mercury removal from wastewater by pistachio wood wastes-derived activated carbon prepared by chemical activation using a novel activating agent. J Environ Manage 223:1001–1009. https://doi.org/10.1016/j.jenvman.2018.06.077

Sande K (2016) Biochar metal sorption and effect on microbial sulfate reduction. Master of Science Dissertation, University of Minnesota, Twin Cities, MN

Shaheen SM, Niazi NK, Hassan NE, Bibi I, Wang H, Tsang DC, Rinklebe J (2019) Wood-based biochar for the removal of potentially toxic elements in water and wastewater: a critical review. Int Mater Rev 64:216–247. https://doi.org/10.1080/09506608.2018.1473096

Shen B, Tian L, Li F, Zhang X, Xu H, Singh S (2017) Elemental mercury removal by the modified bio-char from waste tea. Fuel 187:189–196. https://doi.org/10.1016/j.fuel.2016.09.059

Shen Z, Zhang Z, Zhang M, Rinklebe J, Ma Y, Hou D (2021) Effect of production temperature and particle size of rice husk biochar on mercury immobilization and erosion prevention of a mercury contaminated soil. J Hazard Mater 420:126646. https://doi.org/10.1016/j.jhazmat.2021.126646

Shu R, Wang Y, Zhong H (2016) Biochar amendment reduced methylmercury accumulation in rice plants. J Hazard Mater 313:1–8. https://doi.org/10.1016/j.jhazmat.2016.03.080

Singh B, Singh BP, Cowie AL (2010) Characterization and evaluation of biochars for their application as a soil amendment. Soil Res 48:516–525. https://doi.org/10.1071/SR10058

Sohi SP, Krull E, Lopez-Capel E, Bol R (2010a) A review of biochar and its use and function in soil. Adv Agron 105:47–82. https://doi.org/10.1016/S0065-2113(10)05002-9

Tan G, Sun W, Xu Y, Wang H, Xu N (2016) Sorption of mercury(II) and atrazine by biochar, modified biochars and biochar based activated carbon in aqueous solution. Bioresour Technol 211:727–735. https://doi.org/10.1016/j.biortech.2016.03.147

Tang W, Su Y, Gao Y, Zhong H (2019) Effects of farming activities on the biogeochemistry of mercury in rice–paddy soil systems. Bull Environ Contamin Toxicol 102:635–642. https://doi.org/10.1007/s00128-019-02627-9

Ullrich SM, Tanton TW, Abdrashitova SA (2001) Mercury in the aquatic environment: a review of factors affecting methylation. Crit Rev Environ Sci Tech 31:241–293. https://doi.org/10.1080/20016491089226

Wang T, Liu J, Zhang Y, Zhang H, Chen WY, Norris P, Pan WP (2018a) Use of a non-thermal plasma technique to increase the number of chlorine active sites on biochar for improved mercury removal. J Chem Eng 331:536–544. https://doi.org/10.1016/j.cej.2017.09.017

Wang T, Wu J, Zhang Y, Liu J, Sui Z, Zhang H, Pan WP (2018b) Increasing the chlorine active sites in the micropores of biochar for improved mercury adsorption. Fuel 229:60–67. https://doi.org/10.1016/j.fuel.2018.05.028

Wang Z, Sun T, Driscoll CT, Yin Y, Zhang X (2018c) Mechanism of accumulation of methylmercury in rice (*Oryza sativa* L.) in a mercury mining area. Environ Sci Technol 52:9749–9757. https://doi.org/10.1021/acs.est.8b01783

Wang Y, Dang F, Zheng X, Zhong H (2019) Biochar amendment to further reduce methylmercury accumulation in rice grown in selenium-amended paddy soil. J Hazard Mater 365:590–596. https://doi.org/10.1016/j.jhazmat.2018.11.052

Wang Q, Lai Z, Mu J, Chu D, Zang X (2020) Converting industrial waste cork to biochar as Cu(II) adsorbent via slow pyrolysis. Waste Manage 105:102–109. https://doi.org/10.1016/j.wasman.2020.01.041

Weber K, Quicker P (2018) Properties of biochar. Fuel 217:240–261. https://doi.org/10.1016/j.fuel.2017.12.054

Xiang W, Zhang X, Chen J, Zou W, He F, Hu X, Gao B (2020) Biochar technology in wastewater treatment: a critical review. Chemosphere 252:126539. https://doi.org/10.1016/j.chemosphere.2020.126539

Xie T, Reddy KR, Wang C, Yargicoglu E, Spokas K (2015) Characteristics and applications of biochar for environmental remediation: a review. Crit Rev Environ Sci Technol 45:939–969. https://doi.org/10.1080/10643389.2014.924180

Xing Y, Wang J, Xia J, Liu Z, Zhang Y, Du Y, Wei W (2019) A pilot study on using biochars as sustainable amendments to inhibit rice uptake of Hg from a historically polluted soil in a Karst region of China. Ecotoxicol 170:18–24. https://doi.org/10.1016/j.ecoenv.2018.11.111

Xing Y, Wang J, Shaheen SM, Feng X, Chen Z, Zhang H, Rinklebe J (2020) Mitigation of mercury accumulation in rice using rice hull-derived biochar as soil amendment: a field investigation. J Hazard Mater 388:121747. https://doi.org/10.1016/j.jhazmat.2019.121747

Xu X, Schierz A, Xu N, Cao X (2016) Comparison of the characteristics and mechanisms of Hg(II) sorption by biochars and activated carbon. J Colloid Interface Sci 463:55–60. https://doi.org/10.1016/j.jcis.2015.10.003

Yang W, Liu Y, Wang Q, Pan J (2017) Removal of elemental mercury from flue gas using wheat straw chars modified by Mn-Ce mixed oxides with ultrasonic-assisted impregnation. J Chem Eng 326:169–181. https://doi.org/10.1016/j.cej.2017.05.106

Yang Y, Ye S, Zhang C, Zeng G, Tan X, Song B, Chen Q (2021cb) Application of biochar for the remediation of polluted sediments. J Hazard Mater 404:124052. https://doi.org/10.1016/j.jhazmat.2020.124052

Yang Q, Wang Y, Zhong H (2021) Remediation of mercury-contaminated soils and sediments using biochar: a critical review. Biochar 23–35. https://doi.org/10.1007/s42773-021-00087-1

Yin Q, Lyu P, Wang G, Wang B, Li Y, Zhou Z, Guo Y, Li L, Deng N (2022) Phosphorus-modified biochar cross-linked Mg-Al layered double-hydroxide stabilizer reduced U and Pb uptake by Indian mustard (*Brassica juncea* L.) in uranium contaminated soil. Ecotoxicology 234:113363. https://doi.org/10.1016/j.ecoenv.2022.113363

Yu KL, Lau BF, Show PL, Ong HC, Ling TC, Chen WH, Chang JS (2017) Recent developments on algal biochar production and characterization. Bioresour Technol 24:2–11. https://doi.org/10.1016/j.biortech.2017.08.009

Yuan JH, Xu RK, Zhang H (2011) The forms of alkalis in the biochar produced from crop residues at different temperatures. Bioresour Technol 102:3488–3497. https://doi.org/10.1016/j.biortech.2010.11.018

Yuan H, Lu T, Huang H, Zhao D, Kobayashi N, Chen Y (2015) Influence of pyrolysis temperature on physical and chemical properties of biochar made from sewage sludge. JAAP 112:284–289. https://doi.org/10.1016/j.jaap.2015.01.010

Yuan Y, Bolan N, Prévoteau A, Vithanage M, Biswas JK, Ok YS, Wang H (2017) Applications of biochar in redox-mediated reactions. Bioresour Technol 246:271–281. https://doi.org/10.1016/j.biortech.2017.06.154

Zhang X, Wang H, He L, Lu K, Sarmah A, Li J, Huang H (2013) Using biochar for remediation of soils contaminated with heavy metals and organic pollutants. ESPR 20:8472–8483. https://doi.org/10.1007/s11356-013-1659-0

Zhang H, Wang T, Sui Z, Zhang Y, Norris P, Sun B, Pan WP (2019a) Plasma induced addition of active functional groups to biochar for elemental mercury removal. Plasma Chem 39:1449–1468. https://doi.org/10.1007/s11090-019-10019-4

Zhang H, Wang T, Sui Z, Zhang Y, Sun B, Pan WP (2019b) Enhanced mercury removal by transplanting sulfur-containing functional groups to biochar through plasma. Fuel 253:703–712. https://doi.org/10.1016/j.fuel.2019.05.068

Zhang J, Wu S, Xu Z, Wang M, Man YB, Christie P, Wong MH (2019c) The role of sewage sludge biochar in methylmercury formation and accumulation in rice. Chemosphere 218:527–533. https://doi.org/10.1016/j.chemosphere.2018.11.090

Zhang T, Zhang J, Wei S, Xiong Z, Xiao R, Chuai X, Zhao Y (2023a) Effect of hydrothermal pretreatment on mercury removal performance of modified biochar prepared from corn straw. Fuel 339:126958. https://doi.org/10.1016/j.jece.2023.110048

Zhao B, O'Connor D, Zhang J, Peng T, Shen Z, Tsang DC, Hou D (2018) Effect of pyrolysis temperature, heating rate, and residence time on rapeseed stem derived biochar. J Clean Prod 174:977–987. https://doi.org/10.1016/j.jclepro.2017.11.013

Chapter 11
Application and Development Strategies of Nano-Adsorbents on Mercury Remediation

Visakha Singh, Rabhya Asthana, M. Kiranmai Reddy, Hari Sharan Misra, M. Pradeep, and Saladi Sri Kala Venkata Butchi Raju

11.1 Introduction

With a growing population and the bid to improve living standards, industrialization has shown tremendous growth in this century. However, with this rapid growth came along with the blooming danger of pollution. Various treatments have been developed in the past years to deal with this threat, but they were found lacking in many cases due to their efficiency and sustainability. One of the significant pollutants posing a risk to human life and the surrounding natural environment is Mercury (Hg) (Yang et al. 2019a, b). Mercury is considered one of the most toxic heavy metals since even at lower concentrations, exposure to this heavy metal or its complexes can cause harm to humans and animals. The persistence of mercury in nature, its volatility, and its ability to bioaccumulate have added to the urgency of developing effective removal techniques. Mercury can enter the food chain through both soil and water routes. Its exposure could lead to severe health implications, such as damage to neurocognitive functions (Zheng et al. 2022; Liu et al. 2021). The toxicity of mercury has been observed in its different forms. In aquatic systems, inorganic mercury can be converted into methyl mercury (Feng et al. 2022), which can showcase bioaccumulation and biomagnification in marine life forms and ultimately enter the human

---

V. Singh · R. Asthana · M. K. Reddy (✉) · S. S. K. V. B. Raju
Department of Environmental Science, GITAM School of Science, GITAM (Deemed to Be University), Visakhapatnam, Andhra Pradesh 530045, India
e-mail: kmajji@gitam.edu

H. S. Misra
Department of Microbiology, GITAM School of Science, GITAM (Deemed to Be University), Visakhapatnam, Andhra Pradesh 530045, India

M. Pradeep
Department of Biotechnology, Government Degree College (Men), Srikakulam, Andhra Pradesh 532001, India

N. Kumar (ed.), *Mercury Toxicity*, Environmental Science and Engineering,
https://doi.org/10.1007/978-981-99-7719-2_11

food chain. There are many physical, chemical, and biological processes (Bhowmick et al. 2018) for the removal of mercury, but advanced studies have been carried out on nanotechnology.

A practical method for the remediation of mercury can be attained using nanotechnology. Nanomaterials have gained much fame in the research into bioremediation and monitoring of soil and water pollution. They have particle sizes between 1 and 100 nm. Their unique sizeable specific surface areas, optical properties, and electric properties make them ideal for environmental (Yang et al. 2019a, b) restoration work like removing and detecting heavy pollutants.

Nanoadsorbents are one of the most studied categories of nanomaterials in pollution removal. They can remove pollutants from water, soil, and wastewater, with heavy metals being (Khajeh et al. 2013) one of the contaminants that nano adsorbents are good at eliminating. According to Kumar et al. (2021), adsorption can be defined as a process of mass transfer where a physical or chemical reaction results from the attachment of molecules onto the surface of the adsorbent material. In this case, the adsorbent material is the nano adsorbent that allows for the adsorption of mercury pollutants onto its surface. Adsorbents are considered favorably in the remediation procedures due to their ability to desorb the adsorbed material and be reused after regeneration for multiple cycles. With multiple research projects being conducted into the use of biocompatible materials for the preparation of nano adsorbents, it makes them a valuable alternative to conventional technologies. Nano-adsorbent-based mercury remediation techniques are efficient but cost-effective, feasible, and safe for the environment. They have a variety of distinctive physical and chemical characteristics. One of the most significant is that most of the atoms with high chemical activity and the ability to bind to numerous metal ions are found on the surface of nanomaterials (Kalfa et al. 2009). Since the atoms on the surface are not saturated, static electricity can combine them with ions of different elements. Therefore, a variety of molecules, including polar chemical compounds and trace metals, can be firmly adsorbed by nanoparticles (Wang 2000). The type of nano-objects contained in each nanomaterial determines several important physical properties adding to their effectiveness.

Nano adsorbents commonly studied for the removal of mercury are majorly composed of carbon, zeolites, or polymers such as dendrimer, chitosan, and cellulose, or magnetic in nature. The magnetic nature of nano adsorbents is conferred due to their composition, which can enhance their efficiency of pollutant removal.

The effectiveness of a nano adsorbent is determined by various factors such as its adsorption efficiency, desorption efficiency, and regeneration capacity. An ideal nano adsorbent should be reusable for multiple adsorption and desorption cycles. Without this ability, they will add to the pollution in the environment due to being unsustainable. Various studies pertaining to the removal of mercury have been discussed in this chapter. The current trends in research of nano adsorbents for mercury remediation have been highlighted here, along with their future scope and challenges as an effective remediation technique.

11.2 Nanomaterials for the Removal of Mercury

Mercury (Hg) pollution has become a global problem due to its high levels of ecotoxicity and bioaccumulation. It is a heavy metal present in various chemical forms such as elemental, inorganic, and organic in the environment. Mercury can contaminate air, water, and soil and remain persistent in nature for long periods of time. Yearly around 8000 Mg of mercury accumulates in the atmosphere for various anthropogenic and natural reasons. It can enter the food chain and cause harm to human health due to its carcinogenic, neurotoxic, and genotoxic nature (Singh et al. 2023; Kumari et al. 2020). Hence it has become a vital concern to remove mercury from the environment safely and effectively. Many high-energy-demanding techniques have been developed to remediate mercury from nature (Liu et al. 2022). However, the use of nanomaterials in mercury remediation has gained much interest in the research community due to their effectiveness, ease of manufacture, and regenerative abilities (Emenike et al. 2023). This part of the chapter focuses on removing mercury using nanomaterials synthesized using carbon, zeolite, polymers, metal oxides, and metal, as well as nanoparticles with magnetic properties. In Table 11.1 the adsorption capacities have been listed for the different nanocomposites.

11.2.1 Carbon-Based Nanomaterials for the Removal of Mercury

Carbon-based nanomaterials have become essential in water and soil remediation. These have proven to be quite noteworthy in remediation procedures due to their

Table 11.1 Adsorption capacities of nanocomposites for mercury

Nanocomposite	Adsorption capacity	References
Graphene quantum dots coated carbon cloth	10 ppm	Farhan et al. (2023)
Graphene quantum dots	68 mg/g	Alvand and Shemirani (2017a)
ZIF-8	309.8 mg/g	Yang et al. (2019a, b)
Cysteamine-modified cellulose nano adsorbent	894 mg/g	Li et al. (2019)
Widened MoS_2 nanosheet	1206 mg/g	Fang et al. (2022)
Acetate-Sn(IV) iodophosphate nanocomposite	2500 mg/g	Sharma et al. (2021)
Chitosan crosslinked bismuth sulfide nanoparticles	65.51 mg/g	Han et al. (2022)
SiO_2-FP	0.79 mmol/g	Wang et al. (2023)
Dendrimer-modified mixed oxide nanocomposite	3232 mg/g	Arshadi et al. (2017)

high stability, good surface area and volume ratios, and potential role in catalysis. They have found applications in nanotechnology in various forms like graphene, nanotubes, and nanoparticles (Madima et al. 2020; Madhura et al. 2019).

Adsorption of mercury has been a reliable method for its removal. According to Farhan et al. (2023), while adsorption using carbon-based nanomaterials has been explored, researchers still need to safely dispose of the adsorbed mercury collected by the nano adsorbents. To solve this concern, their study developed a novel carbon cloth coated with a graphene quantum dot (GQD). It remediated Hg(II) to Hg(0). The Hg(0) was then further utilized as a sensor to detect nitrobenzene. During this experiment, the effectiveness of the modified carbon cloth in removing different concentrations of Hg(II) was satisfied through the analysis of SEM and TEM. The removed concentrations were quantified using ICP-AES, which confirmed that the graphene quantum dot-coated carbon cloth took 2 h to remove 10 ppm of the Hg(II). They also studied the removal effectiveness when silver nanoparticles were added to the quantum dots. This addition was able to improve the removal efficiency by three times. The Hg(0) based biosensor was found to be capable of a 30 pM limit of detection (LOD) for nitrobenzene.

In another study involving graphene quantum dots, Alvand and Shemirani (2017a) covalently immobilized the graphene quantum dots on the surface magnetite nanospheres coated with silica. They successfully synthesized a $Fe_3O_4@SiO_2@GQD$ nanocomposite, which displayed fluorescence with excitation at 330 nm and emission peaks at 420 nm. Hg^{2+} selectively quenched the fluorescence due to electron/hole recombination-based annihilation. This nano adsorbent was found to possess abundant Hg^{2+} binding sites and a high surface area leading to an adsorption capacity of 68 mg/g. It also showcased a 30 nM LOD for Hg^{2+} under optimized 0.1–70 μM concentration conditions. The removal of Hg ions was also found to be effective due to its superparamagnetism. This nanocomposite can be recycled using EDTA, making it an ideal mercury detection and removal agent from water.

Yaghmaeian et al. (2015) synthesized a multi-walled carbon nanotube that could remove mercury by adsorption from aqueous solutions. The adsorption was performed in batch technique and favored pH 7. Equilibrium was achieved in 120 min when >85% of the mercury was removed, the adsorbent concentration was 0.5 g/L, and the initial mercury concentration was 0.1 mg/L. The adsorption reaction was found to fit in the Freundlich and Langmuir isotherm models and followed the pseudo-second-order model.

11.2.2 Zeolite Nano Adsorbents

The reactions between mineral resources, sodium aluminate, sodium silicate, and combustion by-products can produce zeolite. In fact, they are hydrated aluminosilicates and porous in nature, mainly sourced from fly ash, making them cost-effective for water purification (Czarna et al. 2018). They were found to be excellent adsorbents of mercury due to their ion exchange mechanism for adsorption.

In a study by Uttam et al. (2022), dansyl appended calix$_4$ arene (DanC_4A) was embedded into a zeolite imidazolate framework (ZIF 8) in order to create a hybrid nanomaterial. This nanomaterial was found to be capable of removing 95% of Hg^{2+} from water. It was also found to be efficient in sensing Hg^{2+} from water through its emission of green fluorescence in HeLa cells, which was confirmed by using fluorescence microscopy and confocal microscopy. It showed 80% quenching of fluorescence titration in the presence of mercury ions. Researchers have extensively studied ZIF 8 due to its efficiency in mercury removal.

Zeolites were found to be applied in the removal of mercury from not only water but also from flue gas. In a study by Yang et al. (2019a, b), it was found that a zeolite-based imidazole framework 8 (ZIF 8) showed excellent adsorption of Hg(0) with the capacity reaching 309.8 mg/g at an average rate of 105.3 μg/g minute at 50 °C. This rate was showcased by the nanocomposite containing copper selenide supported by ZIF 8 prepared by using the two-step surfactant-assisted procedure. The higher adsorption capacity of the nanocomposite can be attributed to the Se-terminated sites, which exhibit an affinity towards Hg(0). Also, to be noted is the importance of the environmental impact of this nanocomposite. Due to the stability of mercury selenide formed in the adsorbent, its impact on the environment is minimal when compared to the natural form of mercury.

In a study by Shirzadi and Nezamzadeh-Ejhieh (2017), a zeolite nanocomposite was synthesized using a natural clinoptilolite. It was modified using hexadecyltrimethylammonium bromide (HDTMABr). The modified nanocomposite showed significant mercury removal due to its complexation and ion exchange procedure. The adsorption reaction occurred within the Langmuir equation, indicating monolayer adsorption. The reaction also fell under the pseudo-second-order model and was found to be spontaneous and exothermic in nature.

11.2.3 Polymer-Based Nano Adsorbents

A polymer-based nanocomposite can be defined as a composite material that contains both organic and inorganic parts. The inorganic part involves nanoparticles that are dispersed in a polymer matrix of organic nature (Khan et al. 2023). Its properties will depend on the nanofiller type and the polymer being used, making it versatile for numerous applications (Amin et al. 2022). As in the case of mercury removal, polymers such as chitosan, cellulose, and dendrimers have been extensively studied due to their biocompatibility, cost-effectiveness, performance, and strength (Zhao et al. 2018).

11.2.3.1 Cellulose-Based Nano Adsorbents

Cellulose is a biological polymer with a linear structure. It is the most abundant polymer of biological origin. Cellulose is a major component of plant cell walls and contains glucose units connected by acetal links. This biopolymer has gained favor as a nanomaterial (Eichhorn et al. 2018) due to its abundance, renewability, better aspect ratio, mechanical strength, and biological compatibility. According to Trache et al. (2020), it also acts as an ideal nanocomposite due to the possibility of various functionalization on it owing to the presence of hydroxyl groups.

In a study by Li et al. (2019), a novel cellulose nanocrystals-based nanocomposite was prepared by modifying it using cysteamine. This synthesis was carried out using mild periodate oxidation on the nanocrystals, following which it was grafted with cysteamine. The nanocomposite was found to have a better adsorbing capacity of mercury ions from aqueous solutions due to the addition of cysteamine. The maximum adsorption was found to be 894 mg/g. It was capable of removing 99% of the mercury from a solution with a concentration of 50 mg/L within 10 min. It also showed good performance after four consecutive cycles. The reaction fit under pseudo-second-order and occurred as a monolayer type of adsorption, falling under the Langmuir model when tested.

Mercury pollution must be effectively removed from the water to improve the water quality available to the public. A widened molybdenum disulfide nanosheet was immobilized on cellulose fibers in order to create a nanocomposite capable of removing 1206 mg/g of mercury ions from water. This nanocomposite showcased high removal efficiency due to the fully exposed binding sites constituting sulfur atoms present in its sizeable interlayer spacing of 0.94 nm (Fang et al. 2022). The Sol–gel technique was used by Sharma et al. (2021) to prepare a new nanocomposite constituting tin iodophosphate and cellulose acetate. Cellulose acetate at 2 w/w% was able to remove 91% of Hg^{2+} from water. The maximum adsorption capacity was found to be 2500 mg/g for the mercury ions.

11.2.3.2 Chitosan-Based Nano Adsorbents

Chitosan is a natural amino saccharide extensively studied for removing heavy metals (Kumar et al. 2020). Chitosan can be adjusted to fit various applications due to its functional groups. Since it is a natural polymer, it has found various applications in nanotechnology due to its efficiency and biocompatible properties (Latha et al. 2023).

In a study by Kumar et al. (2020), a novel nanocomposite was prepared using chitosan nanoparticles embedded in Brassica gongylodes leaf extracts. It was found to be effective in removing 76.67% of mercury at equilibrium in 45 min when the pH was at 10, and the adsorbent concentration was 50 mg. This adsorption occurred as a feasible, spontaneous reaction and followed the pseudo-second-order kinetic model. The removal of mercury from an aqueous solution by using the chitosan-embedded nanocomposite has been described in Fig. 11.1. The figure shows the synthesis of the

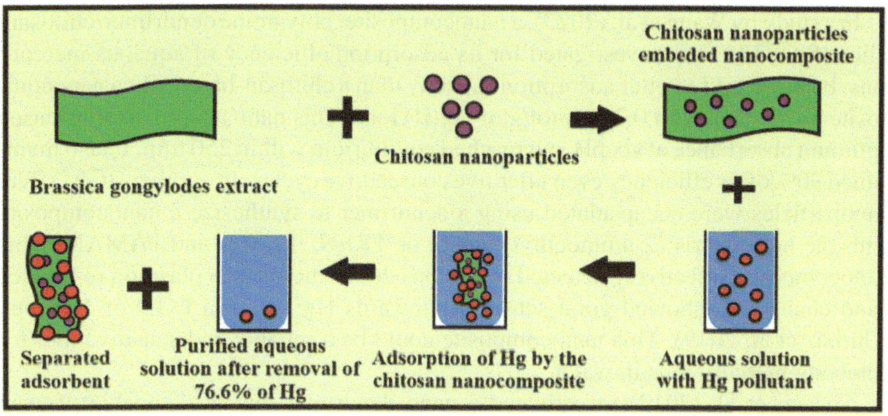

Fig. 11.1 Adsorption of Mercury using a chitosan nanoparticle-embedded nanocomposite

nanocomposite consisting of chitosan nanoparticles and plant extract. The prepared nanocomposite, when added to an aqueous solution containing mercury, adsorbs the pollutant onto its surface and leaves behind the solution with significantly less contamination.

In another study by Han et al. (2022), a nanocomposite was prepared using bismuth sulfide nanoparticles crosslinked with chitosan. The cross-linking was found to improve the adsorption of mercury from water. Its maximum adsorption capacity was found to be at 65.51 mg/g in 45 min when the pH was 5, and the temperature at 30 °C. The adsorption was found to occur through the complexation mechanism, where ions get complexed with the material.

Mahmoud et al. (2022) investigated using SnO_2-formaldehyde—chitosan crosslinking nanocomposite to remove mercury ions with the assistance of microwaves. It was shown to remove 94.2% of mercury from tap water and 98.4% of mercury from wastewater. The maximum adsorption capacity of this nanocomposite was found to be 1050 μmol/g at pH 6. In another study by Eskandari et al. (2020), a silver manganese disulfide/chitosan-polyvinyl alcohol nanocomposite was used to detect the presence of mercury. The LOD was found to be at 9 nM at pH 5.

11.2.3.3 Dendrimer-Based Nano Adsorbents

A dendrimer is a 3-D macromolecule belonging to the family of synthetic polymers. It has tree-like branches and a well-defined structure with a central core (Goel et al. 2022; Umoren et al. 2022). While they can be of various types, such as poly (amidoamine) dendrimers (PAMAM), chiral, or hybrid, they all constitute common structures such as an exterior with functionalities, interior repeating units, and an internal core (Patel et al. 2021) which forms like a matrix to absorb pollutants.

In a study by Wang et al. (2023), a nanocomposite, polyamine dendrimer/chitosan/silica (SiO_2-FP), was investigated for its adsorption efficiency of aqueous mercury ions. It showcased a better adsorption capacity than a chitosan-based silica nanocomposite by 46.30% with 0.79 mmol/g of Hg(II) ions. This nano adsorbent showcased optimum absorbance at six pH and reached equilibrium within 250 min. It also maintained 90% of its efficiency even after five consecutive cycles. In another study, silver nanoparticles were encapsulated using a dendrimer to synthesize a nanocomposite with the help of tris (2-aminoethyl) amine or TREN, $T_4.NH_2$ and PAMAM. This nanocomposite effectively detected metal ions due to the surface plasmon resonance band changes. It showed good sensitivity towards Hg^{2+} with a LOD of 1.18 ppb (Gürbüz et al. 2019). This nanocomposite could be a reliable and sensitive tool for detecting mercury ions in water.

Arshadi et al. (2017) investigated a nano dendrimer that was capable of even removing mercury after five consecutive cycles from the water. This nano adsorbent was prepared by grafting triazine, diethylenetriamine, and L cysteine methyl ester on aluminum silicate mixed oxide. The nano dendrimer showcased an excellent adsorption capacity at 3232 mg/g when equilibrium was achieved at 6 min. It was found to show high reusability even for 15 consecutive cycles. This adsorption occurred as a pseudo-second-order reaction while showcasing exothermic and spontaneous nature. The adsorption mechanism showed high results due to the underlying mechanisms involving ionic exchange, metal chelation, and electrostatic interactions resulting in the removal of mercury ions. Table 11.3 lists the mercury removal efficiencies of the nanocomposites mentioned in this chapter.

11.2.4 Magnetic Nanoadsorbents

Magnetic nano-adsorbents are made of a base adsorbent that has magnetic particles incorporated in it. These particles, oxides of metals like Fe, Co, Ni, and Cu, are easily separable (Mehta et al. 2015) from water when subjected to a magnetic field. According to Maksoud et al. (2020), magnetic nano-adsorbents are highly effective functional materials with remarkable micropollutant sequestration properties at the laboratory scale. With their excellent operational efficiency and low cost, magnetic nano-adsorbents (Liu et al. 2019) have proved to be effective materials for water/wastewater treatment applications. In general, magnetic nano adsorbents are characterized by high specific surface areas (Liu et al. 2020), high pore volumes (Gupta et al. 2018), a sturdy architecture, and extensively interconnected porous networks that jointly support ultrahigh adsorption capabilities for micropollutants (Lingamdinne et al. 2019; Tan et al. 2020). The ability to be separated in situ from adsorption-remediated waters in the form of a magnetic nano-adsorbent(s)-adsorbate(s) sludge by applying a strong enough magnetic field is one of the critical characteristics of magnetic nano-adsorbents, in addition to their redox activity and surface charge properties, low cost of synthesis, non-toxicity, high selectivity, binding specificity, and excellent reusability (Balbino et al. 2020). Recent examples of magnetic nano-adsorbents include polymeric nano adsorbents for removing heavy metals and organic pollutants (Priya et al. 2020) and carbonaceous/spinel ferrite nanocomposite for removing pharmaceutical contaminants from wastewater (Al-Hetlani et al. 2022). When compared to traditional methods, magnetic separation offers certain appealing features. These benefits can be broadly categorized as (i) the ability to carry out an integrated one-step capture and purification of particular species, (ii) the processing of high throughputs, and (iii) the low energy needs, and associated costs entailed by semi-continuous or continuous processes run at relatively low pressure (Schwaminger et al. 2019). According to Mudhoo and Sillanpää (2021), at the tertiary effluent treatment level, where effluent from the upstream secondary treatment units is polished through selective adsorptive sequestration of the target micropollutant(s), the use of magnetic nano adsorbents and the integration of magnetic separation for water purification and wastewater treatment can be envisaged. Besides water purification, magnetic nano adsorbent is also used for environmental remediation and biomedical applications. Magnetic nano adsorbents have proven to be highly effective in biomedical applications. They offer various uses, including drug delivery, hyperthermia treatment, and magnetic resonance imaging (MRI). These applications have been made possible through the unique magnetic properties of the nano adsorbents, which make them ideal for use in a range of biomedical procedures. In addition to their effectiveness, magnetic nano adsorbents have also been found to be safe and non-toxic, making them an attractive option for medical professionals. The potential uses of magnetic nano adsorbents in the biomedical field are game-changing (Schwaminger et al. 2019). Using magnetic nano adsorbents in environmental remediation is a promising yet emerging field that requires further scientific exploration. While extensive research has been conducted, in-depth analysis and investigation are still needed to optimize the performance and efficacy of these materials. Despite their vast potential, magnetic nano adsorbents are still in the early stages of development and require additional experimentation. With more detailed research, we can unlock the produced incapabilities of these materials and utilize them to their maximum potential in environmental remediation (Khan et al. 2020). They can be produced in a variety of forms and utilized in a range of applications. Magnetic nanoparticles, magnetic mesostructured silica, and hydrophilic-lipophilic balanced magnetic nanoparticles are a few nanomagnetic adsorbents (Peralta et al. 2021). In Fig. 11.3, the general structure of a magnetic nanocomposite has been described. Its two main parts are the magnetic core and the surface functional groups. The surface is where the active adsorption sites are located facilitating the adsorption of mercury from contaminated surroundings. In Table 11.2 the adsorption capacities of some of the magnetic nanocomposites have been listed.

Table 11.2 Adsorption capacities of magnetic nanocomposites for mercury

Nanocomposite	Adsorption capacity (mg/g)	References
Fe_3O_4-Ag	71.3	Inglezakis et al. (2020)
Fe_2O_4-SiO_2-Co	504	Zhang et al. (2020)
Fe_3O_4 based nanocomposite	109.5	Behjati et al. (2018)
Fe_3O_4-Chitosan	152	Kyzas and Deliyanni (2013)
Cellulose magnetic nanosheet	469.48	Gao et al. (2021)

Table 11.3 Removal efficiencies of biobased nanocomposites for mercury

Nanocomposite	Removal efficiency (%)	References
Yam peel nanocomposite	75	Marimón-Bolívar et al. (2018)
Fe_3O_4-activated carbon	99.44	Gheitasi et al. (2022)
Carbon dot-based nanocomposite	108.8	Xie et al. (2021)
Chitosan-magnetite	99.91	Rahbar et al. (2014)
Carbon nanotube	85	Yaghmaeian et al. (2015)
ZIF 8	95	Uttam et al. (2022)
Cellulose nanocomposite	99	Li et al. (2019)
Nano cellulose—iodophosphate nanocomposite	91	Sharma et al. (2021)
Chitosan-based nanocomposite	76.67	Kumar et al. (2020)
Chitosan-SnO_2-formaldehyde nanocomposite	94.2	Mahmoud et al. (2022)

11.2.4.1 Nanocomposite Magnetic Nano Adsorbents

A particular class of materials called nanocomposite magnetic nano adsorbents combines magnetic nanoparticles with additional elements, greatly enhancing their adsorption capacity. This technology is groundbreaking in its ability to efficiently remove various pollutants from various substances. Magnetic nanoparticles are highly successful at adsorbing and separating molecules that are too small to be filtered by other methods due to their unique magnetic properties. According to Kharissova et al. (2015), these particles can be effectively concentrated and recovered from various liquids and gases by combining them with other substances, including carbon nanotubes, clay, and activated carbon, further improving their adsorption efficiency. Magnetic nanocomposite adsorbents are being widely used for various purposes. They have easy separation, good stability, and large adsorptive areas. Their most common use is in removing toxic substances from wastewater due to their effectiveness and low cost. They can also be recycled multiple times. According to Liu et al. (2020), different adsorbents have been synthesized, and used to treat various pollutants in solutions, including inorganic, organic, and biological. As industrialization and human demands continue to grow, the production of global goods containing harmful metal particles has also increased. As a result, this has led to an increase in the amount of toxic metal particularly mercury wastewater discharged by industrial activities (Reddy and Yun 2016). Efforts have been made to create functionalized sorbents to remove metal toxins from complex matrices. This has led to the development of magnetic nanocomposite adsorbents, some mentioned in this section, which help remove and recover metal ions from wastewater.

According to a study done by Marimón-Bolívar et al. (2018), a magnetic nanocomposite synthesized by using aminopropyl silane, peptone, and yam peel biomass was found to be effective in removing 75% of mercury at pH 8, with saturation magnetization of 26.6 emu/g. In another study by Inglezakis et al. (2020), mercury ions could be removed from water with a magnetic Fe_3O_4–Ag^0 nanocomposite. When compared with Fe_3O_4 particles with a removal rate for mercury ions at 28 mg/g, the nanocomposite showed a removal rate of 71.3 mg/g. The removal occurred through mechanisms such as adsorption, reduction, and oxidation between the constituents of the nanocomposite. A magnetic nanocomposite Fe_2O_4, Co, and SiO_2, which was improved using amino and thiol groups, was utilized to remove mercury ions. The nanocomposite's maximum adsorption capacity (Zhang et al. 2020) was 504.34 mg/g of Hg (II) at pH 7.2. Behjati et al. (2018) synthesized an effective nanocomposite by modifying Fe_3O_4 nanoparticles with carbon disulfide and tetraethylenepentamine along with dithiocarbamate groups on the surface of the nanocomposite. It had a saturation magnetization of 27 emu/g. Maximum adsorption was observed at 109.5 mg/g when the pH was between 2 and 6.5. This nanocomposite was found to be successful in removing mercury pollutants from groundwater as well.

11.2.4.2 Carbon Materials Magnetic Nanoparticles

There are several uses for organic nanoparticles formed of carbon magnetic nanoparticles in various industries, including engineering, environmental science, and healthcare. Because of their high surface areas and biocompatibility, these nanoparticles can be used for water filtration, biomedical imaging, and drug delivery systems. Additionally, their magnetic and carbon qualities make them perfect for magnetic resonance imaging, magnetic field sensors, nanoparticles (Malhotra et al. 2020), and hyperthermia cancer therapy. Future improvements in numerous fields of study and applications may result from the creation of organic nanoparticles formed by carbon magnetics. According to Pyrzyńska and Bystrzejewski (2010), a type of cutting-edge material known as carbon-encapsulated magnetic nanomaterials consists of metallic cores encased in carbon layers that typically have a thickness of 1 to 5 nm (nm). This thickness and encasing have helped to adsorb mercury onto the surface. A graphene oxide-based nanocomposite was synthesized using silver nanoparticles and Fe_3O_4 with magnetic properties. The silver nanoparticles have the capacity to remove mercury in low temperatures, and after incorporation into the magnetic nano adsorbent, they were able to remove mercury in a gaseous state with an efficiency of 92%. This adsorption resulted from the mechanism of mercury uptake involving an amalgam formation of silver and mercury on the surface of graphene oxide. The nano adsorbent was also reusable after using thermal based desorption method (Ma et al. 2019). Due to the synergistic interaction between the magnetic behavior of the metallic core and the carbon shell's extraordinary stability and functional qualities, these nanomaterials display distinctive features. The metallic cores provide magnetic functionality and can be made of different magnetic materials like iron, cobalt, or nickel. The carbon coatings act as protective shells and contribute to the overall stability and performance of the nanomaterials. The carbon coatings have a crystalline formation of nano for completely separating the nanoparticles within. Interestingly, carbon-encapsulated magnetic nanomaterials-based sorbents are highly mobile, which makes them simple to relocate or separate using external permanent magnets. This feature makes them highly desirable for various scientific applications, including biomedicine, environmental sciences, and catalysis. Researchers can develop innovative solutions to address (Bystrzejewski et al. 2009) significant scientific issues by leveraging their mobility and magnetic properties. As Pyrzyńska and Bystrzejewski (2010) concluded, that a high sorption efficiency ranging from 30 to 60% was displayed by a sorbent based on CEMNP material for mercury removal. The sorption occurred under acidic conditions with pH levels between 5 and 6. As a result, these materials are more promising for the pre-concentration of heavy metals apart from mercury and this method was found viable to remove heavy metals from wastewater under acidic conditions. It was found in an experiment by Bystrzejewski et al. (2009) that the CEMNPS adsorbent worked efficiently when pH was 6. Magnetic carbon nanomaterials can be applied to various processes, such as adsorption, catalysis, super capacitance, immobilization of enzymes, CO_2 capture, and drug delivery. Activated carbon derived from the shells of hazelnut was used during research conducted by Zabihi et al. (2022) to synthesize nanomagnetic activated carbon. Doing X-ray diffraction analysis revealed that the carbon was activated by using zinc chloride and iron oxide crystallites to work efficiently as an adsorbent. The nanocomposite was able to remove 80 mg/g of mercury from a solution.

These materials are adaptable and beneficial in various applications because of their magnetic and carbon properties, which can interact with the desired substances. Thus, it is possible to view magnetic carbon nanomaterials as a viable answer in various scientific and technical fields (Siddiqui et al. 2019). Researchers have looked at several kinds of magnetic nanoparticles and how they affect them to maximize the catalyst's performance. As a result, numerous synthetic techniques have been created to regulate nanoparticle surface characteristics, size, and form Martínez-Prieto et al. (2020). Fe_3O_4 and thiol were used to modify a nanocomposite involving the use of beer barley hust derived activated carbon. This nanocomposite, synthesized by Gheitasi et al. (2022), could effectively remove Hg(II) ions with 99.44% adsorption capacity when the pH was at 6, and the time was 60 min. In another study, a carbon dot-based magnetic nanocomposite was used to remove and detect mercury ions from water samples. This nanocomposite exhibited fluorescence properties which helped in the detection of mercury. It also showed the LOD or lowest detectable limit for mercury ions at 0.3 nm, (Xie et al. 2021) and the maximum recovery of mercury was 98.8%.

11.2.4.3 Polymer Functionalized Magnetic Nanoparticles

Various bio-based or synthetic polymers can be added to nanocomposites to increase adsorption efficiency. In research work done by Rahbar et al. (2014), it was seen that chitosan-coated magnetite nanoparticles could be synthesized to remove mercury ions from water. This nanocomposite was biocompatible and cost-effective in nature. Using the Box-Behnken model, all the parameters were optimized for the reaction. The maximum Hg(II) removal of 99.91% was observed at optimized pH of 5, with the initial mercury ion concentration being 6.2 mg/L. In another study, a magnetic chitosan-based Fe_3O_4 magnetic nanocomposite was cross-linked with glutaraldehyde and compared with a non-magnetic chitosan-glutaraldehyde cross-linked adsorbent. At optimum pH 5, the magnetic nanocomposite had a maximum adsorption capacity of 152 mg/g, whereas the non-magnetic nanocomposite was at 145 mg/g (Kyzas and Deliyanni 2013) which clearly indicates that the nanocomposite has a better side of removal. In a study by Gao et al. (2021), microcrystalline cellulose-based magnetic nanocomposite was synthesized in the form of nanosheets. It showed a maximum adsorption capacity of 469.48 mg/g. The adsorption could occur due to the interactions between mercury ions and the S atoms.

11.2.5 Metal Oxide and Metal-Based Nanomaterials

In order to combat heavy metal pollution, various new nano adsorbents were developed by using metals and metal oxides. They possess highly accessible and active surface area along with modifiable surface chemistry, are easy to synthesize, recyclable (Gupta et al. 2018), and cost-effective in nature. In a study by Vélez et al. (2016), Fe_3O_4 nanoparticles were synthesized by using aloe vera to stabilize them. It was able to remove 70% of the mercury. Similarly, in another study by Liu et al. (2018), graphene oxide adsorbent was modified using three different metal oxide nanoparticles, CuO, MnO_2, and ZnO. Their capacity for adsorbing mercury was measured by using a cold vapor atomic fluorescence spectrophotometer. The adsorbent composed of MnO_2 nanoparticles showed a chemisorption-based adsorption capacity of 85% at 200 °C, whereas the adsorption by the remaining two types of adsorbents did not show any significant removal of mercury. In Fig. 11.2, the types of iron oxides are described which are a well-researched constituent of nanocomposites. The figure also gives examples of the three types: oxides, hydroxides, and oxy-hydroxides.

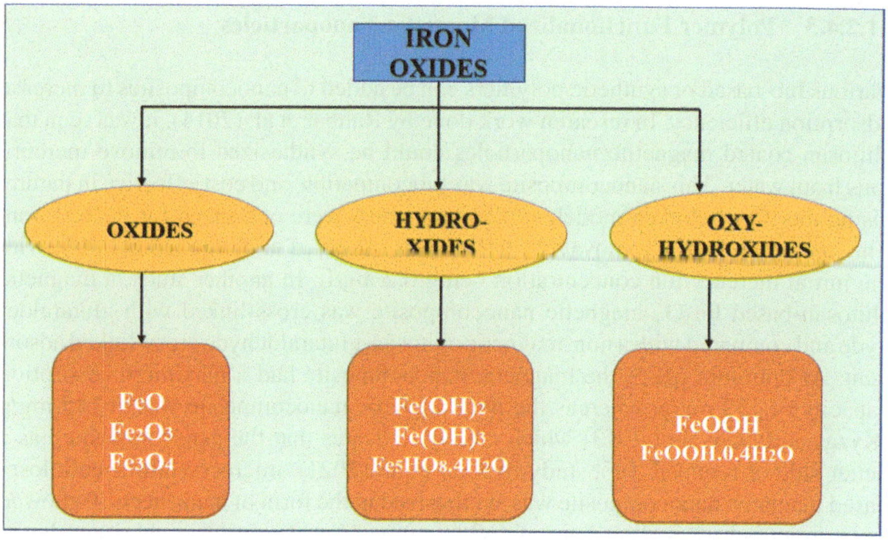

Fig. 11.2 Types of iron oxides

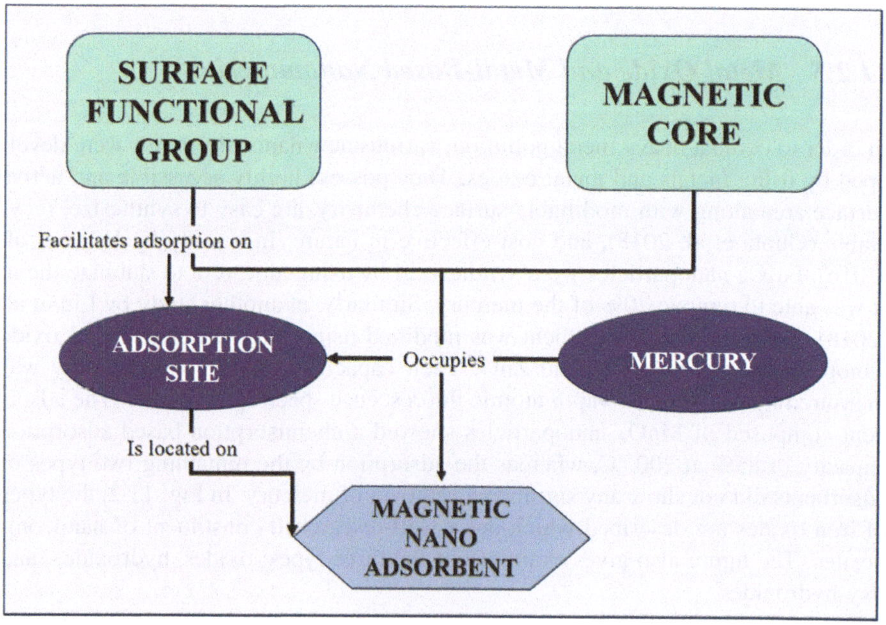

Fig. 11.3 Magnetic nano adsorbent

11.3 Surface Modification for Adsorption Capacity

To advance the nanoparticles' (NPs) dispersibility in a polymer fusion system and augment the binding power of NPs to other components, surface alterations of the NPs are required. NP surface modification involves the modification and coursing of the NPs surface by chemical deep processing and physical methods, thereby regulating the internal stress, improving the revolting force between the NPs, mitigating the gravitation amid the particles, and decisively altering the surface of the NP Physical and chemical features, which give the NP a new function and convene the needs of NP processing and application.

11.3.1 Silica-Based Nano Adsorbents

Silica-based nanoadsorbents are nanoscale materials composed of silica nanoparticles functionalized with specific groups or ligands for enhanced mercury adsorption. These materials offer several advantages for mercury remediation. Nanoparticles synthesized using silica are finding a vast number of uses and scope in heavy metals contaminated wastewater management (Baby et al. 2022). Significant studies have been conducted for remediating heavy metals from contaminated water by using these NPs and their offshoots as adsorbents. They were found to possess excellent properties such as specific pore size, increased surface area, excellent selectivity along with high adsorption ability. Silica has found application in its tailored formation, through functionalized or complex form (Yang et al. 2019a, b). They possess a large surface area, providing ample active sites for mercury adsorption. The high surface area-to-volume ratio allows for efficient mercury capture, leading to improved remediation outcomes. The surface of silica-based nanoadsorbents can be modified to enhance the selectivity and affinity for mercury ions. Functional groups such as thiol (–SH), amino ($–NH_2$), or carboxyl (–COOH) can be introduced to increase the binding capacity and specificity towards mercury. Silica-based nanoadsorbents can be synthesized through relatively simple and scalable processes, making them suitable for large-scale mercury remediation projects. The controllable synthesis methods allow for tuning their physical and chemical properties to optimize their performance. Several studies have demonstrated the effectiveness of silica-based nanoadsorbents for mercury remediation in various environmental matrices.

Wang (2021) developed a thiol-functionalized silica-based nanoadsorbent for efficient mercury removal from wastewater. The nanomaterial exhibited a high mercury adsorption capacity and excellent selectivity, resulting in a significant reduction in mercury concentration. Similarly, Zhu (2022) synthesized a silica-based nano adsorbent functionalized with amino groups for the removal of mercury from contaminated soil. The material effectively immobilized mercury, reducing its bioavailability and minimizing the potential risks to the ecosystem. Li (2023) reported the development of a novel silica-based nanoadsorbent decorated with magnetic nanoparticles for mercury capture in aqueous systems. The material showed rapid and efficient removal of mercury ions and could be easily separated from the solution using a magnetic field.

A silica-based nano adsorbent was prepared using *Saccharomyces cerevisiae* to create a biohybrid composite. It rapidly removed 98% of mercury ions within 30 min of application. The maximum adsorption capacity was found to be 185.19 mg/g at pH 6.5 (Shukla et al. 2020). The adsorption reaction occurred as endothermic and spontaneous in nature. Rabie et al. (2019) studied the adsorption of ethylene diamine and cubic mesoporous-modified silica nanomaterials in another study. The adsorbents composed of ethylene diamine showed an adsorption capacity of 95% of mercury removal compared with 82.5% of the ones with mesoporous.

These studies highlight the versatility and potential applications of silica-based nanoadsorbents in different environmental contexts. Their successful implementation in real-world scenarios offers a promising solution for mercury-contaminated sites. Mercury contamination is a pressing environmental issue, and the development of effective remediation technologies is crucial. Silica-based nanoadsorbents present a promising avenue for mercury removal, offering high surface area, tailored surface chemistry, and scalability. Recent advancements in this field demonstrate the potential of these materials for efficient mercury capture and remediation. Continued research and development efforts in silica-based nanoadsorbents hold promise for mitigating the adverse effects of mercury contamination and ensuring a safer environment.

11.3.2 Iron-Based Nano Adsorbents

Iron-based nanoadsorbents are nanoscale materials composed of iron nanoparticles or iron oxide nanoparticles that possess unique properties for mercury capture and removal. These materials offer several advantages for mercury remediation. Iron-based nanoadsorbents exhibit a strong affinity for mercury ions due to their high surface reactivity. Iron or iron oxide nanoparticles can effectively bind with mercury, facilitating its removal from contaminated environments. Iron-based nanoadsorbents possess redox-active properties, allowing them to participate in oxidation–reduction reactions. This capability is advantageous for the removal of various forms of mercury, including both divalent and elemental mercury species. Iron-based nanoadsorbents demonstrate good stability and can be easily regenerated for repeated use. After mercury capture, they can be treated or regenerated to release the captured mercury ions, making them suitable for multiple remediation cycles. Iron oxide-based nanoparticles consist of maghemite or magnetite particles with diameters that could range from 1 to 100 nm. These nanoparticles have been thoroughly studied due to their superparamagnetic properties and potential biomedical applications that resulted in biocompatibility, low cost, catalysis, high surface area, high degree of functionalization, and no toxic effects (Pankhurst et al. 2003). Recently many advances were made in the synthesis of iron oxide nanoparticles by using the thermal decomposition of iron carboxylate salts leading to significant improvements in the quality of the nanoparticles in terms of crystalline structure, size tunability, and monodispersity. Hydrophobic, organic, and ligand-coated iron oxide nanoparticles were converted into water-soluble and bio-accessible nanoparticles by applying a monolayer polymer coating strategy. Their high stability in harsh conditions such as high pH and temperature allowed for their conjugation with biomolecules. Other biocompatible coatings using polysaccharides like dextran and lipid molecules were developed, resulting in the procurement of safe and approved nanoparticles. Improvements in the quality of organic and soluble iron oxide nanoparticles opened up many opportunities for nanoparticle-based applications.

Iron nanoparticles possess many unique properties when compared with the bulk form. They have a significant amount of applications due to the multitude of research being done into their properties (Lei et al. 2018). Iron oxides consist of oxides, hydroxides, and oxy-hydroxides which could be prepared using cations such as Fe(II) and Fe(III) and anions like O^{2-} and OH–. The different phases of oxides are $Fe(OH)_3$, $Fe(OH)_2$, $Fe_5HO_8{\cdot}4H_2O$, Fe_3O_4, FeO, 5 polymorphs of FeOOH, and 4 of Fe_2O_3. The most known iron oxide forms involve magnetite and hematite. Several types of products could be derived using iron oxide nanoparticles such as goethite, lepidocrocite, magnetite, maghemite, and bluish-green Fe(II)–Fe(III) hydroxyl salts. Their earlier applications involved the removal of metals and dyes from wastewater and surface water through adsorption. Iron oxide-based nano-adsorbents including forms such as magnetite, hematite, or maghemite were significantly studied by researchers for their ability to remove pollutants such as heavy metals ions of mercury, lead, arsenic, chromium from differing industrial effluents (Assefa and Mazengiaw 2021).

The surface properties of nanoparticles can be controlled, allowing for different applications (Fatima and Kim 2018). Iron nanoparticles were found to be very suitable for treating waste water due to high surface-volume ratio, adsorption capacity, sensitivity, and functionalization. The higher surface area led to an increased catalytic activity affecting how efficient the nanoparticles could be in catalyzing the reactions. They could find applications as a catalysis-based adsorbent in wastewater treatment and have found remarkable applications in pollution remediation along with other environmental applications.

11.3.3 Alumina-Based Nano Adsorbents

Alumina-based nano adsorbents are nanosphere particles synthesized from alumina. Methods such as sol–gel, pyrolysis, laser ablation, sputtering, and sol–gel could be used to synthesize these nanoparticles (Yatsui et al. 2000). Laser ablation is usually considered as more convenient as it can be used in different environments like liquid, gas, and even vacuum. Since collecting nanoparticles is easier in liquid than in a gaseous environment, it adds to the advantages of this technique. Aluminum nanoparticles can be found as individual spheres (Kumar and Thareja 2010) or as oriented fibers. Alumina nanoparticles have properties differing from the bulk alumina. The nanoscale fibers and particles have a smaller small diameter of 2–10 nm. Their specific surface area is usually more than 100 m^2/g. The ratio of length and diameter of the nanofibers is 20,000,000:1. The nanofibers have a highly oriented structure but weak interactions among themselves. They have a high surface concentration of the hydroxyl groups and no surface pores. The size of the nanoparticles helps in determining their physio-chemical characteristics and any changes in the size alter the atomic arrangement. They were reported to possess three distinct layers with the surface layer being outermost and the first. It is composed of metal ions which contribute to the functioning of the surface. The shell layer comes as the second layer which is quite distinct from the other layers. The last and third layer is the core which represents the basic chemical formula of the nanoparticles.

In a study by Mahapatra et al. (2013), alumina and iron oxide mixed nanofiber was prepared through electrospinning. This nano adsorbent showed an adsorbing capacity of 63.69 mg/g for mercury ions from an aqueous solution. In another study, a novel gold nanoparticle-based aluminum oxide nano adsorbent was synthesized for the removal of mercury in different forms from water. It showed effective removal efficiency of 97% for water sources from ground and surface areas (Lo et al. 2012).

11.4 Desorption and Recyclability of Nano Adsorbents

In order to be considered an effective nano adsorbent, it not only needs to show efficient adsorption capacity but also an ability to be reused for multiple cycles. Without this characteristic, the adsorbents would be neither sustainable nor cost-effective. It is essential that nano adsorbents should be recyclable after multiple adsorption and desorption cycles while maintaining a high adsorption efficiency.

Recent work used desorption and regeneration in seven recycling rounds for STNTs-Ch beads. Surprisingly, the beads' ability to adsorb substances during these cycles remained unchanged. This highlights the adsorbent's durability and reuse capacity while underscoring its economic potential. To recycle adsorbents, the desorption procedure is essential. According to the study above, the loaded adsorbent could be successfully regenerated without losing its adsorption effectiveness. The desorption procedure successfully eliminated the adsorbed pollutants by treating the adsorbent with diluted sodium hydroxide, enabling the adsorbent to be reused for additional adsorption cycles. In many applications, the capacity to reuse and recycle adsorbents by desorption is a key benefit. It not only lowers the overall cost of buying new adsorbents, but it also lessens the environmental impact by producing less trash. In a magnetic nanocomposite containing graphene quantum dots, the superparamagnetism of the material led to easy separation of the mercury ions with the help of a magnet. This nanocomposite could also be desorbed by using EDTA, allowing for multiple cycles of regeneration (Alvand and Shemirani 2017b). In another case by Zhang et al. (2015), involving a hybrid nanocomposite composed of Au nanoparticles, Fe_2O_3, and graphene oxide, more than 99% of the adsorbed mercury was removed from the surface by using a magnetic field. This nanocomposite was found to be efficient up to 96% even after 15 cycles of regeneration. In one study by Zhang et al. (2020) involving a magnetic nano adsorbent, the maximum adsorption capacity for mercury was found to be 517.4 mg/g and this nano adsorbent retained the adsorption capacity of more than 300 mg/g even after five consecutive cycles.

Based on ceria nanostructures on mesoporous silica beads, another intriguing example of an adsorbent with excellent sorption/desorption performance is shown. This adsorbent demonstrated a stunning 99% removal efficiency in just 60 min and an exceptional adsorption capacity of up to 110 mg/l of mercury. Notably, the adsorbent retained its high performance after numerous adsorption and desorption cycles, proving its appropriateness for long-term and sustainable uses.

In conclusion, the sustainability and economic viability of the adsorption processes depend on the ability to recycle and reuse adsorbents. Adsorbents such STNTs-Ch beads and ceria nanostructures on mesoporous silica beads have been proven in studies to exhibit great reusability and to maintain high adsorption capacity even after numerous cycles of recycling. This emphasizes how crucial the desorption and regeneration processes are in enabling the efficient reuse of adsorbents, which helps to save money and protect the environment.

11.5 Challenges and Future Perspective

Various synthetic and natural compounds have found numerous applications in nanotechnology and remediation technology. However, many modifications are still necessary to improve nano adsorbents as an ideal remediation tool. While the effectiveness of a nanocomposite is always a significant concern, the reusability of the nanocomposite needs to be considered. Nanomaterials could have unintended effects on health and the environment. Hence, a strict protocol must be maintained for the treatment of nanotechnology-based wastes that might enter the environment. There is also the need to evaluate and monitor the nanocomposites due to their combinations, which may have toxic effects. The future of nanotechnology as an efficient remediation tool depends on the synthesis of cost-effective products that could have large-scale applications. They need to be sustainable and practical in large-scale manufacturing in order to meet the demands of the ever-increasing pollution of natural resources (Tahir et al. 2021; Pokrajac et al. 2021; Köhler 2021).

11.6 Conclusion

Nanotechnology is a solution to many of the current pollution-related problems being faced by society today. Nanoadsorbents, in their various forms and types, present a more sustainable approach to remediation techniques. While there are still many challenges required to be overcome by the researchers, there is no doubt about the applications that nano adsorbents have when it comes to the removal of heavy metals such as mercury from the natural environment.

Mercury is a highly harmful pollutant toward human beings and the whole ecosystem. There is a need for extensive research into synthesizing novel nano adsorbents that can effectively remove mercury and its different forms without causing further contamination or damage to the environment. The current research into nanocomposites has advanced from simple formulations to complex compounds, as evidenced by this chapter. There is now a global demand for effective solutions for the bioremediation of mercury by efficient, sustainable, cost-effective, and practical alternatives to conventional methods. Nanotechnology has proven favorable to fulfilling these demands through various nanocomposites with varying compositions and characteristics like magnetism, pollutant detection, and removal. With recent advances in the reusability and biocompatibility of nano adsorbents, the contributions of researchers in nanotechnology are already paving a much-anticipated path towards mercury pollution-free ecosystems.

Acknowledgements The authors would like to thank the Gandhi Institute of Technology and Management (Deemed to be University), Visakhapatnam for the support of resources.

References

Al-Hetlani E, D'Cruz B, Amin MO, Madkour M (2022) An effective magnetic nanoadsorbent based on a carbonaceous/spinel ferrite nanocomposite for the removal of pharmaceutical pollutants from wastewater. Environ Sci Water Res Technol 8:998–1010. https://doi.org/10.1039/D1EW00495F

Alvand M, Shemirani F (2017a) A Fe_3O_4@SiO_2@ graphene quantum dot core-shell structured nanomaterial as a fluorescent probe and for magnetic removal of mercury(II) ion. Microchim Acta 184:1621–1629. https://doi.org/10.1007/s00604-017-2134-2

Alvand M, Shemirani F (2017b) A Fe_3O_4@SiO_2@ graphene quantum dot core-shell structured nanomaterial as a fluorescent probe and for magnetic removal of mercury(II) ion. MCA 1621–1629. https://doi.org/10.3390/ma16020514

Amin KF, Nahin AM, Hoque ME (2022) Polymer nanocomposites for adhesives and coatings. In: Advanced polymer nanocomposites. Woodhead Publishing, pp 235–265

Arshadi M, Mousavinia F, Khalafi-Nezhad A, Firouzabadi H, Abbaspourrad A (2017) Adsorption of mercury ions from wastewater by a hyperbranched and multi-functionalized dendrimer modified mixed-oxides nanoparticles. COCIS 293–306. https://doi.org/10.1016/j.jcis.2017.05.052

Assefa T, Mazengiaw F (2021) Iron-based nanoparticles in wastewater treatment: a review on synthesis methods, applications, and removal mechanisms. J Saudi Chem Soc 25:101280. https://doi.org/10.1016/j.jscs.2021.101280

Baby R, Hussein MZ, Abdullah AH, Zainal Z (2022) Nanomaterials for the treatment of heavy metal contaminated water. Polymers 14:583. https://doi.org/10.3390/polym14030583

Balbino TAC, Bellato CR, da Silva AD, Neto JDOM, Ferreira SO (2020) Preparation and evaluation of iron oxide/hydrotalcite intercalated with dodecylsulfate/β-cyclodextrin magnetic organo-composite for phenolic compounds removal. Appl Clay Sci 193:105659. https://doi.org/10.1016/j.clay.2020.105659

Behjati M, Baghdadi M, Karbassi A (2018) Removal of mercury from contaminated saline wasters using dithiocarbamate functionalized-magnetic nanocomposite. J Environ Manage 213:66–78. https://doi.org/10.1016/j.jenvman.2018.02.052

Bhowmick N, Dadhich AS, Prasuna RG, Reddy MK (2018) Screening and isolation of mercury tolerant bacteria from soil around the industrial areas of Visakhapatnam, India. Pollu Res 37:757–762

Bystrzejewski M, Pyrzyńska K, Huczko A, Lange H (2009) Carbon-encapsulated magnetic nanoparticles as separable and mobile sorbents of heavy metal ions from aqueous solutions. Carbon 47:1201–1204. https://doi.org/10.1016/j.carbon.2009.01.007

Czarna D, Baran P, Kunecki P, Panek R, Żmuda R, Wdowin M (2018) Synthetic zeolites as potential sorbents of mercury from wastewater occurring during wet FGD processes of flue gas. JCLP 172:2636–2645. https://doi.org/10.1016/j.jclepro.2017.11.147

Eichhorn SJ, Rahatekar SS, Vignolini S, Windle AH (2018) New horizons for cellulose nanotechnology. Philos Trans Royal Soc 376:20170200. https://doi.org/10.1098/rsta.2017.0200

Emenike EC, Adeniyi AG, Iwuozor KO, Okorie CJ, Egbemhenghe AU, Omuku PE, Okwu KC, Saliu OD (2023) A critical review on the removal of mercury (Hg^{2+}) from aqueous solution using nano adsorbents. Environ Nanotechnol Monit Manag 100816. https://doi.org/10.1016/j.enmm.2023.100816

Eskandari L, Andalib F, Fakhri A, Jabarabadi MK, Pham B, Gupta VK (2020) Facile colorimetric detection of Hg (II), photocatalytic and antibacterial efficiency based on silver-manganese disulfide/polyvinyl alcohol-chitosan nanocomposites. Int J Biol Macromol 164:4138–4145. https://doi.org/10.1016/j.ijbiomac.2020.09.015

Fang Y, Lu A, Zhao D, Su K, Li Z, Wang Y, Shen XC, Liu K, Ruan C (2022) Facile synthesis of widened MoS_2 nanosheets vertically anchored on natural cellulose fibers for efficient removal of mercury ions from aquatic systems. J Environ Chem Eng 10:108229. https://doi.org/10.1016/j.jece.2022.108229

Farhan A, Zulfiqar M, Rashid EU, Nawaz S, Iqbal HM, Jesionowski T, Bilal M, Zdarta J (2023) Removal of toxic metals from water by nanocomposites through advanced remediation processes and photocatalytic oxidation. Curr Pollut Rep 1–21. https://doi.org/10.1007/s40726-023-00253-y

Fatima H, Kim KS (2018) Iron-based magnetic nanoparticles for magnetic resonance imaging. Adv Powder Technol 29:2678–2685. https://doi.org/10.1016/j.apt.2018.07.017

Feng X, Li P, Fu X, Wang X, Zhang H, Lin CJ (2022) Mercury pollution in China: implications on the implementation of the Minamata convention. Environ Sci Process Impacts 24:634–648. https://doi.org/10.1039/D2EM00039C

Gao P, Lei J, Tan J, Wang G, Liu H, Zhou L (2021) Self-assembled magnetic microcrystalline cellulose/MoS_2/Fe_3O_4 composite for efficient adsorptive removal of mercury ions (Hg^{2+}). Compos Commun 25:100736. https://doi.org/10.1016/j.coco.2021.100736

Gheitasi F, Ghammamy S, Zendehdel M, Semiromi FB (2022) Removal of mercury (II) from aqueous solution by powdered activated carbon nanoparticles prepared from beer barley husk modified with Thiol/Fe_3O_4. J Mol Struct 1267:133555. https://doi.org/10.1016/j.molstruc.2022.133555

Goel H, Saini K, Razdan K, Khurana RK, Elkordy AA, Singh KK (2022) In vitro physicochemical characterization of nanocarriers: a road to optimization. In: Nanoparticle therapeutics. Academic Press, pp 133–179. https://doi.org/10.1016/B978-0-12-820757-4.00018-1

Gupta N, Pant P, Gupta C, Goel P, Jain A, Anand S, Pundir A (2018) Engineered magnetic nanoparticles as efficient sorbents for wastewater treatment: a review. Mater Res Innov 22:434–450. https://doi.org/10.1080/14328917.2017.1334846

Gürbüz MU, Elmacı G, Ertürk AS (2019) Tren-cored PAMAM dendrimer/silver nanocomposites: efficient colorimetric sensors for the determination of mercury ions from aqueous solutions. Chem Select 4:7715–7721. https://doi.org/10.1002/slct.201901538

Han Y, Tao J, Khan A, Ullah R, Ali N, Ali N, Malik S, Yu C, Yang Y, Bilal M (2022) Design and fabrication of chitosan cross-linked bismuth sulfide nanoparticles for sequestration of mercury in river water samples. Environ Res 215:113978. https://doi.org/10.1016/j.envres.2022.113978

Inglezakis VJ, Kurbanova A, Molkenova A, Zorpas AA, Atabaev TS (2020) Magnetic Fe_3O_4-Ag^0 nanocomposites for effective mercury removal from water. Sustainability 12(13):5489. https://doi.org/10.3390/su12135489

Kalfa OM, Yalçınkaya Ö, Türker AR (2009) Synthesis of nano B_2O_3/TiO_2 composite material as a new solid phase extractor and its application to preconcentration and separation of cadmium. JHM Lett 166:455–461. https://doi.org/10.1016/j.jhazmat.2008.11.112

Khajeh M, Laurent S, Dastafkan K (2013) Nanoadsorbents: classification, preparation, and applications (with emphasis on aqueous media). Chem Rev 113:7728–7768. https://doi.org/10.1021/cr400086v

Khan FSA, Mubarak NM, Khalid M, Walvekar R, Abdullah EC, Mazari SA, Karri RR (2020) Magnetic nano adsorbents potential route for heavy metals removal—a review. ESPR 27:24342–24356. https://doi.org/10.1007/s11356-020-08711-6

Khan I, Khan I, Saeed K, Ali N, Zada N, Khan A, Ali F, Bilal M, Akhter MS (2023) Polymer nanocomposites: an overview. Smart Polym Nanocomposites 167–184. https://doi.org/10.1016/B978-0-323-91611-0.00017-7

Kharissova OV, Dias HR, Kharisov BI (2015) Magnetic adsorbents based on micro-and nano-structured materials. RSC Adv 5:6695–6719. https://doi.org/10.1039/C4RA11423J

Köhler JM (2021) Challenges for nanotechnology. Encyclopedia 1(3):618–631

Kumar B, Thareja RK (2010) Synthesis of nanoparticles in laser ablation of aluminum in liquid. J Appl Phys 108:064906. https://doi.org/10.1063/1.3486517

Kumar J, Arland S, Gour P (2020) Mercury removal from simulated waste water by chitosan nano composite embedded with leaf extract of Brassica Gongylodes. Mater Today: Proc 26:728–739. https://doi.org/10.1016/j.matpr.2019.12.412

Kumar R, Rauwel P, Rauwel E (2021) Nanoadsorbants for the removal of heavy metals from contaminated water: current scenario and future directions. Processes 9:1379. https://doi.org/10.3390/pr9081379

Kumari S, Jamwal R, Mishra N, Singh DK (2020) Recent developments in environmental mercury bioremediation and its toxicity: a review. Environ Nanotechnol Monit Manag 13:100283. https://doi.org/10.1016/j.enmm.2020.100283

Kyzas GZ, Deliyanni EA (2013) Mercury (II) removal with modified magnetic chitosan adsorbents. Mol 18(6):6193–6214. https://doi.org/10.3390/molecules18066193

Latha S, Gomathi T, Pavithra S, Sudha PN, Nalluri A, Bhardwaj P (2023) Chitosan-based porous carbon materials for agriculture and agro-waste applications. In: Handbook of porous carbon materials. Springer Nature Singapore, Singapore, pp 763–812

Lei C, Sun Y, Tsang DCW, Lin D (2018) Environmental transformations and ecological effects of iron-based nanoparticles. Environ Pollut 232:10–30. https://doi.org/10.1016/j.envpol.2017.09.052

Li X (2023) Magnetic iron-based nanoadsorbents for highly efficient removal of mercury ions from industrial effluents. J Environ Manage 298:113519

Li W, Ju B, Zhang S (2019) Preparation of cysteamine-modified cellulose nanocrystal adsorbent for removal of mercury ions from aqueous solutions. Cellulose 26:4971–4985. https://doi.org/10.1007/s10570-019-02420-1

Lingamdinne LP, Koduru JR, Karri RR (2019) A comprehensive review of applications of magnetic graphene oxide-based nanocomposites for sustainable water purification. J Environ Manage 231:622–634. https://doi.org/10.1016/j.jenvman.2018.10.063

Liu Y, Chen G, Tian C, Gupta R, Wang X, Zeng H (2018) Metal oxide nanoparticle-modified graphene oxide for removal of elemental mercury. Environ Technol 40:3602–3610. https://doi.org/10.1080/09593330.2018.1482372

Liu S, Wang X, Guo G, Yan Z (2021) Status and environmental management of soil mercury pollution in China: a review. J Environ Manage 277:111442. https://doi.org/10.1016/j.jenvman.2020.111442

Liu Y, Chen H, Zhu N, Zhang J, Li Y, Xu D, Gao Y, Zhao J (2022) Detection and remediation of mercury contaminated environment by nanotechnology: progress and challenges. Environ Pollut 293:118557. https://doi.org/10.1016/j.envpol.2021.118557

Liu S, Ma C, Ma M, Xu F (2019) Magnetic nanocomposite adsorbents. Compos Nanoadsorbents 295–316. https://doi.org/10.1016/B978-0-12-814132-8.00013-7

Liu R, Zenglu QI, Huachun LAN, Zhu L, Liu H, Qu J (2020) U.S. Patent No. 10,569,250. U.S. Patent and Trademark Office, Washington, DC

Lo SI, Chen PC, Huang CC, Chang HT (2012) Gold nanoparticle–aluminum oxide adsorbent for efficient removal of mercury species from natural waters. Environ Sci Technol 46:2724–2730. https://doi.org/10.1021/es203678v

Ma Y, Mu B, Zhang X, Xu H, Qu Z, Gao L, Li B, Tian J (2019) Ag-Fe_3O_4@ rGO ternary magnetic adsorbent for gaseous elemental mercury removal from coal-fired flue gas. Fuel 239:579–586. https://doi.org/10.1016/j.fuel.2018.11.065

Madhura L, Singh S, Kanchi S, Sabela M, Bisetty K, Inamuddin (2019) Nanotechnology-based water quality management for wastewater treatment. Environ Chem Lett 17:65–121. https://doi.org/10.1007/s10311-018-0778-8

Madima N, Mishra SB, Inamuddin I, Mishra AK (2020) Carbon-based nanomaterials for remediation of organic and inorganic pollutants from wastewater. A review. Environ Chem Lett 18:1169–1191. https://doi.org/10.1007/s10311-020-01001-0

Mahapatra A, Mishra BG, Hota G (2013) Electrospun Fe_2O_3–Al_2O_3 nanocomposite fibers as efficient adsorbent for removal of heavy metal ions from aqueous solution. J Hazard Mater 258:116–123. https://doi.org/10.1016/j.jhazmat.2013.04.045

Mahmoud ME, Abdelwahab MS, Ibrahim GA (2022) The design of SnO_2-crosslinked-chitosan nanocomposite for microwave-assisted adsorption of aqueous cadmium and mercury ions. Sustain Chem Pharm 28:100731. https://doi.org/10.1016/j.scp.2022.100731

Maksoud MA, Elgarahy AM, Farrell C, Ala'a H, Rooney DW, Osman AI (2020) Insight on water remediation application using magnetic nanomaterials and biosorbents. Coord Chem Rev 403:213096. https://doi.org/10.1016/j.ccr.2019.213096

Malhotra N, Audira G, Chen JR, Siregar P, Hsu HS, Lee JS, Hsiao CD (2020) Surface modification of magnetic nanoparticles by carbon-coating can increase its biosafety: evidences from biochemical and neurobehavioral tests in zebrafish. Mol 25:2256. https://doi.org/10.3390/molecules25092256

Marimón-Bolívar W, Tejeda-Benítez L, Herrera AP (2018) Removal of mercury (II) from water using magnetic nanoparticles coated with amino organic ligands and yam peel biomass. Environ Nanotechnol Monit Manag 10:486–493. https://doi.org/10.1016/j.enmm.2018.10.001

Martínez-Prieto LM, Marbaix J, Asensio JM, Cerezo-Navarrete C, Fazzini PF, Soulantica K, Corma A (2020) Ultrastable magnetic nanoparticles encapsulated in carbon for magnetically induced catalysis. ACS Appl Nano Mater 3:7076–7087. https://doi.org/10.1021/acsanm.0c01392

Mehta D, Mazumdar S, Singh S (2015) Magnetic adsorbents for the treatment of water/wastewater—a review. JWPE 7:244–265. https://doi.org/10.1016/j.jwpe.2015.07.001

Mudhoo A, Sillanpää M (2021) Magnetic nano adsorbents for micropollutant removal in real water treatment: a review. Environ Chem Lett 19:4393–4413. https://doi.org/10.1007/s10311-021-01289-6

Pankhurst QA, Connolly J, Jones SK, Dobson J (2003) Applications of magnetic nanoparticles in biomedicine. J Phys D: Appl Phys 36:167–181. https://doi.org/10.1088/0022-3727/36/13/201

Patel D, Talele D, Pathak D, Misra A (2021) Polymers used in Peyer's patch targeting. In: Applications of polymers in drug delivery. Elsevier, pp 159–185. https://doi.org/10.1016/B978-0-12-819659-5.00006-9

Peralta ME, Mártire DO, Moreno MS, Parolo ME, Carlos L (2021) Versatile nano adsorbents based on magnetic mesostructured silica nanoparticles with tailored surface properties for organic pollutants removal. J Environ Chem Eng 9:104841. https://doi.org/10.1016/j.jece.2020.104841

Pokrajac L, Abbas A, Chrzanowski W, Dias GM, Eggleton BJ, Maguire S, Mitra S (2021) Nanotechnology for a sustainable future: Addressing global challenges with the international network. Sustain Nanotechnol. https://doi.org/10.1021/acsnano.1c10919

Priya T, Mishra BK, Prasad MNV (2020) Physico-chemical techniques for the removal of disinfection by-products precursors from water. In: Disinfection by-products in drinking water. Butterworth-Heinemann, pp 23–58. https://doi.org/10.1016/B978-0-08-102977-0.00002-0

Pyrzyńska K, Bystrzejewski M (2010) Comparative study of heavy metal ions sorption onto activated carbon, carbon nanotubes, and carbon-encapsulated magnetic nanoparticles. Colloids Surf 362:102–109. https://doi.org/10.1016/j.colsurfa.2010.03.047

Rabie AM, Abd El-Salam HM, Betiha MA, El-Maghrabi HH, Aman D (2019) Mercury removal from aqueous solution via functionalized mesoporous silica nanoparticles with the amine compound. Egypt J Petrol 28:289–296. https://doi.org/10.1016/j.jece.2020.103725

Rahbar N, Jahangiri A, Boumi S, Khodayar MJ (2014) Mercury removal from aqueous solutions with chitosan-coated magnetite nanoparticles optimized using the box-behnken design. Jundishapur J Nat Pharm Prod 9. https://doi.org/10.17795/2Fjjnpp-15913

Reddy DHK, Yun YS (2016) Spinel ferrite magnetic adsorbents: alternative future materials for water purification? Coord Chem Rev 315:90–111. https://doi.org/10.1016/j.ccr.2016.01.012

Schwaminger SP, Fraga-García P, Eigenfeld M, Becker TM, Berensmeier S (2019) Magnetic separation in bioprocessing beyond the analytical scale: from biotechnology to the food industry. Front Bioeng Biotechnol 7:233. https://doi.org/10.3389/fbioe.2019.00233

Sharma A, Singh M, Arora K, Singh PP, Badru R, Kang TS, Kaushal S (2021) Preparation of cellulose acetate-Sn (IV) iodophosphate nanocomposite for efficient and selective removal of Hg^{2+} and Mn^{2+} ions from aqueous solution. Environ Nanotechnol Monit Manag 16:100478. https://doi.org/10.1016/j.enmm.2021.100478

Shirzadi H, Nezamzadeh-Ejhieh A (2017) An efficient modified zeolite for simultaneous removal of Pb (II) and Hg (II) from aqueous solution. J Mol Liq 230:221–229. https://doi.org/10.1016/j.molliq.2017.01.029

Shukla P, Mishra A, Manivannan S, Melo JS, Mandal D (2020) Parametric optimization for adsorption of mercury (II) using self assembled bio-hybrid. J Environ Chem Eng 8:103725

Siddiqui MTH, Nizamuddin S, Baloch HA, Mubarak NM, Al-Ali M, Mazari SA, Griffin G (2019) Fabrication of advance magnetic carbon nano-materials and their potential applications: a review. J Environ Chem Eng 7:102812. https://doi.org/10.1016/j.jece.2018.102812

Singh AD, Khanna K, Kour J, Dhiman S, Bhardwaj T, Devi K, Sharma N, Kumar P, Kapoor N, Sharma P, Arora P, Bhardwaj R (2023) Critical review on biogeochemical dynamics of mercury (Hg) and its abatement strategies. Chemosphere 137917. https://doi.org/10.1016/j.chemosphere.2023.137917

Tahir MB, Sagir M, Rafique M (2021) Challenges and future of nanotechnology. Nanotechnol: Trends Future Appl 141–142

Tan Z, Gao M, Dai J, Ge W, Zhu W, Yan Y (2020) Magnetic Interconnected macroporous imprinted foams for selective recognition and adsorptive removal of phenolic pollution from water. Fibers Polym 21:762–774. https://doi.org/10.1007/s12221-020-8695-4

Trache D, Tarchoun AF, Derradji M, Hamidon TS, Masruchin N, Brosse N, Hussin MH (2020) Nanocellulose: from fundamentals to advanced applications. Front Chem 8:392. https://doi.org/10.3389/fchem.2020.00392

Umoren SA, Solomon MM, Saji VS (2022) Polymeric materials in corrosion inhibition: fundamentals and applications. Elsevier

Uttam B, Polepalli S, Sinha S, Majumder A, Rao CP (2022) Selective sensing and removal of mercury ions by encapsulating dansyl appended calix conjugate in a zeolitic imidazolate framework as an organic-inorganic hybrid nanomaterial. ACS Appl Nano Mater 5(8):11371–11380. https://doi.org/10.1021/acsanm.2c02459

Vélez E, Campillo GE, Morales G, Hincapié C, Osorio J, Arnache O, Uribe JI, Jaramillo F (2016) Mercury removal in wastewater by iron oxide nanoparticles. J Phys Conf Ser 687:012050. https://doi.org/10.1088/1742-6596/687/1/012050

Wang ZL (2000) Characterizing the structure and properties of individual wire-like nanoentities. Adv Mat 12:1295–1298. https://doi.org/10.1002/1521-4095(200009)12:17%3C1295::AID-ADMA1295%3E3.0.CO;2-B

Wang Y (2021) Thiol-functionalized mesoporous silica nanoparticles for efficient removal of mercury from wastewater. ESPR 28(21):27435–27444

Wang B, Wu K, Liu T, Luan H, Xue K, Liu Y, Niu Y (2023) Synthesis of hyperbranched polyamine dendrimer/chitosan/silica composite for efficient adsorption of Hg (II). Int J Biol Macromol 123135. https://doi.org/10.1016/j.ijbiomac.2023.123135

Xie R, Qu Y, Tang M, Zhao J, Chua S, Li T, Zhang F, Wheatley AE, Chai F (2021) Carbon dots-magnetic nanocomposites for the detection and removal of Hg^{2+}. Food Chem 364:130366. https://doi.org/10.1016/j.foodchem.2021.130366

Yaghmaeian K, Khosravi Mashizi R, Nasseri S, Mahvi AH, Alimohammadi M, Nazmara S (2015) Removal of inorganic mercury from aquatic environments by multi-walled carbon nanotubes. J Environ Health Sci Eng 13:1–9. https://doi.org/10.1186/s40201-015-0209-8

Yang Z, Li H, Yang J, Feng S, Liu X, Zhao J, Qu W, Li P, Feng Y, Lee PH, Shih K (2019b) Nanosized copper selenide functionalized zeolitic imidazolate framework-8 (CuSe/ZIF-8) for efficient immobilization of gas-phase elemental mercury. Adv Funct Mater 29:1807191. https://doi.org/10.1002/adfm.201807191

Yang L, Yang L, Ding L, Deng F, Luo XB, Luo SL (2019) Principles for the application of nanomaterials in environmental pollution control and resource reutilization. In: Nanomaterials for the removal of pollutants and resource reutilization. Elsevier, pp 1–23. https://doi.org/10.1016/B978-0-12-814837-2.00001-9

Yatsui K, Yukawa T, Grigoriu C, Hirai M, Jiang W (2000) J Nanopart Res 2:75–83. https://doi.org/10.1023/A:1010090115429

Zabihi M, Omidvar M, Motavalizadehkakhky A, Zhiani R (2022) Competitive adsorption of arsenic and mercury on nano-magnetic activated carbons derived from hazelnut shell. Korean J Chem Eng 1–10. https://doi.org/10.1007/s11814-021-0903-4

Zhang S, Li H, Wang Z, Liu J, Zhang H, Wang B, Yang Z (2015) A strongly coupled Au/Fe_3O_4/GO hybrid material with enhanced nanozyme activity for highly sensitive colorimetric detection, and rapid and efficient removal of Hg^{2+} in aqueous solutions. Nanoscale 7:8495–8502. https://doi.org/10.1039/C5NR00527B

Zhang Z, Xia K, Pan Z, Yang C, Wang X, Zhang G, Guo Y, Bai R (2020) Removal of mercury by magnetic nanomaterial with bifunctional groups and core-shell structure: synthesis, characterization and optimization of adsorption parameters. Appl Surf Sci 500:143970. https://doi.org/10.1016/j.apsusc.2019.143970

Zhao G, Huang X, Tang Z, Huang Q, Niu F, Wang X (2018) Polymer-based nanocomposites for heavy metal ions removal from aqueous solution: a review. Polym Chem 9:3562–3582. https://doi.org/10.1039/C8PY00484F

Zheng X, Cao H, Liu B, Zhang M, Zhang C, Chen P, Yang B (2022) Effects of mercury contamination on microbial diversity of different kinds of soil. Micro 10(5):977. https://doi.org/10.3390/microorganisms10050977

Zhu L (2022) Amino-functionalized silica nanoparticles for efficient mercury immobilization in contaminated soil. Environ Pollut 292:118380

Chapter 12
Amputation of Mercury from Wastewater: Perspectives of Action Methods

Visakha Singh, Sunita Kar, M. Kiranmai Reddy, Y. Satyanarayana, B. Madhava Rao, and Hippolyte Mutanyagwa

12.1 Introduction

As the population of our planet increases, so does the need for urbanization and industrialization. This rapid growth has led to widespread contamination of the natural environment due to a myriad of pollutants being released into the environment. These pollutants often harm not only the ecosystem but also gravely affect the health of human beings as well (Chu and Karr 2017). In such a dire situation, water is one of the natural resources largely compromised. In today's world, water scarcity has become a most pressing issue. Due to the contamination of water with toxic pollutants, the problem is further exacerbated (Kjellstorm et al. 2006). This contamination can occur when toxic effluent is released from various industries into water sources like rivers and oceans. An urgent need is to efficiently treat wastewater to lower the burden on the already stressed resource (Qasem et al. 2021).

One of the major contaminants that is polluting wastewater and hence other water sources is mercury. Mercury is a heavy metal with atomic number 80 and an atomic mass of 200.59 a.m.u. It is a highly toxic and harmful heavy metal polluting water. Mercury can be present in the form of inorganic Hg^{2+} ions, organic complexes such as methyl mercury, and even in its elemental form. It is generally obtained from ores and sediments and is used for various applications such as electrical, clinical, and appliances. It has become a particularly vexing issue due to its ability to bioaccumulate and persist in nature for a very long duration. Exposure to this heavy metal

V. Singh · S. Kar · M. K. Reddy (✉) · Y. Satyanarayana · H. Mutanyagwa
Department of Environmental Science, GITAM School of Science, GITAM (Deemed to Be University), Visakhapatnam, Andhra Pradesh 530045, India
e-mail: kmajji@gitam.edu

B. M. Rao
Department of Chemistry, GITAM School of Science, GITAM (Deemed to Be University), Visakhapatnam, Andhra Pradesh 530045, India

N. Kumar (ed.), *Mercury Toxicity*, Environmental Science and Engineering,
https://doi.org/10.1007/978-981-99-7719-2_12

can cause serious health issues like damage to neurocognitive functions, Minamata disease, and mad hatter's disease. It can contaminate water and enter the food chain of marine animals, ultimately leading to bioaccumulation and reaching human beings as well (Tchounwou et al. 2012). There is an urgent need to develop effective and sustainable tools for the safe removal of mercury from wastewater since it can ultimately pollute other water sources. In recent times, multiple novel techniques have been developed by researchers in order to remove mercury and its different forms from wastewater effectively. Some of these methods have been discussed in this chapter.

Currently, the most viable option for mercury remediation has been based on different types of adsorption and filtration techniques. Adsorbents based on polymers, iron oxides, activated carbon, and hydrogels have found remarkable applications in the field of water treatment. Apart from these, various biosorbents, as well as different membrane filtration techniques, were discussed here. One of the subjects finding much interest in this context is nanotechnology. Nano adsorbents have brought forth a new era for wastewater treatment procedures due to their unique characteristics and sustainable functionality (Vishwakarma et al. 2021).

This chapter aims to discuss the current trends in wastewater treatment technologies focused on the remediation of mercury in all its forms. There are multiple methods to treat wastewater in order to remove mercury, such as physical, chemical, and biological (Bhowmick et al. 2018) methods. The concepts include a description of the latest research being made in adsorption and filtration techniques that could effectively replace conventional wastewater treatment. The novel research being done into mercury removal could pave the way for more effective and sustainable methods without compromising their biocompatibility or economy. This chapter also describes the challenges these novel adsorbents and filters face while highlighting the need for future research.

12.2 Treatment of Wastewater

Wastewater is also known as sewage which contains 99.9% water by mass. It generally contains suspended, dissolved, colloidal, and settling solid matter, biodegradable and soluble organic compounds, inorganic material, nutrients, heavy metals, floating substances, and some pathogens that cause diseases. Wastewater should be treated properly to meet the permissible standards of discharge into the water bodies. It should be disposed of in a manner that does not create any nuisance to the environment. Wastewater can also be utilized (Templeton and Butler 2011) for various purposes such as to recover energy, nutrients, and water that can be reused after treatment, and it also provides some valuable resources.

Wastewater which contains heavy metals such as mercury is lethal to human beings and aquatic species. It can be released from battery manufacturing industries, cement industries, pharmaceuticals, etc. releases mercury into the outlet, which goes to the water bodies. These wastes have to be treated properly before discharging, for which these three types of treatments can be utilized.

12.2.1 Physical Treatments for Mercury Removal

Physical treatments involve coagulation, flocculation, sedimentation, filtration, and disinfection. In one study by Vedrenne et al. (2012), flocculation or coagulation was done by following photo Fenton oxidation treatment. Samples were collected from a matured landfill leachate containing pollutants during drought. This was followed by observation of various parameters like the chemical oxygen demand, ammonia, and total carbon, as well as maximum concentrations of mercury, lead, and arsenic. Whereas other heavy metals such as nickel, cobalt, zinc, cadmium, and manganese were found in trace amounts. After the pollutants' findings, the leachate's lethal concentration was tested on *Artemia salina.* It was found to be 12,161 ± 11 mg/L in COD. This indicated that the leachate constituents showed an antagonistic interaction. In the study, the coagulation-flocculation treatment process was done by using ferric chloride hexahydrate ($FeCl_3{\cdot}6H_2O$) with an optimal dose of 300 mg/L. Then the supernatant was treated with the photo-Fenton process, which was mediated using ferrous chloride tetrahydrate ($FeCl_2{\cdot}4H_2O$) and hydrogen peroxide (H_2O_2) within a compound parabolic concentrator or CPC. In this, a photo-reactor was operating through batch mode by utilizing an R ratio of 114, where $R = [H_2O_2]/[Fe^{2+}]$. The removal efficiency of mercury was found to be 85%.

The author also applied the coagulation method, Lu et al. (2014), in which mercury was removed by using manganese-ferric hydroxides prepared in situ and was analyzed through a reaction between $KMnO_4$ and Fe(II) in a solution along with water. The impact of pH, temperature, and coagulant dosage on mercury removal efficiency was analyzed through a jar test. The result after experimental procedures indicated the efficient removal of mercury by situ Mn-Fe when compared with poly aluminum chloride and ferric chloride, as these coagulants are generally applied in practice. After the experiment, mercury was found in the form of uncharged species such as $Hg(OH)_2$, HgClOH(aq.), and $HgCl_2$(aq.). Later various spectroscopic analyzes were done to know the characterization in which Fourier transform infrared spectroscopy verified that in situ Mn-Fe contained hydroxyl groups that acted as the active sites on the surface, whereas X-ray photoelectron spectroscopy verified that the precipitates contained the mixture of Hg-Mn-Fe and also stated that mercury removal from the solutions was done through a liquid phase to the solid phase. The main removal process of mercury by in situ Mn-Fe were flocculation-precipitation and surface complexation. Mercury was removed with an efficiency of 47 and 70% for a coagulation period of 1 and 7 min. It was also noticed that if the time for the coagulation period increased, mercury removal efficiency was also improved by 80% within 22 min (Table 12.1).

In the context of using a coagulant method for mercury removal, the author, Henneberry et al. (2011), have used metal-based coagulants for removing inorganic

Table 12.1 Physical methods for removal of mercury

Processes	Resources/chemicals used	Efficiency (%)	References
Coagulation/flocculation	$FeCl_3{\cdot}6H_2O$	85	Vedrenne et al. (2012)
Coagulation	In situ Mn-Fe	70–80	Lu et al. (2014)
Coagulation	Ferric chloride, ferric sulphate, polyaluminium chloride	97	Henneberry et al. (2011)
Flocculation	MAPEI	95	Min et al. (2010)
Flocculation	DNA condensed with cetyltrimethyl ammonium bromide	95	Zinchenko et al. (2009)
Disinfection	Chlorine-loaded carbons with RC	75–80	Mochizuki et al. (2021)

mercury and methyl mercury from wastewater. The study provided us with some information regarding the mercury associations with fractions of dissolved organic matter and metal hydroxides. For the study, a water sample was collected from the agricultural drain of the Sacramento-San Joaquin Delta. The water sample was then filtered to isolate dissolved organic matter or DOM. The isolated was then treated with different dosages of the coagulants to see the efficiency of removing all forms of mercury from the solution. In the study, three different industrial coagulants, such as ferric chloride, ferric sulphate, and polyaluminium chloride, were used. The coagulation process removed 85% of DOM from the solution. All the coagulants released inorganic mercury in the solution in the absence of DOM, whereas, in the presence of DOM, the coagulants showed efficacy in removing 97% of inorganic mercury and 80% of methyl mercury. So, the result varies with or without using a DOM solution. It was also found that inorganic methyl bonded with DOM with more aromatic and higher molecular weight fraction whereas DOM did not bind with methyl mercury. This study suggested large amounts of mercury removal from wastewater within environmentally beneficial concentrations.

Mercury can also be removed through the flocculation method, one of the novel works done by author Min et al. (2010), in which a flocculant known as mercaptoacetyl polyethyleneimine (MAPEI) was manufactured by the reaction of polyethyleneimine or PEI with thioglycolic acid (TGA). MAPEI was found to be capable of removing mercury ions (Hg^{2+}) from wastewater. The maximum removal efficiency of Hg^{2+} reached 95%. When the factors related to it were studied then it was found that increasing pH and molecular weight of PEI led to an increase in Hg^{2+} removal. It was also known that alkali metal, alkali earth metal ions, chloride ions (Cl^{-1}), and nitrate ions (NO_3^-) increase the rate of Hg^{2+} removal, whereas sulphate (SO_4^{2-}) decreases the rate. If the wastewater is carrying mercury ions along with turbidity, then both have a synergic removal effect upon each other. When MAPEI compared to TGA, MAPEI was more effective in removing mercury, producing bigger flocs and higher floc sedimentation velocity. Whereas, when compared to PEI, it was noticed that PEI and MAPEI both were not effective in removing turbidity, but when mercury ions were present, they showed efficacy in turbidity removal.

According to the author, Zinchenko et al. (2009), double-stranded DNA can also be used to recover mercury from wastewater, as DNA can combine with different chemical substrates at the molecular level. In a study, DNA was used for its strong bonding nature for removing Hg ions. DNA was condensed by its condensing agent known as cetyltrimethylammonium bromide along with markedly available flocculants produced by using zeolite for precipitating DNA which is bound with mercury ion from wastewater. The mercury solutions with concentrations ranging from 0.02 to 100 ppm with a pH ranging from 2 to 11 were treated with DNA, after which 95% of mercury ions were removed following treatment with sedimentation or filtration.

One of the disinfection processes for the removal of mercury was done by the author Mochizuki et al. (2021), who stated that the mercury adsorption was done by using chlorine-loaded carbons derived by chlorination of rice husk char or RC. The chlorine-loaded carbon was tested for mercury adsorption performance. After this, the effect of chlorine upon mercury adsorption was examined while comparing the chlorine-loaded carbon adsorption performance made from rice husk char (RC) and with demineralized RC (DRC). These RC and DRC reached a maximum level value at 600 °C and then decreased by 1000 °C again. The rise in the chlorination temperature increased RC chlorinated residue's pore properties. It was noticed that the optimum mercury adsorption capacity was 620 mg/g at 1000 °C by RC chlorinated residue. Pseudo-first-order and Langmuir equations expressed the adsorption mechanism and isotherm. Roberts et al. (2005) stated that chlorine disinfectants discharged the maximum mercury ions into the solution.

12.2.2 Chemical Treatment Method

Different chemical treatment methods are used for wastewater treatment, such as ion exchange method, chemical stabilization, chemical precipitation, chemical oxidation, etc.

Ion exchange is a very common and frequently used method for the removal of mercury. In this context Monteagudo and Ortiz (2000) examined a process for the removal of inorganic mercury from mine wastewater by utilizing an ion exchange resin system. In the study, various commercial resins were used as ion exchangers, and the concentration of mercury concentration was 70 to 90 ppm, which was lowered to a permissible level of 34 ppb. Out of the several ion exchangers, Dowex XZS-1, showed a strong cation exchanger in gel form and was best suited for mercury removal. The loaded resin was again regenerated properly by utilizing HCl solutions because of maximum competition within mercury and hydronium ions and also the formation of the $HgCl_4^{2-}$ ionic complex (Table 12.2).

In the ion exchange technique, the author, Oehmen et al. (2014), performed another study, which produced an ion exchange membrane bioreactor (IEMB) to remove

Table 12.2 Chemical methods for removal of mercury

Processes	Resources/chemicals	Efficiency (%)	References
Ion exchange	Commercial resins	99.63	Monteagudo and Ortiz (2000)
Ion exchange	IEMB	98	Oehmen et al. (2014)
Ion exchange	(PGCP-COOH) adsorbent	99	Anirudhan et al. (2008)
Chemical precipitation	Fe_3O_4 and γ-Fe_2O_3	68–87	Vélez et al. (2016)
Advanced oxidation process	AOPs UV/H_2O, UV/Fe^{2+}, UV/O_3, UV/H_2O_2/ Fe^{2+}, UV/Cl	78–98	Ghosh et al. (2022)

mercury from drinking wastewater released from industries. In this process, Hg(II) was transported through the cation exchange membrane and paired with its bioreduction to mercury (Hg°) to remove mercury from the concentrated water bodies. In this process, a very minute amount of residues or by-products were observed. In the process, mercury was removed with greater than 98% efficiency and also the throughput of the mercury-polluted water was optimized by membrane pretreatment. Hence, it can be said that this IEMB technique is a novel technology for treating mercury from water bodies as it generates a minimum amount of polluted waste, ultimately reducing the overall environmental impact of the process.

In another ion exchange process, a novel approach was made for the removal of mercury by the author, Anirudhan et al. (2008), in which a new adsorbent was used. The adsorbent (PGCP-COOH) from the carboxylate functional group at the chain end was produced by grafting it with poly (hydroxyethyl methacrylate) onto coconut coir pith, CP which is a lignocelluloses residue from a coir-based industry. This CP used potassium peroxydisulphate as an initiator and also in the presence of a cross-linking agent *N*, *N′*-methylenebisacrylamide. The batch adsorption process was used under kinetic and equilibrium conditions to see the efficacy of PGCP-COOH for removing mercury from water streams. The adsorbents showed maximum removal efficiency of mercury with 99.0% in the 5.5–8.0 pH range. The process had shown that an increase of medium ionic strength decreased in the amount of metal removal indicated the outer-sphere surface complex mechanism. Freundlich equilibrium model was suited well for collecting equilibrium data. The removal effectiveness was analyzed by using chloralkali wastewater.

Chemical precipitation can be defined as the change of the form of materials into solid particles during water and wastewater treatment. Chemical precipitation is done to reduce the solubility of ions from water bodies. It is generally used for the removal of cations and anions of metals (Wang et al. 2005). In a study, iron oxide nanoparticles were utilized as a chemical precipitation method for the removal of mercury from wastewater (Vélez et al. 2016). They are preferred due to their properties, such as ease of separation, biocompatibility, and magnetic properties. In the study, Fe_3O_4 and γ-Fe_2O_3 nanoparticles were produced by utilizing iron salts along with NaOH as precipitation agents, and aloe vera was also used as their stabilizing agent. After this, the particles used in the process were characterized by three measurements. The iron particles contained narrow size distribution, which is ~100 nm. After the analysis, it was observed that the maximum percentage of mercury removal was 87% for the concentration of 8 mg/L, 73% for 6 mg/L, and 68% for 4 mg/L and was also confirmed by atomic absorption spectroscopy (AAS) measurements.

Mercury can also be removed from wastewater by advanced oxidation processes (AOPs) as discussed by the author, Ghosh et al. (2022), in which AOPs and adsorption processes can do the removal of mercury. The author discussed the adsorption of mercury through batch and column methods. The result stated that the removal of mercury through AOPs was applicable for wastewater appropriately. Mercury removal through available technologies should consider three important things: it should be cost-effective, have maximum efficacy, and be suitable for application in any conditions.

12.2.3 Biological Treatment Method

Various processes have been adopted in the biological treatment method to check mercury pollution from wastewater. One method has been adopted by Wang et al. (2019), where sulphur reduction process was performed by using sulfate-reducing bacteria (SRB) to remove mercury from mercury-contaminated wastewater. The prolonged performance for removing mercury and accumulating methylmercury was also checked. The wastewater used for the study contained 0–50 mg/L of mercury in Hg(II) form, and SRB showed maximum tolerance efficacy in the sulfur-reducing bioreactor with 99.4% efficiency of Hg(II) removal through biogenic sulphide. In the prolonged trials, the sulfidogenic bacteria were always exposed to higher levels of mercury then also methyl mercury was undetected in the bioreactor. Geobacter sulfurreducens PCA and Desulfomicrobium baculatum DSM 4028 showed no or quite fewer mercury methylators were identified in the bioreactor. Hence, this biological reduction process through SRB can reduce Hg(II) from mercury-polluted wastewater without methylmercury accumulation (Table 12.3).

In the biological treatment method, water hyacinth was used to remove mercury and check other parameters. The study by the author, Fazal et al. (2015), demonstrated that the removal of metals and other parameters was conducted by three integrated constructed wetlands (CW) in which industrial wastewater was introduced. These wetlands have shown efficacy in removing pollutants with minimum expenditure while utilizing natural remediation resources. Out of the three wetlands, the first was planted with water hyacinth, the second was constructed with sludge as an anaerobic wetland, and the third was kept as a control. The values generated after analysis showed that mercury was efficiently removed from the two wetlands with minimum cost. The result also proved that the second wetland was more effective in reducing Hg with 100% efficiency, among other parameters. The study also concluded that hydrophytes can survive in extremely polluted conditions while removing harmful nutrients.

Table 12.3 Biological methods for removal of mercury

Resources used	Efficiency (%)	References
Sulfate-reducing bacteria (SRB)	99.4	Wang et al. (2019)
Water hyacinth	99.9	Fazal et al. (2015)
E. coli	80	Chen and Wilson, (1997)
E .coli	99	Deng and Wilson (2001)
Bacillus species	68.1–92	Green-Ruiz (2006)
Bacillus megaterium, MB1	80	Chien et al. (2012)
Malt spent rootless (MSR)	95	Anagnostopoulos et al. (2012)
Eucalyptus camaldulensis bark	69.9	Ghodbane and Hamdaoui, (2008)
Lemna minor powder	92.6	Li et al. (2011)

In the context of bioremediation for the uptake of heavy metals, another study was conducted by the author, Chen and Wilson (1997), in which *E. coli* was used. The author stated that the intracellular bioaccumulation process and the use of microorganisms through genetically engineered techniques can be used for cleaning and removing mercury ions from polluted *water. Escherichia coli* strains were produced by genetically engineered for the control and expression of Hg^{2+} transport system and also to over-express metallothionein (MT), which was used as a carboxyl-terminal fusion with glutathione S-transferase which is together called (GST-MT). In the study, the particular strain capability was examined for Hg^{2+} accumulation. The MTs have low metal binding proteins which are rich in cysteine residues and are effective for binding heavy metals such as Hg^{2+} and Cd^{2+}, etc. In a metal transport system, MT helps microorganism to accumulate metals. The study suggested that Hg^{2+} bioaccumulation was noticed unaffected by Na^{2+}, Mg^{2+}, and Cd^{2+}. The strain can tolerate extreme pH and ionic strength conditions. The particular recombinant strains have shown effective as a bioremediation method to control Hg^{2+} from polluted sources.

E. coli was again used by the author Deng and Wilson (2001), for eliminating mercury from wastewater. Wastewater containing ten different types of ions in which the concentration of mercury was 2.58 mg/L with 9.6 pH levels. The wastewater was compared with distilled water in which it was seen that mercury uptake was higher in wastewater rather than distilled; this was due to high ionic strength, and it also found that higher pH had less effect on the accumulation of mercury. EDTA also showed uptake of mercury instead of inhibiting it. For the thorough uptake of mercury, a reactor with hollow fiber was used for holding induced cells. The induced cells were first prepared in the laboratory with *E. coli* bacterial strains and antibiotics. These cells showed efficacy in removing more than 99% mercury from the wastewater, and the final quantity of mercury accumulated by these cells was 26.8 mg/g in cell dry weight.

Using microorganisms to treat mercury removal from wastewater is an economically effective way for remediation technology. The study done by Green-Ruiz (2006), examined that using nonviable biomass of *Bacillus* sp., which was native to

Urias Estuary, was capable of adsorbing mercury. During the study, some biosorption batch experiments were performed to know the metal adsorption capability of the particular bacterial species in different mercuric concentrations and pH. A total of six various concentrations such as 0.25, 0.5, 1.0, 2.5, 5.0, and 10.0 mg/L were prepared with mercuric chloride and with Milli-Q purified water in which 0.2 g dry weight of the nonviable biomass was utilized to remove from 0.023 mg at 0.25 mg/L of Hg(II) to 0.681 mg at 10.0 mg/L of Hg(II). It has been noticed that mercury adsorption occurred mostly in the first 20 min, and it was also found that changes in the pH had a major impact on the adsorption of metal by the bacteria in the pH range between 4.5 to 6.0 at 25 °C with the solutions 1.0, 5.0, and 10.0 mg/L of Hg(II).

In this matter, another study done by the author, Chien et al. (2012), where selection of *Bacillus megaterium* MB1, bacteria were used. This bacterium was isolated from the preserved sediment of Minamata Bay. The *B. megaterium* MB1 is a non-pathogenic bacteria with a broad spectrum for resistance to mercury. It has the potential to reduce mercuric ions and can tolerate heat as well. It can produce spores and has shown efficacy in controlling mercury pollution in wastewater. The bacterium *B. Megaterium* was recognized as an effective catalytic bio-agent as a safe remediation method to control mercury pollution in wastewater. The result after analysis showed that the alginate gel immobilized *B. Megaterium* MB1 cells reduced 80% mercury from the 10 ml/L mercuric chloride solution in 24 h. These cells also showed effective performance in removing mercury even though repeated nine times. The analysis was done to see the contents of mercury from the alginate beads with or without immobilized *B. megaterium* MB1 and found that maximum quantity of mercury was arrested in the gel beads. The bacteria efficiently degraded organic mercury to ionic or inorganic mercury, reduced the soluble inorganic mercury to insoluble metallic mercury, and also violated the metallic mercury from the medium growth.

Mercury, even at low concentrations, harms the environment greatly (Anagnostopoulos et al. 2012). So, the author proposed removing mercury with minimum cost and eco-friendly technique by using Malt spent rootless (MSR). This MSR is a by-product from a brewery factory that was used as a sorbent for removing mercury from wastewater. The biomass obtained from MSR was dried first at 50 °C overnight and then sieved to make a fraction of 0.150–1.180 mm chosen for the experiment. In this process, pH, the solid to liquid ratio, the contact time between sorbent, and the mercury concentration at the beginning were analyzed to see the effect of mercury removal from wastewater. The pH also played an important role in the biosorption process as it was noticed that the maximum mercury sorption onto MSR was found at 5 pH. The sorption kinetics during the study showed that the sorption of mercury is a fast process in which diffusion of film and intra-particle diffusion played a major role. During the analysis, the kinetic data were discussed by pseudo-second order and through Elovich models whereas, the isotherm data were suited well through the Langmuir model which is determined by a monolayer with a capacity of q_{max} equal to 50 mg/L and stated that the functional group with limited sorption process. So, from this study, we came to know that MSR can remove mercury because of the presence of carboxyl and phosphonate groups on their surfaces. It was also seen that mercury desorption by using MSR was effective with 0.1 M HCl.

One of the novel processes of the biological mercury recovery method was discussed by the author, Ghodbane and Hamdaoui (2008), in which *Eucalyptus camaldulensis* bark was used, which is known as a solid waste and is obtained from forest. This eucalyptus bark was used as a sorbent to remove Hg(II) from wastewater. The mercury solution was prepared with applicable chloride in distilled water for the experiments. This process was highly based on pH and the sorption capacity was based on operating variables shown in the sorption experiments. During the study, it was noticed that the kinetic measurements in the process were faster with uniformity. After the analysis, the highest sorption capacity was found to be 33.11 mg/g at 20 °C. The change was noticed in the negative value of free energy, signifying that the nature of sorption was spontaneous. Hence, these analyses helped determine that eucalyptus bark can be used as an effective remediation tool for removing mercury from wastewater.

Duckweed (*Lemna minor*) was used in powder form to check the adsorption ability upon mercury removal. The study removed both inorganic mercury and organic mercury (methyl and ethyl mercury) (Li et al. 2011). After 40 min when the adsorption was done by *L. minor* powder, and then the concentrations of mercury were compared from the initial level with the final level. The initial mercury concentration was 12 and 50 μg/L. In this study, the removal of inorganic and organic mercury from wastewater was shown together for the first time. The C-O-P and phosphate groups were found to be major adsorption sites. The reaction occurred due to chelation interactions within amine groups and mercury present on the surface of the cells.

12.3 Adsorption Technologies

The process known as adsorption plays a very important role in wastewater treatment. It can be defined as a surface phenomenon where solid particles such as pollutants are able to attach to the adsorbent surface by various chemical and physical forces such as chemisorption, ion exchange, or covalent bond formation. The pollutants like mercury, once attached to an adsorbent, can be easily separated from water sources by desorption. This makes adsorbents a valuable tool for remediation as they can be recycled and be prepared in a biocompatible manner. Adsorbents can be of various types depending on their constituents, like polymers, activated carbon, or even metal oxides (Artioli 2008).

12.3.1 Activated Carbon

Activated carbon is a commonly used adsorbent for treating water by removing various pollutants like organic matter, heavy metals, odor, and other synthetic compounds. It has a porous structure that allows for the physical attachment of liquid or gaseous molecules to the surface of the absorbent (Soni et al. 2020; Johnson 2014).

Activated carbon can be synthesized in either powdered or granulated form through a process involving the pretreatment of a carbon source followed by carbonization and, finally activation. It is a purified form of charcoal capable of adsorption. According to Nille et al. (2021), activated carbon has much promise in the field of pollutant removal due to its lower cost, ease of functionalization, porosity, and high surface area.

In research work by Dou et al. (2023), activated carbon was found to be capable of removing gaseous mercury. Since pure activated carbon may not have sufficient activated sites on its surface, modification can be done to improve its adsorption capacity by using compounds like metal oxides, halogens, or sulfides. But these additions could add further to the pollution problem. In order to tackle this issue, the author used hydroxyl radicals produced by the advanced oxidation process involving UV/H_2O_2 to modify the activated carbon. The study utilized this clean process to procure activated carbon, which showed optimum activity at a temperature of 120 °C when the concentration of H_2O_2 was at 9%. The maximum adsorption activity was observed by the removal of 3636.43 μg/g of mercury. Mercury is capable of causing immense harm to the human body, and it becomes even more difficult to remove from water when it combines with cyanide and forms a stable $HgCN_2$ complex. This complex is observed in the effluent discharge from gold mines. According to Aliprandini et al. (2020), an activated carbon adsorbent was able to remove 81% of the mercury cyanide from a solution containing the mercury complex as well as cyanide in its free form at pH 12. The adsorption occurred without any interference due to the free cyanide. In another study by Mistar et al. (2019), activated carbon was able to remove 218.08 mg/g of mercury from water at its highest adsorption capacity. The activated carbon was prepared from a Bambusa vulgaris variety known as striata using NaOH's chemical activation method. The activated carbon was used in a fixed bed column to remove mercury from a wastewater discharge (Table 12.4).

Table 12.4 Methods for mercury removal

Removal method	Efficiency (%)	References
Activated carbon	81	Aliprandini et al. (2020)
Cao modified activated carbon	81	Budihardjo et al. (2021)
African Palmae shell-activated carbon	93	Egirani et al. (2021)
Zn based polymer	85	Zeng et al. (2022)
Cysteine-based membrane	97	Islam et al. (2020)
Micellar-enhanced ultrafiltration	95	Yaqub and Lee (2020)
Carbon nanotube ultrafiltration	>76	Nayak et al. (2020)
Yam peel nano adsorbent	75	Marimón-Bolívar et al. (2018)
Peptone nano adsorbent	87	Marimón-Bolívar et al. (2018)
Aminopropyl saline nano adsorbent	94	Marimón-Bolívar et al. (2018)
Magnetic chitosan nano adsorbent	99.91	Rahbar et al. (2014)
Au-GO nanocomposite	99	Zhang et al. (2015)

Activated carbon could also be produced by using CaO from clamshells, coal, and peat soil in order to remove mercury from landfill leachate. The activated carbon was able to reach its equilibrium within 100 min when agitated at 500 rpm. The adsorbent was able to remove 81% of the mercury, with its adsorption capacity being 114 mg/g when mixed with coal and 102 mg/g when mixed with peat soil (Budihardjo et al. 2021). Activated carbon, when modified with thiol, was found to have enhanced adsorption and separation capacity. The adsorbent was synthesized using a hydrothermal process without having any toxic effects. It was able to remove mercury at maximum when pH was 7 at 366.3 mg/g. The activated carbon was found to adsorb through ion exchange and electrostatic attraction by a spontaneous and exothermic reaction (Chen et al. 2019). In a study by Egirani et al. (2021), activated carbon was prepared from African palmae shells and tested in both powdered and granular forms. In powdered form, it was able to remove 93% of the mercury, and 92% in granular form.

12.3.2 *Iron Oxides*

Iron oxides are commonly composed of iron and oxygen and are found in nature. They are abundant on Earth, and their surface chemistry is mainly based on the interaction between solvated ions and water (Voloshina 2017). They have applications in various biological and chemical procedures like water treatment. Iron oxides usually have iron in the form of Fe^{2+} and Fe^{+} with unpaired electrons in a 3d subshell, making them magnetic in nature (Siddiqui and Chaudhry 2017; Valezi et al. 2019).

During research by Zhang et al. (2021), magnetic iron oxide was added to sulfur-doped activated carbon to remove mercury. The activated carbon was prepared by the carbonization of thiol polymer leading to the increase in surface area. The adsorbent showed a large 1329 m^2/g surface area. It was able to adsorb a maximum of 187 mg/g of mercury even when added to mixed solutions, making it ideal for water treatment applications. In Fig. 12.3, the action of the magnetic adsorbent can be observed. The adsorbent was able to remove the mercury ions from the contaminated water by its magnetic action and functional groups. A magnetic force was then used to separate the adsorbent from the water after purification. In a study done by Inglezakis et al. (2020), mercury ions were adsorbed from water by synthesizing a nano adsorbent with magnetic Fe_3O_4 and silver nanoparticles. This nanocomposite showed a removal rate of 71.3 mg/g, which was higher than when pure Fe_3O_4 particles with a removal rate for mercury ions at 28 mg/g. The adsorption reaction occurred as a result of mechanisms involving adsorption, reduction, and oxidation among the particles in the nanocomposite and the pollutants. In another study conducted by Behjati et al. (2018), a nanocomposite was synthesized by modifying Fe_3O_4 nanoparticles with carbon disulfide and tetraethylenepentamine. Dithiocarbamate groups were also added to the surface of the nanocomposite. The nano adsorbent showed a saturation magnetization of 27 emu/g. The maximum adsorption for mercury ions was found between a pH of 2 and 6.5 at 109.5 mg/g.

In a study by Chen et al. (2019), $NiFe_2O_4$ was used to prepare a nano adsorbent along with thiol modification. This magnetic nanocomposite was able to remove mercury ions at a maximum capacity of 366.3 mg/g when the pH was 7. The nano adsorbent showed stability and renewability after multiple cycles.

12.3.3 Polymers

Polymeric materials have gained popularity as ideal adsorbent materials in water treatment procedures. They can be synthetic or natural based on their origin and have found application in removing mercury from wastewater. They constitute repeating monomeric units which form a chain-like structure through linking. Depending on their composition, they are capable of a vast number of modifications with different properties.

In a study by Bhatt and Padmaj (2019), mercury in its three forms of elemental Hg^0, its inorganic form Hg^{2+}, and methyl mercury CH_3Hg^+ were tested for removal by using a thiomer synthesized using chitosan and thiobarbituric acid. The maximum adsorption observed in this case was 178.4 mg/g for methyl mercury, 1367 mg/g for elemental mercury, and 2493 mg/g for inorganic mercury. A study by Say et al. (2008) was conducted on the effectiveness of an organo smectite composite modified by a polymer with dithiocarbamate anchoring. The study used carbon disulfide in order to incorporate the dithiocarbamate into the layers of the clay. This adsorbent was capable of removing not only 214.6 mg/g of methyl mercury but also 157.3 mg/g of Hg^{2+} and phenyl mercury at 90.3 mg/g. The removal of mercury occurred in a competitive manner, with their competitive adsorption capacity at 10 ppm at 12.7 mg/g for phenyl mercury being the highest, followed by 9.2 mg/g for methyl mercury, and finally 7.7 mg/g for the inorganic mercury Hg(II).

Chitosan beads were grafted by using polyacrylamide through ATPR or atom transfer radical process by surface initiation (Li et al. 2005). When compared with the pure chitosan beads, the modified form showed better adsorption. While the maximum adsorption for chitosan polyacrylamide beads was observed at 322.6 mg/g within 60 min of equilibrium time, the pure chitosan beads showcased maximum adsorption of 181.8 mg/g over 15 h. The modified adsorbent could also selectively absorb mercury in a mixture with lead ions at a pH less than 6. This selective adsorption was due to the presence of amide groups on the modified beads, which allow for the formation of covalent bonds with mercury ions. The desorption and regeneration were also effective when done with a solution of perchloric acid. In another study, activated carbon was synthesized from waste rubber tires and reacted upon by a diethylenetriamine trimesoyl chloride-based copolymer. This adsorbent was able to remove both Hg^{2+} ions and methyl mercury by 317.3 mg/g and 263.6 mg/g, respectively. Desorption and regeneration were observed for seven cycles of mercury removal.

In a study by Wang et al. (2022), sulfhydryl modified hyperbranched and highly cross-linked polyamide amines to create an adsorbent for eliminating mercury ions

Hg(II). At pH 4.5, this adsorbent showed its highest removal capacity at 282.74 mg/g at temperature 318 K. Adsorption occurred selectively for the mercury ions in a combination of different metal ions. Desorption and regeneration were successfully done for five cycles without significant loss. In another study by Zeng et al. (2022), a novel high-performance zinc-based coordination polymer was prepared. This adsorbent was able to remove 2763.87 mg/g of the ions from an aqueous solution at pH 3 through an endothermic and spontaneous chemisorption reaction. This adsorbent was found to be stable with good repeatability and showed above 85% of mercury ions removal rate. The reaction depended on ion exchange and chelation mechanisms between the mercury ions and the nitrogen and sulfur atoms of the adsorbent. Inverse vulcanization and post-modification were done to functional rubber adsorbent prepared from biomass. Its preparation utilized industrial by-products and renewable materials like sulfur and cotton seed oil respectively. This plant-based rubber adsorbent showed a maximum adsorption capacity of 343.3 mg/g for mercury ions. The rate of removal remained above 80% even after five desorption cycles.

12.3.4 Hydrogels

Hydrogels can be defined as a structure with a three-dimensional network that can capture large amounts of water molecules leading to swelling. Cross-linked or chained molecules in the hydrogel protect it from dissolving in aqueous solutions. They can exist in nature in the form of polymer networks such as collagen. Their ability to change their volume makes them applicable to a variety of water treatment options (Peppas et al. 2012). The three-dimensional network of hydrogels is composed of cross-linked water-soluble polymers, which become hydrophobic in nature. This allows them to be flexible and swell by taking in a large quantity of water. Their applicability is also improved by the availability (Shmeis 2022) of functional groups such as amides, amines, carboxylic acid, and other polar groups.

In a study, a novel nano cellulose hydrogel was developed using cellulose nanocrystals and nanoclusters of gold. This adsorbent hydrogel was not only useful for the removal of mercury but also for the detection of mercury due to its fluorescent nature. The maximum adsorption capacity was 95.7 mg/g for mercury ions. It depended on the functional groups available on the hydrogel and facilitated by the three-dimensional porous network. This porous structure allowed for increasing adsorption and was formed due to the natural structure of cellulose nanocrystals (Lei et al. 2021). The fluorescence was a result of the gold nanoclusters wrapped into the skeletal structure of the hydrogel and also helped in the sorption of mercury ions.

In another study by Luo et al. (2021), nanocellulose and gold nanoclusters were again employed in the form of a hydrogel. The nano cellulose was derived from wood, and the adsorbent was also able to showcase fluorescence along with the sorption of mercury ions. In this case, the maximum adsorption was observed at 234.4 mg/g. The fluorescence was detected to be highly selective towards mercury ions with a 0.09 μg/L detection limit. The wood-based cellulose was not only able to improve the adsorption in this case due to its porous structure but also was found to be cost-effective for mercury removal. In a study by Saberi et al. (2020), another natural polymer, starch, was studied for its adsorption capacity of mercury ions. The hydrogels in this study were prepared by using first starch and PEG—poly acrylic acid in one case and then silver nanoparticles with the starch and PEG—poly acrylic acid in the other case. The maximum adsorption in the case of adsorbent with starch and PEG was observed at 158.21 mg/g when the pH was 7, and for the adsorbent with silver nanoparticles, it was observed at 182.53 mg/g at pH 6. The adsorption in both cases occurred as a homogeneous process on a monolayer. A novel magnetic hydrogel was prepared by using Fe_3O_4 nanoparticles with partially hydrolyzed polyacrylamide. This hydrogel was found to be capable of removing both mercury and lead ions from wastewater (Ebrahimpour and Kazemi 2023). The maximum adsorption capacity was found to be 256.41 mg/g for mercury through an endothermic and spontaneous reaction. In a study by Khozemy et al. (2020), biodegradable wheat flour and acrylamide-based hydrogel were prepared by using gamma radiation to remove Hg^{2+} ions wastewater. This adsorbent was found to showcase an adsorption capacity of 15.6 mg/g at maximum with more than 9000% swelling.

In another study, graphene oxide hydrogel was prepared by encapsulating it with silica gel with the help of ascorbic acid in a chemical reduction reaction. This hydrogel had a maximum adsorption capacity of 266 mg/g for mercury ions. Its adsorption capacity was 32% higher than that of pure silica gel due to graphene oxide (Lu et al. 2019).

12.4 Membrane Technologies

Membrane technologies have been chosen as an effective tool for wastewater reclamation. In membrane technologies, various types of membranes are used with varied pore sizes used to treat wastewater and remove contaminants up to a safe level before discharge. The membrane technology can have various applications in wastewater treatment, such as desalination, removal of dissolved solids and inorganic chemicals. In the context of mercury remediation, two important types of membrane filter technologies are ultrafiltration and membrane separation.

12.4.1 Ultrafiltration

Ultrafiltration is a type of membrane filter that utilizes pressure to separate solid waste from wastewater. The separated water, after ultrafiltration, tends to be free from solid wastes and pathogenic matter due to its pore size. Like other membrane filtration techniques, it also utilizes a semipermeable membrane. The pore size of the ultrafiltration membrane tends to be in the range of 0.01 and 0.1 um (Khan et al.

2022). The separation in this technique depends on the transmembrane pressure to remove the pollutants like mercury from water. Both surfactant and micellar-based semipermeable membranes can be used to eliminate heavy metals from wastewater by using this process (Hussain et al. 2022). The membrane can be of natural or artificial origin, depending on the selected materials and the pressure applied.

Islam et al. (2020) studied the effectiveness of a three-step procedure involving pre-filtering followed by ultrafiltration and finally microfiltration on a solution containing mercury ions and mercury sulfide nanoparticles. The membranes used during this study were synthesized using thiol through the incorporation of cysteine or cysteamine into a polyvinylidene fluoride membrane which was functionalized by using poly acrylic acid. This study reported that the ultrafiltration procedure was able to bring down the concentration of mercury sulfide nanoparticles to below the detection levels of 2 ppb in 12.5 h of contact. The mercury ions were also removed to a level below detection limits. The adsorption efficiency of mercury from wastewater was found to be 97% within 20 h.

There is an urgent requirement for a process to eliminate dangerous heavy metals like mercury from water bodies. To accomplish this goal, Yaqub and Lee (2020) studied the role of a micellar-enhanced ultrafiltration procedure that utilized cetylpyridinium chloride and sodium dodecyl sulfate for chelation. When the parameters were at optimum pressure levels at 2.5 bar, pH at 8, and initial mercury concentration at 10 ppm, 95% mercury removal was observed.

In another study, an ultrafiltration membrane with a mixed matrix was prepared by using multi-walled carbon nanotubes. It was found that the membrane with 0.3 weight % of the nanotubes showcased adsorption of greater than 76% in the case of mercury ions (Nayak et al. 2020).

12.4.2 Membrane Separation

Mercury removal can also be achieved through the microfiltration process, as discussed by author, Liu et al. (2013). Mostly mercury is present in suspended, adsorbed, and also in dissolved forms of methyl mercury and salts. The effluent should contain 1.3 nanogram/L as the target set by the authority, but in some lakes, it was found to be 12 nanogram/L in the effluent of wastewater. Microfiltration was used for the effluent after conventional treatment processes. It reduced the mercury level to 5 ppt without any use of additives. The target level of mercury was reached, and the automatic system performed efficiently. The hollow fiber system was stable and consistent for operating for more than 12 months on real site conditions. This method helped remove the mercury from the wastewater and met the set target levels with high-crystalline poly(vinylidene fluoride) or PVDF hollow fiber which helped to gather solid pollutants on the outer side of the membrane. The same method was applied in a power plant where the wastewater containing mercury pollutants required treatment before discharge into the outlet. The removal of mercury was done from flue gas desulphurization wastewater with or without using chemical precipitation which MF followed to remove the mercury arrested in the precipitate. In both cases, positive and stable results proved that PVDF hollow fiber membranes are long-term cost-effective, simple, and reliable methods for the removal of mercury from wastewater.

In one study suggested by the author, Broom et al. (1994) stated that crossflow microfiltration was used to remove mercury with other metals from wastewater. The wastewater in the study was taken from a metal plating and a battery recycling plant for the operational performance of the crossflow microfiltration plants. The mercury present in the wastewater was reduced to the ppb range. This crossflow microfiltration showed better performance compared to the traditional treatment system. During the study, performance was noticed for settling tanks, settling tanks with sand filters, and also with crossflow MF system. The crossflow system depends upon the use of a dynamic membrane for the formation of a filtration medium. The membrane was placed upon the internal wall of woven fabric cloth and manufactured by naturally present solids in the feed suspension or by deposition materials like diatomaceous earth, metal hydroxides, or kaolin. In the study, the precipitates of heavy metals provided a membrane without any use of filter support.

Mercury removal can also be done through ultrafiltration as stated by the author, Huang et al. (2015). In the study, mercury was removed from wastewater through polymer-enhanced ultrafiltration by utilizing polyvinyl amine. Mercury was removed from the wastewater with 99% efficiency, which was unable to get through conventional techniques. The amine polymer, which was used in the ultrafiltration technique, was used to bind mercury, and the limiting flux followed the gel layer formation model. The surface of the membrane was cleaned regularly with diluted hydrochloric acid for the smooth running of the membrane and to recover its permeability.

As per the author Yaqub and Lee (2020), mercury removal can also be done by using micellar-enhanced ultrafiltration (MEUF) and polyacrylonitrile membrane along with sodium dodecyl sulphate (SDS) as surfactant. In this method, leakage in the surfactant monomer pervaded water causes secondary pollution that can be checked through MEUF along with an activated carbon filter (MEUF-ACF). For optimizing the MEUF technique, the effect of the working parameters such as concentration of mercury, pH of the feed solution, the molar ratio of SDS to Hg, along with retention pressure was examined. In the study, an artificial neural network (ANN) model was used to evaluate mercury removal efficacy and enhance the MEUF process without any physical labor and lengthy time-taking experiments. ANN model was based upon statistical values like mean square error (MSE) and coefficient of determination, i.e. R2. The operating parameters are known to be as 10 ppm of Hg concentration with pH 7.0, along with the molar ratio of SDS to Hg 8:1, and with 1.5 retention pressure. MEUF results indicated that mercury was removed with 95.75% efficiency along with 50.91% SDS whereas 96.83% of Hg and 97.15% of SDS rejection was seen by utilizing MEUF-ACF.

In membrane-based technique, one study was done by the author, Zunita (2021), where Graphene oxide (GO) was used as a nanofiltration technique for mercury removal from wastewater. Graphene oxide and derivatives are certain kinds of materials that have a nanostructure and are used as thin, flexible sheets with maximum chemical stability and mechanical strength. The study noticed that GO-based membranes were used as a shield or barrier to capture mercury vapor in its nanopores and channel. This nanochannel was also used for ion mobility and molecule filtration properties.

The author studied a graphene-based membrane, Makertihartha et al. (2017), where a graphene-based nanofiltration membrane was used for the effective removal of mercury from wastewater. Table 12.5 gives the efficiency of membrane separation process for different filtration types.

Mercury removal from wastewater can also be done through the forward osmosis process. In this method, trace amounts of mercury can be removed from waste effluents. In the study, two inorganic salts, sodium chloride (NaCl) and magnesium chloride ($MgCl_2$), were used with varied valences and examined with different solution concentrations, pH, and temperature to remove mercury from the wastewater efficiently. It was noticed that operating factors like the concentration of influent and temperature directly affected water flux and changed osmotic pressure along with the internal concentration polarization. The temperature directly affects the ICP, which indirectly causes the diffusibility. The results showed that the rejection of mercury by 1 M solutions of NaCl was 98.2%, whereas MgC12 showed 99.9%. When the operation recovery rate reached 50%, the rejection of mercury was done with 98% efficiency (Wu et al. 2016).

Table 12.5 Membrane separation process

Filtration types	Resources/chemicals used	Efficiency (%)	References
Microfiltration	PVDF hollow fiber	90	Liu et al. (2013)
Crossflow microfiltration	Dynamic membrane-woven with fabric cloth, metal hydroxide, and kaolin	99	Broom et al. (1994)
Ultrafiltration (PEUF)	PEUF with ployvinylamine	99	Huang et al. (2015)
Ultrafiltration (MEUF)	MEUF with polyacrylonitrile membrane and SDS	96.83	Yaqub and Lee (2020)
Nanofiltration	Graphene oxide-based membrane	81–99	Zunita (2021)
Forward osmosis	NaCl and $MgCl_2$	98.2–99.9	Wu et al. (2016)

12.5 Applications of Nanomaterials in the Removal of Mercury from Wastewater

Nanomaterials can be defined as materials having their dimensions in nanoscale ranging between 1 and 100 nm. Nanotechnology has gained a lot of interest from researchers in the field of water purification technologies. Their unique properties make them ideal for wastewater treatment involving the removal of mercury. Nanoadsorbents belonging to different categories have been extensively studied due to their unique properties when it comes to absorbance (Kumar et al. 2021). Nanoadsorbents can be of different types, such as nanoparticles, nanocrystals, nanotubes, nanoclusters, or nanocomposites. These nanomaterials used for the adsorption of mercury can be of natural origin or synthetic. Some of the commonly used nanomaterials are composed of organic materials such as cellulose, chitosan, starch, and chitin, or even inorganic in nature, such as those involving metals or metal oxides. Various factors determine the effectiveness of nanomaterials when it comes to adsorption, such as adsorption efficiency, regeneration, desorption, cost-effectiveness, and large-scale production (Yang et al. 2019).

In a study done by Marimón-Bolívar et al. (2018), nanocomposites were synthesized by using magnet nanoparticles along with different materials such as yam peel and peptone, and aminopropyl saline. The adsorption efficiency of the adsorbent synthesized using yam peels was at 75% when the pH was 8. The nano adsorbent containing peptone was at 87% when the pH was 7, and 94% for the nano adsorbent with aminopropyl saline. In a study conducted by Rahbar et al. (2014), magnetic nanocomposites were prepared by using chitosan as a coating. This led to the formation of a biocompatible nanocomposite which was able to remove 99.91% of mercury ions when the pH was 5 and the initial concentration of Hg was 6.2 mg/L.

Another magnetic nanocomposite was prepared by using microcrystalline cellulose with amine-capped Fe_3O_4 nanoparticles and carboxylic nanosheets. This nanocomposite was able to remove 469.48 mg/g of mercury ions within 20 min of adsorption time. The adsorption occurred due to the S atoms interacting with the mercury ions in the solution (Gao et al. 2021). A magnetic nanocomposite was prepared by using activated carbon derived from the shells of hazelnuts with the help of $ZnCl_2$ and KOH. By using hazelnut shells, iron oxide crystallites were introduced into the nanocomposite. It was able to eliminate 80 mg/g of mercury from the aqueous solution (Zabihi et al. 2022). A study done by Alvand and Shemirani (2017) used covalently immobilized graphene quantum dots on magnetite nanospheres coated with silica. This nanocomposite was not only able to remove 68 mg/g of mercury ions but was also able to detect mercury ions due to its property of fluorescence. The limit of detection for mercury was noted to be at 30 nM. The adsorption capacity of this nanocomposite was also influenced by the superparamagnetism exhibited by the nanocomposite. Desorption could occur with the help of EDTA, making it recyclable in nature. In another case, Zhang et al. (2015) used gold nanoparticles along with graphene oxide and iron oxide to prepare a nanocomposite. It was able to adsorb more than 99% of mercury within 30 min of contact. The desorption was also found to be effective since it could retain 96% of its efficiency even after 15 cycles.

A nanocomposite was prepared using Zinc sulfide modified zeolite prepared from fly ash. An ion exchange procedure was used to introduce the ZnS nanoclusters into the zeolite. This nanocomposite showed a mercury removal efficiency of 99.90%. The adsorption occurred due to the formation of the HgS compound after the reaction between the nano adsorbent and mercury ions. It was also found to be efficient after 5 cycles, with a decrease of only 2% in efficiency (Li et al. 2021). In another study involving the use of zeolites by Shirzadi and Nezamzadeh-Ejhieh (2017), a nanocomposite was synthesized by using a natural clinoptilolite obtained from Iran. The nanocomposite was modified by using Hexadecyltrimethylammonium bromide (HDTMABr). This modification allowed for significant improvement in the adsorption efficiency of the due to the complexation and ion exchange procedures. The adsorption occurred in the monolayer and fit within the Langmuir equation. The reaction was also found to fit as a pseudo-second-order model along with a spontaneous and exothermic nature.

12.6 Biosorption

Biosorption is a type of adsorption reaction involving the use of biological material or biomass as the adsorbent. The adsorption is facilitated by the presence of functional groups present on its surface, leading to the adsorption and hence elimination of pollutants from water. It is a cost-effective solution for the remediation of heavy metals like mercury from wastewater. The biosorbent can be used in an immobilized form or in a free form, depending on the application (Kim et al. 2006; Ghaedi 2021).

In a study by Fathollahi et al. (2020), a living biofilm was attached to a geotextile made of polypropylene and polyethylene. This biosorbent has been described in Fig. 12.2, where the biofilm is observed to remove mercury ions from the contaminated water and leave behind treated water. The mercury ions can be observed to be attached to the biofilm developed on the biosorbent. The biofilm for this study was developed by using pavement biofilms in stormwater. The adsorbent was able to remove a maximum of 65.38% of mercury ions from an aqueous solution within 120 min when the pH was at 5.5, and the temperature was 25 °C. The free mean biosorption energy was detected to be 2.36 kJ/mol. The carboxyl, amine, and hydroxyl groups present in the biosorbent were responsible for the reaction.

In a study by Fabre et al. (2020), a biosorbent was prepared by using banana peels to adsorb and remove mercury from solutions. The initial concentration of mercury was kept at 50 $\mu g/dm^3$, and the larger effect on the adsorption was caused by the contact time and the dosage of added adsorbent. The sorption capacity was 0.75 mg/g of mercury. In Fig. 12.1, the action of banana peel-based adsorbent can be observed. The mercury ions are retained by the structure of banana peel consisting of lignin, cellulose, and hemicellulose, which help in removing the mercury ions from the wastewater and leaving behind decontaminated water. In another study by

Fabre et al. (2021), agricultural and industrial wastes were tested for their effectiveness as biosorbents. Lignocellulose, a material found in agricultural wastes, is capable of forming bonds with mercury ions present in wastewater. This will make such materials a cost-effective and green replacement for many of the conventional methods. In this study as well banana peels were found to be an excellent biosorbent for the removal of mercury. Maximum adsorption of 90% was observed for this biosorbent after 72 h of exposure, and more than 80% of the mercury was removed within 10 h. In another study involving the use of biological material as a biosorbent, Pagala (2023) used the Tamarindus indica flower to remove mercury. This biosorbent could remove 76.5% of the contaminating Hg by using a batch process. A maximum adsorption rate of 78.2% was observed when the pH was 5 and the time of contact was 37.65 min.

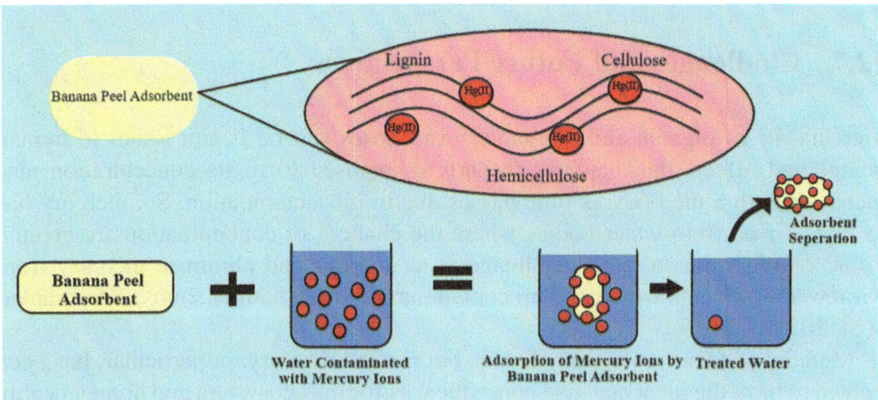

Fig. 12.1 Remediation of mercury ions by using banana peel-based adsorbent

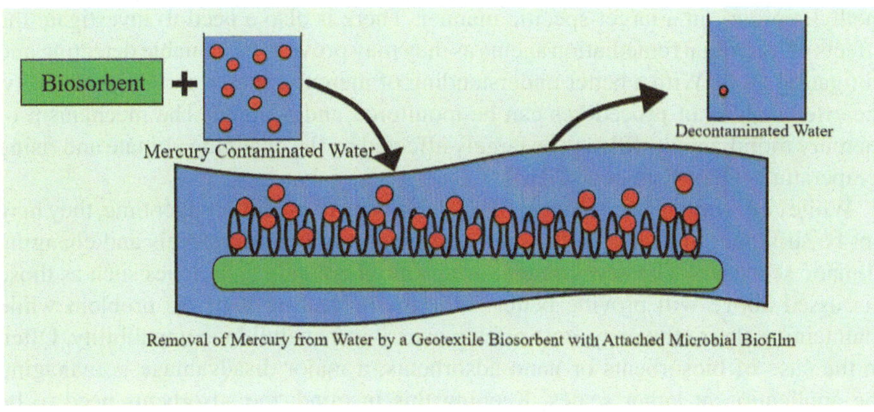

Fig. 12.2 Use of microbial biofilm-based biosorbent to remove mercury ions from wastewater

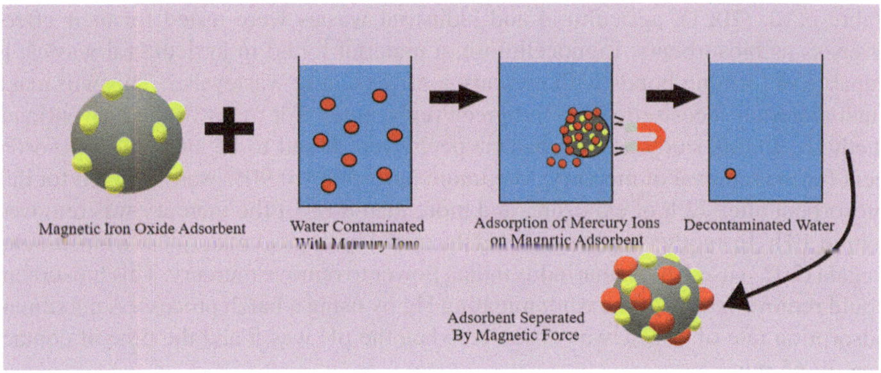

Fig. 12.3 Removal of mercury ions by using a magnetic adsorbent

12.7 Challenges and Future Perspectives

Mercury, in its organic and inorganic form, poses several health issues to human beings and affects the major body parts if exposed to it. Its concentration also increases within the body as time passes due to bioaccumulation. So, mercury has to be removed from water bodies where the chances of contamination are greater. To accomplish this, a major challenge is to monitor and eliminate mercury from wastewater effluents which end up contaminating other natural sources (Jaishankar et al. 2014).

Mercury is toxic in its many forms, but methyl mercury, in particular, has been known to have the most negative consequences for the ecosystem and human health. It has been pointed out in research that mercury remediation largely depends on the time scale of accumulation. Hence a deep understanding of the time scale is necessary for designing appropriate mitigation strategies for dealing with bioaccumulation of methyl mercury in a target-specific manner. There is also a need to investigate the effects of microbial remediation agents as they may prove to be valuable detecting and mitigating tools. With a better understanding of mercury uptake and bioavailability, the effectiveness of procedures can be monitored and adapted. The mechanism of mercury mobilization will also be largely affected by the changing climate and rising temperatures (Hsu-Kim et al. 2018).

While conventional treatment methods have been in use for a longer time, they may not be sufficient to deal with the increasing mercury pollution levels and changing climatic scenarios. The development of novel remediation techniques such as those discussed above will provide better solutions in dealing with the problem while maintaining their efficiency, cost-effectiveness, and overall biocompatibility. Often in the case of biosorbents or nano adsorbents, a major disadvantage is managing the application at larger scales. Keeping this in mind, the adsorbents need to be researched thoroughly for long-term and large-scale application while also keeping their potential toxicity and accumulation in mind (Park et al. 1983; Emenike et al. 2023).

12.8 Conclusion

Water is considered as an essential human right for survival. Keeping this in mind, it has become essential to find proper methods to deal with wastewater pollution as they not only end up contaminating other water sources but also have the potential to be reused if safely treated. To safely manage the threat of highly toxic mercury entering the water systems and food chains, there is a pressing need to find alternative treatment solution systems. The novel systems, such as those discussed above, included a variety of adsorption and filtration systems. Research needs to be done into not only innovating new technologies for wastewater treatment but also into sustainability and future impacts of the novel technologies. There is a bright future in the water treatment field for eco-friendly and cost-effective options provided by nano adsorbents, membrane filtration technologies, and biosorbents. Apart from remediation, research also needs to be done into advanced monitoring systems and better safety checks for industrial effluents. Understanding the procedures and mechanisms behind the bioaccumulation and toxicity of mercury and its different forms, such as methyl mercury, will open up better research opportunities into remediation, monitoring, and safe management practices for treating wastewater contaminated with mercury.

Acknowledgements The authors would like to thank the Gandhi Institute of Technology and Management for the support of resources.

References

Aliprandini P, Veiga MM, Marshall BG, Scarazzato T, Espinosa DC (2020) Investigation of mercury cyanide adsorption from synthetic wastewater aqueous solution on granular activated carbon. J Water Process Eng. 34:101154. https://doi.org/10.1016/j.jwpe.2020.101154

Alvand M, Shemirani F (2017) A $Fe_3O_4@SiO_2@$ graphene quantum dot core-shell structured nanomaterial as a fluorescent probe and for magnetic removal of mercury(II) ion. Microchim Acta 184:1621–1629. https://doi.org/10.1007/s00604-017-2134-2

Anagnostopoulos VA, Manariotis ID, Karapanagioti HK, Chrysikopoulos CV (2012) Removal of mercury from aqueous solutions by malt spent rootlets. Chem Eng J 213:135–141. https://doi.org/10.1016/j.cej.2012.09.074

Anirudhan TS, Divya L, Ramachandran M (2008) Mercury(II) removal from aqueous solutions and wastewaters using a novel cation exchanger derived from coconut coir pith and its recovery. J Hazard Mater 157(2–3):620–627. https://doi.org/10.1016/j.jhazmat.2008.01.030

Artioli Y (2008) Adsorption. Encyclopedia of Ecology Fath. SEJD

Behjati M, Baghdadi M, Karbassi A (2018) Removal of mercury from contaminated saline wasters using dithiocarbamate functionalized-magnetic nanocomposite. J Environ Manage 213:66–78. https://doi.org/10.1016/j.jenvman.2018.02.052

Bernhoft RA (2012) Mercury toxicity and treatment: a review of the literature. J Environ Public Health. https://doi.org/10.1155/2012/460508

Bhatt R, Padmaj P (2019) A chitosan-thiomer polymer for highly efficacious adsorption of mercury. Carbohydr Polym 207:663–674. https://doi.org/10.1016/j.carbpol.2018.12.018

Bhowmick N, Dadhich AS, Prasuna RG, Reddy MK (2018) Screening and isolation of mercury tolerant bacteria from soil around the industrial areas of Visakhapatnam, India. Pollut Res 37:757–762

Broom GP, Squires RC, Simpson MPJ, Martin I (1994) The treatment of heavy metal effluents by crossflow microfiltration. J Membr Sci 87(1–2):219–230. https://doi.org/10.1016/0376-7388(93)E0152-A

Budihardjo MA, Wibowo YG, Ramadan BS, Serunting MA, Yohana E (2021) Mercury removal using modified activated carbon of peat soil and coal in simulated landfill leachate. Environ Technol Innov 24:102022. https://doi.org/10.1016/j.eti.2021.102022

Chen S, Wilson DB (1997) Construction and characterization of *Escherichia coli* genetically engineered for bioremediation of Hg (2+)-contaminated environments. Appl Environ Microbiol 63(6):2442–2445. https://doi.org/10.1128/aem.63.6.2442-2445.1997

Chen K, Zhang Z, Xia K, Zhou X, Guo Y, Huang T (2019) Facile synthesis of thiol-functionalized magnetic activated carbon and application for the removal of mercury(II) from aqueous solution. ACS Omega 4(5):8568–8579. https://doi.org/10.1021/acsomega.9b00572

Chen Y, Yasin A, Zhang Y, Zan X, Liu Y, Zhang L (2020) Preparation and modification of biomass-based functional rubbers for removing mercury(II) from aqueous solution. Materials 13(3):632. https://doi.org/10.3390/ma13030632

Chien M, Nakahata R, Ono T, Miyauchi K, Endo G (2012) Mercury removal and recovery by immobilized *Bacillus megaterium* MB1. Front Chem Sci Eng 6:192–197. https://doi.org/10.1007/s11705-012-1284-3

Chu EW, Karr JR (2017) Environmental impact: concept, consequences, measurement. Ref Module Life Sci. https://doi.org/10.1016/B978-0-12-809633-8.02380-3

Deng X, Wilson D (2001) Bioaccumulation of mercury from wastewater by genetically engineered *Escherichia coli*. Appl Microbiol Biotechnol 56:276–279. https://doi.org/10.1007/s002530100620

Dou Z, Wang Y, Liu Y, Zhao Y, Huang R (2023) Enhanced adsorption of gaseous mercury on activated carbon by a novel clean modification method. Sep Purif Technol 308:122885. https://doi.org/10.1016/j.seppur.2022.122885

Ebrahimpour E, Kazemi A (2023) Mercury(II) and lead(II) ions removal using a novel thiol-rich hydrogel adsorbent; PHPAm/Fe_3O_4@SiO_2-SH polymer nanocomposite. Environ Sci Pollut Res 30:13605–13623. https://doi.org/10.1007/s11356-022-23055-z

Egirani D, Latif MT, Wessey N, Poyi NR, Shehata N (2021) Preparation and characterization of powdered and granular activated carbon from Palmae biomass for mercury removal. Appl Water Sci 11:10. https://doi.org/10.1007/s13201-020-01343-8

Emenike EC, Adeniyi AG, Iwuozor KO, Okorie CJ, Egbemhenghe AU, Omuku PE, Okwu KC, Saliu OD (2023) A critical review on the removal of mercury (Hg^{2+}) from aqueous solution using nanoadsorbents. Environ Nanotechnol Monit Manage 100816. https://doi.org/10.1016/j.enmm.2023.100816

Fabre E, Lopes CB, Vale C, Pereira E, Silva CM (2020) Valuation of banana peels as an effective biosorbent for mercury removal under low environmental concentrations. Sci Total Environ 709:135883. https://doi.org/10.1016/j.scitotenv.2019.135883

Fabre E, Vale C, Pereira E, Silva CM (2021) Sustainable water treatment: use of agricultural and industrial wastes to remove mercury by biosorption. Water Air Soil Pollut 232:284. https://doi.org/10.1007/s11270-021-05165-5

Fathollahi A, Coupe SJ, El-Sheikh AH, Sañudo-Fontaneda LA (2020) The biosorption of mercury by permeable pavement biofilms in stormwater attenuation. Sci Total Environ 741:140411. https://doi.org/10.1016/j.scitotenv.2020.140411

Fazal S, Zhang B, Mehmood Q (2015) Biological treatment of combined industrial wastewater. Ecol Eng 84:551–558. https://doi.org/10.1016/j.ecoleng.2015.09.014

Gao P, Lei J, Tan J, Wang G, Liu H, Zhou L (2021) Self-assembled magnetic microcrystalline cellulose/MoS_2/Fe_3O_4 composite for efficient adsorptive removal of mercury ions (Hg^{2+}). Compos Commun 25:100736. https://doi.org/10.1016/j.coco.2021.100736

Ghaedi M (ed) (2021) Adsorption: fundamental processes and applications. Academic Press

Ghodbane I, Hamdaoui O (2008) Removal of mercury(II) from aqueous media using eucalyptus bark: kinetic and equilibrium studies. J Hazard Mater 160(2–3):301–309. https://doi.org/10.1016/j.jhazmat.2008.02.116

Ghosh S, Othmani A, Malloum A, Christ OK, Onyeaka H, AlKafaas SS, Nnaji ND, Bornman C, Al-Sharify ZT, Ahmadi S, Dehghani MH (2022) Removal of mercury from industrial effluents by adsorption and advanced oxidation processes: A comprehensive review. J Mol Liq 120491. https://doi.org/10.1016/j.molliq.2022.120491

Green-Ruiz C (2006) Mercury(II) removal from aqueous solutions by nonviable Bacillus sp. from a tropical estuary. Bioresour Technol 97(15):1907–1911. https://doi.org/10.1016/j.biortech.2005.08.014

Grimwood K (2013) Thermal degradation. Encyclopedia of forensic sciences. Academic Press

Henneberry YK, Kraus TE, Fleck JA, Krabbenhoft DP, Bachand PM, Horwath WR (2011) Removal of inorganic mercury and methylmercury from surface waters following coagulation of dissolved organic matter with metal-based salts. Sci Total Environ 409(3):631–637. https://doi.org/10.1016/j.scitotenv.2010.10.030

Hsu-Kim H, Eckley CS, Achá D, Feng X, Gilmour CC, Jonsson S, Mitchell CP (2018) Challenges and opportunities for managing aquatic mercury pollution in altered landscapes. Ambio 47:141–169. https://doi.org/10.1007/s13280-017-1006-7

Huang Y, Du JR, Zhang Y, Lawless D, Feng X (2015) Removal of mercury(II) from wastewater by polyvinylamine-enhanced ultrafiltration. Sep Purif Technol 154:1–10. https://doi.org/10.1016/j.seppur.2015.09.003

Hussain CM, Paulraj MS, Nuzhat S (2022) Source reduction, waste minimization, and cleaner technologies. Source Reduction Waste Minimization 23–59. https://doi.org/10.1016/B978-0-12-824320-6.00002-2

Inglezakis VJ, Kurbanova A, Molkenova A, Zorpas AA, Atabaev TS (2020) Magnetic Fe_3O_4-Ag^0 nanocomposites for effective mercury removal from water. Sustainability 12(13):5489. https://doi.org/10.3390/su12135489

Islam MS, Vogler RJ, Abdullah Al Hasnine SM, Hernández S, Malekzadeh N, Hoelen TP, Hatakeyama ES, Bhattacharyya D (2020) Mercury removal from wastewater using cysteamine functionalized membranes. ACS Omega 5(35):22255–22267. https://doi.org/10.1021/acsomega.0c02526

Jaishankar M, Tseten T, Anbalagan N, Mathew BB, Beeregowda KN (2014) Toxicity, mechanism and health effects of some heavy metals. Interdiscip Toxicol 7(2):60–72. https://doi.org/10.2478/intox-2014-0009

Johnson C (2014) Advances in pretreatment and clarification technologies. https://doi.org/10.1016/B978-0-12-382182-9.00029-3

Khan A, Ali J, Jamil SUU, Zahra N, Tayaba TB, Iqbal MJ, Waseem H (2022) Removal of micropollutants. In: Environmental micropollutants. Elsevier, pp 443–461. https://doi.org/10.1016/B978-0-323-90555-8.00012-X

Khozemy EE, Nasef SM, Mohamed TM (2020) Radiation synthesis of superabsorbent hydrogel (wheat flour/acrylamide) for removal of mercury and lead ions from waste solutions. J Inorg Organomet Polym 30:1669–1685. https://doi.org/10.1007/s10904-019-01350-6

Kim SJ, Chung JH, Kim TY, Cho SY (2006) Biosorption of heavy metals and cyanide complexes on biomass. Stud Surf Sci Catal 141–144. https://doi.org/10.1016/S0167-2991(06)81553-3

Kjellstrom T, Lodh M, McMichael T, Ranmuthugala G, Shrestha R, Kingsland S (2006) Air and water pollution: burden and strategies for control. Disease control priorities in developing countries, 2nd edn

Kumar R, Rauwel P, Rauwel E (2021) Nanoadsorbants for the removal of heavy metals from contaminated water: current scenario and future directions. Processes 9(8):1379. https://doi.org/10.3390/pr9081379

Lei X, Li H, Luo Y, Sun X, Guo X, Hu Y, Wen R (2021) Novel fluorescent nanocellulose hydrogel based on gold nanoclusters for the effective adsorption and sensitive detection of mercury ions. J Taiwan Inst Chem Eng 123:79–86. https://doi.org/10.1016/j.jtice.2021.05.044

Li N, Bai R, Liu C (2005) Enhanced and selective adsorption of mercury ions on chitosan beads grafted with polyacrylamide via surface-initiated atom transfer radical polymerization. Langmuir 21(25):11780–11787. https://doi.org/10.1021/la051551b

Li SX, Feng-Ying Z, Yang H, Jian-Cong N (2011) Thorough removal of inorganic and organic mercury from aqueous solutions by adsorption on Lemna minor powder. J Hazard Mater 186(1):423–429. https://doi.org/10.1016/j.jhazmat.2010.11.009

Li Y, Yang L, Li X, Miki T, Nagasaka T (2021) A composite adsorbent of ZnS nanoclusters grown in zeolite NaA synthesized from fly ash with a high mercury ion removal efficiency in solution. J Hazard Mater 411:125044. https://doi.org/10.1016/j.jhazmat.2021.125044

Liu C, Farooq K, Doll B, Venkatadri R (2013) Economical and reliable mercury reduction in refinery and power plant discharge wastewater with robust microfiltration membrane technology. Desalin Water Treat 51(25–27):4980–4986. https://doi.org/10.1080/19443994.2013.809641

Lu X, Huangfu X, Ma J (2014) Removal of trace mercury(II) from aqueous solution by in situ formed Mn–Fe (hydr) oxides. J Hazard Mater 280:71–78. https://doi.org/10.1016/j.jhazmat.2014.07.056

Lu J, Wu X, Li Y, Liang Y, Cui W (2019) Facile fabrication of 3D graphene–silica hydrogel composite for enhanced removal of mercury ions. Nanomater 9(3):314. https://doi.org/10.3390/nano9030314

Luo Q, Huang Y, Lei Z, Peng J, Xu D, Guo X, Wu Y (2021) Wood-derived nanocellulose hydrogel incorporating gold nanoclusters using in situ multistep reactions for efficient sorption and sensitive detection of mercury ion. Ind Crops Prod 173:114142. https://doi.org/10.1016/j.indcrop.2021.114142

Makertihartha IGB, Rizki Z, Zunita M, Dharmawijaya PT (2017) Graphene based nanofiltration for mercury removal from aqueous solutions. Adv Sci Lett 23(6):5684–5686. https://doi.org/10.1166/asl.2017.8802

Marimón-Bolívar W, Tejeda-Benítez L, Herrera AP (2018) Removal of mercury(II) from water using magnetic nanoparticles coated with amino organic ligands and yam peel biomass. Environ Nanotechnol Monit Manage 10:486–493. https://doi.org/10.1016/j.enmm.2018.10.001

Min X, Qing C, Jinjin C (2010) Study on the treatment of wastewater containing mercury by macromolecular heavy metal flocculant mercaptoacetyl polyethyleneimine. Water Environ Res 82(9):790–796. https://doi.org/10.2175/106143010X12609736966766

Mistar EM, Hasmita I, Alfatah T, Muslim A, Supardan MD (2019) Adsorption of mercury(II) using activated carbon produced from *Bambusa vulgaris* var. striata in a fixed-bed column. Sains Malaysiana 48(4):719–725. https://doi.org/10.17576/jsm-2019-4804-03

Mochizuki Y, Bud J, Liu J, Takahashi M, Tsubouchi N (2021) Mercury(II) ion adsorption performance of Cl-loaded carbonaceous material prepared by chlorination of pyrolyzed rice husk char. J Cleaner Prod 305:127176. https://doi.org/10.1016/j.jclepro.2021.127176

Monteagudo JM, Ortiz MJ (2000) Removal of inorganic mercury from mine waste water by ion exchange. J Chem Technol Biotechnol 75:767–772. https://doi.org/10.1002/1097-4660(200009)75:9

Nayak MC, Isloor AM, Lakshmi B, Marwani HM, Khan I (2020) Polyphenylsulfone/multiwalled carbon nanotubes mixed ultrafiltration membranes: fabrication, characterization and removal of heavy metals Pb^{2+}, Hg^{2+}, and Cd^{2+} from aqueous solutions. Arab J Chem 13(3):4661–4672. https://doi.org/10.1016/j.arabjc.2019.10.007

Nille OS, Patil AS, Waghmare RD, Naik VM, Gunjal DB, Kolekar GB, Gore AH (2021) Valorization of tea waste for multifaceted applications: a step toward green and sustainable development. In: Valorization of agri-food wastes and by-products. Academic Press, pp 219–236. https://doi.org/10.1016/B978-0-12-824044-1.00046-5

Oehmen A, Vergel D, Fradinho J, Reis MA, Crespo JG, Velizarov S (2014) Mercury removal from water streams through the ion exchange membrane bioreactor concept. J Hazard Mater 264:65–70. https://doi.org/10.1016/j.jhazmat.2013.10.067

Pagala B (2023) Characteristic evaluation of tamarind flower biomass for mercury biosorption: statistical approach. Biomass Conv Bioref https://doi.org/10.1007/s13399-023-03769-x

Park D, Woo SH, Yun YS, Ahn CK (1983) Jong Moon Park. Education, 1986

Peppas NA, Slaughter BV, Kanzelberger MA, Matyjaszewski K, Möller M (2012) 9.20–hydrogels. Polym Sci A Compr Ref 385–395. https://doi.org/10.1016/B978-0-444-53349-4.00226-0

Qasem NA, Mohammed RH, Lawal DU (2021) Removal of heavy metal ions from wastewater: a comprehensive and critical review. Npj Clean Water 4(1):36. https://doi.org/10.1038/s41545-021-00127-0

Rahbar N, Jahangiri A, Boumi S, Khodayar MJ (2014) Mercury removal from aqueous solutions with chitosan-coated magnetite nanoparticles optimised using the box-behnken design. Jundishapur J Nat Pharm Prod 9(2). https://doi.org/10.17795/jjnpp-15913

Roberts HW, Marek M, Kuehne JC, Ragain JC (2005) Disinfectants' effect on mercury release from amalgam. J Am Dental Assoc 136(7):915–919. https://doi.org/10.14219/jada.archive.2005.0292

Saberi A, Sadeghi M, Alipour E (2020) Design of AgNPs-base starch/PEG-poly (acrylic acid) hydrogel for removal of mercury(II). J Polym Environ 28:906–917. https://doi.org/10.1007/s10924-020-01651-9

Say R, Birlik E, Erdemgil Z, Denizli A, Ersöz A (2008) Removal of mercury species with dithiocarbamate-anchored polymer/organosmectite composites. J Hazard Mater 150(3):560–564. https://doi.org/10.1016/j.jhazmat.2007.03.089

Shirzadi H, Nezamzadeh-Ejhieh A (2017) An efficient modified zeolite for simultaneous removal of Pb(II) and Hg(II) from aqueous solution. J Mol Liq 230:221–229. https://doi.org/10.1016/j.molliq.2017.01.029

Shmeis RMA (2022) Nanotechnology in wastewater treatment. In: Comprehensive analytical chemistry, vol 99. Elsevier, pp 105–134. https://doi.org/10.1016/bs.coac.2021.11.002

Siddiqui SI, Chaudhry SA (2017) Iron oxide and its modified forms as an adsorbent for arsenic removal: a comprehensive recent advancement. Process Saf Environ Prot 111:592–626. https://doi.org/10.1016/j.psep.2017.08.009

Soni R, Bhardwaj S, Shukla DP (2020) Various water-treatment technologies for inorganic contaminants: current status and future aspects. Inorg Pollut Water 273–295. https://doi.org/10.1016/B978-0-12-818965-8.00014-7

Tchounwou PB, Yedjou CG, Patlolla AK, Sutton DJ (2012) Heavy metal toxicity and the environment. Exp Suppl 101:133–164. https://doi.org/10.1007/978-3-7643-8340-4_6

Templeton MR, Butler D (2011) Introduction to wastewater treatment. Bookboon

Tuzen M, Sarı A, Mogaddam MRA, Kaya S, Katin KP, Altunay N (2022) Synthesis of carbon modified with polymer of diethylenetriamine and trimesoyl chloride for the dual removal of Hg(II) and methyl mercury ($[CH_3Hg]^+$) from wastewater: theoretical and experimental analyses. Mater Chem Phys 277:125501. https://doi.org/10.1016/j.matchemphys.2021.125501

Valezi DF, Vicentin BLS, Mantovani ACG, Di Mauro E (2019) Applications of electron magnetic resonance in the study of soils, minerals, and iron oxides. In: Experimental methods in the physical sciences, vol 50. Academic Press, pp 203–217. https://doi.org/10.1016/B978-0-12-814024-6.00010-8

Vedrenne M, Vasquez-Medrano R, Prato-Garcia D, Frontana-Uribe BA, Ibanez JG (2012) Characterization and detoxification of a mature landfill leachate using a combined coagulation–flocculation/photo Fenton treatment. J Hazard Mater 205:208–215. https://doi.org/10.1016/j.jhazmat.2011.12.060

Vélez E, Campillo GE, Morales G, Hincapié C, Osorio J, Arnache O, Uribe JI, Jaramillo F (2016) Mercury removal in wastewater by iron oxide nanoparticles. In: Journal of physics: conference series, vol 687. IOP Publishing, p 012050. https://doi.org/10.1088/1742-6596/687/1/012050

Vishwakarma V (2021) Recovery and recycle of wastewater contaminated with heavy metals using adsorbents incorporated from waste resources and nanomaterials: a review. Chemosphere 273:129677. https://doi.org/10.1016/j.chemosphere.2021.129677

Voloshina E (2017) Hematite, its stable surface terminations and their reactivity toward water. In: Reference module in chemistry, molecular sciences and chemical engineering. Elsevier. https://doi.org/10.1016/B978-0-12-409547-2.14157-5

Wang J, Hong Y, Lin Z, Zhu C, Da J, Chen G, Jiang F (2019) A novel biological sulfur reduction process for mercury-contaminated wastewater treatment. Water Res 160:288–295. https://doi.org/10.1016/j.watres.2019.05.066

Wang Q, Zhu S, Xi C, Jiang B, Zhang F (2022) Adsorption and removal of mercury(II) by a crosslinked hyperbranched polymer modified via sulfhydryl. ACS Omega 7(14):12231–12241. https://doi.org/10.1021/acsomega.2c00622

Wang LK, Vaccari DA, Li Y, Shammas NK (2005) Chemical precipitation. Physicochem Treatment Process 141–197. https://doi.org/10.1385/1-59259-820-x:141

Wu CY, Mouri H, Chen SS, Zhang DZ, Koga M, Kobayashi J (2016) Removal of trace-amount mercury from wastewater by forward osmosis. J Water Process Eng 14:108–116. https://doi.org/10.1016/j.jwpe.2016.10.010

Yang J, Hou B, Wang J, Tian B, Bi J, Wang N, Li X, Huang X (2019) Nanomaterials for the removal of heavy metals from wastewater. Nanomater 9(3):424. https://doi.org/10.3390/nano9030424

Yaqub M, Lee SH (2020) Micellar enhanced ultrafiltration (MEUF) of mercury-contaminated wastewater: experimental and artificial neural network modeling. J Water Process Eng 33:101046. https://doi.org/10.1016/j.jwpe.2019.101046

Zabihi M, Omidvar M, Motavalizadehkakhky A, Zhiani R (2022) Competitive adsorption of arsenic and mercury on nano-magnetic activated carbons derived from hazelnut shell. Korean J Chem Eng 1–10. https://doi.org/10.1007/s11814-021-0903-4

Zeng B, Li J, Xiong C, Lin G, Wang W, Wu Z (2022) High-performance Zn-based coordination polymers selectively adsorb mercury ions from aqueous solutions. J Cleaner Prod 337:130551. https://doi.org/10.1016/j.jclepro.2022.130551

Zhang S, Li H, Wang Z, Liu J, Zhang H, Wang B, Yang Z (2015) A strongly coupled Au/Fe_3O_4/GO hybrid material with enhanced nanozyme activity for highly sensitive colorimetric detection, and rapid and efficient removal of Hg^{2+} in aqueous solutions. Nanoscale 7(18):8495–8502. https://doi.org/10.1039/C5NR00527B

Zhang B, Petcher S, Gao H, Yan P, Cai D, Fleming G, Parker DJ, Chong SY, Hasell T (2021) Magnetic sulfur-doped carbons for mercury adsorption. J Colloid Interface Sci 603:728–737. https://doi.org/10.1016/j.jcis.2021.06.129

Zinchenko AA, Sakai H, Matsuoka S, Murata S (2009) Application of DNA condensation for removal of mercury ions from aqueous solutions. J Hazard Mater 168(1):38–43. https://doi.org/10.1016/j.jhazmat.2009.01.121

Zunita M (2021) Graphene oxide-based nanofiltration for Hg removal from wastewater: a mini review. Membr 11(4):269. https://doi.org/10.3390/membranes11040269

Chapter 13
Environmental Mercury Toxicity and Its Bioremediation

Vikas Menon, Bhairav Prasad, and Himani Sharma

13.1 Introduction

The element mercury (Hg), which has the atomic number 80 and the letter Hg as its chemical symbol, is the only metal that is found in liquid state in the environment. It is distinct because it is the only element that, at ambient temperature, maintains a liquid state and has a characteristic sheen (Beckers and Rinklebe 2017). Mercury is widely used in many different industries, including coal production, mining, pesticide production, smelting, electronics manufacture, and chemical operations, all of which contribute considerably to its emission into the environment. Mercury can also be added to agricultural soils by the use of manures, fertilizers, fungicides, or sludge. When mercury enters the environment, it changes into a neurotoxic methylmercury, which can have a number of negative impacts (Chugh et al. 2022). It exists as elemental (Hg^0), inorganic (Hg^{+1} and Hg^{+2}), and organic methyl mercury (Me-Hg) in the environment and shows its wide distribution in the aquatic to terrestrial ecosystem. As considered one of the main global pollutants, its emission ranges from 6000 to 8000 Mg Hg/year through both anthropogenic as well as natural sources (Singh et al. 2023; Gustin et al. 2020). Hg has high persistence, slow degradation, high toxicity, and potential to contaminate air, water, and soil. Hg enters to the food chain through vegetation, the plants' part biomagnifies the toxicant and linking with the food chain and food web from pioneer group to top-level consumer. Hg exposure in human beings causes gene malfunctioning, neurological disturbance, and cancer. However, it is equally poisonous to plants and adversely distorts plant physiology and metabolism (Singh et al. 2023). Hg is also emitted in gaseous form and transmitted in the atmosphere and travels a long distance before getting settled in the soil. In the year of 2015, the mercury-added by anthropogenic sources was around 220 tons

V. Menon · B. Prasad (✉) · H. Sharma
Department of Biotechnology, Chandigarh College of Technology, CGC, Landran, Mohali, Punjab 140307, India
e-mail: bhairavmicro@gmail.com

N. Kumar (ed.), *Mercury Toxicity*, Environmental Science and Engineering,
https://doi.org/10.1007/978-981-99-7719-2_13

to the atmosphere as reported by UNEP (United Nations Environmental Protection Agency). The common anthropogenic activities that contribute Hg to the environment are mining, industrial activities, coal combustion, mineral extraction, waste discharge, and the use of Hg-based pesticides and fertilizers (Teng et al. 2020). However, approximately 65% of the Hg emission is shared by coal plants to the environment (Yang et al. 2020). The volcanic eruption also emits Hg as elemental, cinnabar (HgS), oxide of Hg, and sulfates of Hg (Castagna et al. 2018). The Hg prevails for a prolonged time in the environment, due to its high volatile nature, it is easily absorbed by plants which is further biomagnified in different trophic levels and gets entered into the food chain (Cai et al. 2020). It is a type of bioaccumulative pollutant that tends to magnify in each trophic level and persist for a long time up to 0.5–2 years in the environment (Varjani et al. 2020). Hg pollution poses a potential threat to both vegetation and animals including human life. It directly hampers the normal metabolism and physiology of the biological entities. High concentration of Hg in plants causes growth, nutrition, and biomass imbalance (Lv et al. 2018). Hg exposure has negative effects on eye irritation, gingivitis, chest pain, interstitial pneumonitis, depression, autoimmune diseases, mental disorders, reproductive disorders, gene expression alterations, nervous system disorders, Minamata disease, and antibiotic resistance, which can be fatal. In addition to affecting the kidneys, exposure to mercury can make people agitated and anxious. Aquatic plants also have the capacity to store mercury inside their tissues (Chugh et al.2022). It also causes oxidative stress, gene malfunctioning, and lipid peroxidation (Pirzadah et al. 2018). Hg shows genotoxic effects in plants such as DNA disfigurement, damage to DNA repair machinery, chromosomal aberration, and mitotic decision aberration (Shahid et al. 2020). It also changes the cell membrane's permeability and prevents normal germination. Additionally, it lowers transpiration rates, hinders photosynthesis, decreases chlorophyll production, and diminishes antioxidant defense mechanisms in plants. The concentrations of magnesium (Mg), potassium (K), and manganese (Mn), are all decreased weather the concentration of iron (Fe) is increased in plants exposed to Hg, which also upsets the balance of critical ions in plants (Chugh et al. 2022). Every person on earth has a certain amount of Hg in their body, which comes from a variety of sources including contaminated food, the use of skin-lightening products that contain mercury, and dental amalgam fillings, according to the UNEP. (UNEP, 2023). Methyl Hg (Me-Hg), the most dangerous form of mercury, is a huge global concern mostly linked to polluted fish and shellfish. Due to its toxicokinetic characteristics, which include a quick uptake into many tissues and organs and a delayed clearance rate, MeHg demonstrates significant toxicity (Crespo-Lopez et al. 2021). In comparison with other forms of Hg, the Me-Hg has potential to enter in all cell membranes including blood–brain barrier and blood–placental membrane (Crespo-Lopez et al. 2021). In the human body this toxic element causes multiple syndromes viz. affect fetus development (He et al. 2015), central nervous system (CNS) problem, urinary system, cardiovascular, respiratory system, immunological disturbance, and hematological abnormalities (Crespo-Lopez et al. 2022). The terrestrial and aquatic deposition of Hg may also show clinical implications like deafness, vision issue, blindness, speech difficulties due to impaired central nervous functioning affected by Hg toxicity. Some other studies also reported that Hg toxicity is associated with neurotoxicity, nephrotoxicity, cardiovascular disease, immunotoxicity, reproductive and developmental toxicity (Genchi et al. 2017). Some recent studies suggested that liver cancer is also induced by Hg (Thompson and Darwish 2019).

To minimize the Hg toxicity in the environment scientists developed many strategies which include physiological, chemical, and biotechnological methods to detoxify and eliminate Hg. The conventional physical methods remediated Hg through membrane filtration, adsorption, precipitation, and soil washing (Mahbub et al. 2017a). Methods including immobilization, vitrification, thermal processes, solidification/stabilization, electrokinetics, and soil washing are all included in the category of physicochemical techniques (Xu et al. 2018). But these methods have several disadvantages viz. producing large quantities of sludges, laborious, cosplay, and time-consuming (Mahbub et al. 2017b). Alternatively, bioremediation which includes microbial remediation and phytoremediation is the most eco-friendly, cost-effective, less laborious development that eliminates Hg pollution (Mosa and Duffin 2017). Microbial remediation involves the application of bacterial and fungal species. Hg contamination in several polluted sites has been remedied at a small scale using a variety of bacteria and fungus that exhibit tolerance to it (Velásquez-Riaño and Benavides-Otaya 2016). Bacterial species have developed various mechanisms to mitigate Hg toxicity from the environment viz. biosequestration, absorption, volatilization, production of exopolysaccharides (EPS), variation in morphology, oxidation, and different efflux mechanism (Priyadarshanee et al. 2022). On the other hand, Phytoremediation methods depend on some plants that sequester and accumulate heavy metals from contaminated zones, metabolize them, and detoxify into less toxic forms. In order to survive in such settings, some plant species have the amazing capacity to hyperaccumulate and sequester hazardous substances. Different processes, including absorption, chelation, regulated distribution, extrusion, and vacuolar sequestration, are used by plants to sequester elements (Podar and Maathuis 2022). Additional biotechnological techniques must be used in order to successfully detoxify Hg. These strategies include whole-cell biosensors, plant-assisted microbial remediation, transposon-mediated in-situ molecular breeding (ISMoB), and nanotechnology. Effective detoxification of Hg can be achieved by using these methods.

13.2 Mercury Toxicity

Hg is a highly toxic chemical pollutant and WHO listed it in the top ten chemicals of public health concern worldwide (WHO 2020). The source of Hg pollution is distributed globally, and its emission continuously increases, majority emission is contributed by anthropogenic activities as documented by international agencies (UNEP 2019). Many studies revealed that most of the people worldwide must have some amount of Hg and that is seen in some vulnerable groups such as artisanal small-scale gold miners and Indigenous people due to their occupational and dietary

(Basu et al. 2018). The severity of Hg toxicity is governed by its chemical form, route of exposure, and quantity. It gains entry to the human being mainly through vegetation and the food chain. Hg enters into the food chain, bioaccumulates, and exerts negative effects on human health. Across many ecosystems, it is still unclear how exactly Hg enters into the food chain (Rice et al. 2014). It is also reported that it also harms the kidney and heart. The main causes of Hg pollution are human-induced activities, such as the burning of fossil fuel, the use of Hg-based fertilizer and fungicides in agriculture, and its use as a catalyst in many industries Refineries and industrial activities also release Hg into the environment (Jain et al. 2022).

Hg is a persistent, bioaccumulative, and toxic element introduced to the environment by many means. Hg enters naturally from volcanoes, forest fire, soil, and weathering of rocks. Many human-induced activities such as burning of coal, mining, municipal or medical waste (thermometer, barometer, and sphygmomanometer). Hg re-entered to the environment naturally from the evaporation of ocean water. National Inventory Report, 2014 of US Environmental Protection Agency reported that the power plant producing electricity using fossil fuel contributes maximum Hg emission of about 42% of all anthropogenic Hg emission (NHDES 2019). The various sources of Hg pollution are depicted in Fig. 13.1.

13.2.1 Effects on Vegetation

Different environmental pollutants can have negative impacts on plants that impede their growth and development. Heavy metals are one of these pollutants that have an effect on plant growth by acting as abiotic stresses. The precise effects of heavy metals on plants significantly varied and depending on the species of plant and the concentration of heavy metal. The amount of heavy metals released into the environment has dramatically increased because of growing industrialization. Trace metal emissions from industrial processes including mining, metalworking, and chemical manufacture significantly contaminate the soil (Chibuike and Obiora 2014). The industrial effluents from these industries contain many trace metals viz. Hg, Cd, Pb, and Cr even in minute quantities show deleterious effects in plants and modulate the physiology and metabolism (Pravin et al. 2012). These metal pollutants reduce plant growth by lowering the performance of chlorophyll, lipid, and protein production (Kim et al. 2014).

Hg is one of the very dangerous trace elements that significantly affect our ecosystem due to its persistent and bioaccumulative long-distance travel ability in nature (Liu et al. 2017). Hg adversely affects many physiological and metabolic processes of plants such as breakdown of photosystem, growth inhibition, reduction in chlorophyll synthesis, change in cell membrane permeability, and also affects antioxidant system (Hassan et al. 2020). Hg also promotes oxidative stress in many plants. Since nitrate is converted to nitrite by the nitrate reductase enzyme, nitrate serves as a frequent source of nitrogen for plants. Plants can assimilate nitrogen

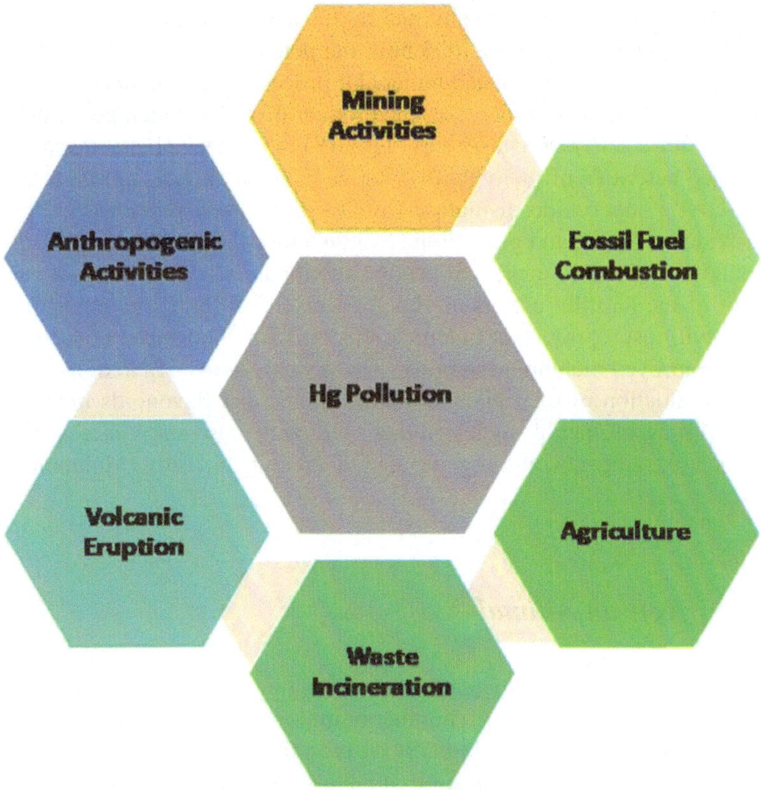

Fig. 13.1 Various sources of mercury pollution

into their systems thanks to this process (Carrasco-Gil et al. 2023). This reduction in nitrate reductase causes accumulation of (ROS) reactive oxygen species and (RNS) reactive nitrogen species which ultimately cause deleterious effects on plant cells, nucleic acid, and proteins (Hafeez et al. 2023). The presence of Hg in agricultural soils is of great concern worldwide as it affects crop yield by inhibiting plant growth and development. Hg causes abnormal seed germination, inhibiting chlorophyll synthesis and photosynthesis, reducing biomass, disturbing protein functioning, nutrient imbalance, and water relation in plants (Sun et al. 2018; Safari et al. 2019). Increased levels of Hg affect chlorophyll production in higher plants and reduce the rate of photosynthesis (Teixeira et al., 2018). Similarly, Hg toxicity also reduces the production of chlorophyll content in lemon plants (Safari et al. 2019). Lemon plant chlorophyll concentration is declining as a result of oxidative stress brought on by mercury. This oxidative stress interferes with the plants' ability to absorb water and minerals as well as their ability to replace metal ions with Hg ions (Hafeez et al. 2023). Hg stress exerts adverse effects on nutrient uptake by plants (Tran et al. 2021). Hg toxicity also prevents plants from producing too many reactive oxygen species (ROS). Nucleic acids, membrane lipids, and proteins are just a few of the biological components that these ROS have the potential to harm. They also disrupt the biochemical, physiological, and molecular processes that occur in plants (Zhang et al. 2016). Hg toxicity also increases the production of hydrogen peroxide (H_2O_2), and thiobarbituric acid reactive species (TBARS) in Chicory plants. this also causes the increased activities of many anti-oxidative enzyme such as peroxidase (POD), catalase (CAT), superoxide dismutase (SOD), glutathione reductase (GR), ascorbate peroxidase (APX), and glutathione S transferase (GST) (Malik et al. 2019). A study found that soluble sugars and proline significantly accumulated in duckweed as a result of the harmful effects of Hg. As a result of these compounds acting as osmolytes with osmoprotectant qualities, this buildup contributes to increased cell turgor and improved membrane stability (Zhang et al. 2016). Hg toxicity also shows increased production of total phenol, anthocyanins, and flavonoids in okra plants. Further, profiling of phenol suggested that okra plants also show increased content of rosmaric acid, apigenin, chlorogenic acid, quercetin, and rutin (Mohammadi et al. 2021).

13.2.2 Effects on Animal

Hg exposure also affects aquatic and terrestrial animals. Hg alters how proteins and enzymes are expressed, affecting a number of important pathways including glucose metabolism and apoptosis. Additionally, it promotes the expression of metallothioneins, which aid in the detoxification of metals and the defense of cells (Rowland et al. 2023). Hg has a tendency to bioabsorb and biomagnify throughout the food chain, having an effect on both animal populations and human populations (Zheng et al. 2019). Me-Hg is the most noxious form of Hg toxicity and approximately 95% of the Hg accumulated in fish as Me-Hg being the most absorbed form by human being when fish is the main protein source in the community (Rodríguez Martín-Doimeadios et al. 2014). Me-Hg exposure causes significant health effects on wildlife. Me-Hg accumulates in a variety of tissues of wildlife and poses detrimental health implications (Chetelat et al. 2020). Long exposure to Hg may cause physiological, anatomical, biochemical, reduced hepatic function, impaired reproduction, and nervous system disorder even at sub-lethal doses (da Silva and de Oliveira Lima 2020). *Danio rerio* (Zebrafish) shows visual deficits and neurological disturbances when exposed to Hg (Cusset et al. 2023). Hg toxicity also causes bone lesions, slow ossification as well as microstructural changes in alveolar bone in pregnant rats (Nuens et al. 2022). Hg accumulation interacts with cortisol levels in adult female northern elephant seals (*Mirounga angustirostris*) and significantly influences the endocrine and immune biomarkers (Peterson et al. 2023). In the American alligator (*Alligator mississippiensis*), the interplay between maternally transferred Hg and incubation temperature has a considerable impact on reproductive success, offspring development, behavior, and eventual survival (Johnson et al. 2023).

13.2.3 Effects on Human

Hg exposure has severe toxic effects on human beings. It causes oxidative stress in the cell and alters the energy metabolism by targeting mitochondria on priority (Zheng et al. 2019). The reactive oxygen species in abundance destroy the antioxidant defense mechanism of the body (Zhao et al. 2021). Many human-induced processes such as industrial activities, coal combustion, mineral extraction, and application of Hg-based pesticide and fertilizer contributing Hg pollution to the atmosphere (Teng et al. 2020). Hg is easily taken up by plants and other vegetation due to its persistent and highly volatile nature, it also allows Hg to pass in the food chain and eventually into the human body (Cai et al. 2020). Hg drastically affects the physiology and metabolism of plants. However, Hg impairs normal functioning in human beings. Hg causes genotoxicity, lipid peroxidation, and oxidative stress (Ahmad et al. 2018). Me-Hg is the highly toxic form among all Hg compounds and a potent neurotoxin that causes Minamata disease (Sakamoto et al. 2018). The main source of Me-Hg poisoning is due to the consumption of shellfish and fish contaminated with Me-Hg in Minamata bay Japan (Harada 1995). It has also been reported that Me-Hg (CH_3-Hg^{+}) entered in the food chain shows toxic effects on human health. It impairs fetus development and neurological functioning (He et al. 2015). Hg also causes adverse effects on hair and is potentially used as a bioindicator of Hg exposure (Ruggieri et al. 2017). Hg exposure also causes preterm delivery in pregnant women (Bello et al. 2023). In a study, it has also been reported that people who consume fish frequently more than once in a week indicated that the blood Hg significantly reduced the performance of lung in young men and impaired the functioning of the respiratory system (Miao et al. 2023). Hg exposure also causes deleterious effects in multiple organ systems viz. immunological system, central nervous system, hematological system, respiratory system, renal system, and cardiovascular system (Crespo-Lopez et al. 2022). Hg even at low concentration causes severe Hg intoxication if exposed chronically. Hg concentration 1–2 ppm in human hair as body burden bioindicator can have non-fatal and fatal cardiovascular complications (Hu et al. 2021a). The acute and repetitive contact of skin with Hg may lead to allergic, pruritic skin rashes, reddening, erythematous soles of the feet, and peeling skin on the palms of the hands (Rauf et al. 2020). Kids have a higher risk of disease than adults due to their different body intake ratio, children have high intake than adults. Children are more vulnerable because of their physiology, immature neurodevelopment, and weak immune system (Du et al. 2021). Serum hs-CRP (highly sensitive C-reactive protein), a marker protein showing persistent low-grade inflammation in people, rises in response to elevated Hg levels in the blood. This implies that a high blood Hg level may have a role in the emergence of a number of inflammatory disorders, such as cardiovascular disease (Kim and Park 2023). In some regions of Iraq and Minamata bay, pregnant women who were exposed to Me-Hg had evidence of MeHg placental transfer and caused neurotoxicity in neonates and infants (Azevedo et al. 2023). Hg also affects joints and other connective tissue studied in mice. The uptake of Hg by articular chondrocytes, synovial cells periosteal, and tracheal cartilage cells. It is also found in dermis, aorta, striated muscles, and esophagus and in blood vessels (Pamphlett and Kum Jew 2019). Hg is also present in CD 44 progenitor cells, renal tubules, and liver periportal cells (Pamphlett and Kum Jew 2019). Hg was found in liver periportal cells and renal tubules, showing that these cells only take it in when they need to. Its existence has been linked to the onset of rheumatoid arthritis and osteoarthritis. Furthermore, it has been demonstrated that exposure to Hg causes immunological, inflammatory, genetic, and epigenetic conditions that affect bone and connective tissue. These results support the idea that Hg exposure may cause bone and connective tissue abnormalities in those with autoimmune diseases or weakened immune systems (Pamphlett and Kum Jew 2019). Hg^{+2} accumulates in the kidney and may cause acute renal failure. It also causes autoimmune disorders and generation of antinuclear antibodies, also triggering Parkinson's and Alzheimer's disease. Therefore, Hg has strong toxic effects on both digestive and nervous systems (Bae et al. 2023). Mercury toxicity and its effects are depicted in Fig. 13.2.

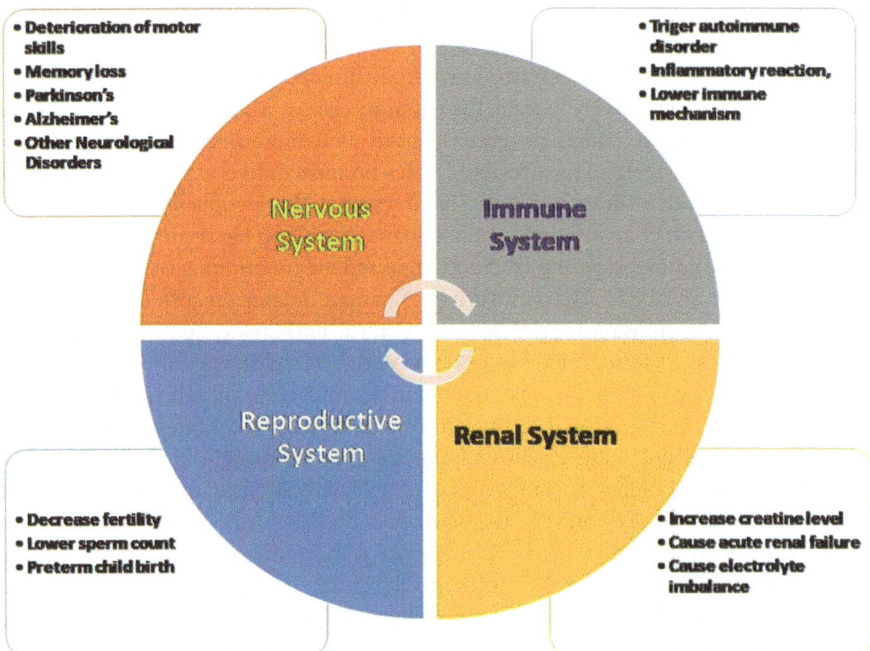

Fig. 13.2 Mercury toxicity and its effects

13.3 Biotechniques for Mercury Removal

Hg is a dangerous heavy metal that exists in the environment and can have detrimental effects on humans, animals, and plants even at low concentrations. Due to different human activities like mining, burning fossil fuels, metal extraction, and the use of Hg-based herbicides and fertilizers in agriculture, its concentration is currently steadily rising. Hg enters the body of a person through the food chain, as well as through medical and dental operations, and causes major health problems such as neurological diseases, vision loss, cardiovascular disease, cancer, problems with the bones and connective tissues, and in some cases, even death. It is essential to prioritize Hg's detoxification as an urgent priority due to its strong toxicity and extensive tissue distribution many traditional approaches have been implicated so far for the elimination and detoxification of Hg but these methods are laborious, expensive, and also generate secondary Hg-based pollutants which are difficult to remove from the environment. The conventional methods include physical and chemical methods mainly based on:

- **separation from contaminated soil**
- **decrease its mobility and solubility and**
- **vaporization and condensation of Hg** (Naguib et al. 2018)

On the other hand, the biological remediation technique famously known as bioremediation is an eye-catching method nowadays for the decontamination of heavy metals including Hg. Many biological populations including bacteria, fungi, actinomyces, and algae successfully detoxify Hg. Many plants expressing metal binding protein are also used for remediation purpose commonly known as phytoremediation. The bioremediation methods are more efficient, economic, and environment friendly. The microbes possess many *mer* operons that support them to survive in metal-contaminated sites and also metabolize them. Recently genetically modified bacteria, plants have been employed for the detoxification of both inorganic and organic Hg (Naguib et al. 2018). The bioremediation of Hg is mainly accomplished by various mechanisms which include biosorption, bioaccumulation, biotransformation, bioleaching, siderophore production, and bacterial encapsulation. Each mechanism is presented in Fig. 13.3.

13.3.1 Biosorption of Mercury

Biosorption is a type of sorption technique in which we use material of biological origin like plants and microorganisms. It is one of the most efficient, economical, and ecologically benign technologies for treating industrial effluents (Kumar et al. 2023). Biosorption is one of the mechanisms by which microbes detoxify Hg. It is safer, economical, and more efficient than other conventional methods (Jariyal et al. 2020). SRB (Sulfate-reducing bacteria) were reported to be more efficient

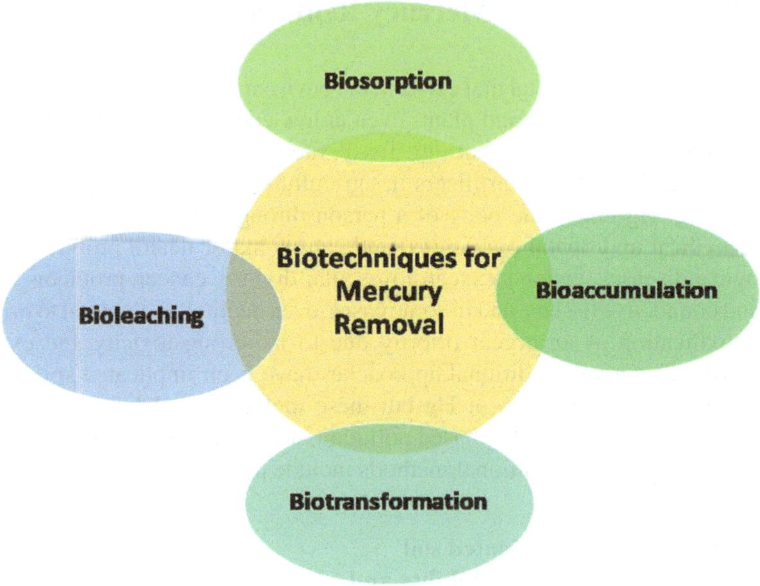

Fig. 13.3 Different mechanisms of mercury bioremediation

in degrading Hg and MeHg (Liu et al. 2022). It was also found that Bacillus sp. secret some extracellular polymers that adsorb Hg up to 123.40 mg/g (Yan et al. 2022). Similarly, *Acinetobacter indicus* yy-1 was reported to remove Hg efficiently (Hu et al. 2021b). Soil fungi either dead or alive have the ability to accumulate large quantities of trace metals by biosorption (Ayele et al. 2021). Fungal fruiting bodies (mushrooms) have the potential to bioaccumulate Hg, while the role of fungal mycelial in Hg bioaccumulation has no significant results (Durand et al., 2020). Biosorption of Hg on fungal necromass (fungal mycelial residues) was studied on lab scale (Amin et al. 2016). Fungal necromass rapidly and passively absorbs Hg from aqueous solution (Martínez-Juárez et al. 2021). Different species of microalgae have the capabilities of removing Hg very effectively from the saline water coasts and estuaries. The three different types of microalgae namely *Ulva vesiculosus, Fucus vesiculosus* and *Grcilaria gracilis* can remove concentration of Hg from 10 to 100 $\mu g\ L^{-1}$ (Henriques et al. 2015). Various fungal strains including *Aspergillus flavus, A. fumigatus, Cladosporium* sp., *Heliminthosporium* sp. *Mucor rouxii, Mucor rouxii IM-80*, and *Candida albicans*, etc., involved in Hg biosorption. It has been reported that the biomass of Mucor was very efficient in removing Hg from 95 to 78% (Martínez-Juárez et al. 2021). The biosorption capabilities of some staple plants for Hg such as millet, groundnut, beans, and corn were evaluated. The tested plants were exposed to Hg concentration of 10–50 mg/L for 7 days and then metal analysis of the solution was evaluated. the metal uptake of the plants was 4.7–9.3 mg/l, 4.6–9.3 mg/l, 3.2–9.6 mg/l and 3.0–7.7 mg/l for groundnut, beans, corn, and millet respectively for the 10 mg/L Hg solution. The effective biosorption of the above crop at 50 mg/L was

26–30.2 mg/L, 18.2–35.7 mg/L, 16.2–42 mg/L, and 17.6–34.2 mg/L for groundnut, beans, corn, and millet respectively. This study confirmed that the above staple crop has good potential of Hg recovery from the contaminated sites (Yahaya et al. 2022).

13.3.2 Bioaccumulation of Mercury

Several microbes have been reported to accumulate Hg. The main mechanism of accumulation of Hg by microbial cells is by converting Hg into MeHg. In a study 32 bacterial strains were screened for Hg bioaccumulation potential, 27 isolates were found to have good potential for Hg tolerance. among the 27 isolates, 2 isolates showed highest Hg removal efficiency and were *Fictibacillus nanhainensis* (SKT-B) and *Bacillus toyonensis* (PJM-F1) with bioaccumulation up to 82.25 and 81.21% respectively (Nurfitriani et al. 2020). Some endophytic bacterial strains also have been reported for Hg bioaccumulation. *Pseudomonas* sp. BacI38 and *Serratia marcescens* BacI56 able to volatilize Hg up to 62. 42% and 47.16% respectively. *Pantoea* sp. BacI23 and *Bacillus* sp. BacI34 with other six endophytic bacterial strains favor Hg accumulation in plant tissue as well (Mello et al. 2020). *Metarhizium robertsii* an Hg-resistant symbiotic fungus inhabiting plants that obtains nourishment from plants and demethylate MeHg via enzyme demethylase and volatilizes Hg divalent using another enzyme reductase. Intermittent removal of Hg from soil in this fashion decreases soil toxicity and enhances plant growth. Recombinant *M. robertsii* has enhanced Hg removal ability than the wild one (Wu et al. 2022). Similarly, genetically engineered *E. coli* have also been reported to remove Hg from wastewater with great efficiency. The *E. coli* cells removed more than 99% of the Hg from wastewater and the bio-accumulated concentration of Hg was 26.8 mg/g cell dry weight. The effect of EDTA was also evaluated on Hg bioaccumulation and found that EDTA stimulates Hg uptake rather than inhibition (Deng and Wilson 2001). Bacillus DC-B2 is reported to volatilize Hg up to 91.6% in Luria agar medium (Chen et al. 2019). In a similar study, *Sphingobium* SA2 was grown in a phosphate minimum medium and demonstrated a stunning 80% volatilization of Hg in just 6 h (Mahbub et al. 2016). Another strain, SE2, displayed a 44% mercury volatilization rate and a 23% mercury accumulation rate inside of its cells, with no volatilization of dead biomass being seen (Mahbub et al. 2017b). Additionally, over a 30-day period in a microcosm under flooding circumstances, Bacillus DC-B2 showed a 17% increase in mercury volatilization. Furthermore, under comparable settings, Hg elimination on cropland with contamination reached an amazing 82.1% within 30 days (Chen et al. 2019).

13.3.3 Biotransformation of Mercury

Biotransformation, a metabolic process which is carried out by a single bacterium or consortia of microorganisms, can be either endogenous or exogenous. Hg is biotransformed into a less harmful form in the environment by a number of different types of microbes. *Pseudomonas* a gram-negative bacillus inhabiting in soil showing Hg resistance and capable of transforming Hg^{+2} to Hg^0, hence have great application in Hg remediation (Li et al. 2022). Several studies have shown that many indigenous microorganisms have the potential to remove heavy metals from contaminated sites and have positive effects on the removal of Hg (Qi et al. 2022). The biotransformation of Hg^{+2} by blue-green algae was evaluated under aerobic and pH-controlled culture conditions. The cyanobacterial species involved in Hg transformation were *Synechococcus leopoldiensis, Limnothrix planctonica*, and *Phormidium limnetica* found to rapidly synthesize β-HgS and Hg^0 (Bellinger et al. 2007). Similarly, *Pseudomonas putida* isolated from Uppanar estuary has the potential to transform Hg from the contaminated soil (Rajadurai and Reddy 2021). *Clostridium cochlearium* has the ability to decompose Me-Hg (Pan-Hou and Imura 1982). The elemental Hg converted to Me-Hg through a complex process and biotransformed in humans is depicted in Fig. 13.4.

13.3.4 Bioleaching of Mercury

Bioleaching is another mechanism by which microbes eliminate heavy metals from contaminated sites. In this process microbes like bacteria, fungi, and others present indigenously in the environment solubilize metal sulfides and oxides from their ore deposits and secondary waste (Jafari et al. 2019). Several techniques, including membrane separation, adsorption, selective precipitation, and ion exchange, are used to recover the solubilized metal. The dissolved metal can be separated from it and recovered using these approaches (Rohwerder et al. 2003). It is eco-friendly and economic technique as it does not produce any gas or secondary pollutants (Asghari et al. 2013). In study 5 bacillus sp. recovered showing resistance 30–40 μg/ml against $HgCl_2$. All the resistant bacterial strains were identified as *Bacillus thuringiensis* AA-35, *B. subtilis* AA-16, *B. cereus* AA-18, *Bacillus* sp. AA-20 and *B. paramycoides*. *B. cereus* AA-18 exhibits significant results of $HgCl_2$ tolerance level up to (40 μg/ml) due to the presence of the merA gene (Al-Amin et al. 2021). *Acidithiobacillus ferrooxidans,* is a Iron-oxidizing bacterium, also used for the bioleaching of gold and copper from their ores. The Hg-reducing activity *A. ferrooxidans* strains SUG 2–2 and MON-1 was evaluated. According to the study, the strains SUG 2–2 and MON-1 have a special ability to volatilize Hg in response to ferrous iron, as well as an activity of Hg reductase that contributes to Hg bioleaching (Takeuchi and Sugio 2006).

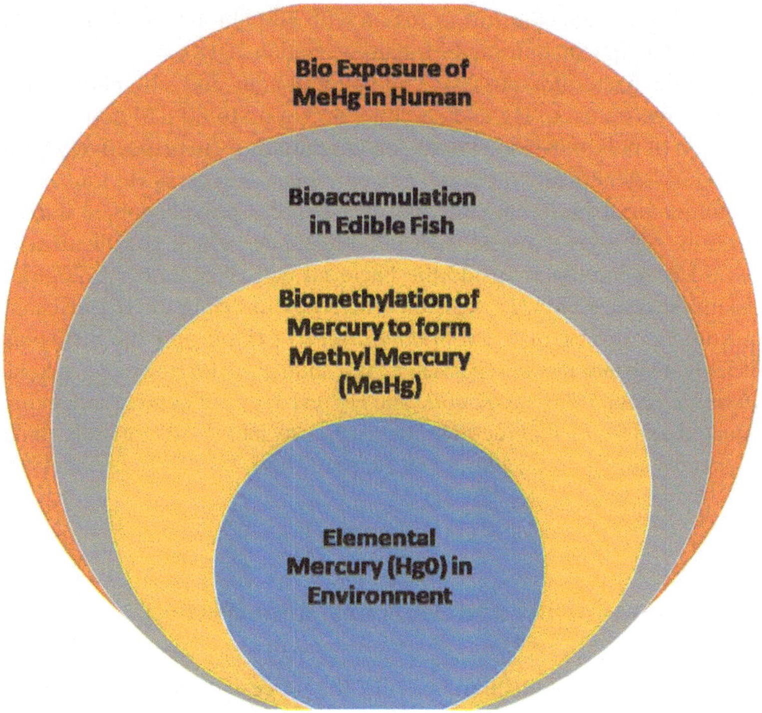

Fig. 13.4 Biotransformation of mercury

13.3.5 Genetically Modified Microorganisms (GEMs) for Mercury Bioremediation

Recombinant DNA technology or Genetic engineering has the potential to design genetically modified microbes to enhance recovery rate of pollutants and increase the rate of detoxification. To improve the removal and recovery of Hg, a method for its bioremediation has been used that makes use of the production of the bacterial mer gene (Jackson 1998). Hg is one of the deadliest pollutants for both flora and fauna. The conventional approaches such as physical, chemical, and physio-chemical methods are expensive, and generate large amounts of secondary waste. Bioremediation of Hg through microbes has gained popularity in recent years but they also have certain limitations, such as being less efficient, unable to tolerate high Hg concentration, and other abiotic factors. Therefore, genetically improved microbes have vast applications in the present scenario as they tolerate high toxicity, adopt new adverse conditions, are highly efficient, etc., on the same theme for phytoremediation genetically improved plants are also being employed in place of their wild variety for phytoremediation of pollutants. In a study genetically engineered *E. coli* express four rice metallothionein isoforms as fusion with glutathione-S-transferase have the ability to remove Hg (Shahpiri and Mohammadzadeh 2018).

A transgenic bacterium possesses metallothionein and polyphosphate kinase provides high resistance to Hg and can tolerate Hg up to 80 μM and 120 μM with the expression of polyphosphate kinase and metallothionein respectively (Ruiz et al. 2011). Similarly, genetically modified *Escherichia coli* express Hg transport system and organomercurial lyase enzyme and over-express polyphosphate during metal detoxification. The over-expression of polyphosphate which is a bivalent chelator serves as a Hg accumulator inside the bacterial cell (Nagata et al. 2006). A new Hg-resistant plasmid pTMJ212 was constructed by inserting the Hg-resistant determinants from *mer* operon of *Acidithiobacillus ferrooxidans* into the IncQ plasmid of pJRD215. Hg-resistant plasmid pTMJ212 is a shuttle vector and can propagate both in *Acidithiobacillus caldus* and *E. coli*. The Hg resistance of control and recombinant was evaluated and found that recombinant *A. caldus* increased markedly under Hg^{2+} stress especially at Hg^{2+} concentrations ranging from 2.0 to 4.5 μg/ml (Chen et al. 2011).

13.4 Microorganisms and Mercury Remediation

By the use of diverse biological treatments, the process of bioremediation converts hazardous and persistent contaminants into innocuous compounds in an inexpensive and environmentally benign manner. Microbes are used in the bioremediation process to break down pollutants through microbial metabolism (Deshmukh et al. 2016). This process involves the biochemical events or pathways that control the activity, growth, and reproduction of the organism. Reactants, pollutants, oxygen, or other electron acceptors are used in chemical reactions in microbial metabolism to change metabolites into certain products. This system gives organisms' access to the carbon, electrons, and other vital elements they need to survive. Many microbe-metal interactions, such as biotransformation, biosorption, bioaccumulation, biomineralization, and bioleaching, are possible (Tayang and Songachan 2021; Jeyakumar et al. 2023; Taha et al. 2023).

Through a process known as bioremediation, microorganisms can detoxify Hg by transforming it into less dangerous forms. It is possible for this process to take place naturally in soil and water systems, but it can also be improved by using particular microbes that are known to be efficient at detoxifying Hg (Bala et al.2022).

Sulfate-reducing bacteria (SRB) are one type of microorganism that is particularly good at detoxifying Hg. These bacteria can convert elemental mercury to less harmful forms like mercury sulphide, which is insoluble and less accessible, by using sulphate as an electron acceptor in their metabolism. Methylotrophic bacteria are a different class of microbes crucial in the detoxification of mercury. Since Me-Hg is a more dangerous and bioavailable form of mercury, these bacteria can methylate inorganic mercury to create it. However, once it has been created, Me-Hg can be further changed into less harmful forms by other microorganisms, such sulfate-reducing bacteria (Zhang et al. 2016).

In addition to these microorganisms, it has been demonstrated that specific kinds of algae and fungi may also detoxify Hg. For instance, several types of green algae can accumulate and detoxify Hg by combining it with the amino acid cysteine, which contains sulfur. Overall, a key component of bioremediation techniques for contaminated environments is the capacity of microorganisms to detoxify Hg. However, many variables, including the type and concentration of Hg, the presence of other pollutants, the particular microorganisms, and environmental circumstances involved, can affect how effective these tactics are (Danouche et al. 2021).

13.4.1 Mercury Bioremediation by Bacteria

The widespread use of mercury, a naturally occurring dangerous heavy metal, with extensive industrial use agriculture and urbanization has significantly increased local pollution (Dash and Das 2012; De et al. 2014; Tayang and Songachan 2021). It has been proposed that mercury levels can be decreased by volatilizing the metal form by bioremediation employing bacteria that are resistant to Hg. In contrast, the main components of physical and chemical approaches entail capturing and gathering mercury from contaminated areas or employing chemical processes to precipitate mercuric compounds for removal (Essa et al. 2002). Mercury can be changed by bacteria using a variety of chemical reactions, including oxidation, reduction, methylation, and demethylation. With the use of plasmids and other extrachromosomal genetic material, they become resistant. The *mer* operon-mediated pathway has been widely explored for bacterial resistance to mercury among the various suggested pathways. Regarding Hg detoxification, various kinds of Hg tolerance mechanisms have been identified. These processes include decreased uptake of mercuric ions, Me-Hg sequestration, methylation of Hg, and enzymatic reduction of divalent Hg to elemental Hg (De et al. 2014).

Hg-tolerant bacteria contain the *mer* operon, which is made up of functional genes, a promoter, a regulator, and an operator, in their genomes. Several kinds of bacteria include the *mer* operon, which is the Hg-resistance determinant. It is found on plasmids of both Gram-negative and Gram-positive bacteria. The operon contains a number of genes, including the regulatory genes *mer*R and *mer*D, the transport genes *mer*T and *mer*P, and the gene *mer*A, which encodes mercuric reductase (Tayang and Songachan 2021). *mer*A and *mer*B are the most frequently detected functional genes among those that are regularly found. The enzymes organomercurial lyase and mercuric ion reductase are produced by these genes, respectively. These enzymes are extremely important for the detoxification of Hg because they work together. *mer*B helps to turn poisonous organomercurial compounds like Me-Hg and phenyl mercuric acetate into combustible elemental mercury, while *mer*A helps change mercuric ions into less dangerous elemental mercury. *mer*A and *mer*B work together to give bacteria broad-spectrum resistance to Hg. Nevertheless, merA alone itself only offers partial resistance to inorganic mercury (Dash and Das 2012). Positively controlled operons with an operator region, a promoter region, and a number of downstream genes can be found in broad-spectrum resistant organisms. Several proteins associated in mercury uptake and resistances are encoded by this operon. There is a unique regulatory protein named *mer*R at the operator-promoter region. These structural genes, which are in charge of absorbing mercury, include *mer*T, *mer*P, and *mer*C. Additional genes that contribute to the overall resistance mechanism include *mer*F, *mer*G, and *mer*E. Additionally, the operon contains genes for the reduction and lyase enzymes *mer*A and *mer*B, which are essential for the detoxification and removal of mercury (Mahbub et al. 2017a, b).

Mycolicibacterium peregrinum (Sanjaya et al. 2021), *Pseudomonas maculicola, P. putida,* and *Enterobacter aerogenes* are used to clean up mercury-contaminated soils in regions where gold mining once occurred and these can also dissolve phosphate and fix nitrogen from the atmosphere (Ginting et al. 2021). Mercury reductase activity and the *mer* operon have been discovered in the mercury-resistant strains of *Pseudomonas aeruginosa* and *Bacillus licheniformis*, enabling them to transform the poisonous form of Hg into a non-toxic form and flourish in Hg-containing environments. This species can efficiently tolerate Hg doses of up to 100 mg/ml (Kotwal et al. 2018). *Vibrio fluvialis* strain can remove Hg ecofriendly (Saranya et al. 2017). *Pseudomonas syringae* can almost completely remove the mercury from a liquid medium by microbial metabolism after 120 h of incubation (Kowalczyk et al. 2016).

Fusobacterium aquatile, Brevundimonas vesicularis, Fusobacterium necrogenes, and *Nitrococcus mobilis* can accumulate 10, 20, and 30 ppm Hg (Chasanah et al. 2018). Mercury-resistant bacteria from the genera Bacillus, Enterobacter, Klebsiella, and Acinetobacter live in the Mithi river in the Bombay area and have the capacity to bioremediate Hg from the river water. It has been discovered that these bacteria can withstand high Hg concentrations after being isolated (Pushkar et al. 2019). Potential Bacterial isolates involved in mercury bioremediation are presented in Table 13.1.

Table 13.1 Potential bacterial isolates involved in mercury bioremediation

Potential bacterial isolates	Target heavy metal	References
Acidithiobacillus caldus	Hg^{+2}	Chen et al. (2011)
Pseudomonas aeruginosa	Hg^{+2}, Pb^{+2}, Cd^{+2}	Imron et al. (2021)
Plenococus sp.	Hg^{+2}	Cheng et al. (2022)
Stenotrophomonas maltophilia ADW10	Hg^{+2}	Naguib et al. (2019)
Pseudomonas otitidis	Hg^{+2},	Naguib et al. (2019)
Herbispirillum huttiense TL36	Hg^{+2}, As^{+2}	Rojas-Solis et al. (2023)
Klebsiella oxytoca TL49	Hg^{+2}, As^{+2}	Rojas-Solis et al. (2023)
Rhizobium radiobacter TL52	Hg^{+2}, As^{+2}	Rojas-Solis et al. (2023)

13.4.2 Mercury Bioremediation by Fungi

Heavy metals can be efficiently absorbed by fungi from the genera, *Streptoverticillium, Aspergillus, Rhizopus,* and *Saccharomyces* (Saraswathi and Sumithra 2018). Recently, bioremediation of contaminants like heavy metals has been demonstrated by using filamentous fungi. *Aspergillus flavus* KRP1 strain has the potential to bioremediate Hg using a biosorption mechanism (Kurniati et al. 2014). The other fungal strains that exhibit excellent Hg removal potential in aquatic environments were identified recently and including *Cladosporium* sp., *Phoma costaricensis, Fusarium oxysporum, Didymella glomerata, Sarocladium kiliense* (Vacar et al. 2021) and *Cladosporium halotolerans* (Sanjaya et al. 2021).

It has been discovered that the fungus *Metarhizium robertsii*, which lives symbiotically with plants, degrades Me-Hg and reduces Hg^{+2}, which lowers the amount of Hg that accumulates in plants and encourages their growth in contaminated soils. This process is achieved by an enzyme methylmercury demethylase (MMD) which converts Me-Hg into Hg^{+2} and causes the demethylation, the Hg^{+2} is further reduced to volatile Hg^0 by another enzyme mercury ion reductase (MIR). Moreover, even without additional nutrients, *M. robertsii* has the capacity to eliminate Me-Hg and Hg^{+2} from salt and freshwater (Wu et al. 2022). The use of fungi that are naturally present in locations that are contaminated with heavy metals (HMs) for bioremediation is a potential. Isolates with high minimum inhibitory concentrations (MIC) for Hg, ranging from 140 to 200 mg/L, include *Phoma costaricensis, Cladosporium* sp., *Fusarium oxysporum, Sarocladium kiliense,* and *Didymella glomerata* (Vacar et al. 2021).

The potential to remove Cd and Hg metals from soil, sewage, and industrial effluent was evaluated for two extremely tolerant fungal isolates, *Alternaria alternata* and *Penicillium aurantiogriseum.* Both isolates were discovered to be resistant to these heavy metals up to 1000 ppm. With the investigation, the best microorganisms for removing heavy metals from liquid media were discovered (Bahobil et al. 2017). The white rot fungus *Phlebia floridensis* could withstand a concentration of mercury up to 100 μM. A recent study reported that some potential fungal isolates significantly reduce Hg up to 97% in laboratory conditions including *Lindgomycetaceae* P87, *Westerdykella* sp. P71, *Aspergillus* sp. A31 and *Curvularia geniculata* P1 (Pietro-Souza et al. 2020). Potential fungal isolates involved in mercury bioremediation is presented in Table 13.2.

13.4.3 Mercury Bioremediation by Algae

Microalgae have drawn a lot of interest because of their extraordinary capacity to react to and accumulate several types of contaminants in ecosystems. Additionally, their potential for use as biocatalysts in environmental applications has been

Table 13.2 Potential Fungal isolates involved in Mercury Bioremediation

Potential fungal isolates	Target heavy metal	References
Penicillium sp.	Hg^{+2}	Cheng et al. (2022)
Westerdykella sp. P7, 1 *Aspergillus* sp. A31, *C. geniculata* P1 and *Lindgomycetaceae* P87	Hg^{+2}	Pietro-Souza et al. (2020)
Westerdykella aquatica P71	Hg^{+2}	Senabio et al. (2023)
Pseudomonodictys pantanalensis nov. A73	Hg^{+2}	Senabio et al. (2023)
Aspergillus hiratsukae LF1	Hg^{+2}	Palanivel et al. (2023)
Aspergillus terreus LF2	Hg^{+2}	Palanivel et al. (2023)

acknowledged. Heavy metals are a significant contributor to environmental pollution since they are known to be harmful and carcinogenic even in minute amounts. The ecosystem, as well as the health of humans and animals, is seriously threatened by these heavy metals. A variety of advantages are provided by microalgae in the context of phytoremediation. They are easily available, affordable, and very effective at removing hazardous metals from the environment. In addition, using microalgae for phytoremediation does not add more toxicity to the ecosystem, making it a secure and sustainable method (Manikandan et al. 2022). Microalgae have been found to be effective detoxification candidates and are being looked at as a less expensive alternative to conventional physicochemical therapies. A microalgae's ability to absorb metals can be broken down into two stages: an initial step that happens quickly and is unrelated to cell metabolism, called "adsorption" onto the cell surface, and a second, more drawn-out stage that depends on cell metabolism, called "absorption" or "intracellular uptake" (Dwivedi 2012). Numerous significant heavy metal pollution need to be managed quickly and effectively. Among them, mercury stands out as a highly corrosive heavy metal that is also particularly hazardous. The industrial sector is significantly in danger from its presence in water bodies since it can damage aluminum-containing machinery. Long-term Hg exposure can cause severe neurological conditions, including paralysis. Strict regulations have been developed in the European Union (EU) to guarantee the safety of drinking water and wastewater with regard to mercury. The permissible limit of Hg in drinking water and wastewater is 0.001 mg/L and 0.005 mg/L, respectively. In order to safeguard public health and maintain the integrity of aquatic ecosystems, these restrictions act as critical benchmarks (Znad et al. 2022).

Diatoms cell walls are essential in their defense against mercury toxicity. Hg undergoes chemical changes when it enters the cell, interacting with sulfur proteins like glutathione to reduce its toxicity. Diatoms have tolerance mechanisms that allow them to store heavy metals in their bodies. These pathways result in the production of active ligands like organic acids, amino acids, metallothioneins, and phytochelatins. These ligands can bind to heavy metals and help them enter vacuoles as complexes by binding to them. These procedures help diatoms effectively control and ensconce heavy metals inside their cellular frameworks (Naseri et al. 2020). Hg and extracellular functional groups on the surface of algae can interact electrostatically thanks

Table 13.3 Potential algal isolates involved in mercury bioremediation

Potential algal isolates	Target heavy metal	References
Chlorella vulgaris	Hg^{+2}	Kumar et al. (2020)
Dunaliella sp.	Hg^{+2}	Imani et al. (2011)
Selenastrum minutum	Hg^{+2}	Rani et al. (2021)
Chlorella fusca var. fusca	Hg^{+2}	Rani et al. (2021)
Navicula pelliculosa	Hg^{+2}	Rani et al. (2021)
Galdieria sulphuraria	Hg^{+2}	Rani et al. (2021)
Microcystis aeruginosa	Hg^{+2}	Tang et al. (2023)
Spirogyra sp.	Hg^{+2}	Almeida et al. (2023)
Chlorella sp.	Hg^{+2}	
Sargassum magnetite	Hg^{+2}	Sandoval-Cárdenas et al. (2023)

to oxygen lone pairs. For instance, it has been discovered that the algae *Chlorella vulgaris* can absorb 17.49 mg/g of Hg in about 2 h at pH 5 and 20 °C by using the carboxylic groups on its surface. *Scenedesmus obtusus* XJ-15, a different species, has shown the capacity to bind mercury via carboxyl (–COOH), phosphate ($-PO_4^{-2}$), and hydroxyl (–OH) containing functional groups. By treating the algal biomass with phosphate, researchers were able to increase the removal efficiency even more. This led to a 180-min sorption time of 95 mg/g at temperature 25 °C and pH 5 (Chugh et al. 2022).

A contact time of 72 h and a concentration of 5,000 were shown to be ideal condition for *Skeletonema costatum* cells in Hg^{+2} contaminated seawater resulting in a removal rate of 79.5% for an starting Hg concentration of 2 mg/L (Soedarti et al. 2017). A recent study suggests that *Chlorella vulgaris* biomass has the potential to remove Hg from aqueous contaminated water. The study was conducted using batch culture at temperature 35 °C, keeping other process parameters variables such as initial concentration of Hg, contact time, desorption rate of Hg, and pH. The pH 6, shows the maximum adsorption of Hg (Kumar et al. 2020). Researchers are continuously looking at algae-based adsorbents for the removal of Hg ions due to the potential negative health hazards even at low concentration. Numerous studies have looked into the detoxification of Hg using various kinds of microalgae. Six potential species of algae, viz. *Osmundea pinnatifida, Ulva lactuca, U. intestinalis, Gracilaria* sp., *Fucus vesiculosus* and *F. spiralis* for Hg adsorption (Znad et al. 2022). Potential algal isolates involved in mercury bioremediation are presented in Table 13.3.

13.4.4 Mercury Bioremediation by Cyanobacteria

Compared to other traditional methods, heavy metal removal mediated by cyanobacteria is a promising technique. It offers in situ operability, cost-effectiveness, and environmentally friendly chemistry. Cyanobacteria are adaptable metal sequestration agents because they can sequester metals both by biosorption and bioaccumulation. Exopolysaccharides (EPS), one of the main components of cyanobacteria's cell wall, are essential for metal ion biosorption. In order to uncover new engineering approaches for creating this extraordinary polymer that improves metal ion adsorption, this paper examines various mechanisms of EPS formation. The study also examines the capacities for heavy metal sequestration of several cyanobacterial species under various environmental circumstances. Strains of *Limnococcus* sp., *Synechococcus* sp. PCC 7942 and *Nostoc muscorum*, show excellent heavy metal removal efficiency, heavy metal detoxification, multi-metal sequestration, biosorption, and bioaccumulation (Al-Amin et al.2021).

Through a variety of metabolic processes, cyanobacteria have the ability to detoxify, remove or degrade a variety of pollutants. They have the capacity to alter cellular processes and generate stress-tolerant metabolites that help with cleanup. When it comes to cleaning up contaminated sites with high concentrations of metals like arsenic (As), copper (Cu), chromium (Cr), cobalt (Co), mercury (Hg), iron (Fe), manganese (Mn), lead (Pb), nickel (Ni), molybdenum (Mo), vanadium (V), cadmium (Cd), and zinc (Zn) cyanobacteria have a remarkable amount of potential. Additionally, cyanobacteria are good at cleaning up pollutants like dyes and polycyclic aromatic hydrocarbons (PAHs). This article gives a brief introduction to bioremediation and emphasizes the important roles that cyanobacteria play in the recovery of damaged ecosystems (Gupta et al. 2012).

The cyanobacterial species Komarek, *Phormidium limnetica* (Lemm.), *Limnothrix planctonica* (Lemm.), and *Synechococcus leopoldiensis* (Racib.) have shown the potential to convert Hg^{+2} ($HgCl_2$) into β-HgS and Hg (0) under controlled aeration and pH. Although there were only slight variations across the various cyanobacterial species, Hg^0 production was quite modest and varied with exposure level. Biotransformation depleted the acid-reducible Hg by converting nearly all of the remaining Hg in the culture into β-HgS. These results indicate that the epilimnion (upper layer) of lakes that receive rainfall contaminated with Hg is where the biotransformation of Hg^{+2} into β -HgS most likely takes place (Lefebvre et al. 2007).

To absorb and minimize the hazards and toxic effects of heavy metals, cyanobacteria use a variety of methods, such as biosorption, bioaccumulation, metal transporter activation, induction of detoxifying enzymes, and biotransformation. In order to find potential genes and proteins that can be altered to enhance the efficiency of cyanobacterial bioremediation, it is important to understand physiological responses and regulatory mechanisms at the gene level. Potential targets for strain improvement and optimization include transcriptional regulators, metal transporters, cellular metabolites, and chaperones such as biosurfactants, extracellular polymers, phytochelatins, and metallothioneins in order to enhance the effectiveness of bioremediation using cyanobacteria (Chakdar et al. 2022). Metallothioneins (MTs) are the metal-binding proteins that can be produced by cyanobacteria. Small in size and abundant in cysteinyl residues, these proteins bind metal ions via cysteine ligands to form cysteinyl thiolate bridges. While Zn and Cu are two metals that are frequently connected with MTs, they can also bind to other hazardous metals like Cd, Hg, and Pb. The presence of cysteine residues and their precise configuration, such as -Cys-Cys-Cys-X-Cys- or -Cys-X-X-Cys (where X represents other amino acids), are what give MTs their ability to bind metals (Yadav et al. 2020).

To combat the difficulties brought on by the buildup of heavy metals (HM), cyanobacterial cells have developed particular defense mechanisms. The three primary ways that cyanobacterial cells catch HMs are described here. First, extracellular polysaccharides (EPS) with anionic groups are found in the cyanobacteria's Gram-negative cell wall's outer layer. Because of the HMs' anionic characteristics, these EPS can bind them. Second, metallothionein enzymes, notably cysteine residues, are abundant in thiol groups and are found in the cytoplasm of cyanobacterial cells. Through their negatively charged thiol groups, these enzymes may efficiently grab HMs, preventing them from interacting with essential biological components like key enzymes. Finally, via membrane transport proteins, cyanobacterial cells can reflux the stored HMs back into the extracellular matrix. The HM homeostasis within the cell is supported by this process.

13.4.5 Mercury Bioremediation by Actinomycetes

Various actinomyces genera have already been reported to be used in bioremediation of Hg. The Hg resistance genes in *Streptomyces lividans* 1326 were identified and exhibit similar properties like transposons encoded by AUD2 and have repeat sequences at its end as insertion sequences. This component connected genetic instability and Hg resistance. The Hg resistance of *S. lividans* is mediated by 6 ORFs (open reading frames) arranged in two different operons. The first operon contains regulatory and transport genes, while the second operon comprises organomercurial lyase and mercuric reductase genes. Diverse promoters situated in the intercistronic region regulate these operons. The organomercurial lyase enzyme, which breaks the Me-Hg link and releases Hg^{2+} ions, is encoded by the *mer*B gene. The Hg^{2+} ions are then transformed into volatile Hg^0 (elemental mercury) by another enzyme, the mercuric reductase, which is encoded by the *mer*A gene. Several reports suggested that together *mer*A and *mer*B genes impart broad spectrum Hg resistance, while alone *mer*A provides narrow-spectrum Hg resistance to the organisms (Ravel et al. 1998). Currently, it is of utmost necessity to apply biological techniques to eliminate dangerous heavy metals and radioactive contaminants. Bioremediation based on microorganisms is regarded as a secure and effective method. Actinobacteria are extremely beneficial for bioremediation because they play a crucial role in the degradation and transformation of metal substrates and organic pollutants. High concentrations of heavy metals such as Cd, Cr, Hg, Pb, Zn, and Cu, chemical compounds, and pesticides can all be eliminated and detoxified by actinobacteria. They also possess the ability to use pollutants as their nutrient sources such as carbon sources and produce industrially important products viz. antibiotic, enzymes, proteins (Kannabiran 2017), and organic acids. Recently two potential Hg tolerant acti nomycetes strains CHR28 and CHR3 recovered from a metal-contaminated zone in Baltimore's Inner Harbour, USA. Streptomyces VITSVK9, another Actinomycete strain with a remarkable potential for metal tolerance, has been used for mercury biosorption. A strain of Streptomyces TY046-017, was recovered from tin tailings, and has also demonstrated possible resistance to Hg, indicating that it can withstand exposure to this heavy metal (Jain et al. 2022).

Three potential strains of *Streptomyces* spp. AS6, AS1, and AS2 grown in Hg supplemented medium and shown growth inhibition zones of 20 mm, 10.3 mm, and 9.6 mm respectively, displayed decreased resistance to $HgCl_2$ at a dosage of 1 mM. Previous studies also noted that these isolates were resistant to $HgCl_2$ at the same dose, with the growth inhibition diameters for Streptomyces spp. AS6, AS1, and AS2, being 8.77 mm, 6.53 mm, and 7.47 mm respectively. Streptomyces spp. BR 28's resistance decreased slightly from 9.23 mm to 9.3 mm, showing that it was still largely stable against $HgCl_2$ (Rahayu et al. 2021).

13.5 Phytoremediation of Mercury

It is a process in which plants and associated microbes are employed for the detoxification of pollutants and heavy metal detoxification. It is an eco-friendly and cost-effective method widely used to clean the environment. Particularly in places affected by gold mining, phytoremediation is the low-cost remedy for remediating Hg-polluted soils. Removing mercury from the soil, may entail phytoextraction, followed by the recovery or secure storage of the mercury that was extracted. Through phytostabilization, phytoremediation not only stops mercury emissions and acid mine drainage but also lowers mercury bioavailability and leaching. Hg can be removed from substrates using ligand-induced phytoextraction, which results in a reduced disposal residue. The site can be used again once the Hg levels in the soil are acceptable. This plant-based remediation approach can also be used in conjunction with lucrative ventures like forestry, the sale of carbon credits, or biomass energy projects (Moreno et al. 2005). Three agricultural crops viz. wheat, barley, and yellow lupin were evaluated for Hg tolerance potential in field experiments. Hg concentrations in the soil were quite low, at 29.17 g/g in the 0–10 cm stratum and 20.32 g/g in the 10–40 cm horizon. The crops were successful at removing Hg, with wheat having the greatest concentration (0.4791 g/g) of all the crops. The low availability of Hg in the soil, however, reduced the overall intake of Hg. The highest phytoextraction yield was achieved by barley, which may reach 719 mg ha1. Future initiatives will boost agricultural biomass yields and increase Hg availability utilizing solubilization agents (Rodriguez et al. 2005). When opposed to removing the soil, phytoremediation is a more affordable option for treating soil that has been contaminated with Hg. Using ammonium thiosulphate to remove Hg, this study tested the efficacy of *Cyperus kyllingia* Endl, *Paspalum conjugatum* L., and *Lindernia crustacea* L. for phytoremediation and evaluated their effects on maize growth. The crops were grown for 63 days in containers with Hg-contaminated soil while receiving two applications of ammonium thiosulfate. Compared to no ammonium thiosulphate, ammonium thiosulphate enhanced Hg accumulation in plant roots and shoots by 47% and 82% respectively. Compared to the control treatment, maize grown on previously remedied soil increased in dry weight by 40% when ammonium thiosulphate was added, but by 62% when it wasn't (Muddarisna et al. 2013).

Phytoremediation is gaining popularity due to its cost-effective and eco-friendly strategy to remediate Hg pollution in soils. Many plant species that have the potential to accumulate Hg have been identified and strategies to improve phytoremediation are being investigated. *Axonopus compressus* is the only species with a BAF (bioaccumulation factor) <1.0, and some plants have a TF (translocation factor) <1.0, showing their capacity to transfer Hg to above-ground tissues. The effectiveness of remediation is influenced by variables such as plant species, soil characteristics, and Hg bioavailability. Compost and chemical accelerators like potassium iodide can encourage plants to absorb Hg (Liu et al. 2020). Although additional fungal species and families should be assessed for comparison, the *L. perenne* and arbuscular mycorrhizal fungus residing symbiotically impacts shoot growth and root development. The GST gene had a distinct response to mycorrhizal treatments, highlighting its significance in toxin removal. A buildup of glutathione without evacuation, however, might have had a harmful effect in the instance of Hg^{2+}. To comprehend how the structures of the plant are impacted by Hg remediation, more research should examine additional genes and enzymes including Hsp70, CAT, GPX, pAPX, and Hsp90. Additionally, the estimated 1.17% Hg mass balance errors caused by sample and equipment accuracy must be taken into consideration (Leudo et al. 2020).

The inclusion of *Sansevieria trifasciata* and *Celosia plumosa* led to Hg removal efficiencies of 75.63% and 66.81%, respectively, with final Hg concentrations of 58 mg/kg and 79 mg/kg in a reactor utilizing growth media made up of 80% contaminated soil and 20% compost. Similar results were obtained in a reactor using *Sansevieria trifasciata* and *Celosia plumosa*, with final Hg concentrations of 60 mg/kg and 82 mg/kg, respectively, and Hg removal efficiencies of 74.79% and 66.55%, respectively (Ratnawati 2020). The potential to rehabilitate unproductive land is a current emphasis for reclaiming Hg-contaminated locations. A developing technique called phytoremediation enables plants to deal with toxins while preserving or enhancing topsoil fertility through root exudation. It provides a low-maintenance, cost-effective option. Despite the lack of Hg-hyperaccumulator species, progress has been made in figuring out which plant species may survive in soil that has been poisoned with Hg (Tiodar et al. 2021).

13.6 Conclusions and Future Research

Hg is a heavy metal and toxic in nature, as it enters the environment by various anthropogenic and industrial processes and passes to the food chain. Many researchers suggested that Hg exerts detrimental toxic effects equally to vegetation, animals, and human beings. Hg causes neurological disorders, kidney failure, liver failure, skin diseases, and other severe diseases in human beings. Therefore, elimination and detoxification of Hg from the environment is an essential task to combat this problem. Many traditional approaches such as physical, chemical, and physicochemical methods have been adopted for the remediation of Hg. Nowadays, bioremediation is gaining popularity for the removal of heavy metals and other xenobiotics due to its least harmful, eco-friendly economic, and high efficacy rate. On the other hand, the traditional approaches such as physical and chemical methods are less effective, expensive, time taking, and laborious. The chemical methods also contribute to environmental pollution as in these methods the chemical reacts with the pollutants and other compounds to produce intermediate compounds and sometimes it is very difficult to detoxify. There are many biological methods such as bioaccumulation, biosorption, bio-adsorption, biosequestration, biotransformation, biomineralization, and bioprecipitation has been successfully employed for the elimination and detoxification of heavy metal including Hg with the help of various microorganisms. A variety of bacterial, fungal, actinomycetes, cyanobacteria, and algal species isolated and identified that effectively eliminate and detoxify Hg, Pb, Cd, Cr, Cu, and other pollutants from contaminated sites and restore their natural condition. However, there are still some constraints and limitations in the wide applicability of bioremediation. As the indigenous microorganisms exposed to the pollutants may sometimes be killed due to the toxicity of the pollutants. Therefore, the more resistant and genetically modified microorganisms need to develop with greater removal and detoxification efficiency. The molecular and enzymatic pathways for the detoxification of the Hg are required to study for the bioaccumulation of Hg by microorganisms. Enzymatic pathways of detoxification, active–passive transport, extra-intracellular sequestration are the main defensive mechanisms of the microorganism to Hg, which tend to resist and detoxify/weaken the toxicity of Hg. Usually, a single microorganism can be resistant to certain pollutants only. Therefore, further research is required for the development of a more efficient recombinant strain (superbug) by applying recombinant DNA technology such as mutation, gene amplification, recombination, and protoplast fusion for bioremediation of Hg and other xenobiotic compounds, which target more than one pollutant. It is also seen that one single approach is not capable of removing pollutants effectively. Therefore, microbial remediation in combination with physical and chemical methods provides greater efficiency than their individual one.

References

Ahmad P, Ahanger MA, Egamberdieva D, Alam P, Alyemeni MN, Ashraf M (2018) Modification of osmolytes and antioxidant enzymes by 24-epibrassinolide in chickpea seedlings under mercury (Hg) toxicity. J Plant Growth Regul 37:309–322

Al-Amin A, Parvin F, Chakraborty J, Kim YI (2021) Cyanobacteria mediated heavy metal removal: a review on mechanism, biosynthesis, and removal capability. Environ Technol Rev 10:44–57

Almeida Â, Cotas J, Pereira L, Carvalho P (2023) Potential Role of *Spirogyra* sp. and *Chlorella* sp. in bioremediation of mine drainage: a review. Phycology 3(1):186–201
Asghari I, Mousavi SM, Amiri F, Tavassoli S (2013) Bioleaching of spent refinery catalysts: a review. Ind Eng Chem Res 19:1069–1081
Ayele A, Haile S, Alemu D, Kamaraj M (2021) Comparative utilization of dead and live fungal biomass for the removal of heavy metal: a concise review. Sci World J
Azevedo LF, Karpova N, Rocha BA, Barbosa Junior F, Gobe GC, Hornos Carneiro MF (2023) Evidence on neurotoxicity after intrauterine and childhood exposure to organomercurials. Int J Environ Health Res 20(2):1070
Bae SJ, Shin KS, Park C, Baek K, Son SY, Sakong J (2023) Risk assessment of heavy metals in tuna from Japanese restaurants in the Republic of Korea. Ann Occup Env Med 35
Bahobil A, Bayoumi RA, Atta HM, El-Sehrawey MM (2017) Fungal biosorption for cadmium and mercury heavy metal ions isolated from some polluted localities in KSA. Int J Curr Microbiol Appl Sci 6(6):2138–2154
Bala S, Garg D, Thirumalesh BV, Sharma M, Sridhar K, Inbaraj BS, Tripathi M (2022) Recent strategies for bioremediation of emerging pollutants: a review for a green and sustainable environment. Toxics. 10(8):484
Basu N, Horvat M, Evers DC, Zastenskaya I, Weihe P, Tempowski J (2018) A state-of-the-science review of mercury biomarkers in human populations worldwide between 2000 and 2018. Environ Health Perspect 126(10):106001
Beckers F, Rinklebe J (2017) Cycling of mercury in the environment: sources, fate, and human health implications: a review. Crit Rev Environ Sci Technol 47(9):693–794
Bellinger DC, Daniel D, Trachtenberg F, Tavares M, McKinlay S (2007) Dental amalgam restorations and children's neuropsychological function: the New England Children's Amalgam Trial. Environ Health Perspect 115(3):440–446
Bello TC, Buralli RJ, Cunha MP, Dórea JG, Diaz-Quijano FA, Guimarães JR, Marques RC (2023) Mercury exposure in women of reproductive age in Rondônia state, Amazon Region, Brazil. Int J Environ Health Res 20(6):5225
Cai L, Fu Z, Cui F (2020) Synthesis of carbon dots and their application as turn off–on fluorescent sensor for mercury(II) and glutathione, 11–20
Carrasco-Gil S, Ortega-Villasante C, Sobrino-Plata J, Barón-Sola Á, Millán R, Hernández LE (2023) Attenuation of mercury phytotoxicity with a high nutritional level of nitrate in alfalfa plants grown hydroponically. Plant Stress 100131
Castagna J, Bencardino M, D'Amore F, Esposito G, Pirrone N, Sprovieri F (2018) Atmospheric mercury species measurements across the Western Mediterranean region: behaviour and variability during a 2015 research cruise campaign. Atmos Environ 173:108–126
Chakdar H, Thapa S, Srivastava A, Shukla P (2022) Genomic and proteomic insights into the heavy metal bioremediation by cyanobacteria. J Hazard Mater 424:127609
Chasanah U, Nuraini Y, Handayanto E (2018) The potential of mercury-resistant bacteria isolated from small-scale gold mine tailings for accumulation of mercury. J Ecol Engg 19(2):236–245
Chen D, Lin J, Che Y, Liu X, Lin J (2011) Construction of recombinant mercury resistant *Acidithiobacillus caldus*. Microbiol Res 166(7):515–520
Chen L, Liang S, Liu M, Yi Y, Mi Z, Zhang Y, Li Y, Qi J, Meng J, Tang X, Zhang H (2019) Trans-provincial health impacts of atmospheric mercury emissions in China. Nat Commun 10(1):1484
Cheng Z, Shi C, Gao X, Wang X, Kan G (2022) Biochemical and Metabolomic responses of Antarctic bacterium P*lanococcus* sp. O5 induced by copper ion. Toxics 10(6):302
Chetelat J, Ackerman JT, Eagles-Smith CA, Hebert CE (2020) Methylmercury exposure in wildlife: a review of the ecological and physiological processes affecting contaminant concentrations and their interpretation. Sci Total Environ 711:135117
Chibuike GU, Obiora SC (2014) Heavy metal polluted soils: effect on plants and bioremediation methods. Appl Environ Soil Sci 1–12. https://doi.org/10.1155/2014/752708

Chugh M, Kumar L, Shah MP, Bharadvaja N (2022) Algal Bioremediation of heavy metals: An insight into removal mechanisms, recovery of by-products, challenges, and future opportunities. Energy Nexus 100129

Crespo-Lopez ME, Augusto-Oliveira M, Lopes-Araujo A, Santos-Sacramento L, Yuki Takeda P, Macchi BM, do Nascimento JLM, Maia CSF, Lima RR, Arrifano GP (2021) Mercury: what can we learn from the Amazon? Environ Int 146:106223

Crespo-Lopez ME, Augusto-Oliveira M, Lopes-Araújo A, Santos-Sacramento L, Souza-Monteiro JR, da Rocha FF, Arrifano GDP (2022) Mercury neurotoxicity in gold miners. In: Advances in neurotoxicology. Academic Press, Cambridge, MA, USA

Cusset F, Reynolds SJ, Carravieri A, Amouroux D, Asensio O, Dickey RC, Fort J, Hughes BJ, Paiva VH, Ramos JA, Shearer L (2023) A century of mercury: Ecosystem-wide changes drive increasing contamination of a tropical seabird species in the South Atlantic Ocean. Environ Poll 323:121187

da Silva SF, de Oliveira Lima M (2020) Mercury in fish marketed in the Amazon triple frontier and health risk assessment. Chemosphere 248:125989

Danouche M, El Ghachtouli N, El Arroussi H (2021) Phycoremediation mechanisms of heavy metals using living green microalgae: physicochemical and molecular approaches for enhancing selectivity and removal capacity. Heliyon 7(7):e07609

Dash HR, Das S (2012) Bioremediation of mercury and the importance of bacterial mer genes. Int Biodeterior Biodegrad 75:207–213

De J, Dash HR, Das S (2014) Mercury pollution and bioremediation—a case study on biosorption by a mercury-resistant marine bacterium. In: Microbial biodegradation and bioremediation. Elsevier, pp 137–166

Deng X, Wilson D (2001) Bioaccumulation of mercury from wastewater by genetically engineered *Escherichia coli*. Appl Microbiol Biotechnol 56:276–279

Deshmukh R, Khardenavis AA, Purohit HJ (2016) Diverse metabolic capacities of fungi for bioremediation. Indian J Micro 56:247–264

Du B, Li P, Feng X, Yin R, Zhou J, Maurice L (2021) Monthly variations in mercury exposure of school children and adults in an industrial area of southwestern China. Environ Res 196:110362

Durand A, Maillard F, Foulon J, Chalot M (2020) Interactions between Hg and soil microbes: microbial diversity and mechanisms, with an emphasis on fungal processes. App Microbiol Biotechnol 104:9855–9876

Dwivedi S (2012) Bioremediation of heavy metal by algae: current and future perspective. J Adv Lab Res Biol 3(3):195–199

Essa AMM, Macaskie LE, Brown NL (2002) Mechanisms of mercury bioremediation, pp 672–674

Genchi G, Sinicropi MS, Carocci A, Lauria G, Catalano A (2017) Mercury exposure and heart diseases. Int J Environ Health Res 14(1):74

Ginting RCB, Solihat N, Hafsari AR (2021) Potential bacteria capable of remediating mercury contaminated soils. In IOP conference series. Earth Environ Sci 648:012136

Gupta RK, Choudhary KK, Kumar M, Negi A, Rai H (2012) Bioremediation and cyanobacteria an overview. BioNano Front 9:190–196

Gustin MS, Bank MS, Bishop K, Bowman K, Branfireun B, Chételat J, Eckley CS, Hammerschmidt CR, Lamborg C, Lyman S, Martínez-Cortizas A (2020) Mercury biogeochemical cycling: a synthesis of recent scientific advances. Sci Total Environ 737:139619

Hafeez A, Rasheed R, Ashraf MA, Qureshi FF, Hussain I, Iqbal M (2023) Effect of heavy metals on growth, physiological and biochemical responses of plants. In: Plants and their interaction to environmental pollution. Elsevier, pp 139–159

Harada M (1995) Minamata disease: methylmercury poisoning in Japan caused by environmental pollution. Crit Rev Toxicol 25(1):1–24

Hassan JM, Ali Raza M, Ur Rehman S, Ansar M, Gitari H, Khan I, Wajid M, Ahmed M, Abbas Shah G, Peng Y, Li Z (2020) Effect of cadmium toxicity on growth, oxidative damage, antioxidant defense system and cadmium accumulation in two sorghum cultivars. Plants 9(11):1575

He F, Gao J, Pierce E, Strong PJ, Wang H, Liang L (2015) In situ remediation technologies for mercury-contaminated soil. Environ Sci Pollut Res 22:8124–8147

Henriques B, Rocha LS, Lopes CB, Figueira P, Monteiro RJ, Duarte ADC, Pardal MA, Pereira E (2015) Study on bioaccumulation and biosorption of mercury by living marine macroalgae: prospecting for a new remediation biotechnology applied to saline waters. Chem Eng J 281:759–770

Hu L, Liu B, Li SZ, Zhong H, He ZZ (2021a) Study on the oxidative stress and transcriptional level in Cr(VI) and Hg(II) reducing strain *Acinetobacter indicus* yy-1 isolated from chromium-contaminated soil. Chemosphere 269:128741

Hu XF, Lowe M, Chan HM (2021b) Mercury exposure, cardiovascular disease, and mortality: a systematic review and dose-response meta-analysis. Environ Res 193:110538

Imani S, Rezaei-Zarchi S, Hashemi M, Borna H, Javid A, Zand AM, Abarghouei HB (2011) Hg, Cd and Pb heavy metal bioremediation by Dunaliella alga. J Med Plants Res 5(13):2775–2780

Imron MF, Kurniawan SB, Abdullah SRS (2021) Resistance of bacteria isolated from leachate to heavy metals and the removal of Hg by *Pseudomonas aeruginosa* strain FZ-2 at different salinity levels in a batch biosorption system. Sustain Environ Res 31:1–13

Jackson TA (1998) Mercury in aquatic ecosystems. Springer US, pp 77–158

Jafari M, Abdollahi H, Shafaei SZ, Gharabaghi M, Jafari H, Akcil A, Panda S (2019) Acidophilic bioleaching: a review on the process and effect of organic–inorganic reagents and materials on its efficiency. Min Proc Ext Met Rev 40:87–107

Jain S, Gupta I, Walia P, Swami S (2022) Application of actinobacteria in agriculture, nanotechnology, and bioremediation. In: Actinobacteria-diversity, applications and medical aspects. IntechOpen

Jariyal M, Yadav M, Singh NK, Yadav S, Sharma I, Dahiya S, Thanki A (2020) Microbial remediation progress and future prospects. In: Bioremediation of pollutants. Elsevier, Amsterdam, The Netherlands, pp 187–214

Jeyakumar P, Debnath C, Vijayaraghavan R, Muthuraj M (2023) Trends in bioremediation of heavy metal contaminations. Environ Eng Res 28(4)

Johnson JM, Bock SL, Smaga CR, Lambert MR, Rainwater TR, Wilkinson PM, Parrott BB (2023) Relationships between maternally-transferred mercury and hatchling development, behavior, and survival in the American alligator (*Alligator mississippiensis*). Sci Total Environ 870:162010

Kannabiran K (2017) Actinobacteria are better bioremediating agents for removal of toxic heavy metals: an overview. Int J Environ Tech Manage 20(3–4):129–138

Kim YH, Khan AL, Kim DH, Lee SY, Kim KM, Waqas M, Jung HY, Shin JH, Kim JG, Lee IJ (2014) Silicon mitigates heavy metal stress by regulating P-type heavy metal ATPases, *Oryza sativalow* silicon genes, and endogenous phytohormones. BMC Plant Biol 14(1):1–13

Kim K, Park H (2023) Association of mercury exposure with the serum high-sensitivity C-reactive protein level in Korean adults. Front Public Health 11

Kotwal DR, Shewale NB, Tambat US, Thakare MJ, Bholay AD (2018) Bioremediation of mercury using mercury resistant bacteria. Res J Life Sci Bioinform Pharm Chem Sci 4(2)

Kowalczyk A, Wilińska M, Chyc M, Bojko M, Latowski D (2016) Isolation, screening and identification of mercury resistant bacteria from mercury contaminated soil. In: E3S web of conferences, p 10

Kumar M, Singh AK, Sikandar M (2020) Biosorption of Hg(II) from aqueous solution using algal biomass: kinetics and isotherm studies. Heliyon 6(1):e03321

Kumar A, Sidharth S, Kandasubramanian B (2023) A review on algal biosorbents for heavy metal remediation with different adsorption isotherm models. Environ Sci Pollut Res 30(14):39474–39493

Kurniati E, Arfarita N, Imai T, Higuchi T, Kanno A, Yamamoto K, Sekine M (2014) Potential bioremediation of mercury-contaminated substrate using filamentous fungi isolated from forest soil. J Environ Sci 26(6):1223–1231

Lefebvre DD, Kelly D, Budd K (2007) Biotransformation of Hg(II) by cyanobacteria. App Environ Micro 73(1):243–249

Leudo AM, Cruz Y, Montoya-Ruiz C, Delgado MDP, Saldarriaga JF (2020) Mercury phytoremediation with Lolium perenne-Mycorrhizae in contaminated soils. Sustainability 12(9):3795

Li DB, Li XJ, Tao Y, Yan ZN, Ao YS (2022) Deciphering the bacterial microbiome in response to long-term mercury contaminated soil. Ecotoxicol Environ Saf 229:113062

Liu Z, Wang LA, Xu J, Ding S, Feng X, Xiao H (2017) Effects of different concentrations of mercury on accumulation of mercury by five plant species. Ecol Eng 106:273–278

Liu Z, Chen B, Wang LA, Urbanovich O, Nagorskaya L, Li X, Tang L (2020) A review on phytoremediation of mercury contaminated soils. J Hazard Mater 400:123138

Liu JL, Liu YP, Duan DD, Peng GG, Li P, Lei P, Zhong H, Tsui MT, Pan K (2022) Effects and mechanisms of organic matter regulating the methylmercury dynamics in mangrove sediments. J Hazard Mater 432:128690

Lv S, Yang B, Kou Y, Zeng J, Wang R, Xiao Y, Li F, Lu Y, Mu Y, Zhao C (2018) Assessing the difference of tolerance and phytoremediation potential in mercury contaminated soil of a non-food energy crop, *Helianthus tuberosus* L. (Jerusalem artichoke). Peer J 6:e4325

Mahbub KR, Krishnan K, Naidu R, Megharaj M (2016) Mercury resistance and volatilization by *Pseudoxanthomonas* sp. SE1 isolated from soil. Environ Technol Innov 6:94–104

Mahbub KR, Bahar MM, Labbate M, Krishnan K, Andrews S, Naidu R, Megharaj M (2017a) Bioremediation of mercury: not properly exploited in contaminated soils! App Micro Biotech 101:963–976

Mahbub KR, Krishnan K, Naidu R, Andrews S, Megharaj M (2017b) Mercury toxicity to terrestrial biota. Ecol Indic 74:451–462

Malik LA, Bashir A, Qureashi A, Pandith AH (2019) Detection and removal of heavy metal ions: a review. Environ Chem Lett 7:1495–1521

Manikandan A, Suresh Babu P, Shyamalagowri S, Kamaraj M, Muthukumaran P, Aravind J (2022) Emerging role of microalgae in heavy metal bioremediation. J Basic Micro 62(3–4):330–347

Martínez-Juárez VM, Cárdenas-González JF, Torre-Bouscoulet ME, Acosta-Rodríguez I (2021) Biosorption of mercury(II) from aqueous solutions onto fungal biomass. Bioinorg Chem Appl 2012(2012):156190

Mello IS, Targanski S, Pietro-Souza W, Stachack FFF, Terezo AJ, Soares MA (2020) Endophytic bacteria stimulate mercury phytoremediation by modulating its bioaccumulation and volatilization. Ecotoxicol Environ Saf 202:110818

Miao J, Feng S, Dou S, Ma Y, Yang L, Yan L, Yu P, Wu Y, Ye T, Wen B, Lu P (2023) Association between mercury exposure and lung function in young adults: a prospective cohort study in Shandong, China. Sci Total Environ 878:162759

Mohammadi S, Pourakbar L, Moghaddam SS, Popović-Djordjević J (2021) The effect of EDTA and citric acid on biochemical processes and changes in phenolic compounds profile of okra (*Abelmoschus esculentus* L.) under mercury stress. Ecotoxicol Environ Saf 208:111607

Moreno FN, Sígolo JB, Anderson CW, Stewart RB, Meech JA, Robinson BH (2005) Phytoremediation of mercury contaminated mine wastes. Massey University, New Zealand

Mosa A, Duffin J (2017) The interwoven history of mercury poisoning in Ontario and Japan. CMAJ 189(5):E213–E215

Muddarisna N, Krisnayanti BD, Utami SR, Handayanto E (2013) Phytoremediation of mercury-contaminated soil using three wild plant species and its effect on maize growth. Appl Ecol Environ Sci 1(3):27–32

Nagata T, Kiyono M, Pan-Hou H (2006) Engineering expression of bacterial polyphosphate kinase in tobacco for mercury remediation. Appl Microbiol Biotechnol 72:777–782

Naguib MM, Khairalla AS, El-Gendy AO, Elkhatib WF (2019) Isolation and characterization of mercury-resistant bacteria from wastewater sources in Egypt. Can J Microbiol 65(4):308–321

Naguib MM, El-Gendy AO, Khairalla AS (2018) Microbial diversity of operon genes and their potential rules in mercury bioremediation and resistance. Open Biotechnol J 12(1)

Naseri A, Saadatm S, Noroozi M, Asri Y, Iranbakhsh A (2020) Study the mercury biosorption by unicellular diatom Nitzschia capitellata Hustedt. Mod phytomorphol 14(2)

NHDES (2019) Mercury: sources, transport, deposition and impacts. New Hemisphere 603:271–3503

Nunes PB, Ferreira MKM, Ribeiro-Frazão D, Bittencourt LO, Chemelo VS, Silva MCF, Pereira-Neto AL, Albuquerque ARL, Paz SPA, Angélica RS et al. (2022) Effects of inorganic mercury exposure in the alveolar bone of rats: an approach of qualitative and morphological aspects. Peer J 26:e12573

Nurfitriani S, Arisoesilaningsih E, Nuraini Y, Handayanto E (2020) Bioaccumulation of mercury by bacteria isolated from small scale gold mining tailings in Lombok, Indonesia. J Ecol Eng 21(6)

Palanivel TM, Pracejus B, Novo LA (2023) Bioremediation of copper using indigenous fungi Aspergillus species isolated from an abandoned copper mine soil. Chemosphere 314:137688

Pamphlett R, Kum Jew S (2019) Mercury is taken up selectively by cells involved in joint, bone, and connective tissue disorders. Front Med 6:168

Pan-Hou HS, Imura N (1982) Involvement of mercury methylation in microbial mercury detoxication. Arch Microbiol 131:176–177

Peterson SH, Ackerman JT, Holser RR, McDonald BI, Costa DP, Crocker DE (2023) Mercury bioaccumulation and cortisol interact to influence endocrine and immune biomarkers in a free-ranging marine mammal. Environ Sci Technol 57(14):5678–5692

Pietro-Souza W, de Campos Pereira F, Mello IS, Stachack FFF, Terezo AJ, da Cunha CN, White JF, Li H, Soares MA (2020) Mercury resistance and bioremediation mediated by endophytic fungi. Chemosphere 240:124874

Pirzadah TB, Malik B, Tahir I, Irfan QM, Rehman RU (2018) Characterization of mercury-induced stress biomarkers in *Fagopyrum tataricum* plants. Int J Phytoremediation 20(3):225–236

Podar D, Maathuis FJ (2022) The role of roots and rhizosphere in providing tolerance to toxic metals and metalloids. Plant Cell Environ 45(3):719–736

Pravin US, Manisha PT, Ravindra MM (2012) Sediment heavy metal contaminants in Vasai Creek of Mumbai: pollution impacts. Am Chem Soc 2(3):171–180

Priyadarshanee M, Chatterjee S, Rath S, Dash HR, Das S (2022) Cellular and genetic mechanisms of bacterial mercury resistance and their role in biogeochemistry and bioremediation. J Hazard Mater 423:126985

Pushkar B, Sevak P, Singh A (2019) Bioremediation treatment process through mercury-resistant bacteria isolated from Mithi river. App Water Sci 9:1–10

Qi R, Xue NN, Wang SZ, Zhou XB, Zhao L, Song WJ, Yang YY (2022) Heavy metal(loid)s shape the soil bacterial community and functional genes of desert grassland in a gold mining area in the semi-arid region. Environ Res 214:113749

Rahayu HM, Putri WA, Khasanah AU, Sembiring L, Purwestri YA (2021) Indigeneous *Streptomyces* spp. isolated from *Cyperus rotundus* rhizosphere indicate high mercuric reductase activity as a pontential bioremediation agent. Biodivers J Bio Diversity 22(3)

Rajadurai M, Reddy ER (2021) Tuning the sensitivity towards mercury via cooperative binding to d-fructose: dual fluorescent chemosensor based on 1, 8-naphthyridine-boronic acid derivative. RSC Adv 11(24):14862–14870

Rani L, Srivastav AL, Kaushal J (2021) Bioremediation: an effective approach of mercury removal from the aqueous solutions. Chemosphere 280:130654

Ratnawati R (2020) Phytoremediation of mercury contaminated soil with the addition of compost. J Eng Technol Sci 52(1)

Rauf AU, Mallongi A, Astuti RDP (2020) Mercury and chromium distribution in soil near maros karst ecosystem. Carpathian J Earth Environ Sci 15(2):453–460

Ravel J, Amoroso MJ, Colwell RR, Hill RT (1998) Mercury-resistant actinomycetes from the Chesapeake Bay. FEMS Microbiol Lett 162(1):177–184

Rice KM, Walker EM Jr, Wu M, Gillette C, Blough ER (2014) Environmental mercury and its toxic effects. J Prev Med Public Health 47(2):74

Rodriguez L, Lopez-Bellido FJ, Carnicer A, Recreo F, Tallos A, Monteagudo JM (2005) Mercury recovery from soils by phytoremediation. Environ Chem Green Chem Pollut Ecosyst 197–204

Rodríguez Martín-Doimeadios RC, Berzas Nevado JJ, Guzmán Bernardo FJ, Jimenez Moreno M, Arrifano GP, Herculano AM, Do Nascimento JL, Crespo-López ME (2014) Comparative study of mercury speciation in commercial fishes of the Brazilian Amazon. Environ Sci Pollut Res 21:7466–7479

Rohwerder T, Gehrke T, Kinzler K, Sand W (2003) Bioleaching review part A. Appl Microbiol Biotechnol 63:239–248

Rojas-Solis D, Larsen J, Lindig-Cisneros R (2023) Arsenic and mercury tolerant rhizobacteria that can improve phytoremediation of heavy metal contaminated soils. PeerJ 11:e14697

Rowland FE, Muths E, Eagles-Smith CA, Stricker CA, Kraus JM, Harrington RA, Walters DM (2023) Complex life histories alter patterns of mercury exposure and accumulation in a pond-breeding amphibian. Environ Sci Technol 57(10):4133–4142

Ruggieri F, Majorani C, Domanico F, Alimonti A (2017) Mercury in children: current state on exposure through human biomonitoring studies. Int J Environ Res Public Health 14:519

Ruiz ON, Alvarez D, Torres C, Roman L, Daniell H (2011) Metallothionein expression in chloroplasts enhances mercury accumulation and phytoremediation capability. Plant Biotechnol J 9(5):609–617

Safari F, Akramian M, Salehi-Arjmand H, Khadivi A (2019) Physiological and molecular mechanisms underlying salicylic acid-mitigated mercury toxicity in lemon balm (*Melissa officinalis* L.). Ecotoxicol Environ Saf 183:109542

Sakamoto M, Tatsuta N, Izumo K, Phan PT, Vu LD, Yamamoto M, Nakamura M, Nakai K, Murata K (2018) Health Impacts and biomarkers of prenatal exposure to methylmercury: lessons from Minamata, Japan. Toxics 6(45):1–9

Sandoval-Cárdenas DI, Pool H, Favela-Camacho SE, Santos-Cruz J, Campos-Guillén J, Ramos-López MA, Rodríguez-deLeón E, Urbina-Arroyo JV, Amaro-Reyes A (2023) Sargassum@magnetite composite EDTA-functionalized for the potential removal of mercury. Polymers 15(6):1405

Sanjaya WTA, Khoirunhisa NS, Ismiani S, Hazra F, Santosa DA (2021) Isolation and characterization of mercury-resistant microbes from the gold mine area in Mount Pongkor, Bogor District, Indonesia. Biodiversitas J Biol Diver 22(7)

Saranya K, Sundaramanickam A, Shekhar S, Swaminathan S, Balasubramanian T (2017) Bioremediation of mercury by *Vibrio fluvialis* screened from industrial effluents. BioMed Res Inter

Saraswathi R, Sumithra P (2018) An assessment of heavy metal bioremediation study using estuarine fungi. Int J Res Ana Rev 5(4):702–704

Senabio JA, de Campos Pereira F, Pietro-Souza W, Sousa TF, Silva GF, Soares MA (2023) Enhanced mercury phytoremediation by Pseudomonodictys pantanalensis sp. nov. A73 and *Westerdykella aquatica* P71. Braz J Microbiol 1–16

Shahid M, Khalid S, Bibi I, Bundschuh J, Niazi NK, Dumat C (2020) A critical review of mercury speciation, bioavailability, toxicity and detoxification in soil-plant environment: ecotoxicology and health risk assessment. Sci Total Environ 711:134749

Shahpiri A, Mohammadzadeh A (2018) Mercury removal by engineered *Escherichia coli* cells expressing different rice metallothionein isoforms. Ann Microbiol 68(3):145–152

Singh AD, Khanna K, Kour J, Dhiman S, Bhardwaj T, Devi K, Sharma N, Kumar P, Kapoor N, Sharma P, Arora P (2023) Critical review on biogeochemical dynamics of mercury (Hg) and its abatement strategies. Chemosphere 137917

Soedarti T, Tini S, Sucipto H, Kuncoro EP (2017) Bioremediation of mercury(II) contaminated seawater using the diatom Skeletonema costatum. KnE Life Sci 62–68

Sun Y, Li Y, Rao J, Liu Z, Chen Q (2018) Effects of inorganic mercury exposure on histological structure, antioxidant status and immune response of immune organs in yellow catfish (*Pelteobagrus fulvidraco*). J App Toxicol 38(6):843–854

Taha A, Hussien W, Gouda SA (2023) Bioremediation of heavy metals in wastewaters: a concise review. Egypt J Aquat Biol Fish 27(1):143–166

Takeuchi F., Sugio T (2006) Volatilization and recovery of mercury from mercury-polluted soils and wastewaters using mercury-resistant *Acidithiobacillus ferrooxidans* strains SUG 2–2 and MON-1. Environ Sci An Int J Environ Physiol Toxicol 13(6):305–316

Tang W, He M, Chen B, Ruan G, Xia Y, Xu P, Song G, Bi Y, Hu B (2023) Investigation of toxic effect of mercury on *Microcystis aeruginosa*: correlation between intracellular mercury content at single cells level and algae physiological responses. Sci Total Environ 858:159894

Tayang A, Songachan LS (2021) Microbial bioremediation of heavy metals. Curr Sci 120(6):00113891

Teixeira FB, De Oliveira AC, Leão LK, Fagundes NC, Fernandes RM, Fernandes LM, Da Silva MC, Amado LL, Sagica FE, De Oliveira EH, Crespo-Lopez ME (2018) Exposure to inorganic mercury causes oxidative stress, cell death, and functional deficits in the motor cortex. Front Mol Neurosci 11:125

Teng D, Mao K, Ali W, Xu G, Huang G, Niazi NK, Feng X, Zhang H (2020) Describing the toxicity and sources and the remediation technologies for mercury-contaminated soil. RSC Adv 10(39):23221–23232

Thompson LA, Darwish WS (2019) Environmental chemical contaminants in food: review of a global problem. J Toxicol

Tiodar ED, Văcar CL, Podar D (2021) Phytoremediation and microorganisms-assisted phytoremediation of mercury-contaminated soils: challenges and perspectives. Int J Environ Res Public Health 18(5):2435

Tran TAT, Dinh QT, Zhou F, Zhai H, Xue M, Du Z, Bañuelos GS, Liang D (2021) Mechanisms underlying mercury detoxification in soil–plant systems after selenium application: a review. Environ Sci Pollut Res 28:46852–46876

UNEP (2019) Global mercury assessment 2018. Chemicals and Health Branch, Geneva, Switzerland

UNEP (2023) Por que o Mercúrio Ainda é uma Ameaça à Saúde Humana e Planetária. Available online: https://www.unep.org/pt-br/noticias-e-reportagens/reportagem/por-que-o-mercurio-ainda-e-uma-ameaca-saude-humana-e-planetaria#:~:text=Todas%20as%20pessoas%20est%C3%A3o%20expostas,resultar%20em%20s%C3%A9rias%20consequ%C3%AAncias%20neurol%C3%B3gicas. Accessed on 28 Jan 2023)

Vacar CL, Covaci E, Chakraborty S, Li B, Weindorf DC, Frențiu T, Pârvu M, Podar D (2021) Heavy metal-resistant filamentous fungi as potential mercury bioremediators. J. Fungi 7(5):386

Varjani S, Upasani VN, Pandey A (2020) Bioremediation of oily sludge polluted soil employing a novel strain of *Pseudomonas aeruginosa* and phytotoxicity of petroleum hydrocarbons for seed germination. ci. Total Environ 737:139766

Velásquez-Riaño M, Benavides-Otaya HD (2016) Bioremediation techniques applied to aqueous media contaminated with mercury. Crit Rev Biotechnol 36(6):1124–1130

WHO (2020) Mercury and health: key facts [website]. WHO, Geneva. https://www.who.int/news-room/factsheets/detail/mercury-and-health. Accessed 1 July 2021

Wu C, Tang D, Dai J, Tang X, Bao Y, Ning J, Zhen Q, Song H, St. Leger RJ, Fang W (2022) Bioremediation of mercury-polluted soil and water by the plant symbiotic fungus *Metarhizium robertsii*. Proc Natl Acad Sci 119(47):e2214513119

Xu W, Hussain A, Liu Y (2018) A review on modification methods of adsorbents for elemental mercury from flue gas. Chem Eng J 346:692–711

Yadav APS, Dwivedi V, Kumar S, Kushwaha A, Goswami L, Reddy BS (2020) Cyanobacterial extracellular polymeric substances for heavy metal removal: a mini review. J Composites Sci 5(1):1

Yahaya I, Amoh AM, Mends EA, Amankwah RK (2022) Biosorption of mercury by selected plants–a preliminary study. Ghana Min J 22(2):32–36

Yan XT, Huang JF, Xiao X, Ma CB, Zhang J, Zhur O, Zhou MY, He HL, Wu CL (2022) A new method for determination of polysaccharides in adsorption of Hg^{2+}. Microchem J 183:107962

Yang L, Zhang Y, Wang F, Luo Z, Guo S, Strähle U (2020) Toxicity of mercury: molecular evidence. Chemosphere 1(245):125586

Zhang L, Lin X, Wang J, Jiang F, Wei L, Chen G, Hao X (2016) Effects of lead and mercury on sulfate-reducing bacterial activity in a biological process for flue gas desulfurization wastewater treatment. Sci Rep 6(1):30455

Zhao Y, Zhou C, Guo X, Hu G, Li G, Zhuang Y, Cao H, Li L, Xing C, Zhang C, Yang F (2021) Exposed to mercury-induced oxidative stress, changes of intestinal microflora, and association between them in mice. Biol Trace Elem Res 199:1900–1907

Zheng NA, Wang S, Dong WU, Hua X, Li Y, Song X, Chu Q, Hou S, Li Y (2019) The toxicological effects of mercury exposure in marine fish. Bull Environ Contam Toxicol 15(102):714–720

Znad H, Awual MR, Martini S (2022) The utilization of algae and seaweed biomass for bioremediation of heavy metal-contaminated wastewater. Molecules 27(4):1275

Chapter 14
Biosorption of Mercury from Aqueous Solutions by Biosorbents

Ismael Acosta, Adriana Rodríguez, Juan Fernando Cárdenas, and Víctor Manuel Martínez

14.1 Introduction

Mercury is the only metal found in nature in the form liquid, and their compounds (as amalgam, associated with metals such as gold and silver), are present in a great variety of places, like artisanal mining, abandoned mines, in artisan ovens, firewood, oil or natural gas, have mercury emissions, in association with cinnabar it was discovered in ceramic vessels, the extraction of gold and silver from mines, auxiliary in the production of chlorine-alkali chemicals, in pressure gauges, thermometers, switches electrical and electronic, fluorescent lamps, dental amalgams, alloyed with other metals, in the production of biocides, in murals, antiseptics in pharmaceutical products, for chemical analysis, and as catalysts, to make more efficient manufacture of other chemicals, pigments and dyes, detergents and explosives (Gazzola 2022; Pavithra et al. 2023). In the aquatic environment, elemental mercury probably binds to the sediment and is then transported by marine or river currents, and a part of the metal contaminates aquatic organisms, and there are microorganisms, that can produce other derivates more toxics (like methylmercury, very dangerous at low doses), which can be subsequently transported by different birds (Bell et al.

I. Acosta (✉)
Laboratorio de Micología Experimental, Facultad de Ciencias Químicas, Universidad Autónoma de San Luís Potosí, Av. Dr. Manuel Nava No. 6, Zona Universitaria, 78320 San Luís Potosí, SLP, México
e-mail: iacosta@uaslp.mx

A. Rodríguez · J. F. Cárdenas
Centro de Investigación y Extensión de la Zona Media, El Balandrán, Universidad Autónoma de San Luís Potosí, Escontría 230, Barrio 3, 79660 Cd. Fernández, SLP, México

V. M. Martínez
Instituto de Ciencias Agropecuarias, Área Académica de Medicina Veterinaria y Zootecnia, Universidad Autónoma del Estado de Hidalgo, Rancho Universitario Km 1, 43600 Tulancingo de Bravo, HGO, México

N. Kumar (ed.), *Mercury Toxicity*, Environmental Science and Engineering,
https://doi.org/10.1007/978-981-99-7719-2_14

2014). Occupational exposure to mercury is a current problem for people in many activities, such as the production of dental amalgams, kitchen scales, watches. laser pointers, pocket calculators, and shoes for children that light up when they walk. All may require button cell batteries, which generally contain very small amounts of mercury, and this, does not pose a threat to the environment or human health if the batteries are used and recycled properly, but if they end up in an incinerator or inappropriate landfill, then the mercury inside can leak out and pollute the air or groundwater (Gazzola 2022; Pavithra et al. 2023). Those who suffer the greatest exposure are the millions of workers in artisanal gold mining and those who extract the metal on a small scale, without adequate protection conditions, therefore for this population, there is a much higher risk of exposure to the metal, with big consequences for health and contamination of the work areas and nearby (Mi et al. 2023), like in Mexican mercury mining workers (Saldaña Villanueva et al. 2022), kidney damage in vulnerable populations due to mercury exposure (Díaz de León-Martínez et al. 2021), the determination of ambiental mercury, and exposure in humans and biota in a mining region, México (Camacho de la Cruz et al. 2021), the evaluation of the metal in surface water and sediments of the "Lunar de Oro stream," La Rinconada, in the high Andes at southern Peru (Loza del Carpio and Ccancapa Salcedo 2020), and this metal change soil arbuscular mycorrhizal fungal community (Mi et al. 2023). In the Mexican Republic, Mexico, mercury mines are located mainly in the states of: Aguascalientes, Chihuahua, Coahuila, Durango, Guanajuato, Hidalgo, México, Morelos, Nuevo León, Querétaro, San Luis Potosí, Sinaloa, Sonora, Tamaulipas, Tlaxcala, and Zacatecas, mainly in San Luis Potosí, Zacatecas, Querétaro, Guanajuato, and Guerrero (Gazzola 2022). The mercury production in Mexico began in the nineteenth century, until the end of the twentieth century (Romero Zepeda 2016). However, due to the process of obtaining gold and silver, in the Zacatecas state, were produced 20 tons/year of mercury.

Among the metallic mineral production plants registered with the Mexican Mining Chamber, four have been described, which are dedicated to the production of mercury from the benefit of minerals and mercury tailings, which are in the states of Durango, San Luis Potosi, and Zacatecas. Likewise, it is known that until a few years ago two more plants operating in the state of Zacatecas, which process tailings through the leaching process, using sodium hyposulfite, to obtain gold and silver as a product and mercury as a by-product, and due to the reduction of the commercial uses of the metal both internationally and in Mexico, and due to the saturation of the national market, its production has been reduced until it has ceased to produce it from 1995 to date (Romero Zepeda 2016).

Due to the above, the situation that exists in Mexico with respect to contamination and/or production of mercury is contradictory (Lindig-Cisneros and Gómez-Orozco 2021), because despite the regulatory provisions formulated, which limit mercury emissions into the atmosphere and water and control the disposal of waste that contains mercury, the element has not been regulated as a marketable product and little has been done to promote adequate awareness among the population regarding exposure to mercury and the reduction of its risks, like the evaluation of assess the possible arsenic an mercury exposure of the population, in

Tlalpujahua, mining zone in Eastern Michoacán, Mexico (Lindig-Cisneros and Gómez-Orozco 2021). And in accordance with Article 7° of the General Law for the Prevention and Integral Management of Waste (NOM-052-SEMARNAT-2005; NOM-053-SEMARNAT-1993), being 0.2 mg/L in the case of mercury.

Recent research on the adsorption of heavy metal ions in wastewater including mercury (II), carried out worldwide, reveals the adsorption capacity of different natural biomass such as some material biological origin such as: microorganisms, fruit and plants, agricultural waste, and others, with highly satisfactory results (Ayele et al. 2021; Namdeti 2023), and this are economics, very abundance, and the obtaining biomass, its easy and fast (Pavithra et al. 2023). The cell wall of the microorganisms, mainly fungi, have different chemical compounds, related to the caption of heavy metals, like mercury (Anuar et al. 2020; Gupta et al. 2020).

Among the biomasses mainly fungal biomasses are the removal of this metal with the isolation of filamentous fungi from textile effluent (Lira et al. 2022), indigenous fungi resistant to mercury(II) and lead(II) isolated from illegal gold mining site (Nofiani et al. 2022), the removal of mercury(II) by cells of *Rhizopus oryzae* and *Aspergillus niger* (Anuar et al. 2020), the identification of fungi and yeast of Papaloapan River, in Tuxtepec, Oaxaca, México (Navarro Moreno et al. 2022), the isolation from soils of fungal mercury-resistant in Querétaro State, México (Hernández-Flores et al. 2018), the determination of resistance and biosorption capacity of contaminants by the fungus *A. niger* (Acosta Rodríguez et al. 2018a, b), the removal of copper(II), zinc(II), and mercury(II) by residues of some macromycetes (Li et al. 2018), indigenous fungi isolated from water contaminated with petroleum (El-Bondkly and El-Gendy 2022), fungi isolated from sites contaminated with this metal (Vacar et al. 2021), the biosorption of lead, cadmium, and mercury for the fungi *Penicillium* sp., isolate contaminated soil (Sánchez-Castellón et al. 2022), the elimination of the same metal of with some fungal biomasses (Marínez-Juárez et al. 2012), the captation of mercury from the wastewater by *Phlebia floridensis* (Sharma et al. 2022), *A. niger* resistant to 200 ppm of Arsenic(III) (Santos et al. 2017), which eliminate efficiently mercury(II) (Acosta Rodríguez et al. 2018a, b), some mineral fungal strains resistant to zinc(II) and mercury(II) (Alzahrani and El-Gendy 2019), the biosorption of this heavy metals by *Lentinus edodes,* and microorganisms (Bayramo and Arıca 2008; Kapahi and Sachdeva 2019), the use of some fungi for mercury removal (Pietro Souza et al. 2020), their removal by *Lactarius acerrimus* (Naeemullah et al. 2020), and for *Lecythophora* sp., DC-F1, (Chang et al. 2019). Therefore, in this work, the objective of this work was to study the biosorption capacity of mercury(II) in aqueous solution, by different fungal biomasses isolated from a place contaminated with heavy metals, and determine the optimal metal removal conditions for the specie of *Purpureocillium lilacinum.*

14.2 Material and Methods

14.2.1 Fungal Isolation and Culture Conditions

The fungal strains analyzed were isolated through open and 10 min of exposure to petri dishes containing Lee's minimal medium (LMM) (Lee et al. 1975) [contained 500 mg/L $K_2Cr_2O_7$, adjusting the pH of the medium with sodium-phosphate buffer], from a zone near of the University zone, San Luis Potosi, México (Cárdenas-Gonzalez et al. 2021). Subsequently, the samples obtained were incubated for one week at 28 °C. The strains were identified by their macro and micro-morphological characteristics (López Martínez et al. 2004). Subsequently, the fungi were grown in thioglycolate broth for the obtaining of the fungal biomass. All experiments were performed in triplicate and two times using the right controls. The purified fungal strains were obtained by cultiving on PDA (Potato Dextrose Agar) and were stored at 4 °C.

14.2.2 Resistant Testing by Dry Wet

Mercury(II)-resistant tests of the different fungal strains isolated (1×10^6 spores/mL), were performed on 100 mL of liquid LMM containing mercury(II) (as $HgCl_2$), at different concentrations, and were incubated at 28 °C, one week, pH 6.2, and 100 rpm. Then the samples were centrifugated to 3000 rpm/5 min, in previously weighed tubes, and the supernatant was eliminated, and the botton was drying at 80 °C for 72 h in a bacteriological oven. Dry weight of the samples was determined by weight difference.

14.2.3 Obtaining the Fungal Biomasses

Of the ten fungal strains isolated, cell biomass was obtained by seeding 1×10^6 spores/600 mL in thioglycolate broth medium, incubating for 7 days at 28 °C, in a metabolic bath at 100 rpm. The biomass obtained is obtained by filtration, washed three times with sterile trideionized water, dried for 72 h at 80 °C, and finally ground and saved in an amber bottle.

14.2.4 Biosorption Tests for Mercury(II) by Using Dry Fungal Biomasses

For biosorption studies, a series of 100 mg/L of mercury ($HgCl_2$) solutions were prepared, the pH was adjusted with nitric acid 1.0 N, and 1.0 g/100 mL of fungal biomass was added to the mercury solution. Samples were incubated at different times, and centrifugate (3000 rpm/5 min), for removed the biomass, and the liquid was analyzed to determine the metal ion concentration by the dithizone method, which formed a mercuric dithizonate complex, of orange color, and this is extracted with chloroform, in acid medium, read at 490 nm (minimum detectable of 0.02 mg/L of dithizone solution) (Greenberg et al. 1992).

14.2.5 Mercury Removal from a Wastewater Lagoon

To 500 mL Erlenmeyer flasks containing 195 mL of wastewater (183 mg/L of mercury) of a wastewater lagoon (Tenorio Tank, ponient zone of this city (S.L.P., México), and add 5 g of fungal biomass, pH 5.5 (adjusted), and kept at 28 °C, at 100 rpm, and every 24 h, the concentration of mercury(II) in the supernatant was determined. All experiments were performed three times and in duplicate.

14.3 Results and Discussion

14.3.1 Isolation and Identification of Fungal Strains Resistant to Mercury(II)

From the samples analyzed, 10 different fungal species were isolated (Table 14.1), in addition, all the strains identified, presented different macroscopic characteristics, among which are: different colors, aspects, and texture (López Martínez et al. 2004), and the isolated strains grow at different concentrations of the metal, which indicates that they may be resistant to it, indicating the presence of mercury(II) in the sampled area, and the different mercury resistance concentration could be explained by the different mechanisms resistance of the fungal strains isolated (Hernández-Flores et al. 2018) (Table 14.1).

The results obtained are like for some of the reports in the literature where microorganisms resistant to heavy metals are isolated, which indicates contamination by them, for example: twenty three fungi isolated from textile effluent were resistant for two metals tested (chromium, and others metals analyzed) (Lira et al. 2022), five fungi isolated from a mining site in Indonesia, which growth at 200 mg/L of $HgCl_2$ (Nofiani et al. 2022), the isolation of *Trichoderma* strains, wich growth in presence of mercury from the soils, of Querétaro State, México (Hernández-Flores et al. 2018), by the fungus *A. niger* (Acosta Rodríguez et al. 2018a, b), the growth in presence of different heavy metals by fungi isolated from soil samples (Vacar et al. 2021), the isolation of *Penicillium* sp., from soil samples by the Alacran mine (Córdoba, Colombia) (Sánchez-Castellón et al. 2022), the tolerance and morphological changes by the fungus *P. floridensis* (Sharma et al. 2022), *A. niger* resistant to

Table 14.1 Fungi identified in the samples analyzed

Fungi identified	Growth in mercury(II) (mg/L)
Purpureocillium lilacinum	2000
Mucor sp	1000
Aspergillus flavus	750
Penicillum sp	500
Aspergillus terreus	500
Alternaria sp	400
Paecelomyces sp	400
Trichoderma sp	300
Scopulariopsis sp	250
Cladosporium sp	200

Arsenic(III) (Santos et al. 2017), *Aspergillus tubingensis* Merv4 resistant to different heavy metals (Alzahrani and El-Gendy 2019), mercury resistance (up 600 μg/mL) in 32 strains of fungi (Pietro Souza et al. 2020), the isolation of *Lecythophora* sp., DC-F1, from soil of a mining zone, and contaminated agricultural soil (Chang et al. 2019; Hindersah et al. 2018), fungi isolated from mercury mining plant (Urik et al. 2014), 90 endophytic fungi mercury resistant (Pietro-Souza et al. 2017), and *Rhodotorula taiwanensis* MD1149 isolated from acid mine drainage (Tkavc et al. 2018).

14.3.2 Growth and Dry Weight of the **Purpureocillium lilacinum** *Strain*

The fungal strain of *P. lilacinum* analyzed, growth at 2 000 mg/L of mercury (7 mg of dry weight, and 10.6% with respect to control) (Fig. 14.1).

These results are similar for two strains of *A. niger* (Acosta-Rodríguez et al. 2018a, b), *A. tubingensis* Merv4 growth at 700 μg/mL of mercury(II) (Alzahrani and El-Gendy 2019), the resistance to 600 μg/mL of mercury concentrations of three fungal strains (Pietro Souza et al. 2020), four mercury-resistant fungi up to 25 mg/kg (Hindersah et al. 2018). But, are different for five fungi isolated from an abandoned illegal gold mining, and mercury-resistant *Trichoderma* strains isolated soils, in Queretaro State, México, which grows at 200 mg/L of $HgCl_2$ (Nofiani et al. 2022; Hernández-Flores et al. 2018), the fungi *Cladosporium* sp., *Didymella glomerata*, and others fungi, which tolerate 200 mg/L of the mercury (Vacar et al. 2021), for the whiter rot fungus *P. floridensis* [100 μM of mercury(II)] (Sharma et al. 2022), for *Lecythophora* sp. DC-F1, which grows at 84.5 mg/L of the metal (Chang et al. 2019), the *Suillus luteus* mushroom contains different concentrations in the samples (Saba et al. 2016), the fungi *Cladosporium cladosporoides* (Urik et al. 2014), *Polygonum acuminatum* and *Aeschynomene flumi* that growth at 30 μg/mL

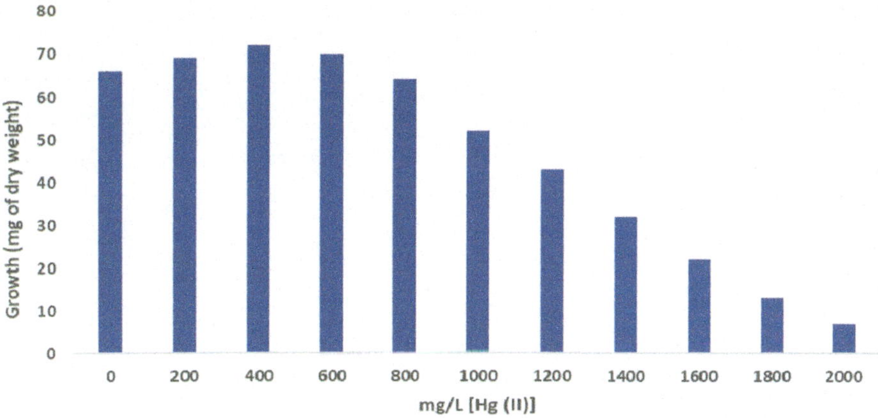

Fig. 14.1 Growth of *Purpureocillium lilacinum*, in LMM. 1×10^6 spores/mL. 28 °C. 1 week of incubation. 100 rpm. pH 6.8

of mercury(II) (Pietro-Souza et al. 2017), and for *R. taiwanensis* MD1149, which growth at 50 μM of $HgCl_2$ (Tkavc et al. 2018).

14.3.3 Removal of Mercury(II) by Dry Cells of the **Purpureocillium lilacinum** *Strain*

14.3.3.1 Effect of Incubation Time

The analyzed biomass presents a better removal percentage (48.1%), in a pH of 5.5, at 24 h of incubation [100 mg/L mercury(II)], 28 °C, 1 g of fungal biomass, and 100 rpm] (Fig. 14.2).

These results are like for the biosorption of the metal (95.4%) by *M. rouxii* biomass IM-80, with the same optimum incubation time (24 h), 100 mg/L of mercury (Martínez-Juárez et al. 2012), the same incubation time for a *A. niger* strain resistant to Arsenic(III) in the same conditions (Acosta-Rodríguez et al. 2018a, b), too for *A. tubingensis* Merv4, the highest mercury(II) biosorption was detected at 24 h of treatment (100%) (Alzahrani and El-Gendy 2019), too the same time of incubation for adsorption of 20 mg/L of mercury on *Aspergillus versicolor* biomass (Das et al. 2007). But, are different for report with the biomasses of *R. oryzae* and *A. niger*, in which the optimum time of removal is 8 h, with 0.1 g of both biomasses (live and dead), and 10 and 100 mg/L of the metal, although, the longer the incubation time, a greater removal is observed (Anuar et al. 2020), for the adsorption of mercury(II) in 2 h by different macromycetes (Li et al. 2018), for the elimination of 100 mg/L at 48 h by different live biomasses (28 and 52% of removal, 48 h, 120 rpm) (Vacar et al. 2021), 3 h for the removal of 15.5% of mercury(II) (51.5 mg/L of initial concentration)

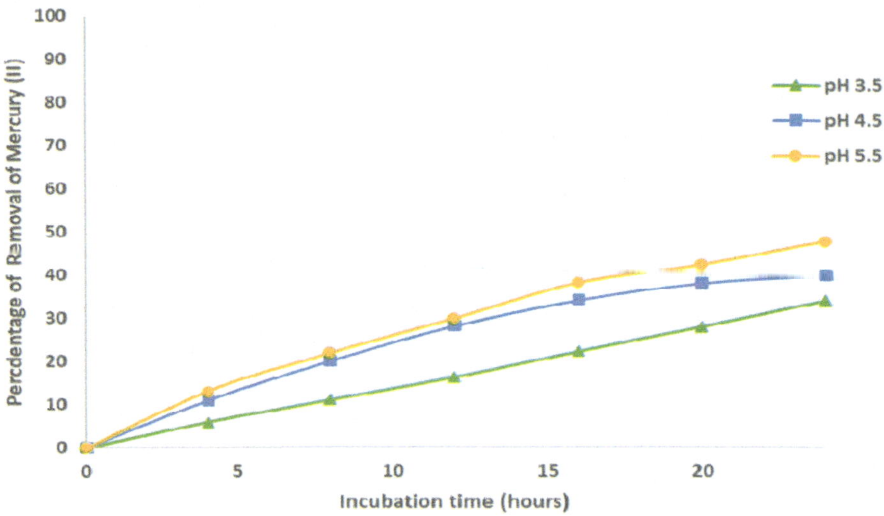

Fig. 14.2 Effect of the incubation time and pH on the biosorption of mercury (II) by *Purpureocillium lilacinum* biomass. 100 mg/L. 1 g of fungal biomass. 28 °C. 100 rpm

by *Penicillium* sp., biomass (Sánchez-Castellón et al. 2022), 7 days for the removal of 100 μM of mercury concentration with whiter rot fungus *P. floridensis* with a percentage of removal between 70 and 84% of the metal (Sharma et al. 2022), for both live and dead cells of *L. edodes*, the biosorption capacity of mercury, cadmium, and zinc ions, increased with increasing contact time until 120 min (Bayramo and Arıca 2008), for *Phanerochaete chrysosporium* most of mercury (100 mg/L), was biosorbed in less than 60 min (Bashardoost et al. 2020), 4 h for the removal of different heavy metals by *Penicillium canescens* (Say et al. 2003), and for the fungus *A. flavus*, was observed a time of 15 days for the elimination of 25 mg/L of the metal (Ghaffar et al. 2023),

14.3.3.2 Effect of the pH

With respect to this parameter, the pH optimum was at 5.5 for the analyzed biomass at the same conditions (Fig. 14.2). These results are similar for the removal of the metal with the biomasses of *R. oryzae* and *A. niger*, dried *M. rouxii* IM-80 biomass, and a *A. niger* strain strain with the same optimum pH value of 5.5 (Anuar et al. 2020; Martínez-Juárez et al. 2012); for a *A. niger* strain resistant to Arsenic(III), in the same conditions (Acosta-Rodríguez et al. 2018a, b), a pH between 5.8 to 6.8 for the removal of 100 μM of the metal with whiter rot fungus *P. floridensis* with a percentage of removal between 70 and 84% (Sharma et al. 2022), for *A. tubingensis* Merv4, the major mercury(II) biosorption was detected at pH 5.5 (Alzahrani and El-Gendy 2019), too the same optimum pH (5.0–6.0) adsorption of mercury on *A. versicolor* biomass (Das et al. 2007), and for *A. flavus* and *Aspergillus fumigatus* (Namdeti 2023), and are different for the removal of the metal by different fungi biomass with a optimum pH of 6.0 (Li et al. 2018; Sánchez-Castellón et al. 2022), for both live and dead biomass of *L. edodes*, the biosorption capacity of some heavy metals, is better at pH value of 6.0 (Bayramo and Arıca 2008), an pH of 5.0 for the removal of this metal by *Lactarius acerrimus* (Naeemullah et al. 2020), for *P. chrysosporium* most of the mercury (100 mg/L), was biosorbed at pH of 6.0 (Bashardoost et al. 2020), an pH of 5.0, by the fungus *P. canescens* for the removal of and others ions (Say et al. 2003), and a pH of 4.8 for the for the elimination of 25 mg/L of the metal by *A. flavus* (Ghaffar et al. 2023).

14.3.3.3 Effect of the Temperature

The optimum temperature of incubation was 28 °C, with a removal percentage of 48.1%, since as the incubation temperature increases, the percentage of metal removal decreases (Fig. 14.3).

These results are like for the removal of the same metal (95.3%) for dried *M. rouxii* IM-80 biomass (30 °C) (Martínez-Juárez et al. 2012), for a *A. niger* strain resistant to Arsenic(III), the optimum was 28 °C (Acosta-Rodríguez et al. 2018a, b), in *A. tubingensis* MERV4 (100% of removal, 12 to 24 h, at 30–35 °C (Alzahrani and El-Gendy 2019), too the same optimum temperature (30 °C) on *A. versicolor* biomass (Das et al. 2007), 28–32 °C with *A. flavus* biomass (Ghaffar et al. 2023), for *A. flavus* and *A. fumigatus* (Namdeti 2023). But, these results are different for *Penicillium* sp., biomass, in which the optimum temperature was 60 °C at 3 h of incubation

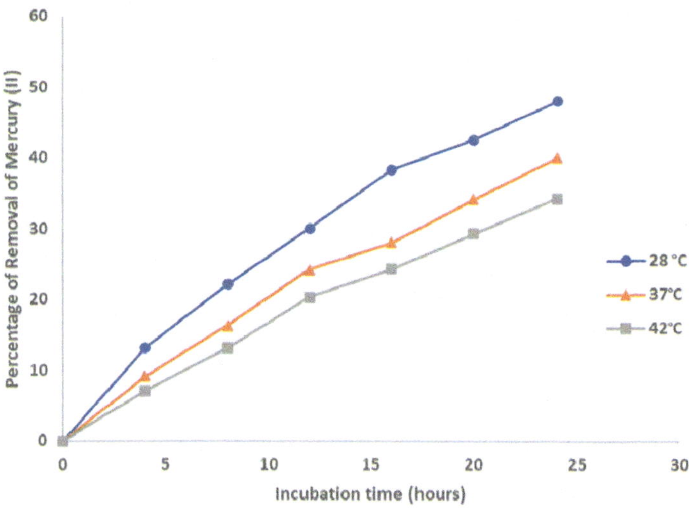

Fig. 14.3 Effect of the temperature on the biosorption of mercury (II). 100 mg/L. 1 g of fungal biomass. pH 5.5. 24 h of incubation. 100 rpm

(Sánchez-Castellón et al. 2022), for *L. edodes*, the biosorption capacity of different heavy metals, the temperature does not influence the bioabsorption (Bayramo and Arıca 2008).

14.3.3.4 Effect of the Initial Concentration of Mercury(II)

On the other hand, a major concentration of the metal, the percentage of removal decrease, at 500 mg/L of the metal the removal percentage decreased by 48.1% (100 mg/L) at 17.1%, for the analyzed biomass (Fig. 14.4). Was report, what a bigger heavy metals concentrations, the elimination is less effective, due to limited free sites of the cells and saturation of the adsorption sites, and at lower concentration, metals ions in the solutions could bind with the binding sites of the cells and caused increasing biosorption (Oves et al. 2013).

These results are similar for the *M. rouxii* IM-80 biomass, and for *A. niger R. Oligosporus*, and *Pleurotus sapidus* where the percentage of adsorption decreased (100 to 500 mg/L) (Martínez-Juárez et al. 2012; Acosta-Rodríguez et al. 2018a, b; Ozsoy 2010; Yalcinkaya et al. 2002). Although, with the biomasses of *R. oryzae* and *A. niger,* in which there is better removal at higher mercury(II) concentration (100 mg/L) (Anuar et al. 2020), for *A. tubingensis* MERV4, was observed a removal (85%–100% and 98%, with 100 and 200 mg/L of mercury, 24 h) (Alzahrani and El-Gendy 2019), for *L. edodes*, the biosorption capacity of different heavy metals, the metal concentration does not influence the bioabsorption (Bayramo and Arıca 2008), for the fungus *P. canescens*, the biosorption capacity of the biomass, increased with increasing initial concentration of metal (Say et al. 2003).

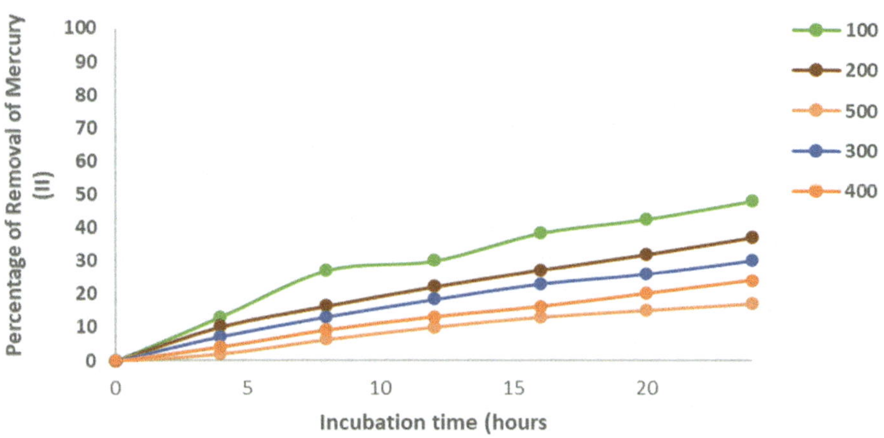

Fig. 14.4 Effect of the concentration of mercury(II) in solution on the biosorption. 1 g of fungal biomass. pH 5.5. 24 h of incubation. 28 °C. 100 rpm

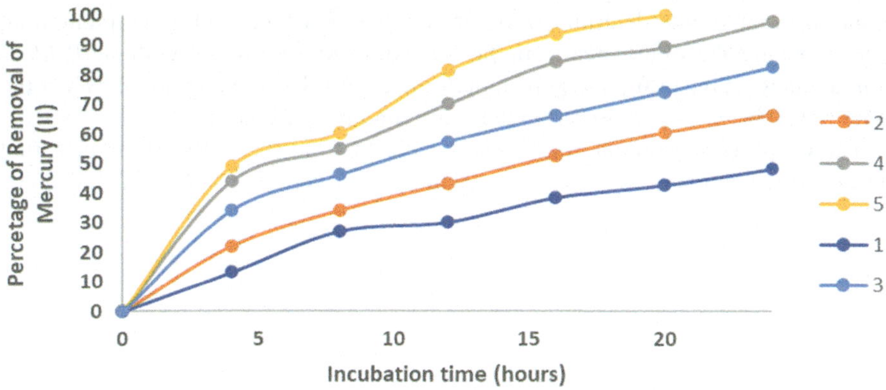

Fig. 14.5 Effect of the fungal biomass concentration on the removal of 100 mg/L of mercury(II). pH 5.5. 24 h of incubation. 28 °C. 100 rpm

14.3.3.5 Effect of the Initial Concentration of Biosorbent

If we increase the biosorbent concentration, too increase the metal removal efficiency, because with 5 g of the bioadsorbent, the elimination is total for this fungal biomass (Fig. 14.5), since there are more bioadsorption sites for it, since the concentration of bioadsorbent added, determines the number of binding sites available for the biosorption of heavy metals (Anuar et al. 2020).

These results are like for *M. rouxii* IM-80 biomass, and *A. niger* strain resustant to arsenic(III) (100% of removal, with 5 g of fungal biomass, at 8 h) (Martínez Juáez et al. 2012; Acosta-Rodríguez et al. 2018a, b), but, with the biomasses live and death of *R. oryzae* and *A. niger,* there are better removal at 0.1 of fungal biomasses (Anuar et al. 2020), for *Penicillium* sp., biomass, in which the biosorbent concentration no influences in the removal of the metal (Sánchez-Castellón et al. 2022), for *A. tubingensis* MERV4, the optimum biosrbent concentration was 200 mg/L of fungal biomass (Alzahrani and El-Gendy 2019),

14.3.4 Removal of Mercury(II) from a Wastewater Lagoon Contaminated

Finally, we analyzed the application of these fungal biomasses, for remediation studies of water and soil contaminated with heavy metals, and for the possible use of these fungal biomasses to eliminate mercury(II) from a wastewater lagoon, a remediation experiment was adapted in aqueous solution. In 500 mL Erlenmeyer flasks containing 195 mL of wastewater contaminated with 183 mg/L of mercury(II) of a wastewater lagoon, were incubated with 5 g of fungal biomass, pH 5.5 (adjusted), and kept at 28 °C, at 100 rpm, and at different times the concentration of mercury(II) in the supernatant was determined, observing that after 8 days of experimentation, 55% of the metal is eliminated from the contaminated wastewater, with the *P. lilacinum* biomass (Fig. 14.6), and percentages of removal between 53 and 32% for the other fungal biomasses analyzed in the same conditions (Table 14.2).

These results are similar for *A. niger* isolated from the polluted air in a fuel station, which removal 94.6% of the metal in the same conditions (Acosta-Rodríguez

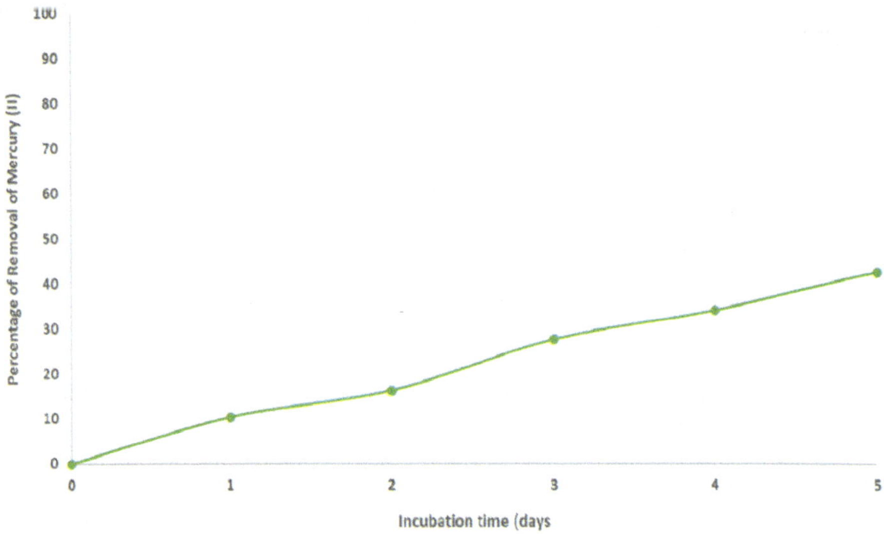

Fig. 14.6 Removal of Mercury(II) from wastewater of the Tenorio Tank, contaminated with 183 mg/L of mercury(II) by the *P. lilacinum* biomass. pH 5.5 (adjusted). 5 g of fungal biomass. 28 °C. 100 rpm

Table 14.2 Removal of mercury(II) from wastewater of the Tenorio Tank, by the different fungal biomasses*

Fungal strain	Percentage of removal (%)
Purpureocillium lilacinum	55
Mucor sp	53
Aspergillus flavus	53
Penicillum sp	48
Aspergillus terreus	43
Alternaria sp	43
Paecelomyces sp	42
Trichoderma sp	36
Scopulariopsis sp	32
Cladosporium sp	32

* 183 mg/L of mercury(II). pH 5.5 (adjusted). 5 g of fungal biomass. 8 days of incubation. 28 °C. 100 rpm

et al. 2018a, b), for the removal of copper, zinc, and mercury in wastewater by *F. velutipes* residue with percentages of removal between 73.11 and 66.67%) (Li et al. 2018), by *F. velutipes, A. polytricha, P. eryngii* and *P. ostreatus* residues (Li et al. 2018), the new fungal what cause disease of maize, *D. glomerata* removal 97% from 100 mg/L aqueous solution (Vacar et al. 2021), for the elimination from wastewater of the same metal ion, with a white rot fungus (Sharma et al. 2022), for *A. niger* (7 days of incubation, the mercury(II) concentration decrease 69% (Acosta Rodríguez et al. 2018a, b), the biosorption of by *A. tubingensis* MERV4 biomass, between 65 and 96% of removal from the real industrial wastewater (Alzahrani and El-Gendy 2019), the bioremediation of contaminated places with mercury by some fungi (Pietro Souza et al. 2020), too, *A. versicolor* biomass removal efficiently mercury from industrial effluent (Das et al. 2007; Sarria Villa et al. 2020), by fungal necromass from *Meliniomyces bicolor* (Maillard et al. 2023), and the bioremediation by fungi, also in terrestrial and aquatic ecosystems (Vaksmaa et al. 2023),

14.3.5 Removal of Mercury(II) by the Different Fungal Biomasses

Subsequently, the mercury(II) removal capacity of the ten isolated strains was analyzed, finding that these biomasses biosorb the metal in different proportions at a pH of 5.5 and 24 h of incubation, being the most efficient: *P. lilacinum*, *Mucor* sp., and *A. flavus* with 100% of removal, and *Penicillium* sp., (97%), *A. terreus* (93%), and *Alternaria* sp., (84%), and the other biomasses in lower percentages (Table 14.3).

The fungi analyzed bioadsorb different percentages of the metal under study, and this is perhaps due to the constitution of the cell wall of the biomasses studied (Anuar et al. 2020), and these results are similar for: the fungal biomasses of *R. oryzae*

Table 14.3 Removal of mercury(II) in solution, by the fungal biomasses*

Fungal Biomass	Percentage of removal (%)
Purpureocillium lilacinum	100
Mucor sp	100
Aspergillus flavus	100
Penicillum sp	97
Aspergillus terreus	93
Alternaria sp	84
Paecelomyces sp	76
Trichoderma sp	68
Scopulariopsis sp	62
Cladosporium sp	57

* 100 mg/L of mercury(II). pH 5.5 (adjusted). 5 g of fungal biomass. 8 days of incubation. 28 °C. 100 rpm

and *A. niger* (Anuar et al. 2020; Acosta-Rodríguez et al. 2018a, b), the removal of heavy metals by *Penicillium* sp., (Sánchez-Castellón et al. 2022), the removal of the same metal by some fungi, and *A. tubingensis* MERV4 biomass (Martínez-Juárez et al. 2012; Alzahrani and El-Gendy 2019), the most strains from endophytic fungi analyzed, growth at higher mercury concentrations like (Pietro Souza et al. 2020), the fungal necromass from *Meliniomyces bicolor* (Maillard et al. 2023), the removal by white rot fungi (Chen et al. 2022), and *P. canescens* for the removal of different heavy metals solutions (Say et al. 2003).

14.4 Conclusions

From the results obtained we can conclude the following:

1. 10 species of fungi were isolated, which present different resistances to mercury(II).
2. *P. lilacinum* removal efficiently the metal in solution (100 mg/L), 24 h of incubation, pH 5.5, 28 °C, 1 g of fungal biomass.
3. The fungal biomass eliminates 55% of the metal of natural water contaminated.
4. All the fungal biomasses efficiently remove mercury (II) in solution, and the more efficient were: *P. lilacinum*, *Mucor* sp., *A. flavus*, and *Penicillium* sp., and this can be an excellent alternative for metal removal from sites contaminated by the same metal.

References

Acosta-Rodríguez I, Cárdenas-González JF, Rodríguez-Pérez AS, Tovar-Oviedo J, Martínez-Juárez VM (2018) Bioremoval of different heavy metals by the resistant fungal strain *Aspergillus niger*. Bioinorg Chem Appl 3457196:7. https://doi.org/10.1155/2018/3457196

Acosta-Rodríguez I, Pacheco-Castillo NC, Cárdenas-González JF, Moctezuma-Zarate MG, Martínez-Juárez VM, Rodríguez-Pérez A (2018) Bioadsorción de mercurio(II) en solución acuosa por la biomasa de *Aspergillus niger*. Mexican J Biotechnol 3(3):15–25. https://doi.org/10.29267/mxjb.2018.3.3.15

Alzahrani NH, El-Gendy MMAA (2019) Torerance and removal of Zinc(II) and mercury(II) by dead biomass of *Aspergillus tubingensis* MERV4. J Indus Pollut Control 35(1):2251–2257. www.icontrolpollution.com

Anuar SNIS, Othman F, Chay TC, Nasir NAHA (2020) Biosorption of mercury ion (Hg^{2+}) using live and dead cells of *Rhizopus oryzae* and *Aspergillus niger*: characterization, kinetic and isotherm studies. J Pure Appl Microbiol 14(3):1749–1760. https://doi.org/10.22207/JPAM.14.3.13

Ayele A, Haile S, Alemu D, Kamaraj (2021) Comparative utilization of dead and live fungal biomass for the removal of heavy metal: a concise review. Hindawi Sci World J 5588111:10. https://doi.org/10.1155/2021/5588111

Bashardoost R, Vahabzadeh F, Shokrollahzadeh S, Monazzami AR (2020) Sorption performance of live and heat-inactivated loofa-immobilized *Phanerochaete chrysosporium* in mercury removal from aqueous solution. Iran J Chem Chem Eng 29(4):79–89. https://doi.org/10.30492/IJCCE.2010.6408

Bayramo G, Arıca MY (2008) Removal of heavy mercury(II), cadmium(II) and zinc(II) metal ions by live and heat inactivated *Lentinus edodes* pellets. Chem Eng J 143(2008):133–140. https://doi.org/10.1016/j.cej.2008.01.002

Bell L, DiGangi J, Weinberg J (2014) Introducción a la contaminación por mercurio y al convenio de Minamata sobre mercurio para las ONG. Red Internacional de Eliminación de los Contaminantes Orgánicos Persistentes (IPEN). Un futuro sin tóxicos. 1ª. Ed. http://ipen.org/sites/default/files/documents/ipen-booklet-hg-update-v1_6a-es-web.pdf

Camacho de la Cruz AA, Espinosa-Reyes G, Rebolloso-Hernández CA, Carrizales-Yáñez L, Ilizaliturri-Hernández CA, Reyes-Arreguín LE, Díaz-Barriga F (2021) Holistic health risk assessment in an artisanal mercury mining region in Mexico. Environ Monit Assess 193(541):1–14. https://doi.org/10.1007/s10661-021-09312-7

Cárdenas González JF, Acosta Rodríguez I, Terán Figueroa Y, Lappe Oliveras P, Martínez Flores R, Rodríguez Pérez AS (2021) Biotransformation of chromium(VI) via a reductant activity from the fungal strain *Purpureocillium lilacinum*. J Fungi 7(1022):1–14. https://doi.org/10.3390/jof7121022

Chang J, Duan Y, Dong J, Shen S, Si G, He F, Yang Q, Chen J (2019) Bioremediation of Hg-contaminated soil by combining a novel Hg-volatilizing *Lecythophora* sp. fungus, DC-F1, with biochar: performance and the response of soil fungal community. Sci Total Environ 67:676–684. https://doi.org/10.1016/j.scitotenv.2019.03.409

Chen L, Zhang X, Zhang M, Zhu Y, Zhuo R (2022) Removal of heavy-metal pollutants by white rot fungi: mechanisms, achievements, and perspectives. J Clean Prod 354:131681. https://doi.org/10.1016/j.jclepro.2022.131681

Das SK, Das AR, Guha AK (2007) A study on the adsorption mechanism of mercury on *Aspergillus versicolor* biomass. Environ Sci Technol 41:8281–8287. https://doi.org/10.1021/es070814g

Díaz de León-Martínez L, Ortega-Romero M, Gavilán-García A, Barbier OC, Carrizalez Yáñez L, Van-Brusel E, Díaz-Barriga F, Flores-Ramírez R (2021) Assessment of biomarkers of early kidney damage and exposure to pollutants in artisanal mercury mining workers from Mexico. Environ Sci Pollut Res 29:13333–13343. https://doi.org/10.1007/s11356-021-16628-x

El-Bondkly AMA, El-Gendy MMAA (2022) Bioremoval of some heavy metals from aqueous solutions by two different indigenous fungi *Aspergillus* sp. AHM69 and *Penicillium* sp. AHM96 isolated from petroleum refining wastewater. Heliyon 8(7):e09854. https://doi.org/10.1016/j.heliyon.2022.e09854

Gazzola J (2022) Cinabrio y mercurio en Teotihuacán y, en particular, en el túnel bajo el templo de la Serpiente Emplumada, México. Journal de la Société des américanistes [En línea]. 108(1):83–115. http://journals.openedition.org/jsa/20694. https://doi.org/10.4000/jsa.20694

Ghaffar I, Hussain A, Javid A, Mehmood S (2023) Impeding effect of polystyrene microplastic pollutants on Hg^{2+} uptake potential of *Aspergillus flavus*. Pakistan J Zool 1–8. https://doi.org/10.17582/journal.pjz/20230130070108

Greenberg AE, Clesceri LS, Eaton AD (1992) Standard methods for the examination of water and wastewater, 18ª. ed. American Public Health Association, Washington, D.C., pp 3–83, 3–107, 1–49, 1–50

Gupta R, Ahuja P, Khan S, Saxena RK, Mohapatra H (2020) Microbial biosorbents: meeting challenges of heavy metal pollution in aqueous solutions. Curr Sci 78(8):967–973. Retrieved from http://www.jstor.org/stable/24103732

Hernández-Flores JL, Melo JGB, Hernández AC, López MAR, Gutiérrez CS, Gomez SR, Moreno VP, Medina RPC, Soto JHV, Hernández SP, Jones GH, Flores SC, Campos-Guillen J (2018) Isolation and characterization of mercury resistant *Trichoderma* Strains from soil with high levels of mercury and its effects on *Arabidopsis thaliana* mercury uptake. Adv Microbiol 2018(8):600–613. https://doi.org/10.4236/aim.2018.87040

Hindersah R, Asda KR, Herdiyantoro D, Kamaluddin NN (2018) Isolation of mercury-resistant fungi from mercury-contaminated agricultural soil. Agriculture 8(33):1–8. https://doi.org/10.3390/agriculture8030033

Kapahi M, Sachdeva S (2019) Bioremediation options for heavy metal pollution. J Health Pollut 9(24):1–20. https://doi.org/10.5696/2156-9614-9.24.191203

Lee K, Buckley L, Campbell CC (1975) An aminoacid liquid synthetic medium for the development of mycelial and yeast forms of *Candida albicans*. Sabouradia 13(2):148–153. https://doi.org/10.1080/00362177585190271

Li X, Zhang D, Sheng F, Qing H (2018) Adsorption characteristics of Copper(II), Zinc(II) and Mercury(II) by four kinds of immobilized fungi residues. Ecotoxicol Environ Saf 147(2018):357–366. https://doi.org/10.1016/j.ecoenv.2017.08.058

Lindig-Cisneros R, Gómez-Orozco SY (2021) Arsénico y mercurio en el paisaje de una zona minera histórica en el oriente de Michoacán, Mexico. Tecnología y Ciencias Del Agua 12(6):480–501. https://doi.org/10.24850/j-tyca-2021-06-12

López Martínez R, Méndez Tovar LJ, Hernández Hernández F, Castañón Olivares R (2004) Micología Médica. Procedimientos para el Diagnóstico de Laboratorio. Ed. Trillas. México. 2ª. Ed

Loza del Carpio AL, Ccancapa Salcedo Y (2020) Mercurio en arroyo altonadino con alto impacto por minería aurifera artisanal, La Rinconada, Puno, Peru. Revista Internacional De Contaminación Ambiental 36(1):33–44. https://doi.org/10.20937/RICA.2020.36.53317

Martínez-Juárez VM, Cárdenas-González JF, Torre Bouscoulet ME, Acosta-Rodrfguez I (2012) Biosorption of mercury(II) from aqueous solutions onto fungal biomass. Bioinorg Chem Appl 5. https://doi.org/10.1155/2012/156190

Mi Y, Bai X, Li X, Zhou M, Liu X, Wang F, Su H, Chen H, Wei Y (2023) Soil mercury pollution changes soil arbuscular mycorrhizal fungal community composition. J Fungi 9(395):1–16. https://doi.org/10.3390/jof9040395

Naeemullah, Tuzen M, Sari A, Turkekul I (2020) Influential bio-removal of mercury using *Lactarius acerrimus* macrofungus as novel low-cost biosorbent from aqueous solution: Isotherm modeling, kinetic and thermodynamic investigations. Mater Chem Phys 29(123168). https://doi.org/10.1016/j.matchemphys.2020.123168

Namdeti R (2023) A review on removal of heavy metals by biosorption: a green technology. Int J Res Rev 10(3): 531–543. https://doi.org/10.52403/ijrr.20230360

Navarro Moreno LG, Vázquez Velasco L, Rangel Cordero A, González JM (2022) Contaminación y hongos: resistencia a metales pesados. LATAM Revista Latinoamericana de Ciencias Sociales y Humanidades. 3(2):215–232. https://doi.org/10.56712/latam.v3i2.76

Nofiani R, Komalasari RK, Puji Ardiningsih P, Santosa SJ (2022) Biosorption of Pb^{2+} using *Fusarium* sp. RS01, a Hg^{2+} and Pb^{2+}-resistant indigenous fungus of an abandoned illegal gold mining site. Sains Malaysiana 51(6):1753–1764. https://doi.org/10.17576/jsm-2022-5106-12

NOM-052-SEMARNAT (2005) Establece las características, el procedimiento de identificación, clasificación y los listados de los residuos peligrosos. 23 de junio de 2006

NOM-053-SEMARNAT-1993. Establece el procedimiento para llevar a cabo la prueba de extracción para determinar los constituyentes que hacen a un residuo peligroso por su toxicidad al ambiente

NOM-117-SSA1-1994, Bienes y servicios. Método de prueba para la determinación de cadmio, arsénico, plomo, estaño, cobre, Fierro, zinc y mercurio en alimentos, agua potable y agua purificada por espectrometría de absorción atómica. -NOM-127-SSA1–1994, Modificación a la Norma Oficial Mexicana, Salud ambiental. Agua para uso y consumo humano. "Límites permisibles de calidad y tratamientos a que debe someterse el agua para su potabilización"

NORMA Oficial Mexicana NOM-147-SEMARNAT/SSA1-2004, Que establece criterios para determinar las concentraciones de remediación de suelos contaminados por arsénico, bario, berilio, cadmio, cromo hexavalente, mercurio, níquel, plata, plomo, selenio, talio y/o vanadio

Lira MMA, Bernal SPF, Castro CCJ, Ramos PMM, Lira MJS, Ottoni JR, Boroski M, Passarini MRZ (2022) Filamentous fungi from textile effluent and their potential application for bioremediation process. Ann Acad Brasil Ciencias 94(2):1–12, e20201718. https://doi.org/10.1590/0001-3765202220201020

Oves M, Khan MS, Zaidi A (2013) Biosorption of heavy metals by *Bacillus thuringiensis* strain OSM29 originating from industrial effluent contaminated north Indian soil. Saudi J Biol Sci 20(2):121–129. https://doi.org/10.1016/j.sjbs.2012.11.006

Ozsoy HD (2010) Biosorptive removal of Hg(II) ions by *Rhizopus oligosporus* produced from corn-processing wastewater. Afr J Biotech 9(51):8791–8799. https://doi.org/10.5897/AJB10.530

Pavithra KG, SundarRajan P, Kumar PS, Rangasam G (2023) Mercury sources, contaminations, mercury cycle, detection and treatment techniques: a review. Chemosphere 312(1):137314. https://doi.org/10.1016/j.chemosphere.2022.137314

Pflender F, Heckman KA, Chalot M, Kennedy PG (2023) Fungal necromass presents a high potential for Mercury immobilization in soil. Chemosphere 311(1):1–5, 136994. https://doi.org/10.1016/j.chemosphere.2022.136994

Pietro-Souza W, Mello IS, Vendruscullo SJ, Silva GF, Cunha CN, White JF, Soares MA (2017) Endophytic fungal communities of *Polygonum acuminatum* and *Aeschynomene fluminensis* are influenced by soil mercury contamination. PLoS ONE 12(7):1–24, e0182017. https://doi.org/10.1371/journal.pone.0182017

Pietro-Souza W, de Campos Pereira F, Souza Mello I, Frutuoso Stachack FF, Terezo AJ, Nunes da Cunha C, White JF, Li H, Soares MA (2020) Mercury resistance and bioremediation mediated by endophytic fungi. Chemosphere 240(124874). https://doi.org/10.1016/j.chemosphere.2019.124874

Romero Zepeda H (2016) Acciones para la protección de la salud humana y el medio ambiente en la identificación de poblaciones vulnerables debida a la exposición de mercurio. Impacto socio-económico en el cumplimiento de las obligaciones del Convenio de Minamata. UAQ – UASLP (2016). Plan de desarrollo humano y tecnológico, sostenible, para la zona productora de mercurio del estado de Querétaro. Universidad Autónoma de Querétaro. Universidad Autónoma de San Luis Potosí. Documento Institucional. https://www.gob.mx/cms/uploads/attachment/file/229177/1600_UNEP_INECC_UAQ_MinamataGe_nero_Dra_Hilda_Romero.pdf

Saba M, Falandysz J, Norom IC (2016) Accumulation and distribution of mercury in fruiting bodies by fungus *Suillus luteus* foraged in Poland, Belarus and Sweden. Environ Sci Pollut Res 23:2749–2757. https://doi.org/10.1007/s11356-015-5513-4

Saldaña-Villanueva K, Pérez-Vazquez FJ, Avila-García IP, Méndez-Rodríguez KB, Carrizalez-Yánez L, Gavilan-García A, Vargas-Morales JM, Van-Brusse E, Diaz-Barriga F (2022) A preliminary study on health impacts of Mexican mercury mining workers in a context of precarious employment. J Trace Elem Med Biol 71(126925):1–10. https://doi.org/10.1016/j.jtemb.2022.126925

Sánchez-Castellón J, Urango-Cárdenas I, Enamorado-Montes G, Burgos-Nuñez S, Marrugo-Negrete J, Díez S (2022) Removal of mercury, cadmium, and lead ions by *Penicillium* sp. Front Environ Chem 2(795632):1–8. https://doi.org/10.3389/fenvc.2021.795632

Santos EE, Cárdenas JF, Torre ME, Martínez VM, Rodríguez A, Acosta I (2017) Bioadsorción de arsénico(III) en solución acuosa por la biomasa modificada de *Aspergillus niger*. Avances En Ciencias e Ingeniería 8(2):1–10. https://doi.org/10.4067/S0718-07642017000600006

Sarria-Villa RA, Gallo-Corredor A, Benítez Benítez R (2020) Tecnologías para remover metales pesados presentes en aguas. Caso Cromo y Mercurio. Journal of Ciencias e Ingeniería 12(1):94–109. https://doi.org/10.46571/JCI.2020.1.8

Say R, Yilmaz N, Denizli A (2003) Removal of heavy metal ions using the fungus *Penicillium canescens*. Adsorpt Sci Technol 21(7):643–650. https://doi.org/10.1260/026361703772776420

Sharma KR, Naruka A, Raja M, Sharma RK (2022) White rot fungus mediated removal of mercury from wastewater. Water Environ Res 94(7):643–650, e10769. https://doi.org/10.1002/wer.10769

Vacar CL, Covaci E, Chakraborty S, Li B, Weindorf DC, Frentiu T, Pârvu M, Podar D (2021) Heavy metal-resistant filamentous fungi as potential mercury bioremediators. J Fungi 7(386):1–21. https://doi.org/10.3390/jof7050386

Vaksmaa A, Guerrero-Cruz S, Ghosh P, Zeghal E, Hernando-Morales V, Niemann H (2023) Role of fungi in bioremediation of emerging pollutants. Front Mar Sci 10:1070905. https://doi.org/10.3389/fmars.2023.1070905

Vallejo Aguilar MLA, Marín Castro MA, Ramos Cassellis ME, Silva Gómez SE, Ibarra Cantún D, Tamariz Flores JV (2021) Biosorption and tolerance of Pb, Cr and Cd by the biomass of *Pleurotus ostreatus* (Jacq. Ex Fr.) P. Kumm. Revista Mexicana Ciencias Agrícolas 12(2):275–289. https://doi.org/10.29312/remexca.v12i2.2687

Yalcinkaya Y, Arica YM, Soysal L, Denizli A, Genc O, Bektas S (2002) Cadmium and mercury uptake by immobilized *Pleurotus sapidus*. Turk J Chem 26(3):441–452. https://journals.tubitak.gov.tr/chem/vol26/iss3/19

SPRINGER NATURE

GPSR Compliance

The European Union's (EU) General Product Safety Regulation (GPSR) is a set of rules that requires consumer products to be safe and our obligations to ensure this.

If you have any concerns about our products, you can contact us on ProductSafety@springernature.com

In case Publisher is established outside the EU, the EU authorized representative is:

Springer Nature Customer Service Center GmbH
Europaplatz 3
69115 Heidelberg, Germany

The manufacturer's authorised representative in the EU is Springer Nature Customer Service Centre GmbH, Europaplatz 3, 69115 Heidelberg, Germany. If you have any concerns regarding our products, please contact ProductSafety@springernature.com

Printed and bound by CPI Group (UK) Ltd, Croydon, CR0 4YY
30/09/2026
02225322-0015